1988 Ordnance Survey Motoring Atlas of Great Britain

Contents

Inside front cover: Key to maps on pages 2-113

II Route Planning Maps
II Limited Interchange Motorway Junctions
1 Legend
2 **Ordnance Survey 3 miles : 1 inch Maps**
108 **Ordnance Survey 5 miles : 1 inch Maps**
114 Town Centre Maps
114 Birmingham, Bristol
115 Cardiff, Edinburgh
116 Glasgow, Leeds
117 Liverpool, Manchester
118 Central London
120 Routes into London
121 Index
132 Distances between Principal Towns

Inside back cover: Motoring Information
Back cover: Motorways and Major Routes

Ordnance Survey · Temple Press

First published 1983 by
Ordnance Survey and Temple Press
Romsey Road, Maybush, Southampton SO9 4DH
Temple Press, an imprint of The Hamlyn Publishing Group Limited, Bridge House, 69 London Road, Twickenham, Middlesex TW1 3SB

Fifth edition 1987
First impression 1987

The representation in this atlas of a road is no evidence of the existence of a right of way.

Temple Press ISBN 0 600 55274 8 (Ordnance Survey ISBN 0 319 00128 8)

Printed and bound by Jarrold & Sons Ltd, England

Route Planning Maps

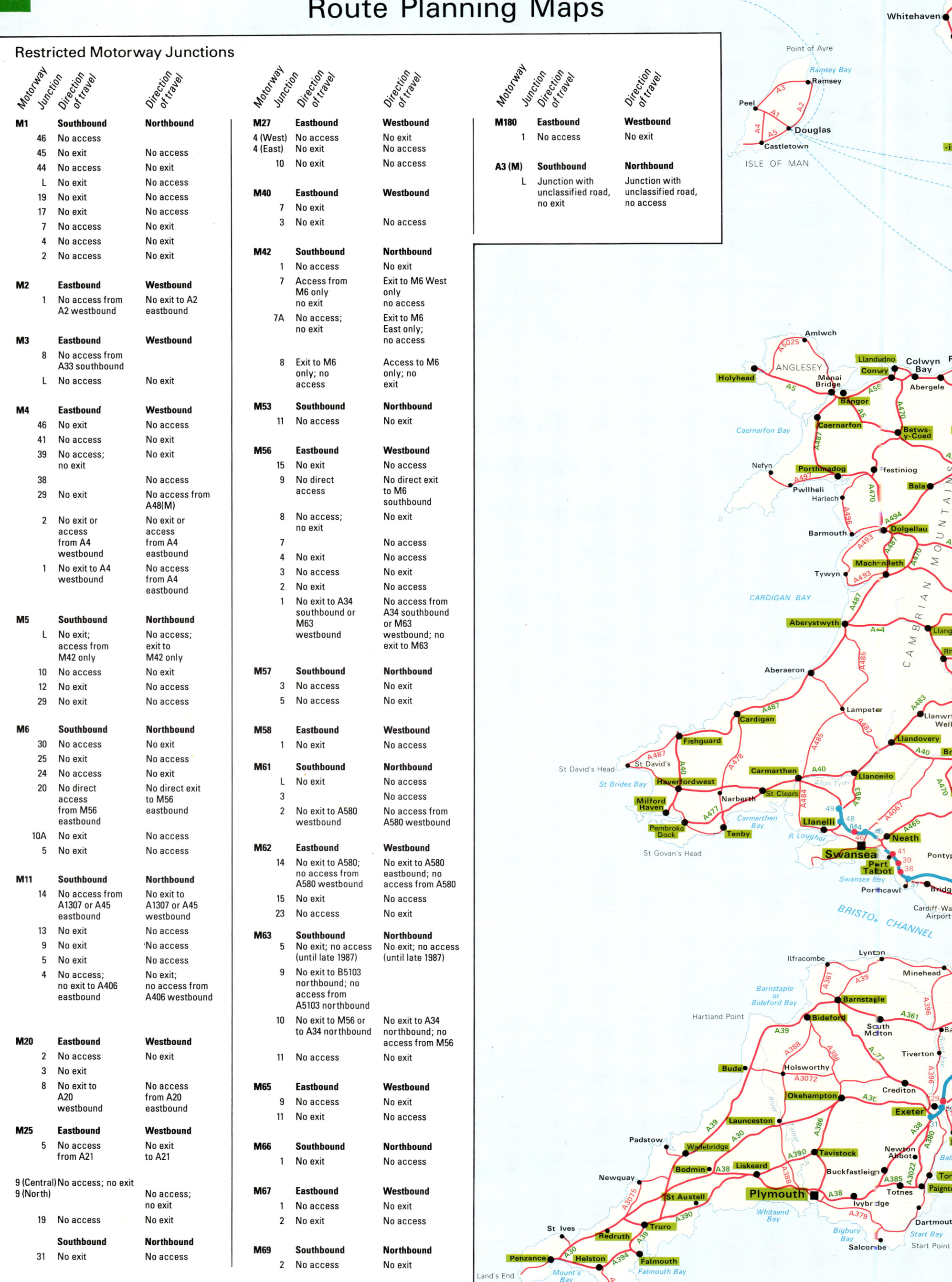

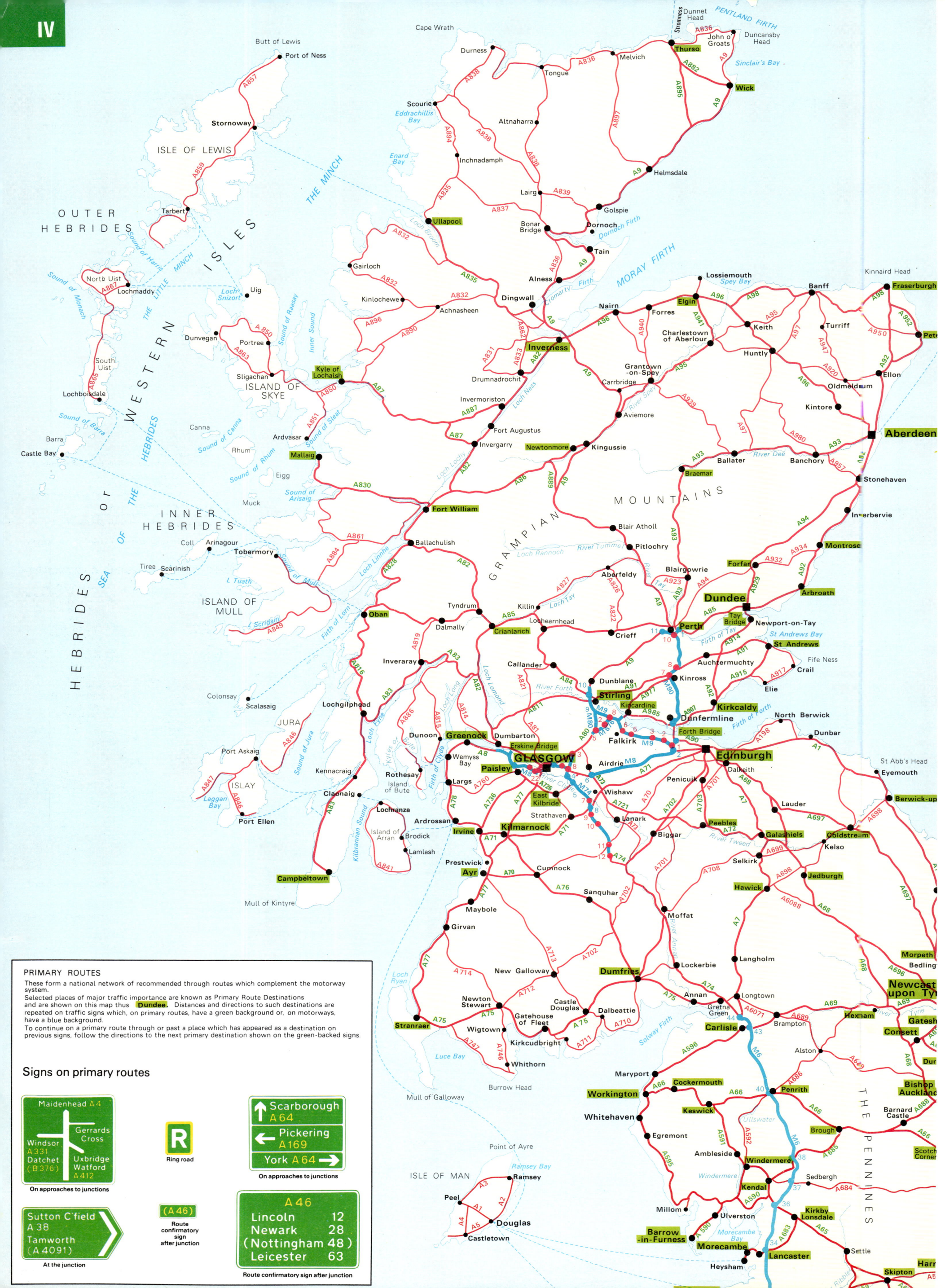

Restricted Motorway Junctions

Motorway	Junction	Direction of travel	Direction of travel
M1		Southbound	Northbound
	46	No access	
	45	No exit	No access
	44	No access	No exit
	L	No exit	No access
	19	No exit	No access
	17	No exit	No access
	7	No access	No exit
	4	No access	No exit
	2	No access	No exit
M2		Eastbound	Westbound
	1	No access from A2 westbound	No exit to A2 eastbound
M3		Eastbound	Westbound
	8	No access from A33 southbound	
	L	No access	No exit
M4		Eastbound	Westbound
	46	No exit	No access
	41	No access	No exit
	39	No access; no exit	No exit
	38		No access
	29	No exit	No access from A48(M)
	2	No exit or access from A4 westbound	No exit or access from A4 eastbound
	1	No exit to A4 westbound	No access from A4 eastbound
M5		Southbound	Northbound
	L	No exit; access from M42 only	No access; exit to M42 only
	10	No access	No exit
	12	No exit	No access
	29	No exit	No access
M6		Southbound	Northbound
	30	No access	No exit
	25	No exit	No access
	24	No access	No exit
	20	No direct access from M56 eastbound	No direct exit to M56 eastbound
	10A	No exit	No access
	5	No exit	No access
M11		Southbound	Northbound
	14	No access from A1307 or A45 eastbound	No exit to A1307 or A45 westbound
	13	No exit	No access
	9	No exit	No access
	5	No exit	No access
	4	No access; no exit to A406 eastbound	No exit; no access from A406 westbound
M20		Eastbound	Westbound
	2	No access	No exit
	3	No exit	
	8	No exit to A20 westbound	No access from A20 eastbound
M25		Eastbound	Westbound
	5	No access from A21	No exit to A21
	9 (Central)	No access; no exit	
	9 (North)		No access; no exit
	19	No access	No exit
		Southbound	Northbound
	31	No exit	No access
M27		Eastbound	Westbound
	4 (West)	No access	No exit
	4 (East)	No exit	No access
	10	No exit	No access
M40		Eastbound	Westbound
	7	No exit	
	3	No exit	No access
M42		Southbound	Northbound
	1	No access	No exit
	7	Access from M6 only no exit	Exit to M6 West only no access
	7A	No access; no exit	Exit to M6 East only; no access
	8	Exit to M6 only; no access	Access to M6 only; no exit
M53		Southbound	Northbound
	11	No access	No exit
M56		Eastbound	Westbound
	15	No exit	No access
	9	No direct access	No direct exit to M6 southbound
	8	No access; no exit	No exit
	7		No access
	4	No exit	No access
	3	No access	No exit
	2	No exit	No access
	1	No exit to A34 southbound or M63 westbound	No access from A34 southbound or M63 westbound; no exit to M63
M57		Southbound	Northbound
	3	No access	No exit
	5	No access	No exit
M58		Eastbound	Westbound
	1	No exit	No access
M61		Southbound	Northbound
	L	No exit	No access
	3		No access
	2	No exit to A580 westbound	No access from A580 westbound
M62		Eastbound	Westbound
	14	No exit to A580; no access from A580 westbound	No exit to A580 eastbound; no access from A580
	15	No exit	No access
	23	No access	No exit
M63		Southbound	Northbound
	5	No exit; no access (until late 1987)	No exit; no access (until late 1987)
	9	No exit to B5103 northbound; no access from A5103 northbound	
	10	No exit to M56 or to A34 northbound	No exit to A34 northbound; no access from M56
	11	No access	No exit
M65		Eastbound	Westbound
	9	No access	No exit
	11	No exit	No access
M66		Southbound	Northbound
	1	No exit	No access
M67		Eastbound	Westbound
	1	No access	No exit
	2	No exit	No access
M69		Southbound	Northbound
	2	No access	No exit
M180		Eastbound	Westbound
	1	No access	No exit
A3 (M)		Southbound	Northbound
	L	Junction with unclassified road, no exit	Junction with unclassified road, no access

NORTH SEA
ENGLISH CHANNEL
STRAIT OF DOVER
THE PENNINES
THE WASH
ISLE OF WIGHT
See Page 120
For Routes Into London
Legend
Motorway with junction
Autoroute avec échangeur
Autobahn mit Anschlußstelle
Motorway junction with limited interchange
Echangeur à possibilités d'intercirculation restreintes
Autobahnanschlußstelle mit begrenztem Richtungswechsel
Motorway under construction
Autoroute en construction
Autobahn im Bau
A3
Primary routes
Itinéraires principaux
Durchgangsstrassen
A358
Other roads
Autres routes
Sonstige Strassen
Exeter
Primary route destinations
Destinations d'itinéraire principal
Durchgangsstrasse Bestimmungsorte
Short ferry routes for vehicles
Les corte traversées en bac pour les véhicules
Kurze Fährstrecken für Kraftfahrzeuge
Scale 1:1 375 000 or about one inch to 22 miles
Miles 20 0 20 40 60
Kilometres 20 0 20 40 60 80 100
LONDON
BIRMINGHAM
MANCHESTER
LIVERPOOL
LEEDS
SHEFFIELD
Cockermouth
Penrith
Keswick
Bishop Auckland
Hartlepool
Middlesbrough
Stockton -on-Tees
Darlington
Whitby
Scarborough
Bridlington
York
Kingston upon Hull
Grimsby
Scunthorpe
Lincoln
Skegness
Boston
Nottingham
Derby
Leicester
Northampton
Coventry
Wolverhampton
Norwich
Great Yarmouth
Lowestoft
Ipswich
Colchester
Cambridge
Chelmsford
Southend-on-Sea
Oxford
Reading
Swindon
Bristol
Bath
Gloucester
Cardiff
Newport
Southampton
Portsmouth
Bournemouth
Brighton
Dover
Folkestone
Canterbury
Maidstone
Hastings
Eastbourne
Blackpool
Preston
Lancaster
Bradford
Huddersfield
Stoke-on-Trent
Shrewsbury
Hereford
Worcester
Salisbury
Winchester
Exeter

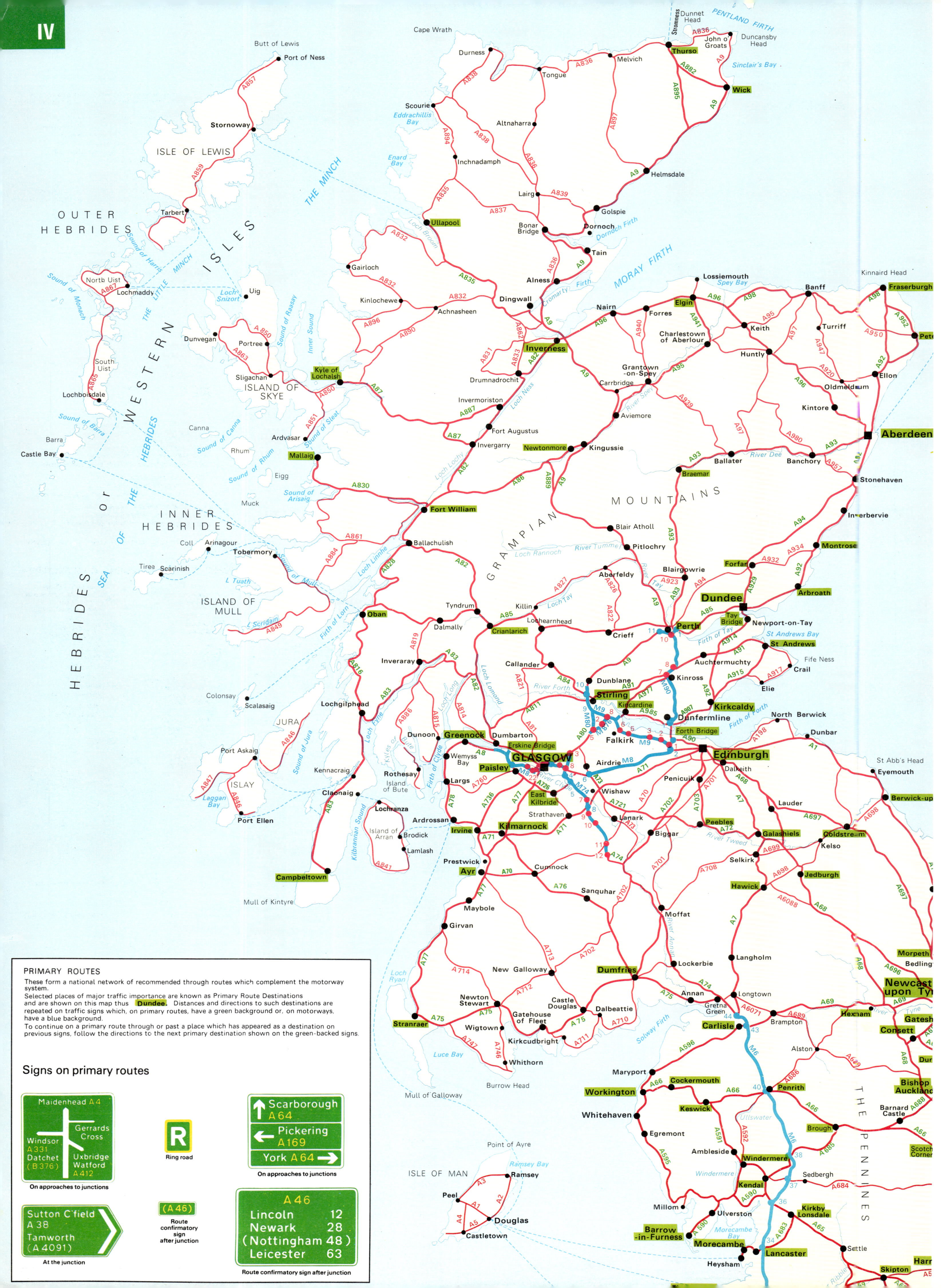

PRIMARY ROUTES

These form a national network of recommended through routes which complement the motorway system.

Selected places of major traffic importance are known as Primary Route Destinations and are shown on this map thus **Dundee.** Distances and directions to such destinations are repeated on traffic signs which, on primary routes, have a green background or, on motorways, have a blue background.

To continue on a primary route through or past a place which has appeared as a destination on previous signs, follow the directions to the next primary destination shown on the green-backed signs.

Signs on primary routes

Maidenhead A4
Gerrards Cross
Windsor A331 Datchet (B376)
Uxbridge Watford A412
R
Scarborough A64
Pickering A169
York A64
Sutton C'field A38
Tamworth (A4091)
(A46)
A46
Lincoln 12
Newark 28
(Nottingham 48)
Leicester 63

On approaches to junctions

Ring road

On approaches to junctions

At the junction

Route confirmatory sign after junction

Route confirmatory sign after junction

Restricted Motorway Junctions

Motorway	Junction	Direction of travel	Direction of travel
M8		Eastbound	Westbound
	25	No access from A739 northbound	No access from A739 northbound
	23	No exit	No access
	22	No exit	No access
	21	No access	No exit
	20	No exit	No access
	19	No access	No exit
	18		No access
	16	No exit	No access
	14	No access	No exit
	9	No access	No exit
	8		No access from A8 eastbound, A89 eastbound or M73 southbound
M9		Eastbound	Westbound
	8	No exit	No access from M876 northbound
	6	No access	No exit
	5	No exit	No access
	3	No exit	No access
	2	No access	No exit
	1	No exit	No access
M73		Southbound	Northbound
	3	No access from A80 northbound	No exit to A80 southbound
	2	No access from A89; no exit to M8 (Junction 8) or A89	No exit to A89. No access from M8 (Junction 8) or A89
M74		Southbound	Northbound
	3	No access	No exit
	7	No access	No exit
	9	No access	No exit; no access
	10	No exit	
	11	No access	No exit
	12	Exit to A74 only, end of motorway	Access from A74 only
M80		Southbound	Northbound
	5	No exit	No access
M90		Southbound	Northbound
	10	No exit to A912	No access from A912
	8	No exit	No access
	7	No access	No exit
M876		Eastbound	Westbound
	2	No access	No exit
A1(M)		Southbound	Northbound
		Junction with A69, no exit; junction with A66(M), no exit; junction with A6129, no access; no exit to A1 northbound	Junction with A69, no access; junction with A66(M), no access from A66(M); junction with A6129, no exit
	5	No exit; no access	No exit
	3	No access	
	2	No exit	

Lerwick
NORTH SEA
Alnwick
Amble-by-the-Sea
Ashington
Blyth
Whitley Bay
Tynemouth
South Shields
Sunderland
Seaham
Hartlepool
Middlesbrough
Stockton -on-Tees
Darlington
Guisborough
Whitby
Northallerton
Thirsk
Pickering
Scarborough
Filey
Flamborough Head
Malton
Bridlington
Ripon
Knaresborough
York
Great Driffield

Legend to Map Pages

ROADS ROUTES STRASSEN

The representation on this map of a road is no evidence of the existence of a right of way

M 1 — Motorway with service area, service area (limited access) and junction with junction number
Autoroute avec aire de service, aire de service (accès restreint) et échangeur avec son numéro
Autobahn mit Servicestation, Servicestation (mit begrenztem Zugang) und Anschlußstelle mit Nummer

M 62 — Motorway junction with limited interchange
Echangeur à possibilités d'intercirculation restreintes
Autobahnanschlußstelle mit begrenzten Richtungswechsel

M 54 — Motorway and junction under construction
Autoroute et échangeur en construction
Autobahn und Anschlussstelle im Bau

A I (T) — Dual carriageway — Trunk road with service area
Route à grande circulation avec aire de service
Fernverkehrsstrasse mit Servicestation

A 15 — Double chaussée — Main road with roundabout or multiple level junction
Route principale avec rond-point, sens giratoire ou échangeur
Hauptstrasse mit Kreisverkehr oder Anschlußstelle

B 676 — Zweibahnige Strasse — Secondary road
Route secondaire
Nebenstrasse

Road under construction
Route en construction
Strasse im Bau

Gradient 1 in 7 and steeper
Pente: 14% et plus
Steigungen: 14% und mehr

Toll — Toll / Peage / Gebühren — Road tunnel / Tunnel routier / Strassentunnel

A 855 B 885 — Narrow road with passing places
Route étroite avec voies de dépassment
Enge Strasse mit Ausweichstelle bzw. Uberholstelle

Other tarred road / Autre route goudronnée / Sonstige asphaltierte Strasse — Other minor road / Autre route / Sonstige Nebenstrasse

18 23 — Distances in miles between markers
Distances en milles entre les marques
Entfernungen in Meilen zwischen den Zeichen

Selected places of major traffic importance are known as Primary Route Destinations and are shown on this map thus DERBY
Distances and directions to such destinations are repeated on traffic signs

TOURIST INFORMATION — RESEIGNEMENTS TOURISTIQUES — ALLGEMEINE TOURISTENANGABEN

- Abbey, Cathedral, Priory / Abbaye, Cathédral, Prieuré / Abtei, Kathedrale, Priorei
- Aquarium / Aquarium / Aquarium
- Camp site / Terrain de camping / Campingplatz
- Caravan site / Terrain pour caravanes / Wohnwagenplatz
- Castle / Château / Schloss
- Cave / Caverne / Höhle
- Country park / Parc naturel / Landschaftspark
- Craft centre / Centre artisanal / Zentrum für Kunsthandwerk
- Garden / Jardin / Garten
- Golf course or links / Terrain de golf / Golfplatz
- Historic house / Manoir, Palais / Historisches Gebäude
- Information centre / Bureau de renseignements / Informationsbüro
- Motor racing / Courses automobiles / Autorennen
- Museum / Musée / Museum
- Nature reserve / Réserve naturelle / Naturschutzgebiet
- Nature or forest trail / Sentier signalisé pour piétons / Natur-oder Waldlehrpfad
- Other tourist feature / Autre site intéressant / Sonstige Sehenswurdigkeit
- Picnic site / Emplacement de pique-nique / Picknickplatz
- Preserved railway / Chemin de fer préservé touristique / Museumseisenbahn
- Racecourse / Hippodrome / Pferderennbahn
- Skiing / Piste de ski / Skilaufen
- Viewpoint / Belvédère / Aussichtspunkt
- Wildlife park / Parc animalier / Wildpark
- Zoo / Zoo / Tiergarten

GENERAL FEATURES

Buildings
Wood
Lighthouse (in use)
Lighthouse (disused)
Windmill
Radio or TV mast
Heliport
Youth hostel
Civil aerodrome — with Customs facilities / without Customs facilities
Public telephone
Motoring organisation telephone

WATER FEATURES

Lake, Bridge, Short ferry routes for vehicles, Canal, Marsh, Slopes, Cliff, Flat rock, Ferry, Transport for vehicles, Light-vessel, Dunes, Low water mark, Foreshore, High water mark

(boat) (hovercraft) Ferry routes for vehicles

RAILWAYS

Standard gauge track
Narrow gauge track
Tunnel
Road crossing under or over
Level crossing
Station

RELIEF — HEIGHTS IN FEET

Feet	Metres
3000	914
2000	610
1400	427
1000	305
600	183
200	61
0	0

·274 Heights in feet above mean sea level

400 Contours at 200ft intervals

To convert feet to metres multiply by 0·3048

ANTIQUITIES

ROMAN ROAD • — Roman antiquity
Castle • — Other antiquities
Native fortress
Site of battle (with date)
Roman road (course of)

BOUNDARIES

National
County, Region or Islands Area

Ancient Monuments and Historic Buildings in the care of the Secretaries of State for the Environment, for Scotland and for Wales and that are open to the public.

Scale 1:190 080
3 miles = 1 inch

1 kilometre = 0·6214 mile — 1 mile = 1·61 kilometres

Kilometres 8 5 0 5 10 15
Miles 5 0 5 10

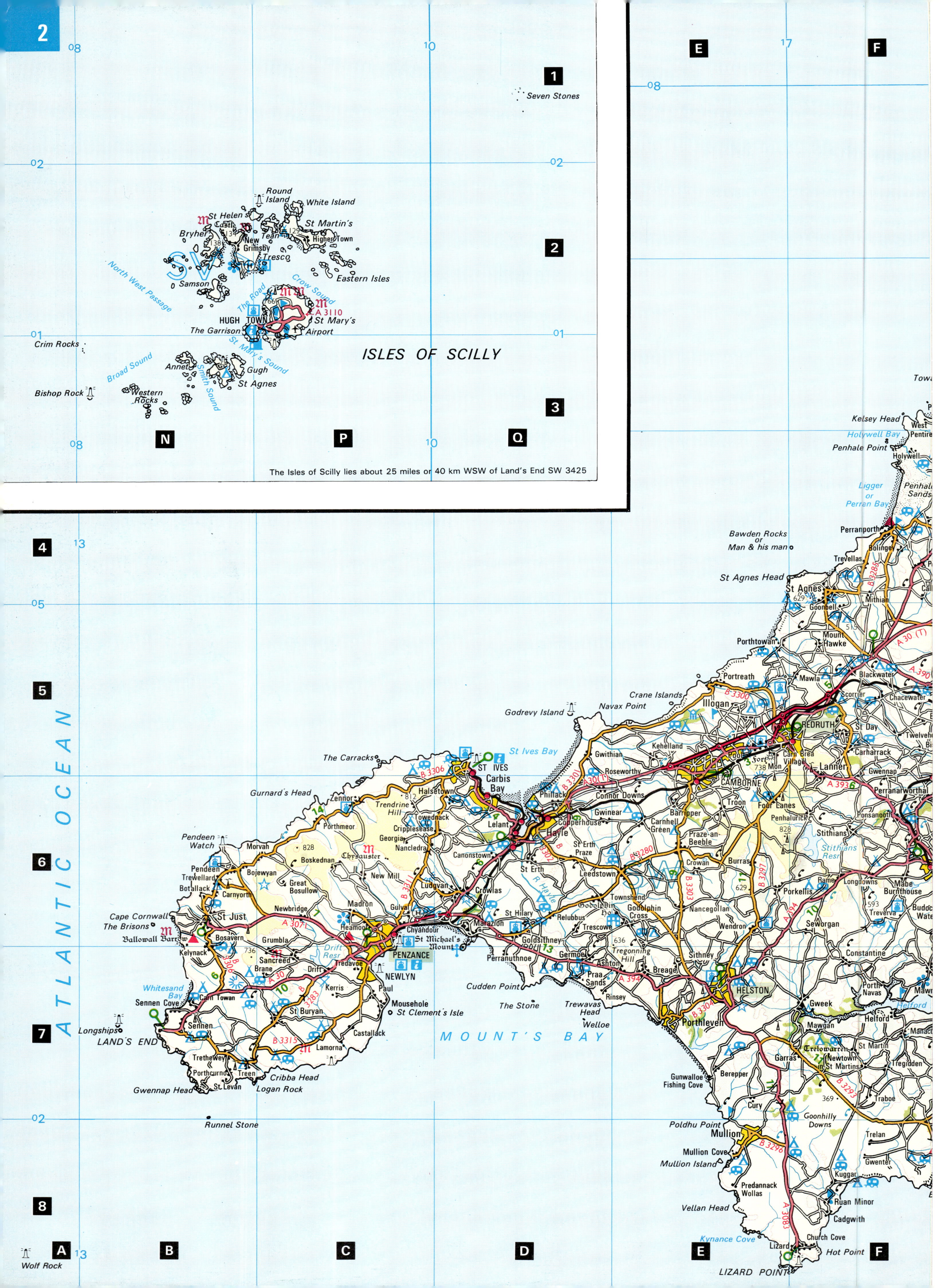
ISLES OF SCILLY
Seven Stones
Round Island
White Island
St Helen's
St Martin's
Bryher
Tean
Higher Town
New Grimsby
Tresco
North West Passage
Samson
Eastern Isles
Crow Sound
The Road
HUGH TOWN
A 3110
St Mary's
The Garrison
Airport
Crim Rocks
St Mary's Sound
Broad Sound
Annet
Gugh
St Agnes
Smith Sound
Bishop Rock
Western Rocks
The Isles of Scilly lies about 25 miles or 40 km WSW of Land's End SW 3425
ATLANTIC OCEAN
MOUNT'S BAY
Kelsey Head
Holywell Bay
Penhale Point
Holywell
Ligger or Perran Bay
Penhale Sands
Perranporth
Bawden Rocks or Man & his man
Bolingey
Trevellas
St Agnes Head
St Agnes
Goonbell
Mithian
Porthtowan
Mount Hawke
A 30 (T)
Portreath
Mawla
Blackwater
A 390
Crane Islands
Navax Point
Illogan
Scorrier
Chacewater
Godrevy Island
REDRUTH
St Day
Twelveheads
St Ives Bay
Gwithian
Kehelland
Carn Brea Village
Lanner
Carharrack
The Carracks
ST IVES
Carbis Bay
Roseworthy
CAMBORNE
Gwennap
Perranarworthal
Gurnard's Head
Halsetown
Phillack
Connor Downs
Troon
Four Lanes
A 393
Zennor
Trendrine Hill
Towednack
Lelant
Gwinear
Carnhell Green
Barripper
Penhalurick
Ponsanooth
Porthmeor
Cripplesease
Georgia
Nancledra
Hayle
Copperhouse
Praze-an-Beeble
Stithians
Stithians Resr
Pendeen Watch
Morvah
Boskednan
Chysauster
Canonstown
St Erth Praze
Crowan
Burras
Pendeen
Trewellard
Bojewyan
New Mill
St Erth
Leedstown
Rame
Longdowns
Mabe Burnthouse
Botallack
Carnyorth
Great Bosullow
Ludgvan
Townshend
Porkellis
Treverva
Budock Water
Cape Cornwall
The Brisons
Ballowall Barrow
St Just
Newbridge
Madron
Gulval
Crowlas
Godolphin Cross
Nancegollan
Wendron
Seworgan
Heamoor
Chyandour
St Michael's Mount
Marazion
St Hilary
Relubbus
Trescowe
Bosavern
Grumbla
Kelynack
Sancreed
Brane
Drift Resr
Drift
PENZANCE
Tredavoe
NEWLYN
Perranuthnoe
Goldsithney
Germoe
Tregonning Hill
Ashton
Praa Sands
Breage
Sithney
Constantine
Whitesand Bay
Carn Towan
Sennen Cove
Kerris
Paul
St Buryan
Mousehole
St Clement's Isle
Cudden Point
The Stone
Trewavas Head
Rinsey
HELSTON
Porthleven
Welloe
Gweek
Porth Navas
Helford
Mawnan
Longships
LAND'S END
Sennen
Castallack
Mawgan
Lamorna
Trethewey
Porthcurno
Treen
Cribba Head
Logan Rock
St Levan
Gwennap Head
Garras
Trelowarren
Newtown in St Martin
St Martin
Tregidden
Gunwalloe Fishing Cove
Berepper
B 3293
Traboe
Runnel Stone
Cury
Goonhilly Downs
Poldhu Point
Mullion
Trelan
Mullion Cove
Mullion Island
B 3296
Gwenter
Kuggar
Predannack Wollas
Ruan Minor
Vellan Head
A 3083
Cadgwith
Kynance Cove
Lizard
Church Cove
Hot Point
LIZARD POINT
Wolf Rock

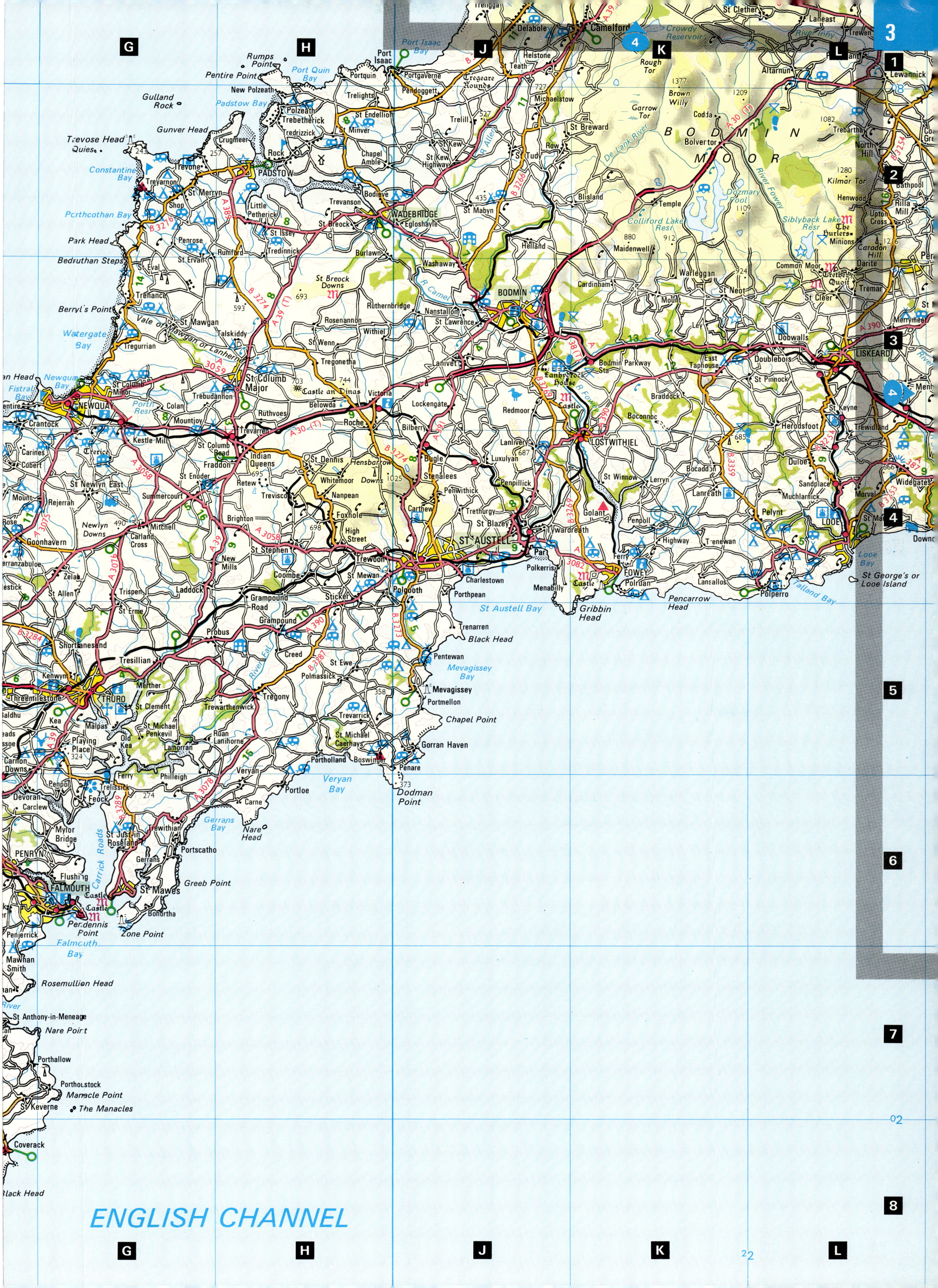
3
G
H
J
K
L
1
2
3
4
5
6
7
8
ENGLISH CHANNEL
BODMIN MOOR
Camelford
Delabole
Helstone
Port Isaac
Port Isaac Bay
Portquin
Port Quin Bay
Rumps Point
Pentire Point
New Polzeath
Polzeath
Trebetherick
Padstow Bay
Gulland Rock
Gunver Head
Trevose Head
Quies
Constantine Bay
Trevone
PADSTOW
Rock
St Minver
St Endellion
Trelights
Pendoggett
Tregeare Rounds
Trelill
St Kew
St Kew Highway
Chapel Amble
Michaelstow
St Teath
St Tudy
St Breward
Rough Tor
Brown Willy
Garrow Tor
Bolventor
Crowdy Reservoir
Altarnun
Laneast
St Clether
River Inny
Trewen
Lewannick
North Hill
Kilmar Tor
Bathpool
Henwood
Rilla Mill
Upton Cross
Caradon Hill
Darite
Tremar
The Hurlers Minions
Siblyback Lake Resr
Dozmary Pool
River Fowey
Temple
Colliford Lake Resr
Maidenwell
Blisland
Warleggan
St Neot
Common Moor
Trethevy Quoit
St Cleer
Mount
Cardinham
Merrymeet
LISKEARD
Dobwalls
Doublebois
East Taphouse
Bodmin Parkway Sta
Lanhydrock House
Castle
Braddock
Boconnoc
St Pinnock
St Keyne
Herodsfoot
Trewidland
Menheniot
Duloe
Sandplace
Widegates
Morval
Muchlarnick
Pelynt
LOOE
Looe Bay
St George's or Looe Island
Talland Bay
Polperro
Lansallos
Pencarrow Head
Trenewan
Highway
Penpoll
Lanreath
Bocaddon
Lerryn
St Winnow
LOSTWITHIEL
Golant
Polruan
FOWEY
Castle
Gribbin Head
Menabilly
Polkerris
Par
Tywardreath
St Blazey
Penpillick
Luxulyan
Lanlivery
Redmoor
Lanivet
Lockengate
Bilberry
Bugle
Stenalees
Penwithick
Trethurgy
Carthew
ST AUSTELL
Charlestown
Porthpean
St Austell Bay
Trenarren
Black Head
Pentewan
Mevagissey Bay
Mevagissey
Portmellon
Chapel Point
Gorran Haven
Penare
Dodman Point
Boswinger
Trevarrick
St Michael Caerhays
Portholland
Veryan Bay
Veryan
Portloe
Carne
Nare Head
Gerrans Bay
Portscatho
Gerrans
Greeb Point
St Mawes
Bohortha
Zone Point
St Just in Roseland
Trewithian
Philleigh
Ruan Lanihorne
Tregony
Trewarthenick
Lamorran
St Michael Penkevil
St Clement
Merther
Tresillian
TRURO
Threemilestone
Kenwyn
Kea
Malpas
Old Kea
Playing Place
Carnon Downs
Penpol
Devoran
Carclew
Feock
Trelissick
Mylor Bridge
Carrick Roads
PENRYN
Flushing
FALMOUTH
Pendennis Point
Castle
Falmouth Bay
Penjerrick
Mawnan Smith
Rosemullion Head
Helford River
St Anthony-in-Meneage
Nare Point
Porthallow
Porthoustock
Manacle Point
The Manacles
St Keverne
Coverack
Black Head
Shortlanesend
Probus
Grampound
Grampound Road
Creed
Polmassick
St Ewe
Sticker
Polgooth
St Mewan
Trewoon
Coombe
St Stephen
High Street
Foxhole
Nanpean
Whitemoor
St Dennis
Hensbarrow Downs
Roche
Victoria
Belowda
Castle an Dinas
Tregonetha
St Wenn
Rosenannon
Withiel
Ruthernbridge
Nanstallon
St Lawrence
BODMIN
Washaway
Helland
St Mabyn
Egloshayle
WADEBRIDGE
Burlawn
Bodieve
Trevanson
St Breock
St Breock Downs
St Issey
Tredinnick
Little Petherick
St Merryn
Shop
Treyarnon
Porthcothan Bay
Park Head
Bedruthan Steps
St Eval
St Ervan
Penrose
Rumford
Trenance
Mawgan Porth
Berryl's Point
Watergate Bay
St Mawgan
Vale of Lanherne
Tregurrian
Talskiddy
St Columb Major
St Columb Minor
NEWQUAY
Newquay Bay
Fistral Bay
Pentire
Crantock
Trebudannon
Colan
Mountjoy
Porth Resr
Ruthvoes
Trevarren
Kestle Mill
St Columb Road
Fraddon
Indian Queens
Retew
Treviscoe
St Enoder
Summercourt
Trerice
Cubert
St Newlyn East
Rejerrah
Newlyn Downs
Carland Cross
Mitchell
Goonhavern
Zelah
St Allen
Trispen
St Erme
Ladock
Brighton
New Mills
A 30
A 39
A 390
A 3058
A 3059
A 3075
A 3076
A 3078
A 389
A 387
B 3266
B 3274
B 3276
B 3287
B 3269
B 3359
B 3254
B 3276
B 3289
B 3284
B 3082
B 3253
B 3268
22
02

A
B
C
D
E
1
2
3
4
5
6
7
8
Widemouth Bay
Marhamchurch
Bridgerule
Pyworthy
Holsworthy
Hollacombe
Graddon Moor
Halwill Junction
Halwill
Beaworthy
Coppathorne
Dizzard Point
Poundstock
Whitstone
North Tamerton
Clawton
Week St Mary
St Gennys
Tregole
Trewint
Cambeak
Crackington
Wainhouse Corner
Jacobstow
Tetcott
Lana
Ashwater
Quoditch
Eworthy
Germansweek
Boasley Cross
Luffincott
Chapmans Well
Northcott
South Wheatley
Clubworthy
West Curry
Canworthy Water
Bennacott
Boyton
East Panson
Virginstow
Bratton Clovelly
Fire Beacon Point
Boscastle
Lesnewth
Marshgate
Otterham
Warbstow
Brazacott
St Giles on the Heath
Broadwoodwidger
Tintagel Head
Castle
Trevalga
Tremaine
North Petherwin
Polapit Tamar
Bossiney
Tintagel
Hallworthy
Treneglos
Yeolmbridge
Werrington
Cross Green
Thrushelton
Trewarmett
Trewassa
Davidstow
Tremail
Tresmeer
Egloskerry
Langore
Stowford
Lewdown
Trekenow
Start Point
Treligga
Tregeare
Red Down
St Stephens
LAUNCESTON
Lifton
Portgate
Lewtrenchard
Lydford
Delabole
Camelford
St Clether
Laneast
Tregadillett
Tinhay
Crowdy Reservoir
River Inny
Trewen
Lawhitton
Coryton
Port Isaac Bay
Halstone
Teath
Polyphant
South Petherwin
Marystow
Chillaton
Kelly
Bradstone
North Brentor
Portgaverne
Rough Tor
Altarnun
Lewannick
Milton Abbot
Pendoggett
Brown Willy
Michaelstow
Garrow Tor
Codda
Trebartha
Lezant
Dunterton
Trelill
St Breward
BODMIN MOOR
Bolventor
Coad's Green
Treburley
Lamerton
St Kew
North Hill
Sydenham Damerel
St Tudy
Row
Kilmar Tor
Bray Shop
St Kew Highway
Blisland
Henwood
Bathpool
Stoke Climsland
Luckett
TAVISTOCK
St Mabyn
Temple
Dozmary Pool
Rilla Mill
Linkinhorne
Downgate
Latchley
Chilsworthy
Whitchurch
WADEBRIDGE
Egloshayle
Colliford Lake Resr
Siblyback Lake Resr
Upton Cross
South Hill
Kelly Bray
Kit Hill
St Ann's Chapel
Gunnislake
Helland
Maidenwell
The Hurlers
Minions
Caradon Hill
Golberdon
Harrowbarrow
Albaston
Washaway
Darite
Pensilva
Trevigro
Haye
Horrabridge
Common Moor
Callington
Cotehele
Calstock
Buckland Monachorum
Yelverton
Crapstone
Buckland Moor
Warleggan
Cardinham
St Neot
Trethurgy
BODMIN
Mount
Tremar
St Cleer
St Dominick
Bere Alston
Nanstallon
St Lawrence
St Ive
Merrymeet
Milton Combe
Dobwalls
St Mellion
Lopwell
Lanivet
Bodmin Parkway Sta
East Taphouse
Doublebois
LISKEARD
Quethiock
Pillaton
Bere Ferrers
Lanhydrock
St Pinnock
Menheniot
Hatt
Cargreen
Tamerton Foliot
Lockengate
Redmoor
Braddock
Botusfleming
Landulph
Bickleigh
Castle
Boconnoc
St Keyne
Trewidland
Landrake
Trematon
Bilberry
Lanlivery
LOSTWITHIEL
Herodsfoot
Tideford
SALTASH
Crownhill
Bugle
Luxulyan
Duloe
Markwell
St Stephens
Budeaux
Stenalees
Lerryn
Bocaddon
Sandplace
Hessenford
St Germans
PLYMOUTH
Penwithick
Penpillick
St Winnow
Lanreath
Widegates
Polbathic
Sheviock
TORPOINT
Devonport
Golant
Muchlarnick
Morval
Penpoll
Pelynt
LOOE
St Martin
Trethurgy
St Blazey
Tywardreath
Highway
Trenewan
Seaton
Crafthole
Antony
ST AUSTELL
Par
Polkerris
FOWEY
Polruan
Downderry
Portwrinkle
St John
Mount Edgcumbe
Drake's Island
Charlestown
Menabilly
Castle
Lansallos
Polperro
Looe Bay
St George's or Looe Island
Talland Bay
Millbrook
Freathy
Polgooth
Porthpean
St Austell Bay
Gribbin Head
Pencarrow Head
Whitsand Bay
Kingsand
Cawsand
The Sound
Plymouth Breakwater
Trenarren
Black Head
Rame
Penlee Point
Heybrook Bay
Wembury Bay
Rame Head
Craft Mew Stone
Pentewan
Mevagissey Bay
PLYMOUTH to
Roscoff .... 6-7 hrs
Santander .... 24 hrs
Mevagissey
Portmellon
Chapel Point
Gorran Haven
Penare
Dodman Point
Due to open Mid 1988
Eddystone Rocks
ENGLISH

F
G
H
J
K
L
1
2
3
4
5
6
7
8
EXETER
CREDITON
OKEHAMPTON
DARTMOOR
FOREST
DARTMOOR
SX
Okehampton Common
Bridestowe and Sourton Common
Meldon Resr
Fernworthy Reservoir
Princetown
HM Prison
Two Bridges
Burrator Reservoir
Avon Dam Resr
Venford Resr
Abbot's Way
Ugborough Moor
Grimspound
Moretonhampstead
Chagford
Bovey Tracey
Chudleigh
NEWTON ABBOT
Kingsteignton
Teignmouth
Shaldon
Dawlish
Dawlish Warren
Starcross
EXMOUTH
Topsham
Exminster
Kenton
ASHBURTON
BUCKFASTLEIGH
Buckfast
Dartington
TOTNES
Ivybridge
South Brent
Kingskerswell
BABBACOMBE BAY
Babbacombe
TORQUAY
Hope's Nose
TORBAY
PAIGNTON
TOR BAY
BRIXHAM
Berry Head
Sharkham Point
Scabbacombe Head
Kingswear
DARTMOUTH
Mew Stone
Stoke Fleming
Strete
START BAY
Slapton
Torcross
Beesands
Hallsands
START POINT
Prawle Point
East Prawle
Bolt Head
SALCOMBE
KINGSBRIDGE
West Charleton
Bolt Tail
Hope
Thurlestone
Bigbury Bay
Burgh Island
Bigbury-on-Sea
Stoke Point
Noss Mayo
Newton Ferrers
Yealmpton
Modbury
CHANNEL
TORQUAY to
Seasonal
Alderney 6 hrs
Jersey 9¼ hrs
Guernsey 6 hrs
Due to open Late 1988
Exeter Airport
30
04

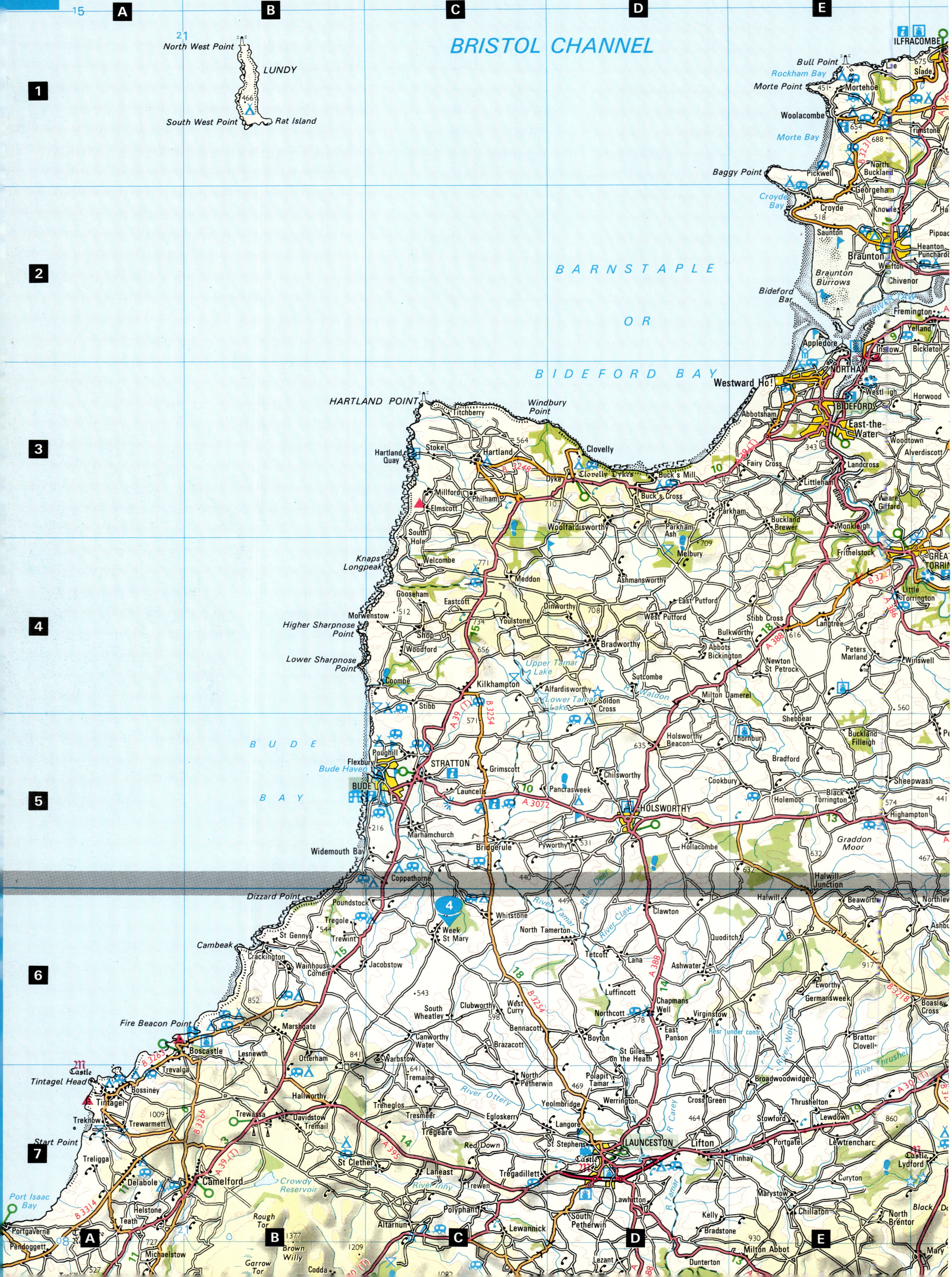
BRISTOL CHANNEL
LUNDY
North West Point
South West Point
Rat Island
BARNSTAPLE OR BIDEFORD BAY
ILFRACOMBE
Bull Point
Rockham Bay
Morte Point
Mortehoe
Woolacombe
Morte Bay
Baggy Point
Croyde Bay
Croyde
Georgeham
Saunton
Braunton
Braunton Burrows
Bideford Bay
Appledore
NORTHAM
Westward Ho!
BIDEFORD
East-the-Water
Abbotsham
HARTLAND POINT
Windbury Point
Hartland Quay
Stoke
Hartland
Clovelly
Clovelly Dykes
Buck's Mill
Buck's Cross
Fairy Cross
Littleham
Landcross
Monkleigh
Frithelstock
Milford
Elmscott
Philham
Woolfardisworthy
Parkham
Buckland Brewer
Parkham Ash
Melbury
South Hole
Knaps Longpeak
Welcombe
Meddon
Ashmansworthy
East Putford
West Putford
Gooseham
Eastcott
Dinworthy
Morwenstow
Higher Sharpnose Point
Lower Sharpnose Point
Shop
Woodford
Youlstone
Bradworthy
Abbots Bickington
Bulkworthy
Stibb Cross
Langtree
Newton St Petrock
Peters Marland
Upper Tamar Lake
Lower Tamar Lake
Alfardisworthy
Soldon Cross
Sutcombe
Kilkhampton
Coombe
Stibb
Milton Damerel
Shebbear
Buckland Filleigh
Holsworthy Beacon
Thornbury
Bradford
BUDE BAY
Bude Haven
Flexbury
Poughill
BUDE
STRATTON
Grimscott
Launcells
Chilsworthy
Pancrasweek
HOLSWORTHY
Cookbury
Holemoor
Black Torrington
Sheepwash
Highampton
Graddon Moor
Marhamchurch
Bridgerule
Pyworthy
Hollacombe
Widemouth Bay
Coppathorne
Halwill Junction
Halwill
Beaworthy
Dizzard Point
Poundstock
Tregole
Trewint
Week St Mary
Whitstone
North Tamerton
Clawton
St Gennys
Cambeak
Crackington
Wainhouse Corner
Jacobstow
Tetcott
Lana
Ashwater
Quodicth
South Wheatley
Clubworthy
West Curry
Luffincott
Chapmans Well
Germansweek
Eworthy
Fire Beacon Point
Boscastle
Marshgate
Bennacott
Northcott
Virginstow
Lesnewth
Otterham
Canworthy Water
Boyton
East Panson
Bratton Clovelly
Warbstow
Tremaine
Brazacott
St Giles on the Heath
Tintagel Head
Castle
Bossiney
Tintagel
Trevalga
Hallworthy
North Petherwin
Polapit Tamar
Broadwoodwidger
Trewarmett
Davidstow
Tremail
Treneglos
Tresmeer
Egloskerry
Yeolmbridge
Werrington
Cross Green
Thrushelton
Trewassa
Tregeare
Langore
Stowford
Lewdown
Start Point
Treligga
St Clether
Red Down
St Stephens
LAUNCESTON
Lifton
Tinhay
Portgate
Lewtrenchard
Lydford
Delabole
Camelford
Crowdy Reservoir
Laneast
Tregadillett
Port Isaac Bay
Helstone
Trewen
Lawhitton
Marystow
Coryton
Chillaton
Portgaverne
St Teath
Rough Tor
Altarnun
Polyphant
Lewannick
South Petherwin
Kelly
Bradstone
North Brentor
Black Down
Pendoggett
Michaelstow
Brown Willy
Garrow Tor
Codda
Lezant
Dunterton
Milton Abbot
River Tamar
River Ottery
River Inny
River Carey
River Thrushel
River Wolf
River Claw
River Deer
River Waldon
A 39 (T)
A 3072
A 388
A 386
A 395
A 30 (T)
B 3254
B 3248
B 3263
B 3266
B 3314
B 3227
B 3231
B 3218
A B C D E
1 2 3 4 5 6 7

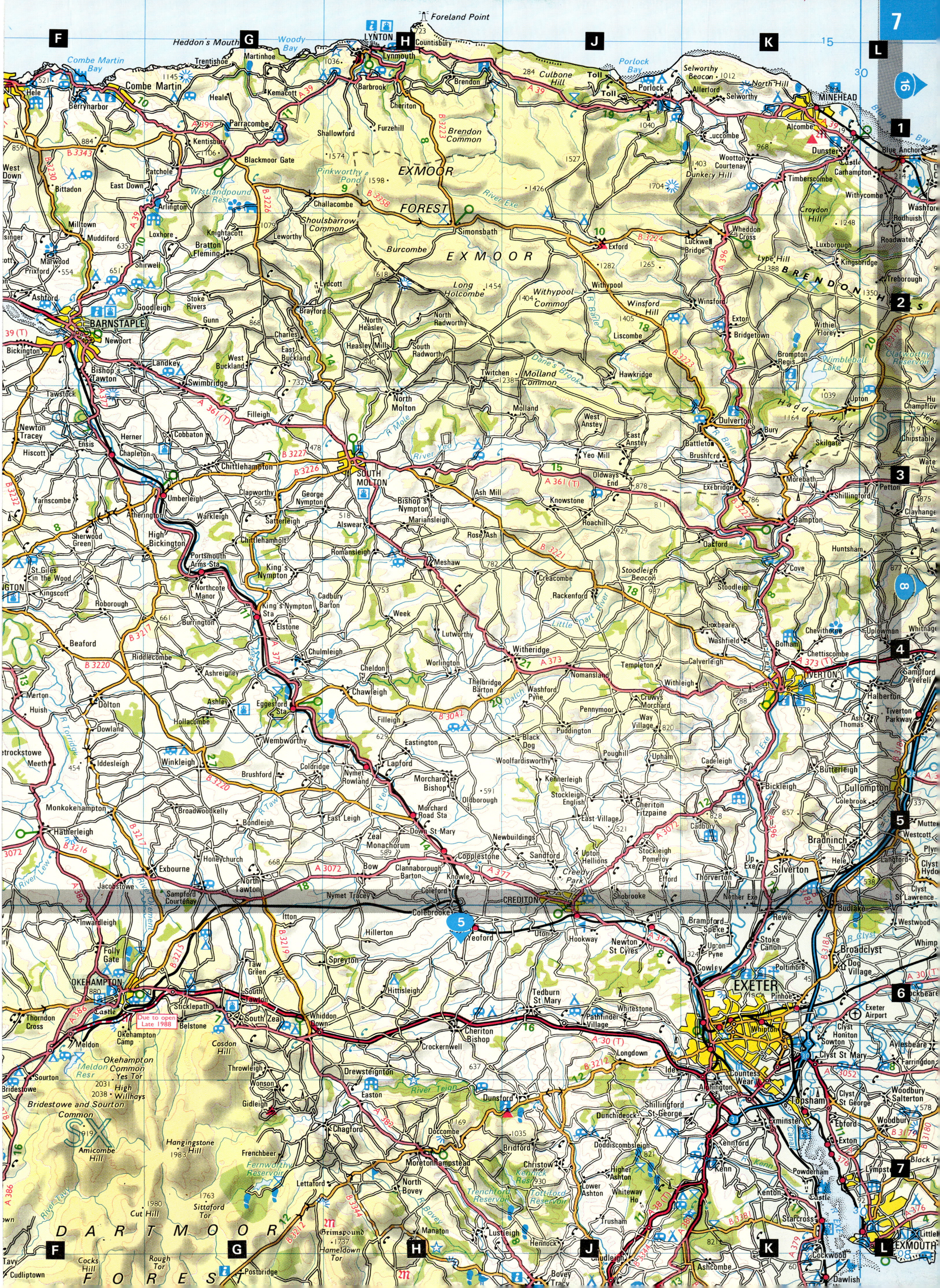

F
G
H
J
K
L
Foreland Point
Heddon's Mouth
Woody Bay
LYNTON
Lynmouth
Countisbury
Combe Martin Bay
Combe Martin
Hele
Berrynarbor
Trentishoe
Martinhoe
Kemacott
Barbrook
Brendon
Culbone Hill
Porlock Bay
Porlock
Selworthy Beacon
Allerford
Selworthy
North Hill
MINEHEAD
Alcombe
Dunster
Blue Anchor
Carhampton
Heale
Parracombe
Kentisbury
Cheriton
Furzehill
Shallowford
Brendon Common
EXMOOR
Pinkworthy Pond
Blackmoor Gate
West Down
Bittadon
East Down
Patchole
Wistlandpound Resr
Challacombe
FOREST
Simonsbath
Exford
Luccombe
Wootton Courtenay
Dunkery Hill
Timberscombe
Withycombe
Washford
Rodhuish
Croydon Hill
Luxborough
Roadwater
Wheddon Cross
Luckwell Bridge
Lype Hill
BRENDON HILLS
Kingsbridge
Treborough
Arlington
Loxhore
Knightacott
Shoulsbarrow Common
Lewortby
Bratton Fleming
Burcombe
Milltown
Muddiford
Marwood
Prixford
Shirwell
Goodleigh
Ashford
BARNSTAPLE
Newport
Bickington
Stoke Rivers
Gunn
Lydcott
Long Holcombe
Withypool Common
Withypool
Winsford Hill
Winsford
Exton
Bridgetown
Withiel Florey
Brompton Regis
Wimbleball Lake
Liscombe
Hawkridge
Brayford
North Heasley
Heasley Mill
Charles
East Buckland
West Buckland
North Radworthy
South Radworthy
Twitchen
Molland Common
Molland
Dulverton
Bury
Haddon Hill
Upton
Skilgate
Landkey
Swimbridge
Bishop's Tawton
Tawstock
Filleigh
North Molton
West Anstey
East Anstey
Yeo Mill
Battleton
Brushford
Newton Tracey
Herner
Cobbaton
Hiscott
Ensis
Chapelton
Chittlehampton
SOUTH MOLTON
Oldways End
Exebridge
Morebath
Shillingford
Petton
Yarnscombe
Umberleigh
Clapworthy
George Nympton
Bishop's Nympton
Mariansleigh
Ash Mill
Knowstone
Roachill
Bampton
Atherington
Warkleigh
Satterleigh
Alswear
Rose Ash
Sherwood Green
High Bickington
Chittlehamholt
Romansleigh
Meshaw
Oakford
Huntsham
St Giles in the Wood
Kingscott
Portsmouth Arms Sta
Northcote Manor
King's Nympton
Creacombe
Rackenford
Stoodleigh Beacon
Stoodleigh
Cove
Roborough
King's Nympton Sta
Cadbury Barton
Week
Little Dart River
Loxbeare
Washfield
Bolham
Chevithorne
Uplowman
Burrington
Elstone
Lutworthy
Beaford
Hiddlecombe
Chulmleigh
Worlington
Witheridge
Templeton
Calverleigh
Chettiscombe
TIVERTON
Sampford Peverell
Ashreigney
Cheldon
Thelbridge Barton
Nomansland
Withleigh
Crown's Morchard
Halberton
Marton
Huish
Dolton
Ashley
Eggesford Sta
Chawleigh
Washford Pyne
Pennymoor
Ash Thomas
Tiverton Parkway
Dowland
Hollacombe
Filleigh
Puddington
Way Village
Black Dog
Wembworthy
Eastington
Poughill
Upham
Cadeleigh
Meeth
Iddesleigh
Winkleigh
Coldridge
Lapford
Woolfardisworthy
Kennerleigh
Bickleigh
Butterleigh
Cullompton
Brushford
Nymet Rowland
Morchard Bishop
Oldborough
Stockleigh English
Cheriton Fitzpaine
Colebrook
Monkokehampton
Broadwoodkelly
East Leigh
Morchard Road Sta
East Village
Cadbury
Hatherleigh
Bondleigh
Down St Mary
Zeal Monachorum
Newbuildings
Upton Hellions
Stockleigh Pomeroy
Bradninch
Westcott
Honeychurch
Bow
Sandford
Copplestone
Clannaborough Barton
Knowle
Creedy Park
Up Exe
Silverton
Hele
Exbourne
Jacobstowe
North Tawton
Nymet Tracey
Coleford
Colebrooke
CREDITON
Shobrooke
Thorverton
Nether Exe
Budlake
Langford
Sampford Courtenay
Inwardleigh
Itton
Hillerton
Yeoford
Uton
Hookway
Newton St Cyres
Brampford Speke
Upton Pyne
Rewe
Stoke Canon
Westwood
Broadclyst
Dog Village
Folly Gate
Spreyton
Cowley
Poltimore
OKEHAMPTON
Sticklepath
Taw Green
South Tawton
Hittisleigh
Tedburn St Mary
EXETER
Pinhoe
Exeter Airport
Clyst Honiton
Sowton
Thorndon Cross
Castle
Due to open Late 1988
Belstone
South Zeal
Whiddon Down
Cheriton Bishop
Pathfinder Village
Whitestone
Whipton
Okehampton Camp
Meldon
Cosdon Hill
Crockernwell
Longdown
Ide
Clyst St Mary
Aylesbeare
Farringdon
Meldon Resr
Okehampton Common
Yes Tor
High Willhays
Sourton
Throwleigh
Wonson
Drewsteignton
River Teign
Easton
Dunsford
Alphington
Countess Wear
Topsham
Clyst St George
Woodbury Salterton
Bridestowe
Bridestowe and Sourton Common
Gidleigh
Chagford
Doccombe
Shillingford St George
Dunchideock
Exminster
Ebford
Woodbury
Exton
Amicombe Hill
Hangingstone Hill
Frenchbeer
Fernworthy Reservoir
Moretonhampstead
Bridford
Christow
Kennick Res
Doddiscombsleigh
Higher Ashton
Kennford
Kenn
Powderham
Lympstone
Cut Hill
Sittaford Tor
Lettaford
North Bovey
Trenchford Reservoir
Tottiford Reservoir
Lower Ashton
Whiteway Ho
Trusham
Kenton
Starcross
EXMOUTH
DARTMOOR
FOREST
Grimspound
Hameldown Tor
Manaton
Lustleigh
Hennock
Chudleigh
Ashcombe
Cockwood
Dawlish
Tavy
Cudlipton
Cocks Hill
Rough Tor
Postbridge
Bovey Tracy
SX
SY
SS
A 39
A 399
A 361 (T)
A 377
A 373
A 3072
A 386
A 30 (T)
A 38 (T)
A 376
A 379
A 382
A 396
A 3052
A 3122
B 3227
B 3226
B 3221
B 3220
B 3217
B 3216
B 3219
B 3215
B 3212
B 3344
B 3358
B 3223
B 3224
B 3230
B 3343
B 3042
B 3190
B 3181
B 3180
B 3178
B 3193
B 3381
5
8
16
1
2
3
4
5
6
7

A
B
C
D
E
1
2
3
4
5
6
7
BRENDON HILLS
QUANTOCK HILLS
VALE OF TAUNTON DEANE
BLACK DOWN HILLS
Williton
Bridgwater
North Petherton
Wiveliscombe
Milverton
TAUNTON
Wellington
Dulverton
Bampton
TIVERTON
Cullompton
Bradninch
Silverton
Willand
Uffculme
Hemyock
Ilminster
CHARD
AXMINSTER
HONITON
OTTERY ST MARY
EXETER
Topsham
Exminster
Broadclyst
Exeter Airport
SIDMOUTH
SEATON
Beer
Beer Head
Seaton Bay
LYME REGIS
Charmouth
EXMOUTH
BUDLEIGH SALTERTON
DAWLISH
TEIGNMOUTH
Shaldon
Kenton
Starcross
Cockwood
Dawlish Warren
Colyton
Axmouth
Kilmington
Shute
Stockland
Dunkeswell
Smeatharpe
Churchstanton
Combe St Nicholas
Chaffcombe
Whitestaunton
Yarcombe
Wambrook
Thorncombe
Hawkchurch
Membury
Clayhidon
Blackborough
Kentisbeare
Broadhembury
Payhembury
Talaton
Feniton
Whimple
Rockbeare
Aylesbeare
Farringdon
Woodbury
Lympstone
Colaton Raleigh
Newton Poppleford
Tipton St John
Sidbury
Sidford
Salcombe Regis
Branscombe
Farway
Offwell
Wilmington
Musbury
Uplyme
Ruishton
Creech St Michael
Durston
Kingston St Mary
Bishops Lydeard
Cheddon Fitzpaine
Monkton Heathfield
Stoke St Mary
Hatch Beauchamp
Curry Mallet
Beer Crocombe
Isle Abbotts
Puckington
Barrington
Shepton Beauchamp
Donyatt
Kingstone
Dowlish Wake
Cudworth
Cricket St Thomas
Winsham
Tatworth
South Chard
Forton
Drimpton
Broadwindsor
Pilsdon Pen
Marshwood
Wootton Fitzpaine
Bettiscombe
Stogumber
Crowcombe
Flaxpool
West Bagborough
Cothelstone
Halse
Oake
Norton Fitzwarren
Bradford-on-Tone
Nynehead
Trull
Pitminster
Corfe
Staple Fitzpaine
Bickenhall
Buckland St Mary
Holcombe Rogus
Burlescombe
Sampford Arundel
Sampford Peverell
Halberton
Tiverton Parkway
Butterleigh
Bickleigh
Cadeleigh
Thorverton
Stoke Canon
Rewe
Brampford Speke
Upton Pyne
Cowley
Pinhoe
Whipton
Countess Wear
Alphington
Clyst St Mary
Clyst St George
Clyst Honiton
Ebford
Exton
Powderham
Kenn
Kennford
Ashcombe
Luton
Little Haldon
Holcombe
Combeinteignhead
Stokeinteignhead
Maidencombe
Babbacombe Bay
LYME B
Due to open Late 1988
Wimbleball Lake
Clatworthy Reservoir
Hawkridge Res
Durleigh Res
Luxhay Reservoir
River Tone
River Parrett
River Axe
River Exe
River Otter
River Culm
West Sedge Moor
North Curry
Thornfalcon
Fivehead
Burrow Bridge
Athelney
Stathe
Langport
Lyng
Middlezoy
Westonzoyland
Chedzoy
Huntworth
Durleigh
Wembdon
Cannington
Nether Stowey
Fiddington
Spaxton
Enmore
Goathurst
Broomfield
Bicknoller
Sampford Brett
Monksilver
Elworthy
Brompton Ralph
Clatworthy
Huish Champflower
Chipstable
Skilgate
Morebath
Shillingford
Petton
Clayhanger
Ashbrittle
Appley
Thorne St Margaret
Runnington
Holywell Lake
Langford Budville
Bathealton
Stawley
Hockworthy
Huntsham
Cove
Stoodleigh
Chevithorne
Cheriton Fitzpaine
Withleigh
Washford
Roadwater
Treborough
Kingsbridge
Luxborough
Exton
Bridgetown
Withiel Florey
Brompton Regis
Upton
Bury
Brushford
Battleton
Exebridge
Oakford
Loxbeare
Calverleigh
Hele
Langford
Plymtree
Clyst Hydon
Dulford
Mutterton
Westcott
Norman's Green
Clyst St Lawrence
Westwood
Fairmile
West Hill
Marsh Green
Venn Ottery
Harpford
Wiggaton
Alfington
Gittisham
Buckerell
Combe Raleigh
Awliscombe
Dumpdon Hill
Upottery
Luppitt
Monkton
Cotleigh
Dalwood
Smallridge
Rousdon
Combpyne
Southleigh
Northleigh
Broad Down
Blackbury Castle
Bowd
Otterton
East Budleigh
Yettington
Black Hill
Hawkerland
Woodbury Salterton
Littleham
Whitford
Colyford
Seatown
Chideock
Morcombelake
Ryall
Whitchurch Canonicorum

F
G
H
J
K
L
1
2
3
4
5
6
7
GLASTONBURY
STREET
SOMERTON
Castle Cary
Bruton
WINCANTON
MERE
GILLINGHAM
SHAFTESBURY
Ilchester
Martock
South Petherton
YEOVIL
SHERBORNE
Milborne Port
Stalbridge
Sturminster Newton
Merriott
CREWKERNE
Beaminster
BLANDFORD FORUM
BRIDPORT
Bothenhampton
DORCHESTER
Puddletown
Wool
WAREHAM
Chickerell
WEYMOUTH
WEYMOUTH BAY
Portland Harbour
FORTUNESWELL
Easton
ISLE OF PORTLAND
BILL OF PORTLAND
Chesil Beach
WEST BAY
ENGLISH CHANNEL
B L A C K M O O R   V A L E
CRANBORNE CHASE
Ringstead Bay
Durdle Door
Lulworth Cove
Worbarrow Bay
Broad Bench
Cerne Abbas
Maiden Castle
WEYMOUTH to
Seasonal
Guernsey 8 hrs
Jersey 10 hrs
Cherbourg 4 hrs

A
B
C
D
E
1
2
3
4
5
6
7
SALISBURY
WILTON
Amesbury
Wilton
Quidhampton
Netherhampton
Harnham
Laverstock
Old Sarum
Stratford Tony
Coombe Bissett
Bishopstone
Broad Chalke
Fovant
Dinton
Teffont Magna
Chilmark
Hindon
Ridge
Sutton Mandeville
Swallowcliffe
Ebbesbourne Wake
Alvediston
Berwick St John
Bowerchalke
Woodminton
Winkelbury
Tollard Royal
Rushmore Ho
Deanland
Sixpenny Handley
Farnham
Chettle
Cashmoor
Stubhampton
Gussage St Michael
Gussage All Saints
Long Crichel
Moor Crichel
Manswood
Witchampton
Hinton Martell
Gaunt's Common
Stanbridge
Badbury Rings
Tadden
Pamphill
WIMBORNE MINSTER
Colehill
Clapgate
Corfe Mullen
Broadstone
Canford Heath
Lytchett Matravers
Lytchett Minster
Upton
Hamworthy
POOLE
Poole Harbour
Brownsea Island
Arne
WAREHAM
Sandbanks
POOLE BAY
POOLE to Seasonal Cherbourg .... 4½ hrs
Studland Bay
Studland
The Foreland or Handfast Point
Ballard Point
Swanage Bay
SWANAGE
Peveril Point
Durlston Bay
DURLSTON HEAD
Tilly Whim Caves
Langton Matravers
Worth Matravers
Seacombe Cliff
ST ALDHELM'S OR ST ALBAN'S HEAD
Kingston
Corfe Castle
Nine Barrow Down
Wytch Heath
Newton Heath
Norden
PURBECK
Kimmeridge
Church Knowle
BOURNEMOUTH
Westbourne
Parkstone
Branksome
Boscombe
Pokesdown
Southbourne
CHRISTCHURCH
Mudeford
Christchurch Bay
HENGISTBURY HEAD
Highcliffe
Barton on Sea
New Milton
Milford on Sea
Hurst Castle
Keyhaven
LYMINGTON
Boldre
Sway
Brockenhurst
Lyndhurst
Ashurst
TOTTON
SOUTHAMPTON
ROMSEY
North Baddesley
Chandler's Ford
Chilworth
Nursling
Marchwood
Hythe
Fawley
Beaulieu
Beaulieu Heath
Holbury
Exbury
N E W F O R E S T
Ringwood
Ringwood Forest
Verwood
St Leonards
St Ives
West Moors
Ferndown
Three Legged Cross
Ashley Heath
Burley
Bransgore
Sopley
Avon
Hurn
Bournemouth (Hurn) Airport
Fordingbridge
Sandleheath
Alderholt
Cranborne
Edmondsham
Woodlands
Downton
Redlynch
Woodgreen
Hale
Breamore
Whitsbury
Rockbourne
Martin
Damerham
Tidpit
Pentridge
Woodyates
West Grimstead
East Grimstead
Alderbury
Whaddon
West Dean
East Dean
Lockerley
Mottisfont
Michelmersh
Timsbury
Awbridge
Sherfield English
Wellow
Plaitford
Landford
Nomansland
Bramshaw
Cadnam
Minstead
Brook
Fritham
Stoney Cross
Emery Down
Bank
Denny Lodge
New Park
Stockbridge
Houghton
Broughton
King's Somborne
Over Wallop
Middle Wallop
Nether Wallop
Longstock
Winterslow
Pitton
Farley
Porton
Idmiston
Boscombe
Allington
Newton Tony
THE NEEDLES
Totland
Freshwater
Yarmouth
Colwell Bay
Totland Bay
Alum Bay
Compton Bay
Brook
Shalcombe
Brighstone
Mottistone
Chilton Chine
The Undercliff
Brighstone Bay
ENGLISH

WINCHESTER
Kings Worthy
New Alresford
Alton
Bordon Camp
Headley
Hindhead
Grayshott
Shottermill
HASLEMERE
Camelsdale
Liphook
Liss
Petersfield
PETERSFIELD
MIDHURST
Twyford
Shawford
Otterbourne
Bishopstoke
EASTLEIGH
Bishop's Waltham
Hedge End
Botley
Wickham
Shirrell Heath
Hambledon
Clanfield
Horndean
Cowplain
Waterlooville
Purbrook
HAVANT
Emsworth
Southbourne
Bosham
CHICHESTER
Westhampnett
Fareham
FAREHAM
Titchfield
Stubbington
Hill Head
Lee-on-the-Solent
GOSPORT
Portchester
Cosham
Wymering
Hilsea
PORTSEA ISLAND
Southsea
PORTSMOUTH
HAYLING ISLAND
South Hayling
West Wittering
East Wittering
Bracklesham Bay
Selsey
SELSEY BILL
Sidlesham
Donnington
Fishbourne
Bognor
Hamble
Netley
Fawley
Calshot
Calshot Spit
Blackfield
SOUTHAMPTON WATER
SOLENT
SPITHEAD
Gilkicker Point
Stokes Bay
Portsmouth Harbour
Langstone Harbour
Chichester Harbour
COWES
East Cowes
Osborne Ho.
Gurnard
Northwood
Parkhurst
NEWPORT
Carisbrooke
Wootton Bridge
RYDE
Puckpool Point
Seaview
St Helens
Bembridge
Bembridge Point
FOREland
Brading
Whitecliff Bay
Culver Cliff
Yaverland
SANDOWN
Sandown Bay
SHANKLIN
Shanklin Chine
Luccombe Chine
DUNNOSE
VENTNOR
St Lawrence
Niton
Whitwell
Godshill
Rookley
Arreton
Chale
Blackgang Chine
Rocken End
ST CATHERINE'S POINT
The Undercliff
Atherfield Point
Whale Chine
Shorwell
Brighstone
ISLE OF WIGHT
FOREST OF BERE
SOUTH DOWNS
CHANNEL
PORTSMOUTH to
Le Havre 5½ hrs
Guernsey 11¾ hrs
St Malo 8½-10 hrs
Jersey 9-9½ hrs
Cherbourg 4¼-6¼ hrs
Caen 5¾ hrs
F
G
H
J
K
L
1
2
3
4
5
6
7
19
12
49
07

CAMBERLEY
FRIMLEY
FARNBOROUGH
FLEET
Church Crookham
ALDERSHOT
FARNHAM
WOKING
Byfleet
Cobham
Oxshott
EPSOM
Woodmansterne
BANSTEAD
Ashtead
LEATHERHEAD
Tadworth
Kingswood
Walton on the Hill
Fetcham
Great Bookham
Effingham
East Horsley
West Horsley
Ripley
Send
Mayford
Worplesdon
Pirbright
GUILDFORD
Shalford
Chilworth
Albury
Shere
Gomshall
Westcott
DORKING
REIGATE
North Holmwood
South Holmwood
Beare Green
Newdigate
Charlwood
Horley
London (Gatwick) Airport
Capel
Ockley
Ewhurst
Cranleigh
Walliswood
Rusper
Ifield
Faygate
Colgate
Warnham
Broadbridge Heath
Roffey
HORSHAM
ST LEONARD'S FOREST
Handcross
Mannings Heath
Lower Beeding
Slaugham
Southwater
Nuthurst
Copsale
Maplehurst
Cowfold
Bolney
Rudgwick
Rowhook
Slinfold
Bucks Green
Alfold
Loxwood
Ifold
Dunsfold
Chiddingfold
Hascombe
Hambledon
Witley
Milford
GODALMING
Busbridge
Hydestile
Farncombe
Bramley
Wonersh
Shamley Green
Compton
Puttenham
Elstead
Tilford
Waverley Abbey
Frensham
Rushmoor
Churt
Hindhead
Grayshott
Headley
Bordon Camp
Lindford
Shottermill
Camelsdale
HASLEMERE
Grayswood
Fernhurst
Kingsley Green
Liphook
Linchmere
Northchapel
Plaistow
Kirdford
Wisborough Green
BILLINGSHURST
Five Oaks
Itchingfield
Christ's Hospital
Barns Green
Coneyhurst
Coolham
Shipley
Dial Post
West Grinstead
Partridge Green
Henfield
Woodmancote
Ashurst
Ashington
West Chiltington
Thakeham
Storrington
Pulborough
Nutbourne
Stopham
Fittleworth
Coldwaltham
Hardham
Wiggonholt
Greatham
Amberley
Rackham
Parham House
Washington
Wiston Park
Steyning
Bramber
Upper Beeding
Small Dole
Poynings
Fulking
Devil's Dyke
Sullington
Kithurst Hill
Harrow Hill
Findon
Cissbury Ring
North Lancing
Upper Cokeham
SHOREHAM-BY-SEA
SOUTHWICK
PORTSLADE-BY-SEA
Kingston by Sea
South Lancing
Sompting
Broadwater
WORTHING
Goring-by-Sea
Ferring
Durrington
High Salvington
Clapham
Patching
Angmering
East Preston
Rustington
LITTLEHAMPTON
Lyminster
Poling
Wick
ARUNDEL
Burpham
South Stoke
North Stoke
Houghton
Bury
Arundel Park
Walberton
Slindon
Madehurst
Eartham
Fontwell
Barnham
Eastergate
Westergate
Aldingbourne
Yapton
Ford
Climping
Flansham
Middleton-on-Sea
Felpham
BOGNOR REGIS
Aldwick
Bersted
Shripney
Woodgate
Colworth
Oving
Tangmere
Boxgrove
Halnaker
Goodwood House
Strettington
Westhampnett
CHICHESTER
Merston
Runcton
Mundham
Hunston
Donnington
Apuldram
Fishbourne
Bosham
Chidham
West Thorney
West Itchenor
Birdham
Earnley
Somerley
Sidlesham
Highleigh
Street End
West Wittering
East Wittering
Bracklesham Bay
Church Norton
Selsey
SELSEY BILL
Chichester Harbour
Southbourne
Hambrook
Funtington
West Stoke
East Ashling
West Ashling
Lavant
The Trundle
West Dean
Singleton
Charlton
East Dean
Upwaltham
Graffham
Heyshott
Bepton
Cocking
Treyford
Elsted
South Harting
East Harting
Nyewood
Rogate
Trotton
Chithurst
Stedham
Iping
Woolbeding
Easebourne
MIDHURST
West Lavington
South Ambersham
Selham
Tillington
Petworth
Byworth
Lodsworth
Lurgashall
Lickfold
Balls Cross
Duncton
Barlavington
Sutton
Bignor
West Burton
East Lavington
SOUTH DOWNS
STANE STREET
Chilgrove
North Marden
East Marden
West Marden
Compton
Uppark
Stoughton
Walderton
Forestside
Milland
Henley
Rake
Hill Brow
Liss
East Liss
Liss Forest
Greatham
Longmoor Camp
Bramshott
Kingsley
Sleaford
Blacknest
Bucks Horn Oak
Dippenhall
Crondall
Ewshot
Upper Hale
Heath End
Crookham Village
Winchfield
Hartley Wintney
Phoenix Green
Minley Manor
Mytchett
Deepcut
Bisley
Knaphill
Brookwood
Normandy
Ash
Tongham
Flexford
Wood Street
Stoughton
Pitch Place
Onslow Village
Blackheath
Peaslake
Holmbury St Mary
Abinger Common
Wotton
Sutton
Coldharbour
Leith Hill
Mickleham
Westhumble
Brockham
Betchworth
Buckland
Leigh
Norwood Hill
Parkgate
Hookwood
Holmwood
Anstiebury
Jayes Park
Forest Green
Ellen's Green
Oakwoodhill
Kingsfold
Lyne House
Lambs Green
Denne Park
Warninglid
Crabtree
Monastery
Wineham
Twineham
Sayers Common
Albourne
Blackstone
Newtimber Place
Edburton
Mock Bridge
Newpound Common
Roundstreet Common
Gay Street
Broomer's Corner
Warnham
Tilford
Pepper Harow
Thursley
Wheelerstreet
Brook
Bowlhead Green
Gibbet Hill
Fisherstreet
Upperton
Petworth House
Crockerhill
Rowly
Thorncombe Street
Loxhill
Eashing
Hurt Wood
Farley Green
Hog's Back
Shackleford
Runfold
Compton
Clandon
West Clandon
East Clandon
Ockham
Effingham Junc Sta
Stoke D'Abernon
Pyrford
Wisley
Burpham
Merrow
Stanwell
Chobham Ridges
Blackwater
Yateley
Hawley
Rowledge
Millbridge
Spreakley
Wrecclesham
Lower Froyle
Bentley
Wyck
East Worldham
Blackmoor
Whitehill
Sheet
Rogate
Nursted
Forestside
Stansted
Woodmancote
Emsworth
Westbourne
Funtington
Walderton
A 272
A 29
A 24
A 283
A 286
A 285
A 281
A 264
A 31
A 3
A 25
A 246
A 287
A 325
A 259
A 27
M 25
B 2133
B 2139
SU
SZ
A B C D E F
1 2 3 4 5 6 7
19
20
48
09

WARLINGHAM
Woldingham
CATERHAM
Godstone
Oxted
Limpsfield
REDHILL
South Nutfield
Salfords
Outwood
Horley
Smallfield
Lingfield
Dormansland
Felbridge
EAST GRINSTEAD
Crawley Down
CRAWLEY
Turners Hill
Balcombe
Lindfield
HAYWARDS HEATH
Cuckfield
BURGESS HILL
Hurstpierpoint
Hassocks
Keymer
Ditchling
Ditchling Beacon
Plumpton
Newick
Chailey
Fletching
Sheffield Park Sta
Uckfield
Maresfield
Nutley
Danehill
ASHDOWN FOREST
Forest Row
Hartfield
Withyham
Groombridge
Eridge Sta
Frant
CROWBOROUGH
Rotherfield
Mayfield
Wadhurst
Ticehurst
Hawkhurst
Flimwell
Lamberhurst
Goudhurst
Bewl Water
Bedgebury Forest
THE WEALD
Edenbridge
Hever
Chiddingstone
Penshurst
Leigh
TONBRIDGE
SOUTHBOROUGH
ROYAL TUNBRIDGE WELLS
Pembury
Paddock Wood
Marden
Horsmonden
Brenchley
Hadlow
East Peckham
West Peckham
Borough Green
SEVENOAKS
Knole
Westerham
Otford
Shoreham
MAIDSTONE
Coxheath
Yalding
Nettlestead
Wateringbury
Mereworth
Etchingham
Hurst Green
Robertsbridge
Burwash
Heathfield
Cross in Hand
Buxted
Framfield
Halland
Horam
Dallington
Netherfield
Whatlington
BATTLE
Catsfield
Ninfield
Hooe
Herstmonceux
Hellingly
HAILSHAM
Chiddingly
Ringmer
LEWES
Mount Caburn
Glynde
Beddingham
Firle Beacon
West Firle
Selmeston
Alciston
Berwick
Alfriston
Wilmington
Folkington
The Long Man
Polegate
Willingdon
Pevensey
Westham
Pevensey Bay
Langney Point
EASTBOURNE
East Dean
Friston
BEACHY HEAD
SEAFORD
NEWHAVEN
Denton
Piddinghoe
Telscombe
Saltdean
Rottingdean
Peacehaven
Woodingdean
Falmer
Kingston near Lewes
BRIGHTON
HOVE
Preston
Patcham
Pyecombe
Clayton
SOUTH DOWNS
PEVENSEY LEVELS
Little Common
Sidley
BEXHILL
Hollington
St Leonards
NEWHAVEN to Dieppe 4-5 hrs
ENGLISH CHANNEL
Royal Sovereign
TQ
G
H
J
K
L
1
2
3
4
5
6
7

ERITH
BEXLEY
DARTFORD
SIDCUP
ORPINGTON
SEVENOAKS
GRAYS
TILBURY
NORTHFLEET
GRAVESEND
ROCHESTER
CHATHAM
GILLINGHAM
SITTINGBOURNE
SHEERNESS
Isle of Grain
River Medway
MAIDSTONE
TONBRIDGE
SOUTHBOROUGH
ROYAL TUNBRIDGE WELLS
CROWBOROUGH
HEATHFIELD
HAILSHAM
BATTLE
HASTINGS
BEXHILL
RYE
TENTERDEN
Cranbrook
Staplehurst
Headcorn
Bearsted
West Malling
Snodland
Paddock Wood
Wadhurst
Ticehurst
Uckfield
Winchelsea
Fairlight
Bedgebury Forest
Isle of Oxney
N O R T H D O W N S
W E A L D
A
B
C
D
E
F
1
2
3
4
5
6
7
8

G
H
23
J
K
64
L
1
2
3
4
5
6
7
8
SHEERNESS to
Vissingen (Flushing).....8·9½ hrs
Warden Point
Warden
Leysdown on Sea
SHEPPEY
Isle of Harty
Shell Ness
WHITSTABLE
Swalecliffe
Seasalter
Chestfield
The Swale
HERNE BAY
REGVLBIVM
Reculver
Beltinge
Hillborough
Herne
Broomfield
Hoath
St Nicholas at Wade
Sarre
Marshside
Chislet
Birchington
Westgate on Sea
MARGATE
South Channel
Long Nose Spit
Foreness Point
White Ness
NORTH FORELAND
BROADSTAIRS
North Goodwin
Acol
Manston Airport
Minster
Monkton
RAMSGATE
Cliffs End
Pegwell Bay
RAMSGATE to
Dunkirk.................2½ hrs
River Stour
Sandwich Bay
Sandwich Flats
SANDWICH
Toll
THE SMALL DOWNS
Goodwin Sands
THE DOWNS
DEAL
Castle
Walmer Castle
Kingsdown
Worth
Sholden
Northbourne
Eastry
Ham
Tilmanstone
Betteshanger
Great Mongeham
Ripple
Sutton
Ringwould
St Margaret's at Cliffe
St Margaret's Bay
SOUTH FORELAND
South Goodwin
West Cliffe
East Langdon
West Langdon
Guston
Whitfield
Military School
DOVER
DVBRIS
DOVER to
Calais.............1¼ hrs
Boulogne.............1¾ hrs
Dunkirk.............2¼ hrs
Ostend.............3½-4½ hrs
Zeebrugge.............4 hrs
Calais.............35 mins
Boulogne.............40 mins
STRAIT OF DOVER
Varne
FAVERSHAM
Goodnestone
Graveney
Oare
Luddenham Court
Uplees
Ospringe
Boughton Street
Hernhill
Dargate
Dunkirk
Yorkletts
Honey Hill
Blean
Rough Common
Tyler Hill
Broad Oak
Sturry
Hales Place
CANTERBURY
DVROVERNVM
Harbledown
Fordwich
Westbere
Stodmarsh
Upstreet
Hersden
Grove
Preston
Elmstone
West Stourmouth
East Stourmouth
Westmarsh
Hoaden
Wickhambreaux
Littlebourne
Ickham
Zoo
Wingham
Ash
Marshborough
Staple
Great Stonar
RVTVPIAE
Woodnesborough
Bekesbourne
Patrixbourne
Bridge
Thanington
Chartham
Nackington
Shalmsford Street
Chartham Hatch
Selling
Sheldwich
Throwley
Badlesmere
Shottenden
Chilham
Old Wives Lees
Leaveland
Challock
Molash
Godmersham
Bilting
Boughton Aluph
Eastwell Park
Boughton Lees
Westwell
Kennington
Wye
Olantigh
Crundale
Hassell Street
Sole Street
Waltham
Petham
Lower Hardres
Bishopsbourne
Upper Hardres Court
Kingston
Barham
Adisham
Aylesham
Goodnestone
Chillenden
Nonington
Knowlton
Easole Street
Elvington
Womenswold
Barfrestone
Eythorne
Waldershare House
Woolage Green
Shepherdswell or Sibertswold
Coldred
Temple Ewell
Lydden
Ewell Minnis
Denton
Wooton
Derringstone
Bossingham
Stelling Minnis
Lyminge Forest
Stone Street
Roman Road
Great Stour
Elmsted
Hastingleigh
Brook
Lymbridge Green
Elham
Acrise Place
Swingfield Minnis
Chapel
Alkham
Buckland
St Radegund's Abbey
West Hougham
Densole
Rhodes Minnis
Stowting
Lyminge
Paddlesworth
Hawkinge
Capel-le-Ferne
Etchinghill
Postling
Beachborough
Newington
Stanford
Sellindge
Brabourne
Brabourne Lees
Smeeth
Hinxhill
Willesborough
Sevington
Mersham
ASHFORD
Hothfield
Chart
Kingsnorth
Shadoxhurst
Bromley Green
Orlestone
Hamstreet
Ruckinge
Bilsington
Bonnington
Aldington Frith
Aldington
Cheeseman's Green
Lympne
Lympne Castle
Saltwood
Castle
HYTHE
Sandgate
FOLKESTONE
Tolls
East Wear Bay
FOLKESTONE to
Boulogne.............1¾ hrs
Royal Military Canal
Warehorne
Burmarsh
Newchurch
ROMNEY MARSH
Snave
Dymchurch
Snargate
Ivychurch
St Mary in the Marsh
St Mary's Bay
Brenzett
Brookland
Old Romney
Littlestone-on-Sea
NEW ROMNEY
Romney Sands
East Road
Greatstone-on-Sea
WALLAND MARSH
LYDD
Lydd Airport
Denge Marsh
Lydd-on-Sea
Denge Beach
Power Station
DUNGENESS
West Road

A
B
C
D
E
1
2
3
4
5
6
7
8
RHONDDA
Ystrad
Tylorstown
ABERCYNON
Ynysybwl
Nelson
Hengoed
YSTRAD MYNACH
PONTLLANFRAITH
ABERCARN
CWMBRAN
Llangeveth
Nant-y-moel
Blaengwynfi
Price Town
Llwynypia
Clydach Vale
Trealaw
TONYPANDY
Penygraig
Wattstown
Ynyshir
PORTH
Trehafod
PONTYCYMER
Wyndham
Ogmore Vale
Ogmore Forest
Gilfach Goch
Cymmer
PONTYPRIDD
Tonyrefail
Treforest
Senghenydd
Abertridwr
Llanbradach
BEDWAS
Trethomas
MACHEN
Pontymister
Crosskeys
RISCA
Caerleon
NEWPORT
CAERPHILLY
Rudry
Church Village
Beddau
Llantrisant
Llantwit Fardre
Efail Isaf
Taff's Well
Tongwynlais
Castellau
Blackmill
Mynydd y Gaer
Heol-y-Cyw
Brynna
Llanharan
Pencoed
BRIDGEND
Coity
Coychurch
Llanharry
Pontyclun
Miskin
Creigiau
Radyr
Pentyrch
Whitchurch
Llanishen
Heath
Roath
Llandaff
Canton
St Fagans
Ely
Peterston-super-Ely
St Bride's-super-Ely
St George's
Welsh St Donats
Pendoylan
Ystradowen
Maendy
COWBRIDGE
Llanblethian
Colwinston
Llysworney
Ewenny
St Brides Major
Wick
Llandow
Broughton
Monknash
Marcross
St Donats
Llantwit Major
Boverton
Castle Ditches
Nash Point
Eglwys Brewis
St Athan
Gileston
Llanmaes
Flemingston
St Mary Church
Llancarfan
Penmark
St Hilary
Llantrithyd
Bonvilston
St Nicholas
St Lythans
Wenvoe
Michaelston-le-Pit
Leckwith
Llandough
Dinas Powys
PENARTH
Lower Penarth
Lavernock Point
Sully
Sully Island
BARRY
Barry Island
Rhoose
Porthkerry
Cardiff Wales Airport
East Aberthaw
Breaksea Point
CARDIFF
Rumney
St Mellons
Peterstone Wentlooge
St Brides Wentlooge
Marshfield
Castleton
Coedkernew
Michaelstone-y-Fedw
Bassaleg
Rogerstone
Bettws
Malpas
Nash
Goldcliff
Whitson
Redwick
Pye Corner
Flat Holm
Steep Holm
MOUTH OF THE SEVERN
Middle Grounds
CLEVEDON
Langford Grounds
Kingston Seymour
Sand Point
Wick St Lawrence
Sand Bay
Worlebury
Kewstoke
St Georges
WESTON-SUPER-MARE
Weston Bay
Weston Airport
West Wick
Locking
Hutton
Uphill
Brean Down
Bleadon
Banwell
Christon
Loxton
Brean
Berrow Flats
Gore Sand
Berrow
Lympsham
Eastertown
East Brent
Brent Knoll
BURNHAM-ON-SEA
Edithmead
Highbridge
Mark
Watchfield
Bason Bridge
East Huntspill
Huntspill
Huntspill Level
BRIDGWATER BAY
Stert Flats
Steart
Stolford
Hinkley Point Power Sta
Shurton
Stockland Bristol
Stogursey
Combwich
Pawlett
Puriton
Woolavington
Cossington
Chilton Polden
Bawdrip
Dunball
Chilton Trinity
Cannington
BRIDGWATER
Wembdon
Durleigh
Chedzoy
Westonzoyland
Middlezoy
North Petherton
North Newton
Huntworth
Selworthy Beacon
Allerford
Selworthy
North Hill
MINEHEAD
Alcombe
Blue Anchor Bay
Blue Anchor
Dunster
Carhampton
Timberscombe
Wootton Courtenay
Dunkery Hill
Withycombe
Washford
Old Cleeve
WATCHET
Williton
Sampford Brett
Bicknoller
West Quantoxhead
East Quantoxhead
Kilve
Kilton
Lilstock
Burton
Stringston
Holford
Nether Stowey
Fiddington
QUANTOCK HILLS
Croydon Hill
Luxborough
Roadwater
Wheddon Cross
Luckwell Bridge
Lype Hill
BRENDON HILLS
Kingsbridge
Treborough
Monksilver
Stogumber
Crowcombe
Elworthy
Flaxpool
West Bagborough
Lower Aisholt
Enmore
Goathurst
Broomfield
Spaxton
Charlynch
Wimbleball Lake
Clatworthy Reservoir
Exton
Bridgetown
Withiel Florey
Brompton Regis
Brompton Ralph
Clatworthy
Upton
Huish Champflower
Langley Marsh
Wiveliscombe
Dulverton
Bury
Skilgate
Haddon Hill
Lydeard St Lawrence
Combe Florey
Tolland
Ash Priors
Bishops Lydeard
Cothelstone
Kingston St Mary
Cheddon Fitzpaine
West Monkton
Monkton Heathfield
Thurloxton
Durston
Hedging
Lyng
Athelney
Burrow Bridge
Stathe
River Tone
Creech St Michael
Ruishton
Thornfalcon
Knapp
North Curry
Stoke St Gregory
Staplegrove
TAUNTON
VALE OF TAUNTON DEANE
Norton Fitzwarren
Bradford-on-Tone
Heathfield
Oake
Milverton
Waterow
Chipstable
Batheaion
Langford Budville
Nynehead
Petton
Shillingford
Clayhanger
Morebath
Exebridge
Brushford
Battleton
Fivehead
Langport
Drayton
King's Sedgemoor
Othery
Greinton
Moorlinch
Sutton Mallet
Stawell
Edington
Catcott
Burtle
Westhay
Wedmore
Compton Bishop
Cross
Badgworth
Biddisham
Stone Allerton
Chapel Allerton
Rooks Bridge
West Stoughton
Puxton
Hewish
Bourton
Yatton

29
30
18
9
F
G
H
J
K
L
1
2
3
4
5
6
7
8
CHEPSTOW
THORNBURY
Wotton-under-Edge
TETBURY
MALMESBURY
CHIPPING SODBURY
YATE
Winterbourne
Frampton Cotterell
Patchway
Filton
Avonmouth
PORTISHEAD
BRISTOL
KINGSWOOD
Nailsea
Backwell
KEYNSHAM
Saltford
BATH
Batheaston
Corsham
CHIPPENHAM
MELKSHAM
BRADFORD ON AVON
TROWBRIDGE
Holt
Congresbury
Bristol Airport
Chew Magna
Chew Valley Lake
Pensford
Farmborough
Peasedown St John
Paulton
MIDSOMER NORTON
RADSTOCK
Axbridge
Cheddar
Cheddar Gorge
MENDIP HILLS
Mendip Forest
Wedmore
WELLS
SHEPTON MALLET
FROME
WESTBURY
WARMINSTER
Longleat House
GLASTONBURY
STREET
Evercreech
Bruton
Castle Cary
WINCANTON
Mere
SOMERTON
GILLINGHAM
SHAFTESBURY
Caldicot
Portskewett
Severn Tunnel Junction Station
Severn Road Bridge
Welsh Grounds
Denny Island
Clevedon Court
Stourhead
White Sheet Hill
Alfred's Tower
Keysley Down
Great Ridge
ROMAN ROAD
M4
M5
M32
A 38
A 420
A 4
A 36
A 37
A 39
A 361
A 303
A 350
A 367
A 368
A 371

A
B
C
D
E
1
2
3
4
5
6
7
8
CRICKLADE
MALMESBURY
Purton
HIGHWORTH
Watchfield
Shrivenham
Uffington
WANTAGE
VALE OF WHITE HORSE
SWINDON
WOOTTON BASSETT
Wroughton
Chiseldon
Lambourn
Aldbourne
Hungerford
CHIPPENHAM
CALNE
Avebury
MARLBOROUGH
Savernake Forest
MELKSHAM
DEVIZES
VALE OF PEWSEY
Pewsey
Upavon
Ludgershall
North Tidworth
ANDOVER
SALISBURY PLAIN
Durrington
Bulford Camp
Amesbury
Larkhill
Stonehenge
WILTON
SALISBURY
Shrewton
Tilshead
Market Lavington
West Lavington
Stockbridge
Wylye Valley
Marlborough Downs
Ridge Way
Kennet and Avon Canal
Wansdyke
M4
A4
A303 (T)
A338
A346
A345
A360
A36 (T)
A30
Barbury Castle
Silbury Hill
Old Sarum
Harewood Forest
Collingbourne Ducis
Everleigh
Enford
Netheravon
Figheldean
Winterbourne Stoke
Porton
Bishops Cannings
Lockeridge
Tidcombe
Faccombe
Kintbury

F
G
H
J
K
L
1
2
3
4
5
6
7
8
HIGH WYCOMBE
BEACONSFIELD
WALLINGFORD
MARLOW
HENLEY-ON-THAMES
MAIDENHEAD
WINDSOR
READING
WOKINGHAM
BRACKNELL
CAMBERLEY
FRIMLEY
FARNBOROUGH
FLEET
ALDERSHOT
FARNHAM
HASLEMERE
BASINGSTOKE
NEWBURY
THATCHAM
WINCHESTER
ALTON
SU
Didcot
Goring
Pangbourne
Tadley
Kingsclere
Overton
Whitchurch
Hook
Odiham
New Alresford
Grayshott
Hindhead
Sandhurst
Crowthorne
Yateley
Twyford
Woodley
Sonning
Theale
Streatley
Cholsey
Newport Pagnell
Ascot
Sunninghill
Windlesham
Bagshot
Godalming
Milford
Liphook
Petersfield

A
B
C
D
E
1
2
3
4
5
6
7
8
35
36
31
19
12
Stratton Audley
Twyford
Swanbourne
Stewkley
Hockliffe
Wingfield
Chalton
Poundon
Charndon
Middle Claydon
East Claydon
Granborough
Hoggeston
Dunton
Linslade
LEIGHTON BUZZARD
Eggington
Stanbridge
Tilsworth
HOUGHTON REGIS
Launton
Marsh Gibbon
Calvert
Edgcott
North Marston
Cublington
Wing
Billington
Totternhoe
DUNSTABLE
Caddington
Kensworth
Ambrosden
Arncott
Blackthorn
Grendon Underwood
Kingswood
Oving
Pitchcott
Whitchurch
Aston Abbotts
Hardwick
Rowsham
Mentmore
Wingrave
Ledburn
Slapton
Eaton Bray
Church End
Edlesborough
Ivinghoe Aston
Whipsnade
Whipsnade Park Zoo
Studham
Markyate
Flamstead
Murcott
Piddington
Ludgershall
Wotton Underwood
Westcott
Waddesdon
Quainton
Weedon
Hulcott
Cheddington
Long Marston
Ivinghoe
Pitstone Green
Dagnall
Boarstall
Brill
Dorton
Ashendon
Upper Winchendon
AYLESBURY
Bierton
Puttenham
Wilstone
Marsworth
Ringshall
Little Gaddesden
Great Gaddesden
Horton-cum-Studley
Oakley
Chilton
Nether Winchendon or Lower Winchendon
Cuddington
Stone
Upton
Dinton
Buckland
Aston Clinton
New Mill
TRING
Aldbury
Ashridge College
Water End
Worminghall
Long Crendon
Easington
Chearsley
Haddenham
Aston Sandford
Ford
Bishopstone
Stoke Mandeville
Weston Turville
Halton
Wigginton
Nettleden
Potten End
Piccotts End
Ickford
Shabbington
Waterperry
Waterstock
Holton
THAME
Kingsey
Towersey
Ilmer
Owlswick
Kimble Wick
Nash Lee
Kimble
WENDOVER
Hastoe
Northchurch
BERKHAMSTED
Ellesborough
St Leonards
Cholesbury
Buckland Common
HEMEL HEMPSTEAD
Bourne End
Bovingdon
Felden
Bedmond
Tiddington
Great Milton
Moreton
Longwick
Monks Risborough
Meadle
Askett
Dunsmore
Little Hampden
The Lee
Lee Clump
Ashley Green
Whelpley Hill
Pitch Green
Henton
PRINCES RISBOROUGH
Great Hampden
Ballinger Common
Chartridge
Lye Green
Botley
CHESHAM
Kings Langley
Chipperfield
Great Haseley
Tetsworth
Emmington
Sydenham
Chinnor
Bledlow
Saunderton
Lacey Green
Prestwood
Great Missenden
South Heath
Hyde Heath
Flaunden
Bucks Hill
Little Milton
Little Haseley
Stoke Talmage
Adwell
Postcombe
Kingston Blount
Speen
Upper North Dean
Little Kingshill
Little Missenden
Chesham Bois
Latimer
Chenies
Sarratt
Chandler's Cross
Drayton St Leonard
Stadhampton
Easington
South Weston
Lewknor
Aston Rowant
Bledlow Ridge
Radnage
The City
Stokenchurch
Walter's Ash
Great Kingshill
Naphill
Bradenham
Holmer Green
AMERSHAM
Little Chalfont
Chalfont and Latimer Sta
CHORLEYWOOD
WATFORD
Croxley Green
Cassiobury Park
Newington
Chalgrove
Pyrton
Cuxham
Shirburn
Watlington
Hughenden Valley
Downley
Hazlemere
Penn Street
Coleshill
Chalfont St Giles
RICKMANSWORTH
Heronsgate
Berrick Salome
Brightwell Baldwin
Christmas Common
Beacon Bottom
West Wycombe
HIGH WYCOMBE
Totteridge
Tylers Green
Winchmore Hill
Knotty Green
Seer Green
Chalfont Common
Jordans
Horn Hill
Maple Cross
West Hyde
Harefield
NORTHWOOD
Benson
Ewelme
Britwell Salome
Greenfield
Northend
Ibstone
Turville Heath
Cadmore End
Lane End
Booker
Wycombe Marsh
Loudwater
Penn
Chalfont St Peter
Ruislip Common
WALLINGFORD
Crowmarsh Gifford
Cookley Green
Russell's Water
Pishill
Turville
Fingest
Frieth
Skirmett
Stonor
Park Corner
Flackwell Heath
BEACONSFIELD
Wooburn Green
Wooburn
GERRARDS CROSS
Cholsey
Nuffield
Nettlebed
Bix
Fawley
Hambleden
Rockwell End
Lower Woodend
Marlow Bottom
Little Marlow
Bourne End
Hedsor
Hedgerley
Fulmer
Denham
UXBRIDGE
HILLINGDON
North Stoke
Ipsden
Middle Assendon
Mill End
Aston
MARLOW
Medmenham
Cookham Dean
Cookham
Cookham Rise
Burnham Beeches
Farnham Common
Stoke Poges
Iver Heath
Iver
Stoke Row
Highmoor Cross
Lower Assendon
Remenham
Remenham Hill
HENLEY-ON-THAMES
Hurley
MAIDENHEAD
Taplow
Burnham
Farnham Royal
SLOUGH
Cowley
YIEWSLEY
South Stoke
Woodcote
Checkendon
Shepherd's Green
Rotherfield Greys
Rotherfield Peppard
Cockpole Green
Warren Row
Knowl Hill
Littlewick Green
Woodlands Park
Bray
Dorney
Upton
Langley
WEST DRAYTON
Goring
Cray's Pond
Sonning Common
Kidmore End
Harpsden
Lower Shiplake
Wargrave
Hare Hatch
Waltham St Lawrence
White Waltham
Boveney
ETON
Datchet
Colnbrook
Poyle
London Heathrow Airport
Lower Basildon
Whitchurch
Whitchurch-on-Thames
Tokers Green
Binfield Heath
Dunsden Green
Shiplake
Twyford
Ruscombe
Holyport
Paley Street
Fifield
Oakley Green
WINDSOR
Frogmore
Old Windsor
Horton
Wraysbury
Stanwell
Upper Basildon
Pangbourne
Mapledurham
Purley on Thames
Emmer Green
Play Hatch
Caversham
Sonning
Whistley Green
Hurst
Shurlock Row
Hawthorn Hill
Windsor Forest
WINDSOR GREAT PARK
Hythe End
STAINES
East Bedfont
Tidmarsh
Sulham
Tilehurst
Calcot
READING
Woodley
Winkfield
Winkfield Row
Cranbourne
Englefield Green
EGHAM
ASHFORD
SUNBURY
Bradfield
Englefield
Theale
Southend
Sheffield Bottom
Whitley
Winnersh
Binfield
Warfield
Newell Green
Woodside
Littleton
Burghfield
Sindlesham
Arborfield
BRACKNELL
North Ascot
Ascot
Sunningdale
Virginia Water
Thorpe
Laleham
Beenham
Ufton Nervet
Sulhamstead
Three Mile Cross
Shinfield
Barkham
WOKINGHAM
Easthampstead
Sunninghill
Lyne
Shepperton
Aldermaston Wharf
Burghfield Common
Wokefield Park
Grazeley
Spencers Wood
Arborfield Cross
Arborfield Garrison
Caesar's Camp
Chobham Common
CHERTSEY
ADDLESTONE
WEYBRIDGE
Padworth
Mortimer
Stratfield Mortimer
Swallowfield
Farley Hill
Finchampstead
Wick Hill
Crowthorne
Surrey Hill
Windlesham
Ottershaw
Silchester
Mortimer West End
Beech Hill
Riseley
Bagshot
Lightwater
Burrowhill
Chobham
Woodham
Pamber Heath
Tadley
Pamber Green
West End Green
Stratfield Saye House
Stratfield Saye Monument
Bramshill Plantn
Eversley
Eversley Cross
Yateley
Sandhurst
Royal Military Academy
CAMBERLEY
Bagshot Heath
West End
Little London
Stratfield Turgis
Heckfield
Hazeley
Bramshill
Blackwater
Frogmore
Chobham Ridges
Bisley
Byfleet
Pamber End
Bramley
Turgis Green
Hound Green
Mattingley
Hawley
Knaphill
Horsell
Pyrford
Charter Alley
Sherfield on Loddon
Hartley Wespall
West Green
Hartley Wintney
Minley Manor
Fox Lane
FRIMLEY
Deepcut
Brookwood
WOKING
Sherborne St John
Church End
Rotherwick
Phoenix Green
Cove
Mytchett
Pirbright
Mayford
Send
Ripley
Ockham
Effingham Junc Sta
Chineham
Old Basing
BASINGSTOKE
Newnham
Hook
Winchfield
FLEET
FARNBOROUGH
Henley Park
Worplesdon
Pitch Place
West Horsley
Wootton
North
Dogmersfield
Church Crookham
Chiltern Hills
Icknield Way
Grand Union Canal
Wendover Arm
Tring Resrs
River Thames
Roman Road
Akeman Street
Blackwater
Basingstoke Canal
M4
M40
M25
M1

LUTON
STEVENAGE
BISHOP'S STORTFORD
HARPENDEN
WELWYN GARDEN CITY
HERTFORD
WARE
SAWBRIDGEWORTH
HARLOW
HATFIELD
ST. ALBANS
HODDESDON
CHESHUNT
WALTHAM ABBEY
EPPING
POTTERS BAR
RADLETT
BOREHAMWOOD
BUSHEY
ENFIELD
BARNET
LOUGHTON
CHIGWELL
EDGWARE
HARROW
WEMBLEY
HENDON
HAMPSTEAD
WILLESDEN
WALTHAMSTOW
ROMFORD
HORNCHURCH
UPMINSTER
BARKING
DAGENHAM
ILFORD
STRATFORD
PADDINGTON
MARYLEBONE
WESTMINSTER
KENSINGTON
HAMMERSMITH
EALING
SOUTHALL
HAYES
HOUNSLOW
ISLEWORTH
TWICKENHAM
RICHMOND
PUTNEY
WANDSWORTH
LONDON
WOOLWICH
GREENWICH
LEWISHAM
BERMONDSEY
CAMBERWELL
BRIXTON
CLAPHAM
BATTERSEA
FULHAM
CHISWICK
ERITH
DARTFORD
CRAYFORD
BEXLEY
SIDCUP
BROMLEY
BECKENHAM
CROYDON
MITCHAM
MERTON
WIMBLEDON
KINGSTON UPON THAMES
SURBITON
ESHER
EPSOM
SUTTON
WALLINGTON
PURLEY
COULSDON
ORPINGTON
SWANLEY
SEVENOAKS
LEATHERHEAD
CATERHAM
REDHILL
NORTH WEALD BASSETT
CHIPPING ONGAR
BRENTWOOD

BRAINTREE
Coggeshall
Kelvedon
WITHAM
GREAT DUNMOW
HARLOW
CHELMSFORD
MALDON
Heybridge
Tiptree
Great Baddow
Danbury
CHIPPING ONGAR
North Weald Bassett
Writtle
Ingatestone
BRENTWOOD
BILLERICAY
WICKFORD
RAYLEIGH
Hockley
Rochford
ROMFORD
HORNCHURCH
UPMINSTER
DAGENHAM
BASILDON
SOUTH BENFLEET
Hadleigh
SOUTHEND-ON-SEA
Shoeburyness
CANVEY ISLAND
Corringham
Stanford-le-Hope
LONDON
ERITH
GRAYS
TILBURY
Thurrock
BEXLEY
DARTFORD
SIDCUP
SWANSCOMBE
NORTHFLEET
GRAVESEND
ROCHESTER
CHATHAM
GILLINGHAM
SHEERNESS
SITTINGBOURNE
ORPINGTON
Swanley
Isle of Grain
River Medway
RIVER THAMES
RIVER Crouch
Osea
Northey Island
Due to open Mid 1989
SHEERNESS to Vlissingen (Flushing) 8·9½ h
NORTH DOWNS

NORTH SEA
COLCHESTER
CAMVLODVNVM
WIVENHOE
BRIGHTLINGSEA
CLACTON-ON-SEA
Holland-on-Sea
FRINTON-ON-SEA
WALTON-ON-THE-NAZE
The Naze
Kirby Cross
Thorpe le-Soken
Weeley
Great Bentley
Alresford
Thorrington
Little Clacton
Great Clacton
Great Holland
St Osyth
St Osyth Marsh
Jaywick
Colne Point
Point Clear
Priory
MERSEA ISLAND
East Mersea
Mersea Flats
WEST MERSEA
Virley Channel
The Nass
Great Wigborough
Layer de la Haye
Abberton Resr
Peldon
Langenhoe
Rowhedge
Fingringhoe
Abberton
Blackheath
Old Heath
Shrub End
Mile End
Horkesley
Ardleigh
Little Bromley
Horsley Cross
Horsleycross Street
Bradfield
Wix
Ramsey
Little Oakley
Great Oakley
Stones Green
Beaumont
Horsey Island
Hamford Water
Kirby le Soken
Tendring
Little Bentley
Frating Green
Aingers Green
Weeley Heath
Elmstead Market
Crockleford Heath
Great Bromley
Rough Tower
Gunfleet Sand
Barrow Deep
Sunk Sand
Midbarrow
Tongue
Buxey Sand
Dengie Flat
Ray Sand
Foulness Sands
Foulness Point
Holliwell Point
MAPLIN SANDS
FOULNESS ISLAND
Potton Island
Havengore Island
Courtsend
Churchend
River Roach
River Crouch
BURNHAM-ON-CROUCH
Southminster
Montsale
Deal Hall
Asheldham
Dengie
Tillingham
St Lawrence
Bradwell-on-Sea
Bradwell Waterside
BLACKWATER
Sales Point
St Peter's Flat
South Channel
MARGATE
Long Nose Spit
Foreness Point
White Ness
NORTH FORELAND
BROADSTAIRS
RAMSGATE
RAMSGATE to Dunkirk 2½ hrs
North Goodwin
Goodwin Sands
Pegwell Bay
Sandwich Bay
Cliffs End
Manston Airport
Minster
Monkton
Westgate on Sea
Birchington
Acol
Sarre
St Nicholas at Wade
Reculver
REGVLBIVM
HERNE BAY
Beltinge
Hillborough
Broomfield
Herne
Marshside
Chislet
Upstreet
Hersden
West Stourmouth
East Stourmouth
Westmarsh
Preston
Elmstone
Hoaden
Grove
Westbere
Stodmarsh
Wickhambreaux
Fordwich
Sturry
CANTERBURY
DVRVERNVM
Littlebourne
Ickham
Wingham
Ash
Marshborough
Staple
Woodnesborough
Great Stonar
RVTVPIAE
River Stour
WHITSTABLE
Swalecliffe
Chestfield
Seasalter
Yorkletts
Dargate
Hernhill
Boughton Street
Dunkirk
Blean
Rough Common
Tyler Hill
Honey Hill
Broad Oak
Hales Place
Harbledown
Chartham Hatch
Thanington
Selling
Sheldwich
Eastling
Ospringe
FAVERSHAM
Goodnestone
Graveney
Oare
Uplees
Luddenham Court
Teynham
Conyer
The Swale
Shell Ness
Isle of Harty
ISLE OF SHEPPEY
Minster
Eastchurch
Warden
Warden Point
Leysdown on Sea
THE SMALL DOWNS
Sandwich Flats
TM
TR

CARMARTHEN
CARMARTHEN BAY
BRISTOL CHANNEL
SWANSEA BAY
SWANSEA
LLANELLI
GORSEINON
AMMANFORD
LLANDEILO
KIDWELLY
BURRY PORT
PORT TALBOT
St Clears
Whitland
Laugharne
Pendine
Pendine Sands
Laugharne Sands
Cefn Sidan Sands
Pembrey Forest
Pembrey
Ferryside
Llansteffan
Trimsaran
Pontyberem
Pontyates
Cross Hands
Tumble
Llannon
Hendy
Pontardulais
Pontardawe
Clydach
Glanaman
Garnant
Gwaun-Cae-Gurwen
Brynamman
Llangennech
Loughor
Gowerton
Dunvant
Killay
Sketty
Morriston
Landore
Llansamlet
Skewen
The Mumbles
Mumbles Head
Bishopston
Pennard
Penmaen
Oxwich
Oxwich Bay
Horton
Port-Eynon
Port-Eynon Point
Port-Eynon Bay
Pwlldu Head
Rhossili
Rhossili Bay
WORMS HEAD
(Penrhyn-gwyr)
Llangennith
Burry Holms
Whiteford Point
Llanrhidian Sands
Llanrhidian
Cheriton
Llanmadoc
Reynoldston
Knelston
Scurlage
Llanddewi
Nicholaston
Penrice
Parkmill
Ilston
Three Crosses
Crofty
Penclawdd
Helwick
Scarweather
Abergwili
Llangunnor
Nantgaredig
Dryslwyn
Dynevor Castle
Llangadog
Llansawel
Talley
Abergorlech
Brechfa
Alltwalis
Pontarsais
Newchurch
Bronwydd Arms
Cwmffrwd
Llangain
Llangynog
Llanboidy
Meidrim
Cwmbach
Blaenwaun
Llanwinio
Trelech
Tegryn
Llanfyrnach
Dinas
Ffairfach
Bethlehem
Trichrug
Carreg Cennen
Llandybie
Pen-y-groes
Saron
Gorslas
Cefneithin
Drefach
Crwbin
Llangendeirne
Pontantwn
Llandefaelog
Meinciau
Pont Henri
Carway
Five Roads
Pontyates
Sylen
Felinfoel
Dafen
Pwll
Llwynhendy
Cefn Padrig
Grovesend
Pontlliw
Penllergaer
Fforest-fach
Llangyfelach
Craigcefnparc
Felindre
Cadle
Waunarlwydd
Upper Killay
Black Pill
West Cross
Newton
Southgate
GOWER
Afon Tywi
Afon Taf
Afon Cywyn
Gwendraeth
Gwendraeth Fach
Gwendraeth Fawr
River Loughor
Afon Llwchwr
Mynydd y Betws
Mynydd y Gwair
THE BLACK MOUNTAIN
Dyffryn Tywi
SN
SS
SWANSEA to Cork 10 hrs
A B C D E
1 2 3 4 5 6 7 8
17 23
27

F
G
H
J
K
L
1
2
3
4
5
6
7
8
28
29
16
LLANDOVERY
MYNYDD BWLCH-Y-GROES
Llanfihangel Nant Bran
Merthyr Cynog
Llandyfalle Hill
Talgarth
Bronllys
Felinfach
Llanfilo
Trefeca
Castell Dinas
Myddfai
Mynydd Myddfai
Usk Resr
Trecastle
Sennybridge
Defynnog
Libanus
BRECON
Llanfrynach
Llangorse
Llangorse Lake
Bwlch
Llanddeusant
MOUNTAIN
FFOREST FAWR
Fan Hir
Fan Llia
Fan Fawr
Heol Senni
Pen Milan
BRECON BEACONS
Talybont Resr
Torpantau
Neuadd Resrs
Pentwyn Resr
Pontsticill Resr
Ystradfellte
Dan-yr-Ogof
Cantref Resr
Beacons Resr
Llwyn-on Resr
Garn Ddu
Cefn Mawr
Abercraf
Ystradgynlais
Ystalyfera
Seven Sisters
Glyn Neath
Penderyn
Penderyn Resr
Hirwaun
Hirwaun Common
Rhigos
Pontneddfechan
Cefn-coed-y-cymmer
MERTHYR TYDFIL
Pentrebach
Troedyrhiw
Dowlais
Tafarnaubach
TREDEGAR
RHYMNEY
EBBW VALE
BRYNMAWR
Beaufort
Crickhowell
NANTYGLO
BLAINA
Abertillery
Mynydd Llangynidr
Mynydd Llangatwg
ABERDARE
Aberaman
Cwmdare
Trecynon
Abercanaid
Aberfan
Merthyr Vale
MOUNTAIN ASH
Penrhiwceiber
Treharris
Nelson
Bedlinog
Fochriw
New Tredegar
BARGOED
Gelli-gaer Common
Gelligaer
Trelewis
Penybryn
Hengoed
YSTRAD MYNACH
Llanbradach
BEDWAS
CAERPHILLY
BLACKWOOD
Oakdale
Crumlin
NEWBRIDGE
Abercarn
Risca
Machen
Rudry
Thornhill
Cwmbran
TREHERBERT
TREORCHY
RHONDDA
Ferndale
Maerdy
Tylorstown
Ystrad
Llwynypia
Trealaw
TONYPANDY
Penygraig
Porth
Ynysybwl
ABERCYNON
Trehafod
Gilfach Goch
Ogmore Vale
Nant-y-moel
Blaengarw
PONTYCYMER
PONTYPRIDD
Tonyrefail
Beddau
Llantrisant
Church Village
Llantwit Fardre
Taff's Well
Pentyrch
Radyr
Whitchurch
Llandaff
CARDIFF
Canton
St Fagans
Pontyclun
Llanharan
Llanharry
NEATH
Tonna
Cwmafan
Glyncorrwg
MAESTEG
Cymmer
Caerau
Port Talbot
Margam
Margam Burrows
Kenfig
Kenfig Burrows
Pyle
Cornelly
Cefn Cribwr
Bryncethin
Sarn
Pencoed
BRIDGEND
Coity
Laleston
Coychurch
Ewenny
Tythegston
Merthyr Mawr
Ogmore
Ogmore-by-Sea
PORTHCAWL
Nottage
Newton
Tusker Rock
Southerndown
St Bride's Major
Wick
Colwinston
Llysworney
COWBRIDGE
Llanblethian
St Hilary
Llantwit Major
Boverton
St Donats
Marcross
Monknash
Nash Point
Llantrithyd
Bonvilston
Peterston-super-Ely
St Nicholas
Dyffryn
Wenvoe
Dinas Powys
Michaelston-le-Pit
Llandough
Leckwith
Penarth
Lower Penarth
Lavernock Point
Sully
Sully Island
BARRY
Barry Island
Rhoose
Porthkerry
Cardiff Airport
East Aberthaw
Gileston
St Athan
Flemingston
Llancarfan
Penmark
Breaksea Point
Castle Ditches
Vale of Neath
Resolven
Melincourt
Morfa Glas
Crynant
Cilybebyll
Rhos
Swansea Valley

ST GEORGE'S CHANNEL
ST BRIDES BAY
FISHGUARD to Rosslare..............3½ hrs
STRUMBLE HEAD
DINAS HEAD
Cemaes Head
Cardigan Island
Pen Brush
Penbwchdy
Carregwastad Point
Fishguard Bay
Newport Bay
GOODWICK
FISHGUARD
Newport
Dinas Cross
Mynydd Carningli
MYNYDD PRESELI
Foel Eryr
Foel Cwmcerwyn
Ynys Deullyn
Penclegyr
Carreg-gwylan-fach
Penllechwen
ST DAVID'S HEAD
Whitesand Bay or Porth-mawr
Bishops and Clerks
Ramsey Island
Ramsey Sound
St David's
Solva
Green Scar
SM
Rickets Head
Stack Rocks
Skomer Island
Wooltack Point
Tower Point
BROAD SOUND
Gateholm Island
Skokholm Island
St Ann's Head
MILFORD HAVEN
Dale Point
HAVERFORDWEST
NARBERTH
Johnston
NEYLAND
Pembroke Dock
PEMBROKE
Angle Bay
Sheep Island
Freshwater West
SR
Linney Head
Crow Rock
Toes
The Wash
ST GOVAN'S HEAD
Stackpole Head
Trewent Point
Old Castle Head
Saundersfoot
Monkstone Point
TENBY
Caldey Sound
Caldey Island
Chapel Point
Giltar Point
Amroth
Mathry
Crymych
Clunderwen
Whitland

G
H
J
K
L
1
2
3
4
5
6
7
8
ABERAERON
NEW QUAY
Cwmtydu
Ynys-Lochtyn
Llangrannog
Aberporth
Parcllyn
Penbryn
Plwmp
Synod Inn
Cross Inn
Llanarth
Oakford
Mydroilyn
Dihewyd
Ystrad Aeron
Cilcennin
Monachty
Pennant
Cross Inn
Bethania
Penuwch
Llangeitho
Tregaron
Llanddewi Brefi
Bwlch-Llan
Trefilan
Tal-sarn
Gartheli
Temple Bar
Cribyn
Pont Creuddyn
Betws Bledrws
Silian
Llangybi
Llanfair Clydogau
Cellan
LAMPETER
Cwmann
Pentre-bach
Llanwnnen
Cwrtnewydd
Gorsgoch
Talgarreg
Bwlch-y-fadfa
Castell Howell
Capel Cynon
Brynhoffnant
Pentregat
Sarnau
Tan-y-groes
Glynarthen
Blaenannerch
Tresaith
Rhydlewis
Ffostrasol
Moel y Mor
Pont-Siân
Cwmsychbant
Llanwenog
Dre-fach
Rhydowen
Tre-groes
Pren-gwyn
Capel Dewi
Maescrugiau
Llanybydder
Pencarreg
Mynydd Pencarreg
Ffaldybrenin
Pumsaint
Caio
Rhydcymerau
Llansawel
Edwinsford
Abbey
Talley
Halfway
Cwmdu
Mynydd Figyn
Pen-y-garn
Abergorlech
Gwernogle
Llidiad-Nenog
Mynydd Llanybyther
Mynydd Llanllwni
Llanllwni
New Inn
Gwyddgrug
Llanfihangel-ar-arth
Llandysul
Llangeler
Pentre-cwrt
Bancyffordd
Pencader
Dolgran
Alltwalis
SN
Cwmduad
Llanpumsaint
Pontarsais
Plâs
Horeb
Brechfa
Rhydargaeau
Llanfynydd
Capel Isaac
Pantglas Hall
Salem
Manordeilo
Bethlehem
Felingwmuchaf
Penybanc
Llanfihangel uwch-Gwili
Court Henry
LLANDEILO
Dynevor Castle
Ffairfach
Llangathen
Nantgaredig
Cape Dewi
Llanegwad
ROMAN ROAD
Llanarthney
Golden Grove
Afon Tywi
Abergwili
CARMARTHEN
Llangunnor
Bronwydd Arms
Newchurch
Merthyr
Ffynnon-ddrain
Nant Cynnen
Afon Gwili
Cynwyl Elfed
Pen-y-bont
Talog
Trelech
Dinas
Hermon
Llanfyrnach
Blaenwaun
Llanwinio
Cwmbach
Cwmfelin Mynach
Gellywen
Abernant
Meidrim
Llangynin
Llanboidy
Cefn-y-pant
Llanglydwen
Henllan Amgoes
Cwmfelin Boeth
Whitland
St Clears
Llanddowror
Red Roses
Llansadurnen
Llanmiloe
East Marsh
Pendine
Pendine Sands
Marros
Laugharne
Laugharne Sands
Ginst Point
Llanybri
Llansteffan
Llangynog
Bancyfelin
Sarnau
Llangain
Llanllwch
Ferryside
Llansaint
Cefn Sidan Sands
Pembrey Forest
Pembrey
KIDWELLY
Mynydd Garreg
Croesyceiliog
Cwmffrwd
Llandyfaelog
Pontantwn
Crwbin
Llangendeirne
Pontyberem
Meinciau
Pont Henri
Pontyates
Carway
Five Roads
Trimsaran
Mynydd Pen-bre
BURRY PORT
Pwll
LLANELLI
Llwynhendy
Dafen
Felinfoel
Llangennech
Bynea
Hendy
Pontarddulais
Fforest
Tumble
Llannon
Cross Hands
Gorslas
Cefneithin
Drefach
Porthyrhyd
Llanddarog
Cwmisfael
Gwendraeth Fach
Gwendraeth Fawr
Maesybont
Carmel
Pen-y-groes
AMMANFORD
Betws
Pantyffynnon
Tycroes
Llandybie
Llandyfan
Pentre Gwenlais
Glanaman
Garnant
Capel Hendre
Saron
Mynydd y Gwair
Mynydd Garn-fach
Cefn Drum
Graig Fawr
River Loughor
Felindre
Craigcefnparc
Clydach
Pontlliw
Grovesend
GORSEINON
Penllergaer
Loughor
Gowerton
Llangyfelach
Morriston
Fforest Fach
Landore
SWANSEA
Sketty
Killay
Dunvant
Waunarlwydd
Three Crosses
Upper Killay
Pen-clawdd
Crofty
Llanmorlais
Llanrhidian Sands
Whiteford Point
Landimore
Llanmadoc
Cheriton
Burry Holms
Llangennith
Llanrhidian
Ilston
CARMARTHEN BAY
Afon Taf
Afon Cywyn
Afon Teifi
Afon Cothi
Afon Cennen
Afon Duad
Afon Dewi Fawr
Afon Cynin
Afon Gronw
Afon Aeron
Afon Twrch
NEWCASTLE EMLYN
Cenarth
Cwm-cou
Aber-banc
Henllan
Penrhiw-llan
Horeb
Drefach
Felindre
Drefelin
Cwmpengraig
Saron
Rhos
Capel Iwan
Moelfre
Penrherber
Newchapel
Cilwendeg
Boncath
Blaenffos
Bwlchygroes
Clydey
Cilrhedyn
Star
Tegryn
Crymych
Freni-fawr
Abercych
Llechryd
Llangoedmor
CARDIGAN
Penparc
Tremain
Blaenporth
Pantgwyn
Ponthirwaun
Troedyraur
Beulah
Betws Ifan
Penrhiwpal
Maesllyn
Llandyfriog
Llandygwydd
Verwig
Castle
Cross Inn
Nanternis
Llwyndafydd
Gilfachreda
Llwyncelyn
Foss-y-ffin
Cilau-Aeron
Aberarth
Aberreurig
Llwyn-y-groes
Bryn Brawd
Craig Twrch
Llanycrwys
Ffarmers
Afon Cothi
Glan Duar
Penuwch
Caledfwlch
Cwmgwili
Llanelli
Afon Lliw
Trapp
Castle
Gwendraeth
Llanddarog
Lloughor
Afon Gwili
Mynydd Sylen
Sylen
Mynydd Cerrig
Graig Fawr
Llanarthney
A 40 (T)
A 48 (T)
A 477 (T)
A 487 (T)
A 484
A 485
A 486
A 482
A 475
A 476
A 483 (T)
A 474
A 4070
B 4300
B 4310
B 4336
B 4301
B 4309
B 4306
B 4317
B 4308
B 4312
B 4298
B 4299
B 4328
B 4314
B 4342
B 4337
B 4338
B 4459
B 4333
B 4334
B 4570
B 4578
B 4576
B 4577
B 4302
B 4297
B 4295
40
41
28
24
26
20

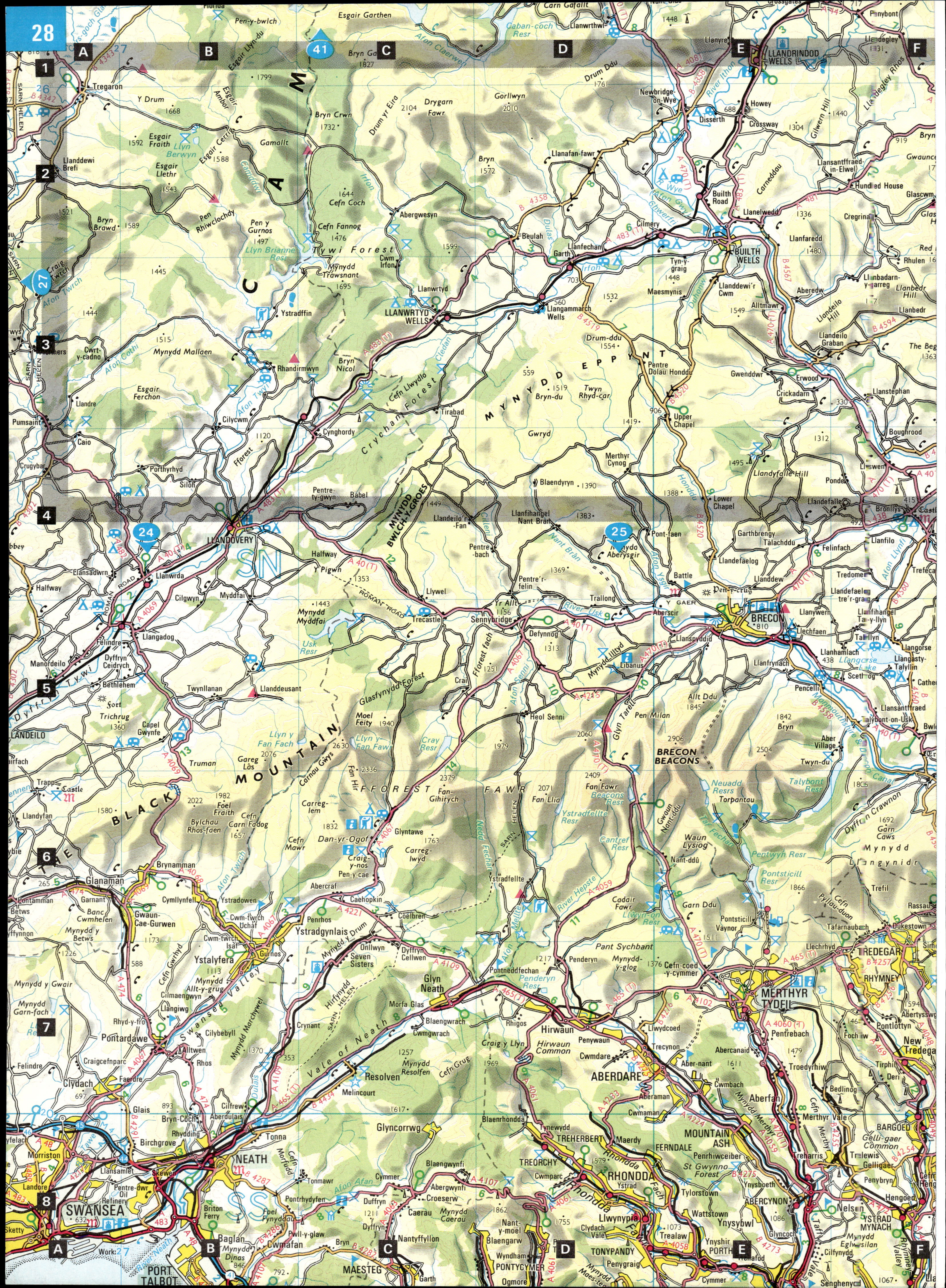
Tregaron
Llanddewi Brefi
Pumsaint
Llandovery
Llanwrtyd Wells
Beulah
Builth Wells
Llandrindod Wells
Newbridge-on-Wye
Llangammarch Wells
Rhandirmwyn
Cynghordy
Mynydd Eppynt
Tirabad
Upper Chapel
Lower Chapel
Crickadarn
Erwood
Brecon
Sennybridge
Trecastle
Llywel
Defynnog
Libanus
Brecon Beacons
Llangadog
Llanwrda
Myddfai
Llanddeusant
Bethlehem
Trichrug
Fan Foel
Black Mountain
Fforest Fawr
Fan Hir
Dan-yr-Ogof
Craig-y-nos
Glyntawe
Ystradfellte
Pont Nedd Fechan
Hirwaun
Glyn Neath
Resolven
Vale of Neath
Neath
Swansea
Port Talbot
Pontardawe
Clydach
Ystalyfera
Ystradgynlais
Brynamman
Glanaman
Aberdare
Merthyr Tydfil
Mountain Ash
Treherbert
Treorchy
Rhondda
Tonypandy
Pontycymer
Maesteg
Cwmafan
Glyncorrwg
Abercynon
Treharris
Tredegar
Rhymney
Bargoed
Ystrad Mynach
Talybont-on-Usk
Llangorse
Llangynidr
Pen-y-fan
Taf Fechan
Crychan Forest
Tywi Forest
Llyn Brianne Resr
Usk Resr
Cantref Resr
Llwyn-on Resr
Pontsticill Resr
Talybont Resr
Beacons Resr
Neuadd Resrs
Cray Resr
Llyn y Fan Fach
Llyn y Fan Fawr
River Usk
River Wye
Afon Tywi
Mynydd Bwlch-y-Groes
Sarn Helen
Roman Road
Llandeilo
Llandefalle
Glasbury
Talgarth
Felinfach
Llanfihangel Nant Brân
Pentre-bach
Heol Senni
Penderyn
Cefn-coed-y-cymmer
Rhigos
Treorchy
Ferndale
Llwynypia
Porth
Cymmer
Nantyffyllon
Caerau
Bryn
Baglan
Briton Ferry
Skewen
Morriston
Landore
Glais
Crynant
Seven Sisters
Onllwyn
Abercraf
Penycae
Cwm-twrch
Gwaun-Cae-Gurwen
Garnant
Cwmllynfell
Rhos
Alltwen
Llangiwg
Ystradowen
Capel Gwynfe
Twynllanan
Llangadog
Manordeilo
Felindre
Dyffryn Ceidrych
Llansadwrn
Halfway
Cilycwm
Porthyrhyd
Siloh
Caio
Crugybar
Llandre
Ffarmers
Cwrt-y-cadno
Mynydd Mallaen
Esgair Ferchon
Ystradffin
Abergwesyn
Cefn Coch
Cefn Fannog
Drygarn Fawr
Gorllwyn
Llanafan-fawr
Garth
Cilmery
Llanelwedd
Disserth
Howey
Crossway
Carneddau
Gwenddwr
Llandeilo Graban
Llanstephan
Boughrood
Llyswen
Pentre Dolau Honddu
Merthyr Cynog
Blaendryn
Gwryd
Llandefaelog
Battle
Pont-faen
Aberyscir
Llanspyddid
Llanfrynach
Pencelli
Llanhamlach
Scethrog
Aber Village
Torpantau
Vaynor
Pontsticill
Trefil
Dukestown
Pontlottyn
Troedyrhiw
Aberfan
Merthyr Vale
Trelewis
Gelligaer
Penybryn
Hengoed
Nelson
Ynysybwl
Trealaw
Wattstown
Ynyshir
Senghenydd
Penrhiwceiber
Abercanaid
Cwmbach
Llwydcoed
Trecynon
Aber-nant
Penywaun
Cwmdare
Blaenrhondda
Tynewydd
Ystrad
Clydach Vale
Penygraig
Blaengarw
Wyndham
Nant-y-moel
Ogmore
Garth
Cwmparc
Croeserw
Abergwynfi
Blaengwynfi
Duffryn
Tonna
Cilfrew
Aberdulais
Melincourt
Rheola
Blaengwrach
Cwmgwrach
Morfa Glas
Penywaun

Radnor Forest
PRESTEIGNE
New Radnor
Kinnerton
Offa's Dyke
KINGTON
Hergest Ridge
Gladestry
Huntington
Kingswood
Michaelchurch-on-Arrow
Newchurch
Brilley
Whitney-on-Wye
Clifford
Clyro
HAY-ON-WYE
Cusop
Llowes
Glasbury
Hay Bluff
Twmpa
Talgarth
THE BLACK MOUNTAINS
Waun Fâch
Mynydd Llysiau
Capel-y-ffin
Llanthony Priory
Castell Dinas
Cwmdu
Tretower
Crickhowell
Llangattock
Sugar Loaf
Llanwenarth
Gilwern
Govilon
Llanfoist
ABERGAVENNY
Llanellen
BLAENAVON
Blorenge
BRYNMAWR
Beaufort
EBBW VALE
NANTYGLO
BLAINA
ABERTILLERY
ABERSYCHAN
PONTYPOOL
Trevethin
Griffithstown
Sebastopol
Crumlin
NEWBRIDGE
Abercarn
Cwmbran
CWMBRAN
BLACKWOOD
Aberbargoed
Pontypool
USK
Glascoed
Monkswood
Llandegfedd Resr
Caerleon
Llangibby
Llantrisant
Usk
Raglan
Llanarth
Bettws Newydd
Clytha
Llanvapley
Llanvihangel-Ystern-Llewern
Dingestow
Penrhos
Wonastow
Tregare
Mitchel Troy
MONMOUTH
Dixton
Rockfield
Llangattock-Vibon-Avel
Newcastle
Cross Ash
Skenfrith
Grosmont
Kentchurch
Garway
Broad Oak
Welsh Newton
Llanrothal
White Castle
Llantilio Crossenny
Pandy
Llanvihangel Crucorney
Walterstone
Longtown
Clodock
Ewyas Harold
Pontrilas
Abbey Dore
Kilpeck
Wormbridge
Much Dewchurch
Orcop
St Weonards
Llangarron
Peterstow
ROSS-ON-WYE
Weston under Penyard
Walford
Goodrich
Whitchurch
Symonds Yat
English Bicknor
Coleford
FOREST OF DEAN
Cinderford
Mitcheldean
Drybrook
Ruardean
Lydbrook
Newland
Parkend
Whitecroft
Bream
Yorkley
Blakeney
Lydney
Aylburton
Alvington
Woolaston
St Briavels
Hewelsfield
Tintern Parva
Tintern Abbey
Trellech
Llandogo
Penallt
Devauden
Itton
Chepstow Park Wood
St Arvans
Tutshill
CHEPSTOW
Mathern
Sedbury
Beachley
Oldbury Sands
Lydney Sand
Shirenewton
Mounton
Caerwent
Llanvair Discoed
Penhow
Wentwood
Earlswood
Newchurch
Kilgwrrwg Common
Wolvesnewton
Llangwm
Gwernesney
Llansoy
Llanishen
Trelleck Grange
Kingcoed
Llandenny
HEREFORD
Holmer
Lugwardine
Bartestree
Tupsley
Lower Bullingham
Dinedor
Holme Lacy
Fownhope
Mordiford
Woolhope
Capler Camp
Brockhampton
How Caple
Fawley Chapel
Kings Caple
Sellack
Hoarwithy
Llandinabo
Harewood End
Pencoyd
Michaelchurch
Tretire
Glewstone
Marstow
Pencraig
Little Birch
Much Birch
Aconbury
Kingsthorne
Dewsall Court
Callow
Grafton
Twyford Common
Allensmore
Kingstone
Thruxton
Cockyard
Didley
Madley
Clehonger
Eaton Bishop
Canon Bridge
Swainshill
Stretton Sugwas
Credenhill
Kenchester
Bishopstone
Byford
Mansel Gamage
Mansel Lacy
Brinsop
Tillington
Wellington
Moreton on Lugg
Marden
Sutton Walls
Sutton St Nicholas
Withington
Shucknall
Westhide
Preston Wynne
Dormington
Weston Beggard
Tarrington
Hampton Bishop
Much Marcle
Putley
Aylton
Stoke Edith
Newtown
Stretton Grandison
Yarkhill
Ashperton
Canon Frome
Castle Frome
Bishop's Frome
Much Cowarne
Moreton Jeffries
Stoke Lacy
Burley Gate
Ocle Pychard
Felton
Ullingswick
Little Cowarne
Maund Bryan
Bodenham
Pencombe
Hegdon Hill
Marston Stannett
Docklow
Humber Court
Risbury
Steen's Bridge
Pudleston
Hatfield
Stoke Prior
LEOMINSTER
Ivington
Kimbolton
Stockton
Bach Camp
Whyle
Grafton
Leysters
Kyre Park
BROMYARD
Edwyn Ralph
Collington
Thornbury
Wall Hills
Bredenbury
Grendon Green
Munderfield Row
Munderfield Stocks
Stanford Bishop
Upper Hill
Dilwyn
Weobley
Weobley Marsh
King's Pyon
Canon Pyon
Westhope
Hope under Dinmore
Bowley
Ledgemoor
Wormsley
Yarsop
Yazor
Norton Canon
Moorhampton
Staunton on Wye
Monnington on Wye
Bredwardine
Letton
Willersley
Winforton
Eardisley
Almeley
Kinnersley
Sarnesfield
Upcott
Lower Welson
Lyonshall
Holme Marsh
Pembridge
Marston
Barewood
Moorcot
Broxwood
Luntley
Sollers Dilwyn
Eardisland
Shirl Heath
Lawton
Monkland
Stretford Court
Aulden
Brierley
Wharton
Staunton on Arrow
Titley
Byton
Combe
Lucton
Eyton
Cobnash
Yarpole
Orleton
Mortimer Cross
Ashton
Luston
Eye
The Hundred
Preston on Wye
Blakemere
Moccas
Tyberton
Peterchurch
Vowchurch
Turnastone
Dorstone
Snodhill
Urishay Common
Golden Valley
Upper Maes-coed
Middle Maes-coed
Michaelchurch Escley
St Margarets
Newton
Bacton
Craswall
Llanveynoe
Rowlestone
Llancillo
Kentchurch Court
Due to open Late 1988

DROITWICH
WORCESTER
GREAT MALVERN
Bromyard
LEDBURY
Pershore
EVESHAM
Alcester
Studley
Broadway
TEWKESBURY
Bishop's Cleeve
CHELTENHAM
CHARLTON KINGS
Winchcombe
Newent
GLOUCESTER
Mitcheldean
Cinderford
FOREST OF DEAN
Lydney
STROUD
Stonehouse
Dursley
Wotton-under-Edge
NAILSWORTH
Minchinhampton
TETBURY
CIRENCESTER
CRICKLADE
Berkeley
Upton upon Severn
Leckhampton
Painswick
Chalford
Source of the River Thames
C O T S W O L D
V A L E
A B C D E
1 2 3 4 5 6 7 8

F
G
H
J
K
L
1
2
3
4
5
6
7
8
DAVENTRY
Daventry Resr
STRATFORD-UPON-AVON
BANBURY
Chipping Campden
Moreton-in-Marsh
STOW-ON-THE-WOLD
Bourton-on-the-Water
CHIPPING NORTON
Shipston-on-Stour
CHARLBURY
WOODSTOCK
Blenheim Palace
KIDLINGTON
BICESTER
BRACKLEY
Middleton Cheney
Deddington
Aynho
BURFORD
WITNEY
Carterton
OXFORD
ABINGDON
Kennington
Wheatley
FARINGDON
LECHLADE
FAIRFORD
HIGHWORTH
Watchfield
Bampton
Eynsham
Southam
Woodstock
Weedon Bec
Kineton
Wellesbourne
Ettington
OT MOOR
Due to open early 1988
34
35
18
19
20

NEWTOWN
BISHOP'S CASTLE
Church Stretton
Clun Forest
Clun
Craven Arms
LUDLOW
KNIGHTON
Radnor Forest
PRESTEIGNE
LANDRINDOD WELLS
LEOMINSTER
KINGTON
HAY-ON-WYE
HEREFORD
Talgarth
THE BLACK MOUNTAINS
Golden Valley
River Wye
River Lugg
River Teme
Llangorse Lake
Offa's Dyke
Due to open Late 1988

BRIDGNORTH
WOMBOURNE
TIPTON
WEST BROMWICH
DUDLEY
OLDBURY
SMETHWICK
BRIERLEY HILL
BLACKHEATH
HALESOWEN
STOURBRIDGE
Kinver
Hagley
KIDDERMINSTER
BEWDLEY
Wyre Forest
STOURPORT ON SEVERN
BROMSGROVE
REDDITCH
Alvechurch
DROITWICH
WORCESTER
Bromyard
GREAT MALVERN
Malvern Link
Malvern Wells
Little Malvern
LEDBURY
Upton upon Severn
Pershore
EVESHAM
TEWKESBURY
Winchcombe
Bishop's Cleeve
CHELTENHAM
CHARLTON KINGS
ROSS-ON-WYE
Newent
Hanley Castle
Kempsey
Ombersley
Stourport
Hartlebury
Alcester
Studley
Astwood Bank
Inkberrow
Fladbury
Broadway
Bredon
Twyning
Ashchurch
Gotherington
Sudeley Castle
Longdon
Eastnor
Colwall
Cradley
Leigh
Knightwick
Martley
Clifton upon Teme
Tenbury
Cleobury Mortimer
Highley
Alveley
Quatford
Kingswinford
Amblecote
Blakedown
Belbroughton
Catshill
Lickey
Rubery
Northfield
Bournville
Selly Oak
Edgbaston
Moseley
Harborne
Handsworth
Aston
Hollywood
Wythall
Hanbury
Feckenham
Hadzor
Martin Hussingtree
Fernhill Heath
Claines
Hallow
Powick
Callow End
Madresfield
Guarlford
Welland
Castlemorton
Birtsmorton
Ripple
Kinnersley
Defford
Eckington
Bredon's Norton
Kemerton
Overbury
Conderton
Beckford
Ashton under Hill
Hinton on the Green
Sedgeberrow
Wickhamford
Badsey
Bretforton
Offenham
Harvington
Norton
Salford Priors
Abbots Morton
Rous Lench
Church Lench
Flyford Flavell
Upton Snodsbury
Spetchley
Whittington
Littleworth
Stoulton
Wadborough
Drakes Broughton
Wyre Piddle
Cropthorne
Charlton
Elmley Castle
Bricklehampton
Netherton
Little Comberton
Great Comberton
Birlingham
Croome Court
Severn Stoke
Clifton
Rhydd
Earl's Croome
Naunton
Strensham
Twyning Green
Shuthonger
Bushley
Forthampton
Walton Cardiff
Fiddington
Oxenton
Dixton
Alderton
Dumbleton
Great Washbourne
Teddington
Alstone
Gretton
Toddington
Stanway
Didbrook
Hailes Abbey
Beckbury
Farmcote
Cleeve Hill
Southam
Woodmancote
Prestbury
Elmstone Hardwicke
Hardwicke
Boddington
Uckington
Staverton
Down Hatherley
Norton
Sandhurst
Ashleworth
Hasfield
Tirley
Apperley
Deerhurst
Chaceley
Odda's Chapel
Eldersfield
Staunton
Pendock
Redmarley D'Abitot
Ryton
Dymock
Kempley
Donnington
Preston
Much Marcle
Upton Bishop
Gorsley
Linton
Kilcot
Upleadon
Hartpury
Highleadon
Tibberton
Taynton
Clifford's Mesne
May Hill
Aston Ingham
Bromsash
Aston Crews
Weston under Penyard
Rushall
Woolhope
Putley
Aylton
Little Marcle
Trumpet
Munsley
Tarrington
Yarkhill
Stretton Grandison
Canon Frome
Ashperton
Bosbury
Coddington
Colwall Green
Colwall Stone
Wellington Heath
Fromes Hill
Mathon
Evesbatch
Acton Beauchamp
Ridgeway Cross
Castle Frome
Bishop's Frome
Munderfield Row
Munderfield Stocks
Stanford Bishop
Little Cowarne
Moreton Jeffries
Cowarne
Bredenbury
Wall Hills
Thornbury
Edvin Loach
Edwyn Ralph
Tedstone Wafre
Tedstone Delamere
Whitbourne
Bromyard Downs
Linley Green
Suckley
Suckley Hills
Alfrick
Longley Green
Storridge
Acton Green
Stifford's Bridge
West Malvern
Wyche
Malvern Hills
Newland
Bransford
Leigh Sinton
Rushwick
Bowling Green
Dines Green
Broadwas
Lulsley
Broad Green
Berrow Green
Horsham
Wichenford
Sinton Green
Holt Heath
Grimley
Moseley
Shoulton
Hawford
Salwarpe
Oddingley
Tibberton
Crowle
Broughton Hackett
White Ladies Aston
Peopleton
Pinvin
Throckmorton
Bishampton
Naunton Beauchamp
North Piddle
Abberton
Kington
Grafton Flyford
Himbleton
Huddington
Sale Green
Stock Green
Stock Wood
Holberrow Green
Bradley Green
Ham Green
Hunt End
Headless Cross
Crabbs Cross
Webheath
Bentley
Woodgate
Stoke Prior
Aston Fields
Tutnall
Blackwell
Barnt Green
Cofton Hackett
Hopwood
Longbridge
Frankley
Romsley
Illey
Clent
Clent Hills
Holy Cross
Fairfield
Bournheath
Dodford
Drayton
Chaddesley Corbett
Mustow Green
Stone
Shenstone
Rushock
Elmley Lovett
Cutnall Green
Elmbridge
Upton Warren
Wychbold
Hampton Lovell
Dunhampton
Doverdale
Oldfield
Crossway Green
Lincomb
Sytchampton
Astley
Astley Cross
Noutard's Green
Shrawley
Great Witley
Little Witley
Woodbury Hill
Abberley
Abberley Hill
The Village
Dunley
Areley Kings
Wilden
Ribbesford
Callow Hill
Bliss Gate
Rock
Clows Top
Pound Bank
Far Forest
Buttonoak
Trimpley
Upper Arley
Shatterford
Kingsford
Wolverley
Franche
Habberley
Blakebrook
Wribbenhall
Cookley
Caunsall
Churchill
Broome
Blakeshall
Whittington
Stourton
Potter's Cross
Enville
Bobbington
Six Ashes
Halfpenny Green
Swindon
Himley
Wordsley
Quarry Bank
Lye
Cradley
Pedmore
Woodgate
Chad Valley
Quatt
Chelmarsh
Hampton
Billingsley
Woodhill
Kinlet
Stottesdon
Chorley
Sidbury
Middleton Scriven
Glazeley
Eardington
Oldbury
Chetton
Upton Cressett
Middleton Priors
Ditton Priors
Cleobury North
Burwarton
Aston Botterell
Loughton
Wheathill
Farlow
Oreton
Cleeton St Mary
Catherton
Neen Savage
Hopton Wafers
Doddington
Hints
Knowle
Milson
Neen Sollars
Bayton
Mamble
Nash
Boraston
Newnham
Lindridge
Frith Common
Eardiston
Pensax
Stockton on Teme
Stanford Bridge
Orleton
Stanford on Teme
Hanley William
Hanley Child
Broad Heath
Rochford
Callows Grave
Kyre Park
Stoke Bliss
Upper Sapey
Sapey Common
Shelsley Walsh
Shelsley Beauchamp
Garmsley Camp
Bank Street
Wolferlow
High Lane
Harpley
Collington
Brown Clee Hill
Vale of Evesham
Long Barrow
Charlton Abbots
Langley Hill
Guiting Power
Cutsdean
Snowshill
Stanton
Laverton
Wormington
Aston Somerville
Childswickham
Willersey
Kersoe
Middle Littleton
North Littleton
South Littleton
Aldington
Cleeve Prior
Dunnington
Iron Cross
Broom
Arrow
Coughton
Sambourne
New End
Beoley
Holt End
Tanworth in Arden
Mappleborough Green
Portway
Rowney Green
King's Heath
King's Norton
Springfield
Mosley
Hamstead
Perry Barr
Gloucester Cheltenham (Staverton) Airport
Golden Valley
Leckhampton
Syreford
Whittington
Sevenhampton
Hawling
Stoke Orchard
The Leigh
Coombe Hill
Tithe Barn
Longford
Maisemore
Glasshouse Hill
Pontshill
Coughton
Brampton Abbotts
How Caple
Hole in the Wall
Foy
Sollers Hope
Crow Hill
Woolhope
A 44
A 449
A 38
A 456
A 442
A 4133
A 4104
A 438
M 5
M 50
M 42

BIRMINGHAM
WEST BROMWICH
DUDLEY
SMETHWICK
HALESOWEN
SOLIHULL
COVENTRY
NUNEATON
BEDWORTH
HINCKLEY
KENILWORTH
WARWICK
ROYAL LEAMINGTON SPA
BROMSGROVE
REDDITCH
STRATFORD-UPON-AVON
EVESHAM
CHELTENHAM
BANBURY
CHIPPING NORTON
STOW-ON-THE-WOLD
Moreton-in-Marsh
Chipping Campden
Broadway
Winchcombe
Alcester
Henley-in-Arden
Southam
Shipston-on-Stour
Bidford-on-Avon
Pershore
Coleshill
Nether Whitacre
Over Whitacre
Meriden
Knowle
Dorridge
Balsall Common
Berkswell
Tile Hill
Coventry Airport
National Exhibition Centre
Birmingham International Airport
Stratford-upon-Avon Canal
Grand Union Canal
Due to open Mid 1990
ROMAN ROAD
RYKNILD STREET
VALE OF EVESHAM
A
B
C
D
E
1
2
3
4
5
6
7
8
30
31
33
44

F
G
H
J
K
L
1
2
3
4
5
6
7
8
LUTTERWORTH
RUGBY
DAVENTRY
MARKET HARBOROUGH
CORBY
KETTERING
DESBOROUGH
ROTHWELL
WELLINGBOROUGH
NORTHAMPTON
Towcester
BRACKLEY
BUCKINGHAM
MILTON KEYNES
NEWPORT PAGNELL
Olney
Winslow
BICESTER
LEIGHTON BUZZARD
Middleton Cheney
Long Buckby
Weedon Bec
Bugbrooke
Roade
Potterspury
Stony Stratford
Wolverton
Woburn Sands
Irchester
Wollaston
Bozeat
Finedon
BURTON LATIMER
Rushden
Weldon
YARDLEY CHASE
WHITTLEWOOD FOREST
Salcey Forest
Dunchurch
Crick
Kilsby
Brixworth
Guilsborough
Naseby
Pitsford Resr
Hollowell Resr
Ravensthorpe Resr
Daventry Resr
Grand Union Canal
M1
M 45
A 5
A 45
A 43
A 508
A 422
A 421
A 413
A 6
A 14
A 427
A 428
A 4500
Due to open early 1988

MARKET HARBOROUGH
CORBY
Weldon
KETTERING
BURTON LATIMER
Finedon
IRTHLINGBOROUGH
WELLINGBOROUGH
RUSHDEN
HIGHAM FERRERS
RAUNDS
THRAPSTON
OUNDLE
DESBOROUGH
ROTHWELL
NORTHAMPTON
Wollaston
Bozeat
Olney
NEWPORT PAGNELL
MILTON KEYNES
Woburn Sands
BEDFORD
KEMPSTON
AMPTHILL
Flitwick
LEIGHTON BUZZARD
DUNSTABLE
LUTON
HITCHIN
LETCHWORTH
BIGGLESWADE
SANDY
ST NEOTS
Brampton
BUCKINGHAM
Winslow
Potterspury
Yardley Gobion
Roade
Earls Barton
Irchester
Great Doddington
Sharnbrook
Clapham
Kimbolton
Grafham Water
Little Paxton
Great Barford
Shefford
Stotfold
Barton-le-clay
Toddington
Woburn
Stony Stratford
Wolverton
YARDLEY CHASE
WHITTLEWOOD FOREST
Salcey Forest
ROCKINGHAM FOREST
ROMAN ROAD
Due to open Early 1988
Luton International Airport
A
B
C
D
E
1
2
3
4
5
6
7
8
35
46
20

BEDFORD LEVEL
(MIDDLE LEVEL)
Ranson Moor
Pondersbridge
White Fen
Wimblington
Doddington
Benwick
The Herne
Middle Moor
Ramsey Mereside
Ramsey St Mary's
Forty Foot or Vermuden's Drain
Sixteen Foot Drain
Wimblington Fen
Manea
Purls Bridge
Welches Dam
Horseway
CHATTERIS
Ramsey Forty Foot
Ramsey Hollow
RAMSEY
Bury
Tick Fen
Langwood Fen
Pymore
Fodder Fen
Littleport
Apes Hall
Mare Fen
Southery
Methwold Fens
Queen's Ground
Brandon Creek
Brandon Bank
Feltwell Anchor
Hockwold Fens
Burnt Fen
Little Ouse River
Lakenheath Sta
BEDFORD LEVEL
SOUTH LEVEL
Shippea Hill Sta
Lakenheath
The Delph
Prickwillow
Queen Adelaide
ELY
Middle Fen
Stuntney
Chettisham
Little Downham
West Fen
Coveney
Wardy Hill
Mepal
Witcham
Witchford
Wentworth
Sutton
Somersham High North Fen
Horseley Fen
Pidley Fen
Chatteris Fen
Upwood
Great Raveley
Wistow
Little Raveley
Warboys
Fenton
Wennington
Abbots Ripton
Broughton
Pidley
Somersham
Old Hurst
Woodhurst
Kings Ripton
Colne
Earith
Bluntisham
Hill Row Doles
Haddenham
North Hill
Aldreth
Wilburton
Stretham
Little Thetford
Barway
Great Fen
Broad Hill
Isleham Fen
Mildenhall Fen
Kenny Hill
Beck Row
Thistley Green
West Row
Worlington
Eriswell
Holywell Row
Mildenhall
Barton Mills
Isleham
Soham
Soham Mere
Prior
Great Stukeley
Due to open Early 1988
HUNTINGDON
Hartford
Houghton
Hemingford Abbots
Hemingford Grey
ST IVES
Needingworth
Holywell
Over
Willingham
Smithey Fen
R Great Ouse
GODMANCHESTER
Fenstanton
Fen Drayton
Swavesey
Chittering
North Fen
Abbey
Upware
Wicken
Fordham
Freckenham
Fordham Abbey
Chippenham
Herringswell
Kennett
Rampton
Cottenham
River Cam
River Bank
Adventurers Fen
Little Fen
Landwade
Snailwell
Offord Cluny
Offord D'Arcy
Hilton
Conington
Boxworth
Lolworth
Longstanton
Oakington
Westwick
Landbeach
Waterbeach
Reach
Burwell
Exning
Kentford
Papworth St Agnes
Graveley
Elsworth
Bar Hill
Knapwell
Histon
Impington
Girton
Dry Drayton
Milton
Horningsea
Lode
Swaffham Prior
Swaffham Bulbeck
Newmarket Heath
NEWMARKET
Moulton
Gazeley
Dalham
Toseland
Yelling
Papworth Everard
Madingley
Chesterton
Fen Ditton
Stow cum Quy
Bottisham
Little Wilbraham
Ashley
Cheveley
Croxton
Eltisley
Caxton
Highfields
Hardwick
Coton
Cambridge Airport
Teversham
Great Wilbraham
Stetchworth
Dullingham
Woodditton
Ditton Green
Saxon Street
Ousden
Lidgate
Upend
Kirtling
Kirtling Green
Abbotsley
Great Gransden
Little Gransden
Bourn
Caldecote
Comberton
Toft
Grantchester
Barton
Sand
Cherry Hinton
Fulbourn
Six Mile Bottom
Westley Waterless
Burrough Green
Brinkley
Waresley
Longstowe
Kingston
Trumpington
CAMBRIDGE
Gog Magog Hills
Wandlebury
Fleam Dyke
Weston Colville
Great Bradley
Cowlinge
Gamlingay
Great Eversden
Little Eversden
Harlton
Haslingfield
Hauxton
Great Shelford
Stapleford
Carlton
Little Bradley
Weston Green
Hatley St George
Wimpole Hall
Arrington
Harston
Little Shelford
Babraham
West Wratting
Balsham
Little Thurlow
Great Thurlow
Potton
East Hatley
Croydon
New Wimpole
Orwell
Barrington
Newton
Sawston
Little Abington
Hildersham
West Wickham
Great Wratting
Barnardiston
Cockayne Hatley
Tadlow
Wendy
Shepreth
Foxton
Thriplow
Whittlesford
Pampisford
Great Abington
Linton
Horseheath
Withersfield
Kedington
Sutton
Wrestlingworth
Whaddon
Meldreth
Fowlmere
Duxford
Hinxton
Bartlow
HAVERHILL
Eyeworth
Guilden Morden
Due to open Late 1988
Kneesworth
Melbourn
Ickleton
Stump Cross
Hadstock
Shudy Camps
Castle Camps
Sturmer
Wixoe
Dunton
Abington Pigotts
Bassingbourn
Steeple Morden
Litlington
Great Chesterford
Little Chesterford
Ashdon
Helions Bumpstead
Birdbrook
Edworth
Hinxworth
ROYSTON
Heydon
Little Walden
Church End
Steeple Bumpstead
Ashwell
Great Chishill
Elmdon
Strethall
Littlebury
Stambourne Green
Caldecote
Barley
Chrishall
Little Chishill
Littlebury Green
Wendens Ambo
Sewards End
Radwinter
Hempstead
Stambourne
Newnham
Therfield
Duddenhoe End
Bridge Green
Audley End
SAFFRON WALDEN
Audley End Sta
Wimbish
Toppesfield
Bygrave
Kelshall
Reed
Barkway
Langley
Great Sampford
Cornish Hall End
Wimbish Green
Gainsford End
BALDOCK
Sandon
Buckland
Nuthampstead
Arkesden
Newport
Howlett End
Little Sampford
Howe Street
Wallington
Mill End
Chipping
Anstey
Clavering
Wicken Bonhunt
Debden
Debden Green
Finchingfield
Clothall
Rushden
Wyddial
Meesden
Brent Pelham
Rickling
Widdington
Thaxted
Little Bardfield
Wethersfield
Weston
Throcking
Cottered
Hare Street
Great Hormead
Berden
Quendon
Great Bardfield
Hall's Green
Cromer
Buntingford
Stocking Pelham
Ugley
Shalford
Graveley
Ardeley
Aspenden
Furneux Pelham
East End
Ugley Green
Henham
Broxted
Duton Hill
Walkern
Westmill
Lindsell
Bardfield Saling
Great Saling
Shalford Green
Wood End
Braughing
Patmore Heath
Manuden
Farnham Green
STANSTED MOUNTFITCHET
Elsenham
Great Easton
Bran End
Stebbing
STEVENAGE
Aston End
Aston
Benington
Nasty
Albury
Farnham
Molehill Green
Butcher's Pasture
Panfield
Puckeridge
Little Hadham
Birchanger
London (Stansted)
Little Easton
GREAT DUNMOW
Churchend
Rayne
Knebworth
Bragbury End
Dane End
Standon
Hadham Ford
Bury Green
Colliers End
BISHOP'S STORTFORD
Stane Street
Takeley
Little Dunmow
Bannister Green
Felsted
Watton at Stone
R Chelmer
R Stort
R Cam or Rhee
R Granta
R Pant
Roman Road
Ermine Street
Icknield Way
Devil's Ditch
New River
Brandon

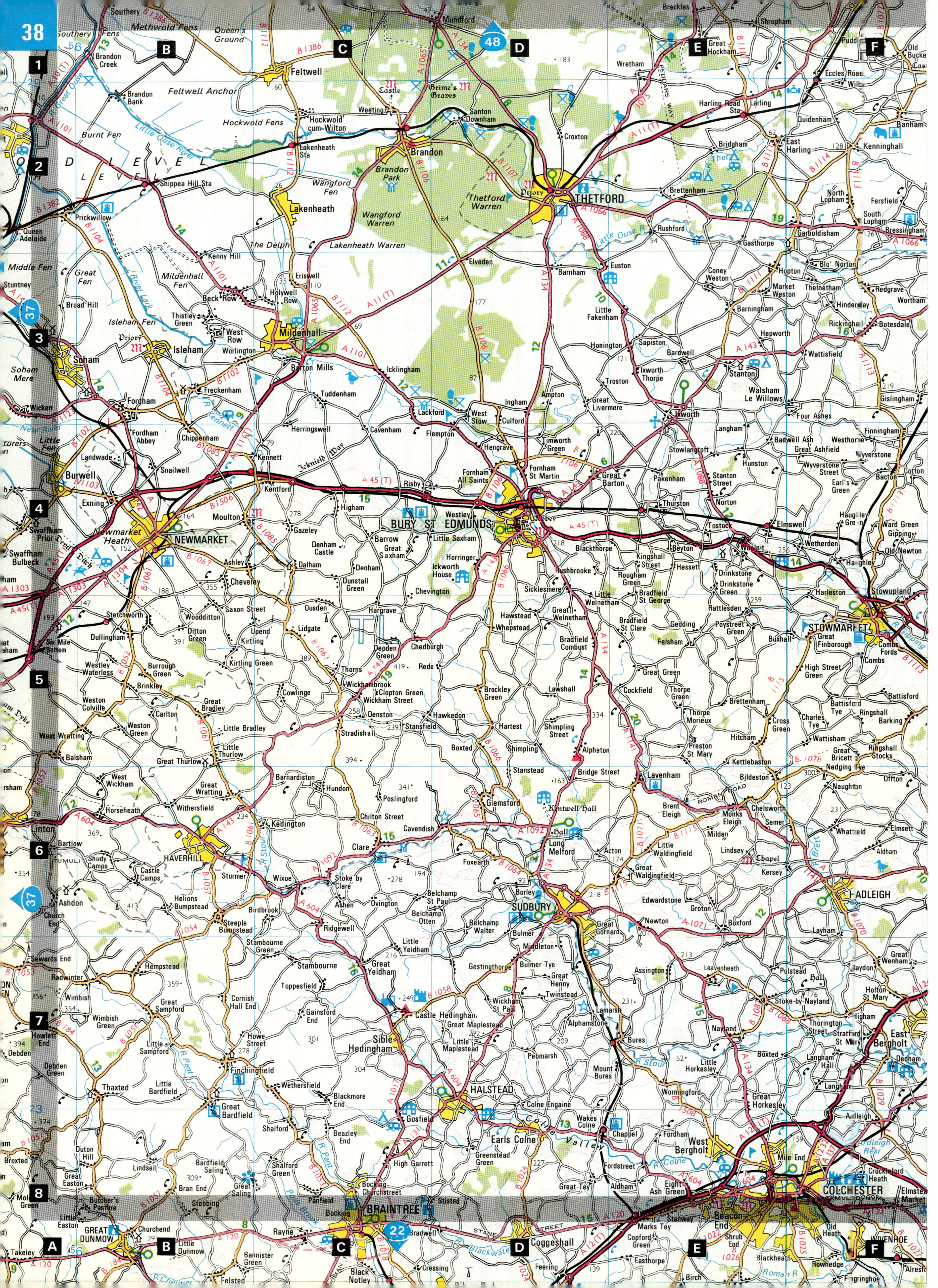
B
C
D
E
F
48
Southery
Methwold Fens
Queen's Ground
Mundford
Breckles
Shropham
Great Hockham
Feltwell
Brandon Creek
Brandon Bank
Feltwell Anchor
Hockwold Fens
Hockwold cum Wilton
Weeting
Grime's Graves
Santon Downham
Wretham
Harling Road Sta
Larling
Eccles Road
Quidenham
Banham
East Harling
Kenninghall
Burnt Fen
Lakenheath Sta
Brandon
Brandon Park
Croxton
Bridgham
Little Ouse River
Shippea Hill Sta
Wangford Fen
Thetford Warren
THETFORD
Brettenham
North Lopham
Fersfield
Lakenheath
Wangford Warren
South Lopham
Bressingham
Prickwillow
Queen Adelaide
The Delph
Lakenheath Warren
Rushford
Garboldisham
Gasthorpe
Kenny Hill
Elveden
Euston
Blo' Norton
Middle Fen
Great Fen
Mildenhall Fen
Eriswell
Barnham
Coney Weston
Hopton
Thelnetham
Stuntney
Broad Hill
Beck Row
Holywell Row
Market Weston
Redgrave
Wortham
Isleham Fen
Thistley Green
West Row
Mildenhall
Little Fakenham
Barningham
Hinderclay
Rickinghall
Botesdale
Prior
Isleham
Worlington
Hepworth
Soham
Soham Mere
Barton Mills
Icklingham
Honington
Sapiston
Bardwell
Wattisfield
Freckenham
Ixworth Thorpe
Stanton
Tuddenham
Troston
Walsham Le Willows
Wicken
Fordham
Ingham
Ampton
Great Livermere
Gislingham
Lackford
West Stow
Culford
Ixworth
Four Ashes
New River
Fordham Abbey
Chippenham
Herringswell
Cavenham
Flempton
Langham
Finningham
Timworth Green
Badwell Ash
Westhorpe
Little Fen
Landwade
Kennett
Hengrave
Stowlangtoft
Great Ashfield
Wyverstone
Snailwell
Icknield Way
Fornham All Saints
Fornham St Martin
Hunston
Wyverstone Street
Burwell
Great Barton
Pakenham
Stanton Street
Bacton
Kentford
Risby
Earl's Green
Cotton
Exning
Higham
Thurston
Norton
Moulton
BURY ST EDMUNDS
Westley
Tostock
Elmswell
Haughley
Haughley Green
Swaffham Prior
Newmarket Heath
NEWMARKET
Gazeley
Barrow
Little Saxham
Great Saxham
Blackthorpe
Beyton
Woolpit
Wetherden
Ward Green
Gipping
Old Newton
Swaffham Bulbeck
Denham Castle
Dalham
Denham
Horringer
Ickworth House
Kingshall Street
Hessett
Drinkstone
Rougham Green
Ashley
Cheveley
Dunstall Green
Chevington
Rushbrooke
Sicklesmere
Drinkstone Green
Harleston
Stowupland
Saxon Street
Ousden
Hargrave
Little Welnetham
Bradfield St George
Rattlesden
Stetchworth
Woodditton
Lidgate
Hawstead
Great Welnetham
Bradfield St Clare
Gedding
Poystreet Green
STOWMARKET
Ditton Green
Upend
Whepstead
Bradfield Combust
Felsham
Buxhall
Great Finborough
Combs Ford
Dullingham
Kirtling
Depden Green
Chedburgh
Six Mile Bottom
Kirtling Green
Combs
Burrough Green
Thorns
Rede
Great Green
High Street Green
Westley Waterless
Wickhambrook
Clopton Green
Wickham Street
Cockfield
Thorpe Green
Brinkley
Cowlinge
Brockley Green
Lawshall
Battisford
Weston Colville
Great Bradley
Denston
Hawkedon
Brettenham
Battisford Tye
Ringshall
Carlton
Stradishall
Stansfield
Hartest
Shimpling Street
Thorpe Morieux
Cross Green
Charles Tye
Barking
West Wratting
Weston Green
Little Bradley
Boxted
Shimpling
Alpheton
Preston St Mary
Hitcham
Wattisham
Great Bricett
Ringshall Stocks
Balsham
Great Thurlow
Little Thurlow
Stanstead
Kettlebaston
Nedging Tye
Offton
West Wickham
Barnardiston
Hundon
Bridge Street
Lavenham
Bildeston
Naughton
Great Wratting
Poslingford
Glemsford
Horseheath
Withersfield
Chilton Street
Kentwell Hall
Brent Eleigh
Monks Eleigh
Chelsworth
Semer
Linton
Kedington
Cavendish
Whatfield
Elmsett
Bartlow
HAVERHILL
Clare
Long Melford
Little Waldingfield
Milden
Lindsey
Chapel
Aldham
Shudy Camps
Castle Camps
Sturmer
Stoke by Clare
Foxearth
Acton
Great Waldingfield
Kersey
Helions Bumpstead
Wixoe
Belchamp St Paul
Belchamp Otten
SUDBURY
Edwardstone
HADLEIGH
Ashdon
Ashen
Ovington
Belchamp Walter
Borley
Groton
Church End
Steeple Bumpstead
Birdbrook
Bulmer
Great Cornard
Newton
Boxford
Layham
Ridgewell
Middleton
Stambourne Green
Little Yeldham
Sewards End
Hempstead
Stambourne
Great Yeldham
Gestingthorpe
Bulmer Tye
Assington
Leavenheath
Polstead
Great Wenham
Radwinter
Toppesfield
Great Henny
Stoke-by-Nayland
Holton St Mary
Wimbish
Great Sampford
Cornish Hall End
Twinstead
Wickham St Paul
Lamarsh
Castle Hedingham
Great Maplestead
Alphamstone
Nayland
Thorington Street
Stratford St Mary
East Bergholt
Howlett End
Wimbish Green
Little Sampford
Gainsford End
Howe Street
Sible Hedingham
Little Maplestead
Bures
Little Horkesley
Langham
Debden
Debden Green
Finchingfield
Pebmarsh
Dedham
Thaxted
Little Bardfield
Wethersfield
Blackmore End
HALSTEAD
Mount Bures
Wormingford
Great Horkesley
Great Bardfield
Colne Engaine
Ardleigh
Shalford
Beazley End
Gosfield
Wakes Colne
Chappel
Fordham
West Bergholt
Mile End
Duton Hill
Earls Colne
Colne Valley
Ardleigh Resr
Lindsell
Greenstead Green
Broxted
Bardfield Saling
Shalford Green
High Garrett
Fordstreet
Crockleford Heath
Great Easton
Great Saling
Bran End
Panfield
Bocking Churchstreet
Great Tey
Aldham
Eight Ash Green
COLCHESTER
Elmstead Market
Butcher's Pasture
Stebbing
Bocking
BRAINTREE
Stisted
Stanway
Beacon End
Little Easton
GREAT DUNMOW
Churchend
Rayne
22
Bradwell
Coggeshall
Marks Tey
Copford Green
Shrub End
Old Heath
WIVENHOE
Little Dunmow
Bannister Green
Black Notley
Cressing
Blackwater
Feering
Easthorpe
Blackheath
Rowhedge
Takeley
Felsted
Stane Street
Birch
Roman R
Fingringhoe
Alresford
A 11(T)
A 45(T)
A 134
A 143
A 1065
A 1066
A 1088
A 1101
A 604
A 1092
A 1017
A 131
A 120
A 12(T)
A 1071
B 1106
B 1107
B 1112
B 1386
B 1085
B 1063
B 1061
B 1057
B 1054
B 1053
B 1058
B 1064
B 1066
B 1115
B 1113
B 1070
B 1068
B 1508
B 1029
B 1025
B 1022
B 1018
B 1024

LOWESTOFT
Kessingland
BECCLES
BUNGAY
HARLESTON
DISS
HALESWORTH
SOUTHWOLD
Walberswick
Dunwich
EYE
FRAMLINGHAM
SAXMUNDHAM
LEISTON
Sizewell
Thorpeness
ALDEBURGH
Aldeburgh Bay
WICKHAM MARKET
Orford
Orford Ness
WOODBRIDGE
Rendlesham Forest
Tunstall Forest
IPSWICH
Hollesley Bay
Shipwash
FELIXSTOWE
Old Felixstowe
HARWICH
Manningtree
Mistley
RIVER STOUR
R. Orwell
Rough Tower
The Naze
Sunk
WALTON-ON-THE-NAZE
Kirby Cross
NORTH SEA
FELIXSTOWE to
Zeebrugge 5-8 hrs
Rotterdam 7-10 hrs
HARWICH to
Hook of Holland 6½-8½ hrs
Esbjerg 20 hrs
Kristiansand 22-24 hrs
Hamburg 21 hrs
Gothenburg 24 hrs
Hirtshals 27½ hrs
Oslo 39½ hrs

Penrhyn Mawr
Bryncroes
Botwnnog
Rhedyn
Nanhoron
Llanbedrog
Mynytho
Trwyn Llanbedrog
50
A
B
C
D
E
Harlech
Rhydlios
Rhoshirwaun
Capel Carmel
Castell Odo
Llawr Dref
Llangian
Abersoch
St Tudwal's Road
Llanfair
Llandanwg
Pen-sarn
Llanbedr
Braich Anelog
Llwchmynydd
Aberdaron
Rhiw
Porth Neigwl or Hell's Mouth
Llanengan
Sarn Bach
Bwlchtocyn
St Tudwal's Islands
Braich y Pwll
Bardsey Sound
Uwchmynydd
Ynys Gwylan-fawr
Pen y Cil
Cilan Uchaf
Trwyn yr Wylfa
Trwyn Cilan
Mochras or Shell Island
Morfa Dyffryn
Coed Ystumgwern
Burial Chamber
Dyffryn Ardudwy
Llanenddwyn
Tal-y-bont
Bardsey Island (Ynys Enlli)
1
2
3
4
5
6
7
Llanaber
BARMOUTH
The Bar
Barmouth Bay
Fairbourne
Llwyngwril
Llangelynnin
Rhoslefain
Llanfendigaid
Bryncrug
Aber Dysynni
TYWYN
Caethle
Aberdyfi
CARDIGAN
BAY
Aberdovey Bar
Ynyslas
Borth
Upper Borth
Llangorwen
ABERYSTWYTH
Llanbadarn Fawr
The Bar
Pen Dinas
Penparcau
Rhydyfelin
Llanfarian
Blaenplwyf
Rhadmad
Llanddeiniol
Carreg Ti-pw
Llanrhystud
Llansantffraed
Llanon
Rhyd-Rosser
Trefenter
Nebo
Cross Inn
Bethania
Blaenpennal
Aberarth
ABERAERON
Pennant
Monachty
Penuwch
Foss-y-ffin
NEW QUAY
27
Gilfachreda
Llwyncelyn
Oakford
Cilcennin
Llangeitho
Bwlch-Llan
A 499
A 496
A 493
A 487
A 482
B 4413
B 4572
B 4577
B 4337
B 4576
22
26
33

F
G
H
J
K
L
1
2
3
4
5
6
7
51
42
32
28
SH
SJ
SN
SO
CAMBRIAN MOUNTAINS
CADAIR IDRIS
PLYNLIMON (Pumlumon Fawr)
Bala
Llangower
Llanuwchllyn
Cwm Pydew
Bronaber
Mynydd Bach
Dyrysgol
Foel Figenau
Foel y Geifr
Y Clogydd
Pistyll Rhaeadr
Craig Rhiwarth
Llangynog
Cefn Coch
Llanrhaeadr-ym-Mochnant
Penybontfawr
Pennant Melangell
Pencraig
Rhobell Fawr
Cefndeuddwr
Ganllwyd
Aran Benllyn
Aran Fawddwy
Foel Rhudd
Bwlch y Groes
Ty-nant
Alltforgan
Cedig
Hirnant
Afon Tanat
Llyn Efyrnwy
Ty-uchaf
Penygarnedd
Abernant
Llanfachreth
Rhydymain
Afon Wnion
Llanelltyd
Cymmer Abbey
Precipice Walk
Nannau
Brithdir
Torrent Walk
DOLGELLAU
Penmaenpool
Abergwynant
Islawr-dref
Arthog
Llanymawddwy
Hen Gerrig
Pen y Ffrid Gownwy
Ddol Cownwy
Llanwddyn
Hotel
Abertridwr
Tycrwyn
LLANFYLLIN
Llanfihangel-yng-Ngwynfa
Pont Llogel
Dolwar Hall
Dolanog
Pontrobert
Dyfnant Forest
Mynydd y Gadfa
Cribin Fawr
Waun-oer
Mynydd Ceiswyn
Mynydd Dolgoed
Cae Afon
Dinas Mawddwy
Minllyn
Mallwyd
Cwm-Cewydd
Afon Dugoed
Esgair Ddu
Tir Rhiwiog
Moel y Llyn
Afon Twrch
Foel
Llangadfan
Llanerfyl
Afon Banwy
Glascoed
Four Crosses
Melin-y-grug
Meifod
Foel-y-ffridd
Aberangell
Llyn Coch-hwyad
Tafolog
Pen Coed
Dolwen
Corris Uchaf
Corris
Aberllefenni
Esgairgeiliog
Cwm-Llinau
Dol-fôr
Mynydd y Cemais
Mynydd Rhiw-Saeson
Pentre-celyn
Llyn Gwyddior
Nant yr Eira
Bryn y Castell
Mynydd Waun Fawr
Llanfair Caereinion
Hotel
Tal-y-llyn
Llanfihangel-y-pennant
Castell y Bere
Mynydd Pennant
Craig Goch
Abergynolwyn
Mynydd Tan-y-coed
Tarrenhendre
Dolgoch Falls
Bryn-glas Station
Pen Trum-gwr
Trum Gelli
Pennal
Pantperthog
Plâs Llwyngwern
Llanwrin
Cemmaes
Cemmaes Road
Abercegir
Commins Coch
Darowen
Tafolwern
Llanbrynmair
Pandy
Esgair Priciau
Ffridd-Rhyd-Ddu
Tirymynach
Talerddig
Rhyd
Llanllugan
Llanwyddelan
Adfa
Manafon
New Mills
Esgair Cwmowen
Mynydd Clogau
Tregynon
Afon Rhiw
MACHYNLLETH
Penegoes
Dovey Valley
Corlan-fraith
Tal-y-Wern
Bont Dolgadfan
Moelfre
Pennant
Bryn Amlwg
Trannon
Plas Llysyn
Carno
Llyn Mawr
Llyn y Tarw
Bwlch-y-ffridd
Bettws Cedewain
Glaspwll
Pant-glâs
Aberhosan
Afon Dulas
Ysgubor-y-coed
Pencarreg-gopa
Cwm Einion
Cors Fochno
Tre'r-ddôl
Taliesin
Llancynfelyn
Moel-y-Llyn
Glaslyn
New Pool
Bugeilyn
Dylife
Staylittle
Gors Goch
Glyntrefnant
Llawryglyn
Afon Trannon
Trefeglwys
Clatter
Llanwnog
Caersws
ROMAN ROAD
Llanllwchaiarn
NEWTOWN
Cwm Ceulan
Tal-y-bont
Cefnyresgair
Drosgol
Source of River Severn
Pen Pumlumon-Arwystli
Llyn Clywedog
Bryn y Fan
Y Fan
Cefn Carnedd
Llandinam
Mochdre
Dolfor
Afon Leri
Nant-y-moch Resr
Bont-goch (Elerch)
Drum Peithnant
Source of River Wye (Afon Gwy)
Esgair y Maesnant
Geufron
Afon Clywedog
LLANIDLOES
Oakley Park
Coed-y-gaer
Pentre
Kerry Hill
Llyn Syfydrin
Salem
Bow Street
Pen-bont Rhydybeddau
Penrhyn-coch
Plas Gogerddan
Y Foel
Pant Mawr
Cwmbelan
Detheryd
Bryn Llyndwr
Capel Bangor
Goginan
Ponterwyd
Afon Rheidol
Aberffrwd
Llanfihangel-y-Creuddyn
Devil's Bridge
Banc Nant-rhys
Esgair Ychion
Llangurig
Foel Gurig
Tylwch
Oldchapel Hill
Hirddywel
Llanbadarn Fynydd
Gors Lydan
Llaniwared
Dolfach
Bryn Titli
Drysgol
Cefn Cenarth
Pant-y-dwr
Afon Marteg
Red Lion Hill
Ddyle
Llanbister
Cnwch Coch
Cwmystwyth
Bryn Garw
Yr Allt
Geifas
Bwlch-y-sarnau
Cefn-crin
Abbeycwmhir
Llanafan
Pont-rhyd-y-groes
Ysbyty Ystwyth
Mynydd Bach
Lledrod
Tynygraig
Ystradmeurig
Marchnant
Trawsallt
Llyn Teifi
Clawdd-du-bach
Craig Goch Resr
Penygarreg Resr
Trumau
St Harmon
Moel Hywel
Gamallt
Camlo Hill
Llanddewi Ystradenni
Bronnant
Swyddffynnon
Pontrhydfendigaid
Strata Florida Abbey
Llyn Egnant
Dibyn Du
Claerwen Resr
Garreg-ddu Resr
Elan Village
Carn Gafallt
Llansantffraed-Cwmdeuddwr
Rhayader
Gwastedyn Hill
Nantmel
Nant Glâs
Crossgates
Penybont
Llandegley
Pen-y-bwlch
Esgair Garthen
Caban-coch Resr
Afon Claerwen
Llanwrthwl
Llanyre
LLANDRINDOD WELLS
Cors goch Glan Teifi
Tregaron
Y Drum
Bryn Garw
Drygarn Fawr
Gorllwyn
Drum Ddu
Newbridge on Wye
Howey
Llandegley Rhos

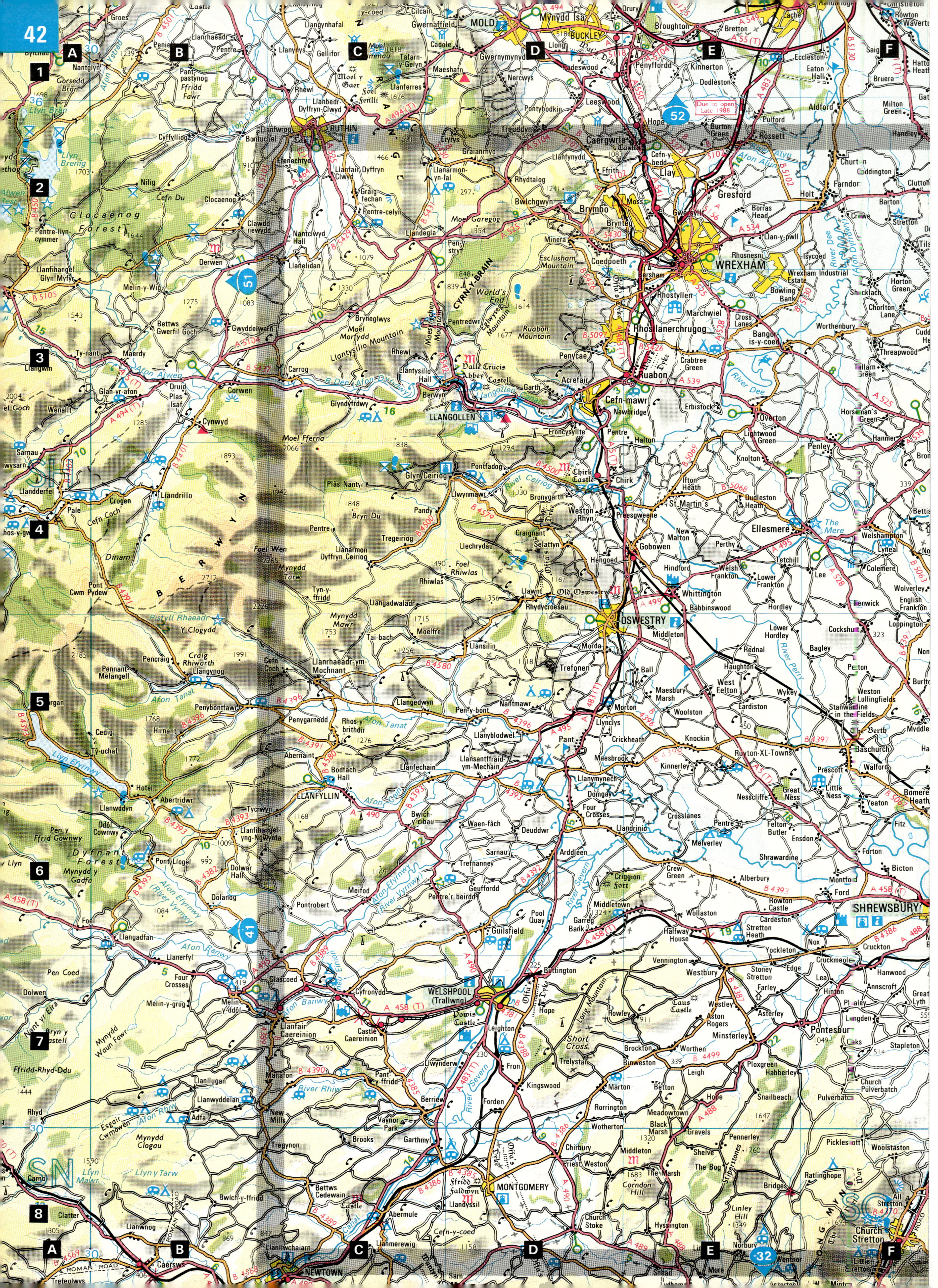
MOLD
BUCKLEY
RUTHIN
WREXHAM
Caergwrle
Gresford
Brymbo
Rhosllanerchrugog
Ruabon
Cefn-mawr
LLANGOLLEN
Chirk
Ellesmere
OSWESTRY
Gobowen
Whittington
Clocaenog Forest
Llanfihangel Glyn Myfyr
Corwen
Cynwyd
Llandrillo
Bala
Berwyn
Pistyll Rhaeadr
Llangynog
Llyn Efyrnwy
Llanwddyn
Llanfyllin
Llansantffraid-ym-Mechain
Llanymynech
Four Crosses
Ruyton-XI-Towns
Baschurch
SHREWSBURY
Dyfnant Forest
Llanfair Caereinion
WELSHPOOL (Trallwng)
Powis Castle
Guilsfield
Pontesbury
Minsterley
Church Stretton
MONTGOMERY
Chirbury
Bishop's Castle
NEWTOWN
Caersws
Carno
Offa's Dyke
World's End
Eglwyseg Mountain
Ruabon Mountain
Esclusham Mountain
Llantysilio Mountain
Moel Fferna
Cyrn-y-Brain
Long Mountain
Stiperstones
Long Mynd
Llyn Brenig
The Mere
Afon Tanat
Afon Banwy
River Severn
River Vyrnwy
River Dee

G
H
J
K
L
1
2
3
4
5
6
7
8
53
44
33
CONGLETON
SANDBACH
Elworth
Wheelock
CREWE
Haslington
ALSAGER
KIDSGROVE
BIDDULPH
LEEK
Leek
NANTWICH
Shavington
Willaston
Audley
Tunstall
Burslem
Hanley
STOKE-ON-TRENT
NEWCASTLE UNDER-LYME
Fenton
Longton
Trentham
Barlaston
Blythe Bridge
STONE
Eccleshall
STAFFORD
MALPAS
WHITCHURCH
Wem
MARKET DRAYTON
Hodnet
Shawbury
NEWPORT
Penkridge
Gnosall
Wellington
Hadley
Oakengates
TELFORD
Dawley
Shifnal
Codsall
Albrighton
Brewood
Iron-Bridge
Broseley
Coalport
MUCH WENLOCK
WOLVERHAMPTON
Pattingham
Bilbrook
WEDNESFIELD
WILLENHALL
BILSTON
DARLASTON
SEDGLEY
COSELEY
Wombourne
TIPTON
Bayston Hill
Condover
Burwardsley
Bunbury
Tattenhall
Audlem
Woore
Madeley
Keele
Loggerheads
Ashley
Bishop's Offley
Wrenbury
Prees
Weston
Stanton upon Hine Heath
High Ercall
Atcham
Wroxeter
Cressage
Kemberton
Beckbury
Stockton
Worfield
Badger
Claverley
Bridgnorth
Morville
Shipton
Much Wenlock
Cardington
Church Stretton
Cound
Acton Burnell
Kenley
Harley
Homer
Wyke
Barrow
Willey
Nordley
Astley Abbots
Tasley
Shropshire Union Canal
Llangollen Canal
Trent and Mersey Canal
Caldon Canal
River Trent
River Sow
River Tern
River Severn
River Dane
Aqualate Mere
Belvide Resr
Rudyard Resr
Tittesworth Resr
WATLING STREET
ROMAN ROAD
M6
M54
A5
A34
A41
A49
A51
A53
A442
A449
A454
A518
A500
A525
A529
A534
B5026
B5027
B5415
B5405
B5062
B5063
B5065
B5476
B5069
B5395
Whitmore
Hanchurch
Swynnerton
Standon
Croxton
Sugnall
Norbury
Woodseaves
Gnosall Heath
Haughton
Derrington
Seighford
Great Bridgeford
Norton Bridge
Yarnfield
Milwich
Sandon
Weston
Hixon
Tixall
Milford
Brocton
Bednall
Acton Trussell
Dunston
Coppenhall
Bradley
Church Eaton
Wheaton Aston
Blymhill
Lapley
Stretton
Gailey
Huntington
Hatherton
Cheslyn Hay
Landywood
Shareshill
Featherstone
Essington
Coven
Chillington Hall
Boscobel House
Kiddemore Green
Weston-under-Lizard
Tong
Cosford
Boningale
Perton
Trysull
Seisdon
Lower Penn
Hilton
Swindon
Himley
Halfpenny Green
Heathton
Roughton
Chetwynd Aston
Edgmond
Longford
Church Aston
Lilleshall
Donnington
Woodcote
Sheriffhales
Great Chatwell
Moreton
High Onn
Forton
Sutton
Puleston
Howle
Tibberton
Cherrington
Kynnersley
Sleapford
Rowton
Crudgington
Waters Upton
Eyton upon the Weald Moors
Preston upon the Weald Moors
Admaston
Wrockwardine
Walcot
Withington
Rodington
Upton Magna
Uffington
Battlefield
Haughton
Roden
Poynton
Astley
Hadnall
Preston Gubbals
Albrighton
Merrington
Grinshill
Clive
Moreton Corbet
Edgebolton
Muckleton
Great Wytheford
Booley
Stanton upon Hine Heath
Peplow
Ellerdine Heath
High Hatton
Ollerton
Child's Ercall
Stoke Heath
Tern Hill
Sutton
Woodseaves
Rosehill
Wistanswick
Lockleywood
Mill Green
Hinstock
Ellerton
Sambrook
Cheswardine
Soudley
Chipnall
Almington
Hales
Knighton
Pipe Gate
Norton in Hales
Mucklestone
Betton
Adderley
Shavington Park
Calverhall
Ightfield
Ash Magna
Broughall
Alkington
Tilstock
Fenn's Moss
Hollinwood
Prees Higher Heath
Prees Green
Darliston
Fauls
Edstaston
Paddolgreen
Newtown
Wolverley
Whixall
Welsh End
Marchamley
Weston
Lee Brockhurst
Preston Brockhurst
Alderton
Yorton
Bury Walls
Moreton Say
Bletchley
Longslow
Stoke on Tern
Burleydam
Wilkesley
Coxbank
Dodcott
Hankelow
Buerton
Hatherton
Walgherton
Stapeley
Wybunbury
Hough
Chorlton
Betley
Wrinehill
Blakenhall
Checkley
Onneley
Baldwin's Gate
Maer
Blackbrook
Chapel Chorlton
Podmore
Hookgate
Ashley Heath
Wetwood
Fairoak
Chatcull
Bishop's Offley
Adbaston
Knighton
Shebdon
High Offley
Lawnhead
Ellenhall
Ranton
Cotes
Millmeece
Slindon
Sturbridge
Walton-on-the-Hill
Weeping Cross
Doxey
Salt
Gayton
Burston
Aston-By-Stone
Oulton
Moddershall
Tittensor
Beech
Meaford
Meir
Meir Heath
Rough Close
Forsbrook
Dilhorne
Caverswall
Cellarhead
Werrington
Wetley Rocks
Cheddleton
Bagnall
Baddeley Green
Milton
Abbey Hulton
Bucknall
Endon
Stanley
Brown Edge
Knypersley
Biddulph Moor
Horton
Rudyard
Longsdon
Basford Green
Harriseahead
Mow Cop
Gillow Heath
Scholar Green
Church Lawton
Rode Heath
Hassall Green
Hassall
Talke
Red Street
Chesterton
Silverdale
Clayton
Hanford
Halmer End
Alsagers Bank
Leycett
Madeley Heath
Wrinehill
Barthomley
Balterley
Weston
Radway Green
Hardings Wood
Winterley
Brereton Heath
Brereton Green
Arclid Green
Brookhouse Green
Fourlanes End
Brownlow Heath
Astbury
Hulme Walfield
Key Green
Timbersbrook
Heaton
Rushton Spencer
Meerbrook
Newtown
Hightown
Occlestone Green
Warmingham
Bradfield Green
Church Minshull
Wettenhall
Hebden Green
Budworth
Bottom Flash
Clotton
Duddon
Burton
Tarporley
Eaton
Rushton
Hargrave
Huxley
Tiverton
Tilstone Fearnall
Alpraham
Calveley
Beeston Castle
Beeston
Newton
Spurstow
Radio Telescope
Wardle
Haughton Moss
Aston juxta Mondrum
Barbridge
Worleston
Rease Heath
Acton
Burland
Faddiley
Ravensmoor
Chorley
Cholmondeley Castle
Bickerton
Bulkeley
Peckforton
Chowley
Harthill
Fuller's Moor
Hampton Heath
No Man's Heath
Bickley Moss
Norbury
Marbury
Wirswall
Higher Wych
Grindley Brook
Marley Green
Aston
Newhall
Newtown
Butt Green
Weston
Cotwalton
Hilderstone
Middleton Green
Garshall Green
Fradswell
Yarlet
Marston
Whitgreave
Chebsey
Cotton Clanford
Long Compton
Allimore Green
Levedale
Mitton
Whiston
Marston
Bishop's Wood
Horsebrook
Oaken
Wergs
Tettenhall
Trescott
Shipley
Burnhill Green
Patshull Hall
Ackleton
Rudge
Ryton
Grindle
Brockton
Sutton Maddock
Norton
Kemberton
Sutton Hill
Madeley
Coalbrookdale
Little Wenlock
Leighton
The Wrekin
Upper Longwood
Rushton
Eaton Constantine
Donnington
Cross Houses
Berrington
Cantlop
Uppington
Norton
Pitchford
Golding
Acton Pigott
Frodesley
Dorrington
Longnor
Ruckley
Leebotwood
Hughley
Presthope
Bourton
Easthope
Stretton Westwood
Church Preen
Plaish
Acton Round
Haughton
Brockton
Longville in the Dale
Eaton
East Wall
Gretton
Enchmarsh
Hope Bowdler
Rushbury
Wall under Heywood
Tickerton
Monkhopton
Aston Eyre
Upton Cressett
Oldbury
Chetton
Westonbridge
Morville
Stanton
Viroconium
Horsehay
Lawley
Arleston
Old Park
Crackleybank
Priorslee
Trench
Leegomery
Lilleshall
Abbey
Castle
Wem
Moreton
Woodcote
The Walls
Kenley
Atcham
Attingham

A
B
C
D
E
1
2
3
4
5
6
7
8
54
55
43
34
LEEK
MATLOCK
Matlock Bath
WIRKSWORTH
Ambergate
ASHBOURNE
BELPER
Duffield
HEANOR
RIPLEY
Somercotes
South Normanton
Chaddesden
DERBY
Alvaston
Chellaston
Mickleover
Cheadle
Blythe Bridge
Werrington
UTTOXETER
Sudbury
Hatton
Tutbury
Rolleston
Stretton
Repton
Willington
Melbourne
STAFFORD
Needwood Forest
BURTON UPON TRENT
SWADLINCOTE
Woodville
ASHBY-DE-LA-ZOUCH
Measham
RUGELEY
Cannock Chase
Hednesford
CANNOCK
Heath Hayes
Burntwood
Chase Terrace
LICHFIELD
Alrewas
BROWNHILLS
TAMWORTH
Polesworth
ATHERSTONE
Ibstock
Market Bosworth
Shenstone
ALDRIDGE
BLOXWICH
WALSALL
WEDNESFIELD
WOLVERHAMPTON
WILLENHALL
BILSTON
DARLASTON
WEDNESBURY
SUTTON COLDFIELD
HINCKLEY
NUNEATON

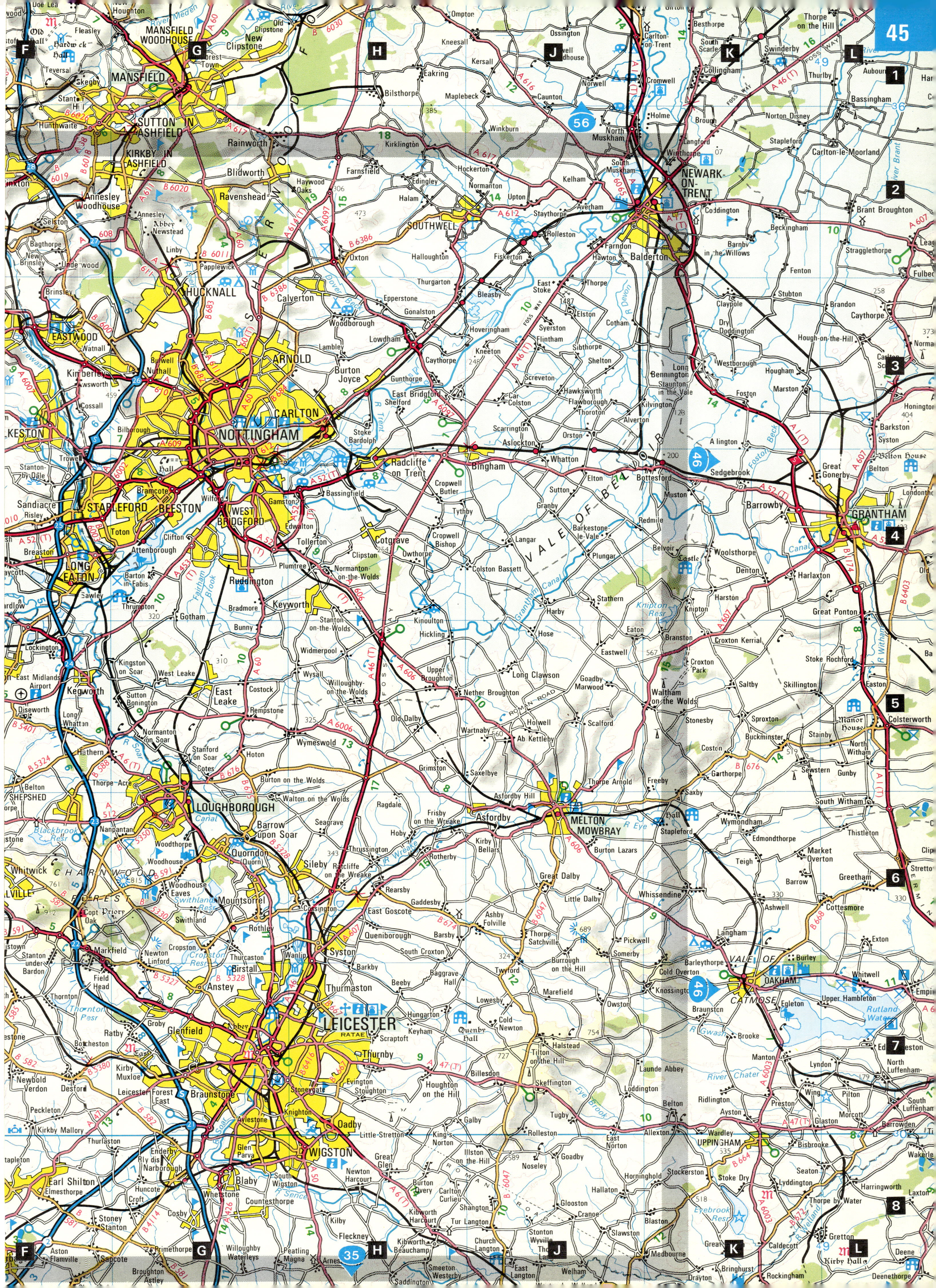
MANSFIELD
MANSFIELD WOODHOUSE
SUTTON IN ASHFIELD
KIRKBY IN ASHFIELD
HUCKNALL
EASTWOOD
ARNOLD
CARLTON
NOTTINGHAM
STAPLEFORD
BEESTON
WEST BRIDGFORD
LONG EATON
LOUGHBOROUGH
MELTON MOWBRAY
LEICESTER
WIGSTON
NEWARK-ON-TRENT
SOUTHWELL
GRANTHAM
OAKHAM
UPPINGHAM
Rutland Water
VALE OF BELVOIR
CHARNWOOD FOREST
Radcliffe on Trent
Bingham
Cotgrave
Ruddington
Keyworth
Kegworth
Shepshed
Quorndon
Mountsorrel
Syston
Birstall
Anstey
Glenfield
Braunstone
Oadby
Blaby
Earl Shilton
Asfordby
Bottesford
Balderton
Collingham
Rainworth
Ravenshead
Calverton
Bestwood
Kirklington
Kibworth Harcourt
Market Overton
Cottesmore
Colsterworth
Bassingham
Brant Broughton
Ollerton
Sileby
Syston
Thurmaston
Harby
Long Clawson
Wymeswold
Barrowby
Sedgebrook
Whatton
Aslockton
Elton

A
B
C
D
E
1
2
3
4
5
6
7
8
Ossington
Carlton on Trent
Besthorpe
South Scarle
Thorpe on the Hill
North Hykeham
Swinderby
Waddington
Nocton
Dunston
Sots Hole
WOODHALL SPA
Kersall
Cromwell
Collingham
Thurlby
Aubourn
Harmston
Metheringham
Martin
Martin Dales
Caunton
Norwell
Bassingham
Coleby
Blankney
Tattershall Thorpe
Tattershall
Winkburn
Bathley
Holme
North Muskham
Brough
Norton Disney
Boothby Graffoe
Scopwick
Timberland
Timberland Delph
South Muskham
Langford
Winthorpe
Stapleford
Carlton-le-Moorland
Navenby
Wellingore
Kirkby Green
Rowston
Walcott
Tattershall Bridge
Kelham
NEWARK-ON-TRENT
Digby
Dogdyke
Upton
Averham
Coddington
Brant Broughton
Welbourn
Ashby de la Laundé
Billinghay
Chapel Hill
Staythorpe
Rolleston
Farndon
Beckingham
Leadenham
Dorrington
North Kyme
Fiskerton
Hawton
Balderton
Barnby in the Willows
Stragglethorpe
Fulbeck
Temple Bruer
Cranwell
Ruskington
Anwick
South Kyme
East Stoke
Thorpe
Fenton
RAF College
Bleasby
Elston
Stubton
Brandon
Caythorpe
Leasingham
Evedon
South Kyme Fen
Syerston
Cotham
Claypole
Dry Doddington
Holdingham
Ewerby
Hoveringham
Flintham
Sibthorpe
Westborough
Hough-on-the-Hill
Normanton
Rauceby
SLEAFORD
Kirkby la Thorpe
Howell
Kneeton
Shelton
Long Bennington
Hougham
Carlton Scroop
Rauceby Sta
Asgarby
Quarrington
Heckington
East Heckington
Screveton
Hawksworth
Staunton in the Vale
Marston
Ancaster
Wilsford
Silk Willoughby
Burton Pedwardine
Hale
Car Colston
Flawborough
Kilvington
Foston
Honington
Thoroton
Alverton
Scarrington
Orston
Barkston
Kelby
Swarby
Aswarby
Scredington
Aslockton
Allington
Syston
Culverthorpe
Heydour
Helpringham
Bingham
Whatton
Sedgebrook
Great Gonerby
Belton
Belton House
Welby
Oasby
Aisby
Aunsby
Osbournby
Swaton
Elton
Bottesford
Muston
Londonthorpe
Dembleby
Sutton
Barrowby
GRANTHAM
Haceby
Newton
Bridge End
Granby
VALE OF BELVOIR
Redmile
Braceby
Walcot
Threekingham
Horbling
Barkestone-le-Vale
Langar
Belvoir
Woolsthorpe
Ropsley
Sapperton
Pickworth
Billingborough
Plungar
Castle
Denton
Old Somerby
Folkingham
Colston Bassett
Harlaxton
Boothby Pagnell
Pointon
Stathern
Harston
Great Ponton
Lenton
Aslackby
Harby
Knipton
Knipton Res
Eaton
Ingoldsby
Kirkby Underwood
Rippingale
Hose
Branston
Croxton Kerrial
Bassingthorpe
Bitchfield
Risegate
Eastwell
Croxton Park
Stoke Rochford
Burton-le-Coggles
Irnham
Bulby
Dunsby
Casswell's Bridge
Long Clawson
Goadby Marwood
Saltby
Skillington
Easton
Corby Glen
Stainfield
Haconby
Pinchbeck West
Nether Broughton
Waltham on the Wolds
Colsterworth
Hanthorpe
Holwell
Scalford
Stonesby
Sproxton
Manor House
Grimsthorpe
Morton
Ab Kettleby
Buckminster
Stainby
North Witham
Swayfield
Swinstead
Dyke
Pode Hole
Coston
Gunby
Edenham
Saxelbye
Garthorpe
Sewstern
BOURNE
Asfordby Hill
Thorpe Arnold
Freeby
Saxby
Creeton
Twenty
Asfordby
MELTON MOWBRAY
Stapleford
Wymondham
South Witham
Castle Bytham
Little Bytham
Lound
Northorpe
Thurlby
Kirby Bellars
Burton Lazars
Edmondthorpe
Thistleton
Witham on the Hill
Toft
Deeping Fen
Market Overton
Clipsham
Careby
Great Dalby
Teigh
Stretton
Manthorpe
Little Dalby
Barrow
Greetham
Pickworth
Carlby
Wilsthorpe
Baston
Ashby Folville
Whissendine
Ashwell
Cottesmore
Braceborough
Langtoft
Thorpe Satchville
Pickwell
Langham
Exton
Essendine
Greatford
Somerby
Ryhall
Barholm
MARKET DEEPING
Deeping St James
Burrough on the Hill
Barleythorpe
VALE OF
Burley
Little Casterton
Belmesthorpe
West Deeping
Twyford
Cold Overton
OAKHAM
CATMOSE
Whitwell
Tickencote
Great Casterton
Deeping Gate
Marefield
Knossington
Egleton
Upper Hambleton
Empingham
Uffington
Tallington
Maxey
Northborough
Lowesby
Owston
Braunston
Rutland Water
Tinwell
STAMFORD
Bainton
Etton
Newborough
Cold Newton
Halstead
Brooke
Edith Weston
Ketton
Burghley Ho
Barnack
Peakirk
Glinton
Tilton on the Hill
Manton
North Luffenham
Easton on the Hill
Ufford
Helpston
Billesdon
Launde Abbey
Lyndon
Collyweston
Southorpe
Werrington
Skeffington
River Chater
Wing
Pilton
Wittering
Marholm
Loddington
Ridlington
Preston
South Luffenham
Thornhaugh
Galby
Tugby
Belton
Ayston
Morcott
Barrowden
Wansford
Upton
Rolleston
Allexton
Wardley
Glaston
Tixover
Duddington
Ailsworth
Castor
East Norton
UPPINGHAM
Bisbrooke
Wakerley
Stibbington
Sutton
Illston on the Hill
Noseley
Goadby
Stockerston
Stoke Dry
Seaton
King's Cliffe
Yarwell
Longthorpe
Horninghold
Lyddington
Harringworth
Laxton
Sibson
Water Newton
Alwalton
Hallaton
Eyebrook Res
Thorpe by Water
Nassington
Orton Longueville
Glooston
Cranoe
Blaston
Blatherwycke
Apethorpe
Chesterton
Langton
Stonton Wyville
Slawston
Gretton
Bulwick
Woodnewton
Farcet
Church Langton
Medbourne
Great Easton
Caldecott
Kirby Hall
Fotheringhay
Elton
East Langton
Welham
Bringhurst
Rockingham
Deenethorpe
Southwick
Haddon
Drayton
Tansor
Warmington
Yaxley
Stilton
36
45
56

THE WASH
SKEGNESS
BOSTON
HUNSTANTON
KING'S LYNN
WISBECH
SPALDING
HOLBEACH
PETERBOROUGH
MARCH
DOWNHAM MARKET
WHITTLESEY
Long Sutton
Sutton Bridge
Coningsby
Wainfleet All Saints
Kirton
Heacham
Snettisham
Dersingham
Sandringham
Wolferton
Terrington St Clement
Gedney Drove End
Crowland
Thorney
Wrangle
Friskney
Sibsey
Stickney
Old Leake
Leverton
Fishtoft
Frampton
Wyberton
Sutterton
Fosdyke
Pinchbeck
Moulton
Whaplode
Fleet
Gedney
Tydd St Giles
Leverington
Walsoken
Emneth
Upwell
Outwell
Nordelph
Stow Bardolph
Hilgay
Southery
Methwold Hythe
Marshland Fen
Terrington Marsh
Friskney Flats
Long Sand
Roger Sand
Toft Sand
Black Buoy Sand
Gat Sand
Old South
Seal Sand
Peter Black Sand
Bulldog Sand
Breast Sand
Wainfleet Sand
BOSTON DEEPS
LYNN DEEPS
Old Lynn Channel
River Witham
River Welland
River Nene
Great Ouse
BEDFORD LEVEL
TF
TL

NORTH SEA
THE WASH
LYNN DEEPS
BRANCASTER ROADS
Brancaster Bay
Holkham Bay
Blakeney Point
Peter Black Sand
Bulldog Sand
Seal Sand
SKEGNESS
Gibraltar
HUNSTANTON
Holme next the Sea
Titchwell
Thornham
Brancaster
Burnham Deepdale
Burnham Norton
Burnham Market
Burnham Overy Staithe
Burnham Overy Town
Burnham Thorpe
Holkham Hall
WELLS-NEXT-THE-SEA
Salt Marshes
Stiffkey
Morston
Blakeney
Cley next the Sea
Newgate
Cockthorpe
Langham
Wiveton
Glandford
Warham
Binham
Westgate
Priory
Wighton
Saxlingham
Field Dalling
Bale
Lower Green
Hindringham
Sharrington
Thornage
Brinton
Gunthorpe
Stody
Briningham
Ringstead
Docking
Stanhoe
North Creake
South Creake
Waterden
Little Walsingham
Great Walsingham
Houghton St Giles
East Barsham
North Barsham
West Barsham
Great Snoring
Little Snoring
Thursford
Barney
Swanton Novers
Melton Constable
Kettlestone
Fulmodeston
Heacham
Sedgeford
Fring
Bircham Newton
Bircham Tofts
Great Bircham
Barmer
Syderstone
Bagthorpe
Snettisham
Southgate
Ingoldisthorpe
Shernborne
Dersingham
Anmer
Sandringham
West Newton
Wolferton
Babingley River
Castle Rising
North Wootton
South Wootton
Gaywood
Tattersett
Sculthorpe
Dunton
Shereford
Tatterford
FAKENHAM
Hempton
Little Ryburgh
Great Ryburgh
Stibbard
Wood Norton
Hindolveston
Guestwick
Foulsham
Themelthorpe
Toftrees
Colkirk
Oxwick
Helhoughton
East Rudham
West Rudham
New Houghton
Harpley
Flitcham
Hillington
Congham
Roydon
Grimston
Pott Row
Little Massingham
Great Massingham
West Raynham
East Raynham
South Raynham
Weasenham St Peter
Weasenham All Saints
Whissonsett
Horningtoft
Gateley
Brisley
North Elmham
Cathedral
Foxley
Bintree
Twyford
Guist
Bawdeswell
Billingford
Wellingham
Tittleshall
Stanfield
Massingham Heath
Rougham
Mileham
Litcham
East Bilney
Beetley
Worthing
Mill Street
Swanton Morley
Lyng
Elsing
Woodgate
River Wensum
Clenchwarton
West Lynn
KING'S LYNN
Tilney All Saints
Ashwicken
Gayton
Gayton Thorpe
East Winch
West Winch
Middleton
Saddle Bow
Wiggenhall St Germans
Wiggenhall St Mary the Virgin
Wiggenhall St Mary Magdalen
Great Ouse
North Runcton
Setchey
Blackborough End
West Bilney
East Walton
Pentney
West Acre
Castle Acre
Priory
South Acre
Newton
West Lexham
East Lexham
Great Dunham
Little Dunham
Great Fransham
Little Fransham
Beeston
Longham
Bittering
Gressenhall
Wendling
Scarning
EAST DEREHAM
Toftwood
North Tuddenham
Mattishall
Mattishall Burgh
Yaxham
Whinburgh
Westfield
Daffy Green
Runcton Holme
Wormegay
Narborough
River Nar
Tottenhill
Watlington
South Runcton
Shouldham
Shouldham Thorpe
Marham
Wallington Hall
Stowbridge
Stow Bardolph
Fincham
Barton Bendish
Beachamwell
Beachamwell Warren
Devil's Dyke
ROMAN ROAD
SWAFFHAM
Great Palgrave
Sporle
Necton
Bradenham
Shipdham
Holme Hale
North Pickenham
South Pickenham
Ashill
Cockley Cley
Letton Hall
Cranworth
Garvestone
Reymerston
Southburgh
Hardingham
Runhall
Woodrising
Carbrooke
Ovington
Saham Toney
WATTON
Griston
Caston
Rockland St Peter
Rockland All Saints
Stow Bedon
Thompson
Breckles
Great Hockham
Shropham
Little Ellingham
Great Ellingham
Deopham
Deopham Green
Hingham
Scoulton
Wicklewood
Hackford
ATTLEBOROUGH
Besthorpe
Wimbotsham
Crimplesham
Stradsett
Bexwell
DOWNHAM MARKET
Denver
Wereham
Boughton
Oxborough Hall
Gooderstone
Hilborough
Great Cressingham
Little Cressingham
Bodney
Merton
Stoke Ferry
Whittington
Wretton
West Dereham
Foulden
Northwold
Methwold
Methwold Hythe
Cranwich
Ickburgh
Mundford
Fordham
Hilgay
River Wissey
Ten Mile Bank
Southery
Southery Fens
Methwold Fens
Queen's Ground
Brandon
Nordelph
Barroway Drove
Stow Bardolph
PEDDARS WAY
TF
TL

G
H
J
K
L
65
36
1
2
3
4
5
6
7
8
Haisborough
SHERINGHAM
West Runton
CROMER
Foulness
Overstrand
Sidestrand
Trimingham
Mundesley
Paston
Bacton
Keswick
Walcott
Happisburgh
Whimpwell Green
Happisburgh Common
Lessingham
Hempstead
Sea Palling
Waxham
Horsey
Horsey Stream
Somerton
Winterton-on-Sea
Hemsby Hole
Martham
Hemsby
Newport
Scratby
California
Ormesby St Margaret
Ormesby St Michael
Caister-on-Sea
West Caister
Rollesby
Potter Heigham
Bastwick
Hickling
Hickling Green
Hickling Heath
Hundred Stream
Ludham
Repps
Thurne
River Thurne
Catfield
Sutton
Stalham
Stalham Green
Ingham
Smallburgh
Dilham
East Ruston
Honing
Crostwight
Ridlington
Witton Bridge
Spa Common
NORTH WALSHAM
Edingthorpe
Swafield
Knapton
Trunch
Gimingham
Southrepps
Northrepps
Roughton
Felbrigg
Crossdale Street
Aylmerton
East Runton
Beeston Regis
Upper Sheringham
Weybourne
Kelling
HOLT
Bodham
West Beckham
East Beckham
Gresham
Baconsthorpe
Hempstead
Edgefield
Plumstead
Matlaske
Bessingham
Sustead
Hanworth
Thurgarton
Alby Hill
Aldborough
Wickmere
Little Barningham
Calthorpe
Erpingham
Itteringham
Saxthorpe
Corpusty
Oulton
Oulton Street
Ingworth
Blickling
AYLSHAM
Burgh next Aylsham
Tuttington
Felmingham
Banningham
Colby
Suffield
Antingham
Gunton Hall
Thorpe Market
Westwick
Skeyton
Swanton Abbot
Worstead
Brampton
Marsham
Cawston
Southgate
Reepham
Eastgate
Brandiston
The Heath
Hevingham
Buxton
Lamas
Little Hautbois
Scottow
Sloley
Sco Ruston
Tunstead
Coltishall
Horstead
Belaugh
Hoveton
Wroxham
Horning
Irstead
Barton Turf
Neatishead
Stratton Strawless
River Bure
Swannington
Alderford
Felthorpe
Horsford
Newton St Faith
Frettenham
Crostwick
Hainford
Horsham St Faith
Spixworth
Rackheath
Salhouse
Woodbastwick
Upper Street
Ranworth
South Walsham
Panxworth
Morton
Attlebridge
Weston Longville
Taverham
Ringland
Drayton
Norwich Airport
Hellesdon
Catton
Costessey
New Costessey
Easton
Honingham
East Tuddenham
Hockering
New Rackheath
Sprowston
Thorpe End
Thorpe St Andrew
Great Plumstead
Little Plumstead
Hemblington
Blofield
Brundall
North Burlingham
Lingwood
Beighton
South Burlingham
Upton
Acle
Stokesby
Billockby
Thrigby
Filby
Mautby
Clippesby
Burgh St Margaret (Fleggburgh)
Runham
River Bure
Damgate
Tunstall
Halvergate
Halvergate Marshes
Freethorpe
Berney Arms Sta
Wickhampton
Breydon Water
Burgh Castle
GREAT YARMOUTH
YARMOUTH ROADS
GREAT YARMOUTH to Scheveningen 9 hrs
Gorleston-on-Sea
Bradwell
Belton
Hobland Hall
Fritton
Hopton-on-Sea
Lound
Herringfleet
Blundeston
Corton
Somerleyton
Oulton
Oulton Broad
LOWESTOFT
Wheatacre
Burgh St Peter
River Waveney
Haddiscoe
Toft Monks
Thorpe
Thurlton
Lower Thurlton
Norton Subcourse
Norton Marshes
Priory
Reedham
Limpenhoe
Cantley
Hassingham
Southwood
Buckenham
Strumpshaw
Postwick
NORWICH
Trowse Newton
Lakenham
Eaton
Earlham
Bowthorpe
Colney
Bawburgh
Marlingford
Colton
Barford
Barnham Broom
Little Melton
Cringleford
Great Melton
Wramplingham
High Green
Kimberley
Hethersett
Keswick
Arminghall
Caistor St Edmund
Framingham Pigot
Framingham Earl
Kirby Bedon
Bramerton
Surlingham
Rockland St Mary
Claxton
Hellington
Ashby St Mary
Yelverton
Poringland
Bergh Apton
Thurton
Hardley Street
Langley Street
Chedgrave
Loddon
Hales
Raveningham
Ferry
River Chet
River Yare
Ketteringham
Swardeston
East Carleton
Swainsthorpe
Dunston
Stoke Holy Cross
Shotesham
Howe
Brooke
Mundham
Seething
Kirstead Green
Thwaite St Mary
Woodton
Hedenham
Kirby Cane
Ellingham
Stockton
Gillingham
Geldeston
Aldeby
Maypole Green
WYMONDHAM
Crownthorpe
Mulbarton
Bracon Ash
Wreningham
Newton Flotman
Saxlingham Nethergate
Flordon
Hapton
Tasburgh
Ashwellthorpe
Spooner Row
Tacolneston
Forncett End
Forncett St Mary
Forncett St Peter
Tharston
Stratton St Michael
Long Stratton
Hempnall
Hempnall Green
Morningthorpe
Fritton
Topcroft Street
Bunwell
Carleton Rode
Suton
ROMAN ROAD
TG
TM
39

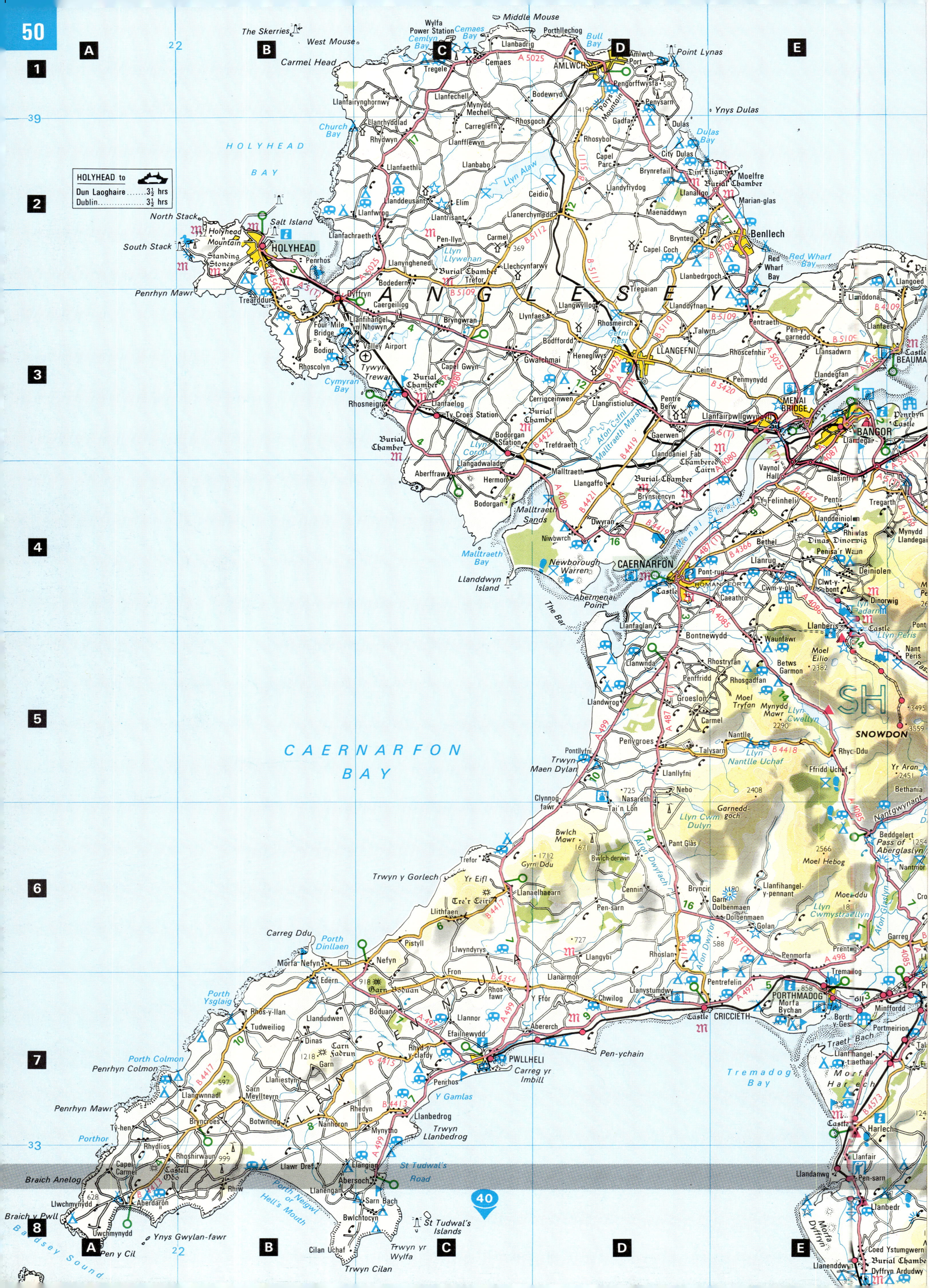
The Skerries
West Mouse
Carmel Head
Wylfa Power Station
Cemaes Bay
Cemlyn Bay
Middle Mouse
Porthllechog
Bull Bay
Point Lynas
Llanbadrig
Cemaes
AMLWCH
Amlwch Port
Tregele
Llanfechell
Llanfairynghornwy
Bodewryd
Penysarn
Ynys Dulas
Mynydd Mechell
Parys Mountain
Gadfa
Dulas
Dulas Bay
Rhosgoch
Carreglefn
Llanrhyddlad
Church Bay
Rhydwyn
Llanfflewyn
Rhosybol
Capel Parc
City Dulas
HOLYHEAD BAY
Llanfaethlu
Llanbabo
Llyn Alaw
Brynrefail
Din Lligwy
Moelfre
Burial Chamber
HOLYHEAD to
Dun Laoghaire 3½ hrs
Dublin 3½ hrs
Llanddeusant
Elim
Ceidio
Llandyfrydog
Llanallgo
Marian-glas
North Stack
Salt Island
Llanfwrog
Llantrisant
Llanerchymedd
Maenaddwyn
Holyhead Mountain
HOLYHEAD
Llanfachraeth
Carmel
Benllech
South Stack
Standing Stones
Penrhos
Pen-llyn
Llyn Llywenan
Capel Coch
Brynteg
Red Wharf Bay
Llechcynfarwy
Llynfaes
Llanbedrgoch
Llanynghenedl
Burial Chamber
Trefor
Tregaian
Llangoed
Penrhyn Mawr
Bodedern
A N G L E S E Y
Trearddur
Dyffryn
Caergeiliog
Llangwyllog
Llanddyfnan
Llanddona
Llanfihangel yn Nhowyn
Bryngwran
Rhosmeirch
Talwrn
Pentraeth
Llanfaes
Four Mile Bridge
Cefni Res
Pen-y-garnedd
Bodffordd
LLANGEFNI
Rhoscefnhir
Llansadwrn
Bodior
Valley Airport
Heneglwys
Castle
BEAUMARIS
Rhoscolyn
Tywyn Trewan
Capel Gwyn
Gwalchmai
Ceint
Penmynydd
Llandegfan
Cymyran Bay
Burial Chamber
Rhosneigr
Llanfaelog
Cerrigceinwen
Llangristiolus
Pentre Berw
MENAI BRIDGE
Penrhyn Castle
Ty Croes Station
Burial Chamber
Llanfairpwllgwyngyll
BANGOR
Bodorgan Station
Afon Cefni
Malltraeth Marsh
Gaerwen
Llandegai
Trefdraeth
Llanddaniel Fab
Chambered Cairn
Llyn Coron
Llangadwaladr
Aberffraw
Malltraeth
Vaynol Hall
Hermon
Llangaffo
Burial Chamber
Glasinfryn
Bodorgan
Brynsiencyn
Y Felinheli
Pentir
Tregarth
Malltraeth Sands
Dwyran
Llanddeiniolen
Rhiwlas
Niwbwrch
Dinas Dinorwig
Mynydd Llandegai
Malltraeth Bay
Bethel
Penisa'r Waun
Newborough Warren
CAERNARFON
Pont-rug
Deiniolen
Llanddwyn Island
Llanrug
Castle
Caeathro
Cwm-y-glo
Clwt-y-bont
Abermenai Point
Dinorwig
The Bar
Llanfaglan
Llyn Padarn
Llanberis
Castle
Llyn Peris
Bontnewydd
Waunfawr
Nant Peris
Moel Eilio
Rhostryfan
Betws Garmon
Llanwnda
Penffridd
Rhosgadfan
SH
Groeslon
Moel Tryfan
Mynydd Mawr
Llyn Cwellyn
Llandwrog
Carmel
SNOWDON
CAERNARFON BAY
Penygroes
Nantlle
Talysarn
Llyn Nantlle Uchaf
Rhyd-Ddu
Pontllyfni
Trwyn Maen Dylan
Llanllyfni
Ffridd Uchaf
Yr Aran
Nebo
Bethania
Clynnog-fawr
Nasareth
Tai'n Lon
Nantgwynant
Llyn Cwm Dulyn
Garnedd-goch
Bwlch Mawr
Pant Glas
Beddgelert
Pass of Aberglaslyn
Trefor
Gyrn Ddu
Bwlch-derwin
Moel Hebog
Nantmor
Trwyn y Gorlech
Yr Eifl
Llanaelhaearn
Cennin
Bryncir
Tre'r Ceiri
Llanfihangel-y-pennant
Moel-ddu
Llithfaen
Pen-sarn
Garn Dolbenmaen
Dolbenmaen
Llyn Cwmystradllyn
Afon Glaslyn
Carreg Ddu
Porth Dinllaen
Pistyll
Golan
Garreg
Afon Dwyfach
Afon Dwyfor
Llwyndyrys
Llangybi
Rhoslan
Penmorfa
Prenteg
Morfa Nefyn
Nefyn
Edern
Fron
Tremadog
Garn Boduan
Llanarmon
Pentrefelin
PORTHMADOG
Porth Ysgolaig
Rhos-fawr
Y Ffor
Llanystumdwy
Morfa Bychan
Minffordd
Rhos-y-llan
Boduan
Chwilog
Tudweiliog
Llandudwen
Llannor
Castle
CRICCIETH
Borth y Gest
Portmeirion
Efailnewydd
Abererch
Traeth Bach
Dinas
Carn Fadrun
Pen-ychain
Porth Colmon
Penrhyn Colmon
Garn
Rhyd-y-clafdy
PWLLHELI
Carreg yr Imbill
Tremadog Bay
Llanfihangel-y-traethau
Llaniestyn
Penrhos
Y Gamlas
Sarn Meyllteyrn
Llangwnnadl
Morfa Harlech
Penrhyn Mawr
Rhedyn
Llanbedrog
Bryncroes
Botwnnog
Nanhoron
Trwyn Llanbedrog
Ty-hen
Castle
Harlech
Porthor
Mynytho
Rhydlios
Rhoshirwaun
Llanfair
Capel Carmel
Castell Odo
Llawr Dref
Llangian
St Tudwal's Road
Braich Anelog
Llandanwg
Pen-sarn
Abersoch
Rhiw
Llanengan
Porth Neigwl or Hell's Mouth
Llwchmynydd
Aberdaron
Sarn Bach
Llanbedr
Braich y Pwll
Bwlchtocyn
St Tudwal's Islands
Uwchmynydd
Ynys Gwylan-fawr
Morfa Dyffryn
Bardsey Sound
Pen y Cil
Cilan Uchaf
Trwyn yr Wylfa
Trwyn Cilan
Coed Ystumgwern Burial Chamber
Llanenddwyn
Dyffryn Ardudwy
40

F
G
H
J
K
1
2
3
4
5
6
7
8
West Hoyle Bank
Welsh Channel
Point of Ayr
Great Ormes Head
(Pen-y-Gogarth)
LLANDUDNO
Little Ormes Head
Penrhyn Bay
Rhôs-on-Sea
Puffin Island
CONWY
BAY
Dutchman Bank
Penmaen Swatch
Lavan Sands
Traeth Lafan
PENMAENMAWR
LLANFAIRFECHAN
Dwygyfylchi
CONWY
COLWYN BAY
Old Colwyn
Mochdre
Abergele Roads
Llanddulas
ABERGELE
Kinmel Bay
Towyn
Pensarn
RHYL
PRESTATYN
Gronant
Gwaenysgor
Meliden
Dyserth
Rhuddlan
Bodelwyddan
St Asaph (Llanelwy)
Trefnant
DENBIGH
Henllan
Llannefydd
Llanfair Talhaiarn
Llangernyw
Llansannan
Betws-yn-Rhôs
Moelfre Uchaf
Dolwen
Trofarth
Eglwysbach
Tal-y-cafn
Tal-y-Bont
Dolgarrog
Llanbedr-y-cennin
Rowen
Ty'n-y-groes
Henryd
Aber Falls
Moel Wnion
FOEL-FRAS
Drum
Dulyn Resr
CARNEDD LLEWELYN
CARNEDD DAFYDD
Ffynnon Llugwy Resr
Llyn Eigiau Resr
Llyn Cowlyd Resr
Llyn Crafnant Resr
BETHESDA
Llyn Ogwen
GLYDER FAWR
GLYDER FACH
Capel Curig
Pont Cyfyng
Swallow Falls
BETWS-Y-COED
Trefriw
LLANRWST
Llanddoged
Melin-y-coed
Pentre-tafarn-y-fedw
Gwytherin
Pandy Tudur
Rhos-y-mawn
Mwdwl-Eithin
Moel Seisiog
Llyn Alwen
Aled Isaf Resr
Llyn Aled
Bryn Trillyn
Llyn Brân
Gorsedd Brân
Nantglyn
Bylchau
Tan-y-fron
Llyn Brenig
Mynydd Hiraethog
Alwen Resr
Clocaenog Forest
Cyffylliog
Clocaenog
Clawdd-newydd
Derwen
Melin-y-Wig
Bettws Gwerfil Goch
Gwyddelwern
Corwen
Cynwyd
Llandrillo
Capel Garmon
Nebo
Fairy Glen
Glan-Conwy
Pentrefoelas
Hafod-Dinbych
Rhydlydan
Cefn-brith
Glasfryn
Cerrigydrudion
Llanfihangel Glyn Myfyr
Pentre-llyn-cymmer
Ysbyty Ifan
Penmachno
Dolwyddelan
Pont-y-pant
Carnedd Moel-siabod
Pen y Bedw
Llyn Conwy
Afon Conwy
Migneint
BLAENAU FFESTINIOG
FFESTINIOG
Maentwrog
Tanygrisiau
Moelwyn Mawr
Cnicht
Llyn Trawsfynydd
Trawsfynydd
Moel Ysgyfarnogod
Carnedd Iago
Arenig Fach
Llyn Celyn
Frongoch
Arenig Fawr
Moel Llyfnant
BALA
Llanfor
Rhos-y-gwaliau
Llangower
Llyn Tegid or Bala Lake
Llanuwchllyn
Gellioedd
Llangwm
Ty-nant
Maerdy
Glan-yr-afon
Foel Goch
Sarnau
Cefn-ddwysarn
Rhiwlas
Llandderfel
Pale
Crogen
Cefn Coch
Dinam
BERWYN
Pistyll Rhaeadr
Y Clogydd
Moel Fferna
Cadair Berwyn
Foel Wen
Mynydd Tarw
Rhobell Fawr
Dyrysgol
Foel Figenau
Foel y Geifr
Cwm Cynllwyd
Aran Benllyn
Foel Rhudd
Bwlch y Groes
Ganllwyd
Llanfachreth
Cefndeuddwr
Trawsfynydd
Llanrhaeadr-ym-Mochnant
Penybontfawr
Llangynog
Pennant Melangell
Hirnant
Alltforgan
Cedig
Ruthin
Llanfwrog
Efenechtyd
Llanbedr-Dyffryn-Clwyd
Rhewl
Llanynys
Gellifor
Llangynhafal
Llandyrnog
Bodfari
Tremeirchion
Afonwen
Caerwys
Rhuallt
Cwm
Trelawnyd
Gop Hill
Llanasa
Talacre
Mostyn
Whitford
Trelogan
Gwespyr
Vale of Clwyd
Clwydian Range
Llanelidan
Nantclwyd Hall
Clocaenog
Llanfair Dyffryn Clwyd

A
B
C
D
E
F
1
2
3
4
5
6
7
8
LIVERPOOL to
Belfast 9 hrs
Dublin 8-8¾ hrs
Seasonal
Douglas 3 hrs
LIVERPOOL BAY
CROSBY CHANNEL
SOUTHPORT
Birkdale
Ainsdale
FORMBY
Freshfield
Great Altcar
ORMSKIRK
Burscough Bridge
Rufford
Banks
Crossens
Churchtown
Marshside
Marshside Sands
Angry Brow
Great Bank
Hightown
Ince Blundell
Lydiate
MAGHULL
Melling
Thornton
Sefton
Netherton
Great Crosby
Blundellsands
CROSBY
Waterloo
Seaforth
LITHERLAND
Aintree
KIRKBY
Knowsley
RAINFORD
BOOTLE
New Brighton
WALLASEY
East Hoyle Bank
Mockbeggar Wharf
Moreton
HOYLAKE
West Kirby
Greasby
Upton
BIRKENHEAD
Frankby
Grange
Caldy
Thurstaston
Irby
Pensby
HESWALL
Barnston
Gayton
Parkgate
NESTON
Ness
Burton
Willaston
Raby
Thornton Hough
Bromborough
BEBINGTON
Rock Ferry
New Ferry
Port Sunlight
Eastham
Eastham Sands
LIVERPOOL
West Derby
Childwall
Woolton
Allerton
Garston
Speke
Halewood
Hale
Liverpool Airport
HUYTON-WITH-ROBY
Knowsley Hall
ELLESMERE PORT
Whitby
Little Stanney
Stoak
Ince Banks
Ince
Thornton-le-Moors
Dungeon Banks
Capenhurst
Ledsham
Puddington
Shotwick
Mollington
Saughall
Backford
Upton
Blacon
CHESTER
Hoole
Guilden Sutton
Mickle Trafford
Bridge Trafford
Picton
Christleton
Rowton
Littleton
Handbridge
Saighton
Bruera
Eccleston
Eaton Hall
Aldford
Pulford
Rossett
Dodleston
Kinnerton
Bretton
Broughton
Hawarden
Mancot
Sandycroft
Sealand
Shotton
Queensferry
CONNAH'S QUAY
Northop Hall
Northop
Ewloe
BUCKLEY
Mynydd Isa
MOLD
Soughton (Sychdyn)
FLINT
Bagillt
Bagillt Bank
Holywell Bank
Greenfield
HOLYWELL
Brynford
Milwr
Halkyn
Halkyn Mountain
Rhes-y-cae
Pentre Halkyn
Lixwm
Nannerch
Rhosesmor
Rhydymwyn
Cilcain
Gwernaffield
Cadole
Gwernymynydd
Maeshafn
Nercwys
Llanferres
Tafarn-y-Gelyn
Moel Famau
Moel Llys-y-coed
Pontybodkin
Leeswood
Padeswood
Pen-y-ffordd
Hope
Caergwrle
Treuddyn
Cefn-y-bedd
Llay
Gresford
Burton
Marford
Gwersyllt
Brymbo
Bwlchgwyn
Llanfynydd
Rhydtalog
Minera
WREXHAM
Coedpoeth
Esclusham Mountain
Moel Gareg
Llandegla
Graigfechan
Pentre-celyn
Llanfair Dyffryn Clwyd
RUTHIN
Llanbedr-Dyffryn-Clwyd
Llanrhydd
Rhewl
Llangynhafal
Llandyrnog
Gellifor
Llangwyfan
Pentre
Bodfari
Tremeirchion
Trefnant
Llannerch Hall
DENBIGH
Henllan
Bont-newydd
St Asaph (Llanelwy)
Rhuallt
Cwm
Dyserth
Rhuddlan
Meliden
PRESTATYN
Gronant
Gwaenysgor
Talacre
Point of Ayr
West Hoyle Bank
Welsh Channel
Mostyn Bank
Ffynnongroyw
Mostyn
Glan-y-don
Llanasa
Trelogan
Gwespyr
Whitford
Carmel
Downing
Gorsedd
Pen-y-cefn
Babell
Ysceifiog
Afon-wen
Caerwys
RHYL
Kinmel Bay
Towyn
ABERGELE
St George
Pengwern
Bodelwyddan
Lowther College
Glascoed
Moelfre Isaf
Llanefydd
Groes
Nantglyn
Bylchau
Gorsedd Brân
Llyn Brân
Cyffylliog
Bontuchel
Efenechtyd
Clocaenog Forest
Clocaenog
Clawdd-newydd
Nantclwyd Hall
Derwen
Cefn Du
Nilig
RIVER DEE (AFON DYFRDWY)
RIVER MERSEY
Great Burbo Bank
Mad Wharf
Formby Hills
Due to open Late 1988

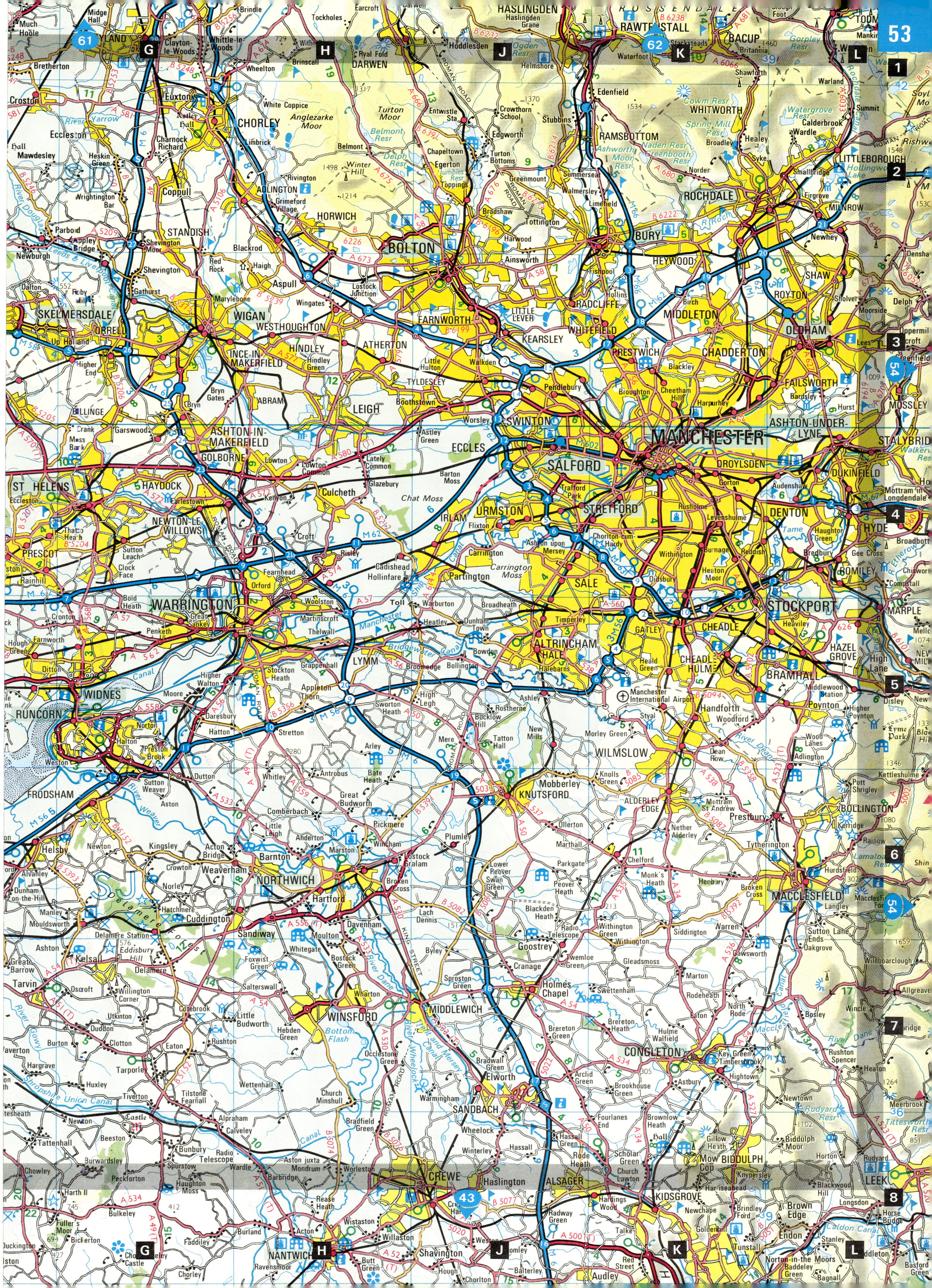
G
H
J
K
L
1
2
3
4
5
6
7
8
61
62
54
43
HASLINGDEN
RAWTENSTALL
BACUP
DARWEN
CHORLEY
WHITWORTH
RAMSBOTTOM
ROCHDALE
LITTLEBOROUGH
MILNROW
HORWICH
BOLTON
BURY
HEYWOOD
SHAW
ROYTON
STANDISH
SKELMERSDALE
WIGAN
WESTHOUGHTON
FARNWORTH
RADCLIFFE
WHITEFIELD
MIDDLETON
OLDHAM
CHADDERTON
PRESTWICH
KEARSLEY
ATHERTON
HINDLEY
INCE-IN-MAKERFIELD
ABRAM
LEIGH
TYLDESLEY
FAILSWORTH
ASHTON-UNDER-LYNE
MOSSLEY
STALYBRIDGE
MANCHESTER
SALFORD
SWINTON
ECCLES
ASHTON-IN-MAKERFIELD
GOLBORNE
ST HELENS
HAYDOCK
NEWTON-LE-WILLOWS
DROYLSDEN
DUKINFIELD
DENTON
HYDE
URMSTON
STRETFORD
IRLAM
PRESCOT
SALE
BREDBURY
ROMILEY
MARPLE
STOCKPORT
WARRINGTON
ALTRINCHAM
HALE
GATLEY
CHEADLE
CHEADLE HULME
HAZEL GROVE
BRAMHALL
LYMM
WIDNES
RUNCORN
WILMSLOW
FRODSHAM
KNUTSFORD
ALDERLEY EDGE
BOLLINGTON
NORTHWICH
MACCLESFIELD
WINSFORD
MIDDLEWICH
HOLMES CHAPEL
CONGLETON
SANDBACH
BIDDULPH
CREWE
ALSAGER
KIDSGROVE
LEEK
NANTWICH
Manchester International Airport
Shropshire Union Canal

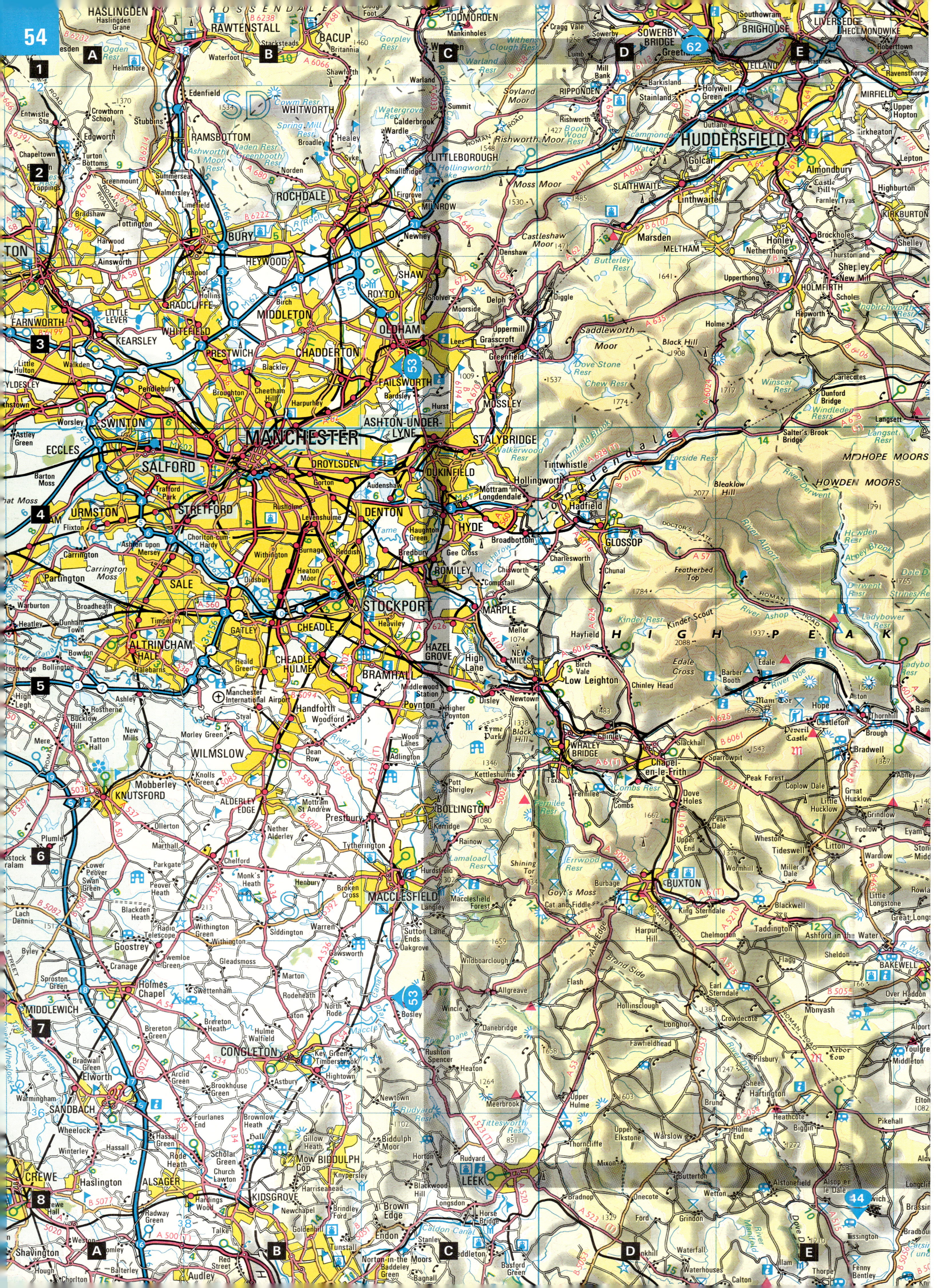
A
B
C
D
E
1
2
3
4
5
6
7
8
HASLINGDEN
RAWTENSTALL
BACUP
TODMORDEN
SOWERBY BRIDGE
BRIGHOUSE
LIVERSEDGE
HECKMONDWIKE
RAMSBOTTOM
WHITWORTH
LITTLEBOROUGH
HUDDERSFIELD
ROCHDALE
BURY
HEYWOOD
MILNROW
SLAITHWAITE
Marsden
MELTHAM
Honley
HOLMFIRTH
SHAW
ROYTON
RADCLIFFE
MIDDLETON
OLDHAM
FARNWORTH
KEARSLEY
WHITEFIELD
PRESTWICH
CHADDERTON
FAILSWORTH
MOSSLEY
SWINTON
ECCLES
MANCHESTER
ASHTON-UNDER-LYNE
STALYBRIDGE
SALFORD
DROYLSDEN
DUKINFIELD
URMSTON
STRETFORD
DENTON
HYDE
Hadfield
GLOSSOP
SALE
ROMILEY
STOCKPORT
MARPLE
ALTRINCHAM
CHEADLE
HAZEL GROVE
BRAMHALL
CHEADLE HULME
NEW MILLS
WHALEY BRIDGE
Chapel-en-le-Frith
WILMSLOW
KNUTSFORD
ALDERLEY EDGE
BOLLINGTON
MACCLESFIELD
BUXTON
BAKEWELL
MIDDLEWICH
Holmes Chapel
CONGLETON
SANDBACH
BIDDULPH
LEEK
CREWE
ALSAGER
KIDSGROVE
HIGH PEAK
HOWDEN MOORS
MIDHOPE MOORS
62
53
44

DEWSBURY
WAKEFIELD
NORMANTON
FEATHERSTONE
PONTEFRACT
KNOTTINGLEY
OSSETT
HORBURY
Crigglestone
Hemsworth
South Elmsall
South Kirkby
Ackworth Moor Top
ROYSTON
BARNSLEY
Cudworth
WOMBWELL
WORSBROUGH
HOYLAND NETHER
DEARNE
WATH UPON DEARNE
SWINTON
MEXBOROUGH
CONISBROUGH
RAWMARSH
ROTHERHAM
STOCKSBRIDGE
Chapeltown
Ecclesfield
SHEFFIELD
Fulwood
Dore
Totley
DRONFIELD
Eckington
Killamarsh
STAVELEY
CHESTERFIELD
Brimington
BOLSOVER
Clowne
Whitwell
WORKSOP
Carlton in Lindrick
Langold
Blyth
Harworth
Bircotes
Bawtry
Tickhill
MALTBY
Thurcroft
Dinnington
ADWICK LE STREET
BENTLEY
DONCASTER
Armthorpe
Bessacarr
Warmsworth
New Edlington
Rossington
Hatfield
Stainforth
Thorne
Moorends
Dunscroft
Hatfield Moors
Askern
Campsall
Carcroft
EAST RETFORD
Ollerton
Edwinstowe
WARSOP
SHIREBROOK
MANSFIELD
MANSFIELD WOODHOUSE
SUTTON IN ASHFIELD
KIRKBY IN ASHFIELD
Clipstone
Rainworth
Blidworth
Ravenshead
Tibshelf
CLAY CROSS
MATLOCK
Wirksworth
Bakewell
Hathersage
Alfreton
Pinxton
South Normanton
Selston
Underwood
Annesley Woodhouse
Southwell
SHERWOOD FOREST
River Don
River Dearne
River Rother
River Idle
River Torne
River Went
River Aire
River Derwent
Ladybower Res
Damflask Res
Broomhead Res
Redmires Resrs
Barbrook Res
Linacre Resrs
Ogston Res
F
G
H
J
K
L
1
2
3
4
5
6
7
8
45
56
58
63

GOOLE
Snaith
Rawcliffe
Reedness
Swinefleet
Ousefleet
Whitgift
Adlingfleet
Alkborough
Whitton
Winteringham
BARTON-UPON-HUMBER
Barrow upon Humber
South Ferriby
Marshland
Goole Fields
Pollington
Balne
River Went
Fenwick
Sykehouse
Moorends
Thorne Waste or Moors
Thorne
Fishlake
Stainforth
Hatfield
Crowle
Eastoft
Luddington
Amcotts
Garthorpe
Fockerby
RIVER TRENT
Burton upon Stather
Normanby
Flixborough
Winterton
Roxby
Risby
Appleby
Santon
Gunness
Keadby
Crosby
SCUNTHORPE
Broughton
BRIGG
Wrawby
Barnetby le Wold
Elsham
Melton Ross
Bigby
Searby
Grasby
Howsham
Cadney
North Kelsey
South Kelsey
Moortown
Holton le Moor
Horkstow
Saxby All Saints
Bonby
Worlaby
Kingsforth
Thornton Curtis
Wootton
Burnham
ERMINE STREET
Old River Ancholme
New River Ancholme
Dunscroft
Barnby Dun
Dunsville
Kirk Sandall
Edenthorpe
Armthorpe
Nutwell
Cantley
DONCASTER
Bessacarr
Branton
Auckley
Blaxton
Hatfield Woodhouse
Hatfield Chase
Hatfield Moors
Sandtoft
Westgate
Belton
Epworth
ISLE OF AXHOLME
Wroot
Low Burnham
Haxey
Owston Ferry
Westwoodside
East Lound
Graizelound
Derrythorpe
Burringham
Brumby
Ashby
Bottesford
Yaddlethorpe
Althorpe
Belton
Butterwick
Messingham
Scotterthorpe
Scotter
Scotton
Manton
Hibaldstow
Redbourne
Kirton in Lindsey
Gainsthorpe Village
East Ferry
Wildsworth
Laughton
Blyton
Northorpe
Grayingham
Blyborough
Willoughton
Hemswell
Harpswell
Glentham
Waddingham
Snitterby
Bishop Norton
Thornton le Moor
North Owersby
Usselby
Kirkby
Osgodby
Middle Rasen
West Rasen
Normanby-by-Spital
Owmby-by-Spital
Toft next Newton
Newton by Toft
Saxby
Spridlington
Cold Hanworth
Hackthorn
Friesthorpe
Faldingworth
Wickenby
Snelland
Welton
Dunholme
Scothern
Stainton by Langworth
Langworth
Sudbrooke
Nettleham
Reepham
Cherry Willingham
LINCOLN
Washingborough
Fiskerton
Heighington
Canwick
Branston
Bracebridge Heath
Potterhanworth
Waddington
Nocton
Dunston
Metheringham
Blankney
Scopwick
Harmston
Coleby
Boothby Graffoe
Navenby
Wellingore
Digby
Asgarby
Welbourn
Leadenham
Fulbeck
Temple Bruer
Cranwell
Ruskington
Hossington
Loversall
New Rossington
Tickhill
Bawtry
Harworth
Bircotes
Austerfield
Finningley
Misson
Misterton
West Stockwith
East Stockwith
Walkeringham
Walkerith
Morton
GAINSBOROUGH
Gringley on the Hill
Beckingham
Everton
Scaftworth
Scrooby
Wiseton
Mattersey
Priory
Ranskill
Clayworth
Saundby
Bole
North Wheatley
Torworth
Blyth
Oldcotes
Carlton in Lindrick
Barnby Moor
Lound
Sutton
Hayton
Clarborough
Ranby
Babworth
EAST RETFORD
Grove
WORKSOP
North Leverton with Habblesthorpe
Sturton le Steeple
Littleborough
Knaith
Lea
Upton
Kexby
Willingham by Stow
Stow
Sturton by Stow
Gate Burton
Marton
Ingham
Cammeringham
Fillingham
Glentworth
Heapham
Springthorpe
Corringham
Aisby
Pilham
South Leverton
Cottam
Treswell
Rampton
Woodbeck
Headon
Stokenham
Eaton
Gamston
Upton
Askham
East Drayton
Laneham
Torksey
Fenton
Brampton
Thorpe le Fallows
Aisthorpe
Brattleby
Scampton
North Carlton
South Carlton
Burton
Saxilby
Kettlethorpe
Laughterton
Dunham on Trent
Newton on Trent
Ragnall
Darlton
Thorney
Skellingthorpe
Doddington
Birchwood
Boultham
North Hykeham
Harby
North Clifton
South Clifton
Wigsley
Spalford
North Scarle
Eagle
Thorpe on the Hill
Swinderby
Thurlby
Aubourn
Bassingham
Norton Disney
Stapleford
Carlton-le-Moorland
Brant Broughton
Stragglethorpe
Beckingham
Fenton
Elkesley
West Drayton
Bothamsall
Haughton
Bevercotes
West Markham
East Markham
Tuxford
Normanton on Trent
Weston
Grassthorpe
Sutton on Trent
Girton
Besthorpe
South Scarle
Collingham
Carlton-on-Trent
Cromwell
Holme
North Muskham
South Muskham
Langford
Winthorpe
NEWARK-ON-TRENT
Coddington
Balderton
Farndon
Hawton
Averham
Kelham
Staythorpe
Rolleston
Fiskerton
SOUTHWELL
Halloughton
Upton
Normanton
Halam
Edingley
Hockerton
Kirklington
Farnsfield
Bilsthorpe
Maplebeck
Winkburn
Eakring
Kersall
Caunton
Kneesall
Ompton
Ossington
Norwell
Norwell Woodhouse
Bathley
Moorhouse
Laxton
Egmanton
Wellow
Ollerton
New Ollerton
Boughton
Walesby
Edwinstowe
Budby
Perlethorpe
Clumber Park
Carburton
Meden Vale
WARSOP
Clipstone
New Clipstone
Rainworth
Blidworth
Haywood Oaks
Oxton
Ravenshead
SHERWOOD FOREST
SE
SK
Whitton
Great Heck
Gowdall
Cowick
Courtney
Southfield Resr
Rawcliffe Bridge
Dutch River
River Aire
River Idle
River Torne
River Witham
River Ryton
River Poulter
River Maun
Fossdyke Navigation
ROMAN ROAD
FOSS WAY
Stainforth & Keadby Canal
Chesterfield Canal
North Kelsey Beck
Read's I.
Reads Channel
Humber Bridge
A B C D E
1 2 3 4 5 6 7 8
45 46 55 58

RIVER HUMBER
MOUTH OF THE HUMBER
NORTH SEA
SPURN HEAD
GRIMSBY
CLEETHORPES
Immingham
MARKET RASEN
LOUTH
MABLETHORPE
Sutton on Sea
ALFORD
Chapel St Leonards
Ingoldmells
SKEGNESS
SPILSBY
HORNCASTLE
WOODHALL SPA
Tattershall
Coningsby
Caistor
Binbrook
THE WOLDS
KINGSTON UPON HULL to
Rotterdam (Europoort)..........14 hrs
Zeebrugge..........14½-15 hrs
Humberston
Waltham
Scartho
Tetney
North Somercotes
Saltfleet
Donna Nook
Gas Terminal
Trusthorpe
Sandilands
Huttoft
Anderby Creek
Hogsthorpe
Burgh le Marsh
Wainfleet All Saints
Gibraltar
Wainfleet Sand
Friskney Flats
Friskney
Sibsey
Stickney
East Fen
West Fen
Wragby
Keelby
Habrough
Ulceby
Killingholme
Patrington
Easington
Kilnsea
Sunk Island Sands
Holmpton
Withernsea
Welwick
Skeffling
Tetford
Revesby
Stickford
Midville
New Bolingbroke
Martin
Timberland
Billinghay
North Kyme
Anwick

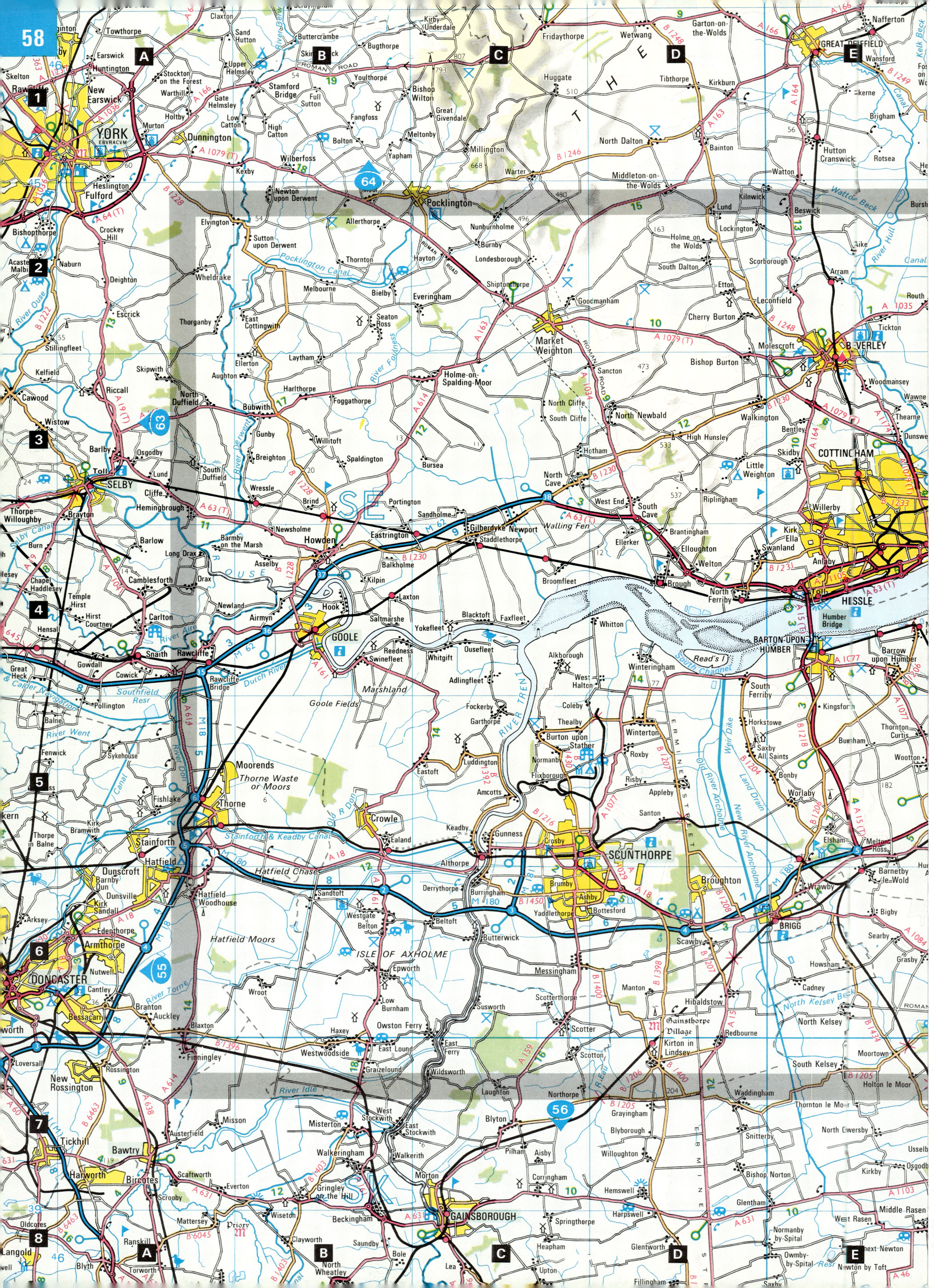

York
EBVRACVM
New Earswick
Huntington
Towthorpe
Earswick
Stockton on the Forest
Warthill
Holtby
Murton
Dunnington
Heslington
Fulford
Bishopthorpe
Crockey Hill
Naburn
Deighton
Escrick
Stillingfleet
Kelfield
Riccall
Skipwith
Cawood
Wistow
Barlby
Osgodby
Selby
Brayton
Thorpe Willoughby
Burn
Chapel Haddlesey
Temple Hirst
Hirst Courtney
Hensall
Gowdall
Great Heck
Cowick
Snaith
Rawcliffe
Rawcliffe Bridge
Pollington
Balne
Fenwick
Sykehouse
Fishlake
Kirk Bramwith
Thorpe in Balne
Stainforth
Hatfield
Hatfield Woodhouse
Dunscroft
Barnby Dun
Dunsville
Kirk Sandall
Edenthorpe
Armthorpe
Doncaster
Cantley
Bessacarr
Branton
Auckley
Blaxton
Finningley
Rossington
New Rossington
Tickhill
Bawtry
Harworth
Bircotes
Scaftworth
Everton
Scrooby
Mattersey
Ranskill
Torworth
Blyth
Oldcotes
Langold
Misson
Austerfield
Misterton
West Stockwith
East Stockwith
Walkeringham
Walkerith
Gringley on the Hill
Wiseton
Clayworth
Beckingham
Saundby
North Wheatley
Morton
Gainsborough
Bole
Lea
Claxton
Sand Hutton
Upper Helmsley
Gate Helmsley
Low Catton
High Catton
Stamford Bridge
Full Sutton
Buttercrambe
Bugthorpe
Kirby Underdale
Youlthorpe
Bishop Wilton
Great Givendale
Fangfoss
Bolton
Meltonby
Yapham
Millington
Wilberfoss
Kexby
Newton upon Derwent
Pocklington
Allerthorpe
Elvington
Sutton upon Derwent
Thornton
Hayton
Nunburnholme
Burnby
Londesborough
Shiptonthorpe
Melbourne
Bielby
Everingham
Seaton Ross
East Cottingwith
Wheldrake
Thorganby
Laytham
Ellerton
Aughton
Harlthorpe
Foggathorpe
Bubwith
Holme-on-Spalding-Moor
North Duffield
South Duffield
Gunby
Willitoft
Breighton
Spaldington
Bursea
Wressle
Brind
Cliffe
Hemingbrough
Lund
Barlow
Long Drax
Drax
Camblesforth
Carlton
Barmby on the Marsh
Asselby
Newsholme
Howden
Portington
Sandholme
Eastrington
Gilberdyke
Newport
Staddlethorpe
Balkholme
Kilpin
Laxton
Newland
Airmyn
Hook
Goole
Saltmarshe
Yokefleet
Blacktoft
Faxfleet
Reedness
Swinefleet
Whitgift
Ousefleet
Adlingfleet
Marshland
Goole Fields
Moorends
Thorne Waste or Moors
Thorne
Crowle
Ealand
Eastoft
Luddington
Garthorpe
Fockerby
Amcotts
Keadby
Gunness
Althorpe
Derrythorpe
Burringham
Sandtoft
Westgate
Belton
Beltoft
Butterwick
Epworth
Isle of Axholme
Hatfield Chase
Hatfield Moors
Wroot
Low Burnham
Haxey
Owston Ferry
Westwoodside
East Lound
Graizelound
East Ferry
Wildsworth
Laughton
Blyton
Pilham
Aisby
Corringham
Springthorpe
Heapham
Upton
Kirkby Underdale
Fridaythorpe
Wetwang
Garton-on-the-Wolds
Huggate
Tibthorpe
Kirkburn
Bainton
North Dalton
Warter
Middleton-on-the-Wolds
Lund
Lockington
Kilnwick
Holme on the Wolds
South Dalton
Etton
Goodmanham
Market Weighton
Cherry Burton
Leconfield
Scorborough
Bishop Burton
Sancton
North Cliffe
South Cliffe
North Newbald
Hotham
North Cave
South Cave
West End
Walkington
High Hunsley
Little Weighton
Riplingham
Brantingham
Ellerker
Elloughton
Welton
Brough
Broomfleet
North Ferriby
Swanland
Kirk Ella
Anlaby
Willerby
Cottingham
Skidby
Bentley
Beverley
Molescroft
Woodmansey
Tickton
Arram
Beswick
Watton
Hutton Cranswick
Rotsea
Brigham
Skerne
Wansford
Great Driffield
Nafferton
Hessle
Humber Bridge
Barton-upon-Humber
Barrow upon Humber
Whitton
Alkborough
West Halton
Winteringham
Coleby
Thealby
Burton upon Stather
Normanby
Flixborough
Winterton
Roxby
Risby
Appleby
Santon
Scunthorpe
Crosby
Brumby
Ashby
Yaddlethorpe
Bottesford
Broughton
Scawby
Brigg
Wrawby
Elsham
Melton Ross
Barnetby le Wold
Bigby
Searby
Grasby
Howsham
Cadney
North Kelsey
South Kelsey
Moortown
Holton le Moor
Thornton le Moor
North Owersby
Usselby
Kirkby
Osgodby
Bishop Norton
Glentham
Middle Rasen
West Rasen
Normanby by Spital
Owmby by Spital
Newton by Toft
Snitterby
Waddingham
Hibaldstow
Redbourne
Kirton in Lindsey
Gainsthorpe Village
Manton
Messingham
Scotterthorpe
Scotter
Scotton
Northorpe
Grayingham
Blyborough
Willoughton
Hemswell
Harpswell
Glentworth
Fillingham
Saxby
Saxby All Saints
Bonby
Worlaby
Horkstow
South Ferriby
Kingsforth
Burnham
Thornton Curtis
Wootton
Read's I
River Ouse
River Derwent
River Aire
River Don
River Trent
River Idle
River Torne
River Went
River Foulness
River Hull
River Ancholme
Old River Ancholme
New River Ancholme
Pocklington Canal
Stainforth & Keadby Canal
Dutch River
Southfield Resr
Walling Fen
Weir Dike
Land Drain
Watton Beck
Kelk Beck
The Wolds
Ermine Street
Roman Road
M 62
M 18
M 180
A 63
A 19
A 614
A 1079
A 1034
A 163
A 164
A 166
A 161
A 18
A 15
A 159
A 631
A 638
A 1077
A 1035
B 1228
B 1230
B 1246
B 1248
B 1206
B 1207
B 1398
B 1400
B 1396
B 1205
B 1403
B 6045
B 1216
B 1450
B 1204
B 1218
B 1434
A
B
C
D
E
1
2
3
4
5
6
7
8
55
56
63
64
SE
SK

G
H
J
K
F
55
45
1
2
3
4
5
6
7
8
HORNSEA
Hornsea Mere
Brandesburton
Seaton
Sigglesthorne
Leven
Catwick
Rolston
Mappleton
Goxhill
Great Hatfield
Long Riston
Rise
Withernwick
Meaux
New Ellerby
South Skirlaugh
Old Ellerby
West Newton
Aldbrough
Burton Constable
Coniston
Swine
Flinton
Garton
Ganstead
Sproatley
Humbleton
Sutton on Hull
Bilton
Lelley
Owstwick
Tunstall
Elstronwick
TA
Burton Pidsea
Preston
Roos
Marfleet
HEDON
Waxholme
KINGSTON UPON HULL
Burstwick
Rimswell
WITHERNSEA
Halsham
Thorngumbald
Paull
Hollym
Keyingham
Ottringham
Winestead
Holmpton
East Halton Skitter
Paull Holme Sands
Patrington
Cherry Cobb Sands
Salthaugh Grange
Out Newton
Welwick
Goxhill
Foulholme Sands
Skeffling
Easington
East Halton
Sunk Island
The Old Hall
Sunk Island Sands
Kilnsea
RIVER HUMBER
Oil Refinery
Killingholme
Immingham
Ulceby
Habrough
Croxton
Kirmington
Brocklesby
Stallingborough
GRIMSBY
SPURN HEAD
Healing
Keelby
Great Coates
CLEETHORPES
Bull
Spurn
MOUTH OF THE HUMBER
Humber
NORTH SEA
Great Limber
Aylesby
Riby
Laceby
R. Freshney
Bradley
Scartho
Humberston
Irby Upon Humber
Barnoldby le Beck
New Waltham
Holton le Clay
Swallow
Cabourne
Beelsby
Waltham
Brigsley
Caistor
Cuxwold
Ashby cum Fenby
Hatcliffe
Tetney Lock
Tetney
North Cotes
Marshchapel
Donna Nook
Nettleton
Rothwell
East Ravendale
Grainsby
North Thoresby
Grainthorpe
North Somercotes
Thoresway
Thorganby
Wold Newton
Fulstow
Covenham Resr
Swinhope
Ludborough
Covenham St Bartholomew
South Somercotes
Normanby le Wold
Stainton le Vale
Covenham St Mary
Saltfleet
Claxby
Binbrook
Utterby
North Ormsby
Yarburgh
Skidbrooke
Walesby
Kirmond le Mire
Fotherby
Saltfleetby St Clement
Alvingham
Tealby
North Elkington
Saltfleetby All Saints
MARKET RASEN
Ludford
Kelstern
Cockerington
South Elkington
Keddington
Saltfleetby St Peter
Theddlethorpe St Helen
North Willingham
Welton le Wold
Grimoldby
Gas Terminal
Linwood
Sixhills
Stewton
Manby
Little Carlton
TF
KINGSTON UPON HULL to
Rotterdam (Europoort) 14 hrs
Zeebrugge 14½-15 hrs

A
B
C
D
E
F
1
2
3
IRISH SEA
Selker Bay
Corney
Prior Park
Broughton Mills
Hycemoor
Bootle Sta
Bootle
Bootle Fell
Duddon Bridge
Broughton in Furness
Annaside
Black Combe
Whitbeck
The Green
Hallthwaites
Foxfield
Grizebeck
Green Road Sta
The Hill
Kirkby-in-Furness
Beck Side
Shooting House Hill
Whicham
Silecroft
Kirksanton
MILLOM
Soutergate
Haverigg
Askam in Furness
Duddon Sands
Ireleth
Pennington
Lindal in Furness
ULVERSTON
DALTON-IN-FURNESS
Stainton with Adgarley
Newton
BARROW-IN-FURNESS
Isle of Walney
North Scale
Vickerstown
Tummer Hill Scar
Leece
Roose
Gleaston
Biggar
Rampside
Roa Island
Sheep Island
Piel Island
Foulney Island
South End
Piel Bar
66

N
P
Q
R
1
2
3
4
5
POINT OF AYRE
The Ayres
Rue Point
The Lhen
Cranstal
Bride
Dhowin
Andreas
Shellag Point
Jurby West
Jurby East
Jurby Head
Ballasalla
Civil War Fort
St Judes
Regaby
Sandygate
Dhoor
The Cronk
The Curragh's
Sulby
RAMSEY BAY
RAMSEY
Churchtown
Port e Vullen
Glen Auldyn
Orrisdale Head
Ballaugh
Maughold
Maughold Head
Ravensdale
North Barrule
Ballajora
Port Mooar
Michael
Kirk Michael
Slieau Dhoo
Corrany
Clagh Ouyr
Cashtal yn Ard
Port Cornaa
Lady Port
Gob y Deigan
Ballacarnane Beg
Barregarrow
Druidale
SNAEFELL
Dhoon
Knocksharry
Shoughlaige-e-Caine
Cronk-y-Voddy
Bulgham Bay
St Patrick's Isle
Castle
PEEL
Ballagyr
Lambfell Moar
Colden
King Orry's Grave
Laxey
Laxey Head
Contrary Head
Patrick
Ballig
St John's
Slieau Ruy
Baldwin
Ballacannell
Laxey Bay
ISLE OF MAN
Baldrine
Clay Head
Glenmaye
Crosby
Glen Vine
Union Mills
Hillberry
Onchan
Dalby Point
Dalby
Foxdale
Garth
Port Groudle
Onchan Head
Niarbyl Bay
Eairy
DOUGLAS
Douglas Bay
Fort
South Barrule
Close Clark
Braaid
Douglas Head
Stroin Vuigh
Ballamodha
St Mark's
Newtown
Quine's Hill
Little Ness
Fleshwick Bay
Lingague
Ronague
Port Soderick
Bradda Head
Bradda
Colby
Ballabeg
Ballasalla
Santon Head
Port Erin
Isle of Man (Ronaldsway) Airport
The Howe
Cregneish
CASTLETOWN
Derbyhaven
Port St Mary
Castle
St Michael's Island
Langness
Calf of Man
SPANISH HEAD
Dreswick Point
Chicken Rock
DOUGLAS to
Heysham 4-5 hrs
Seasonal
Liverpool 3 hrs
Stranraer 6 hrs
Belfast 4½ hrs
Dublin 4½ hrs
Fleetwood 3 hrs
The Isle of Man lies about 36 miles or 58 km W of Tarn Bay SD 0790
52
F

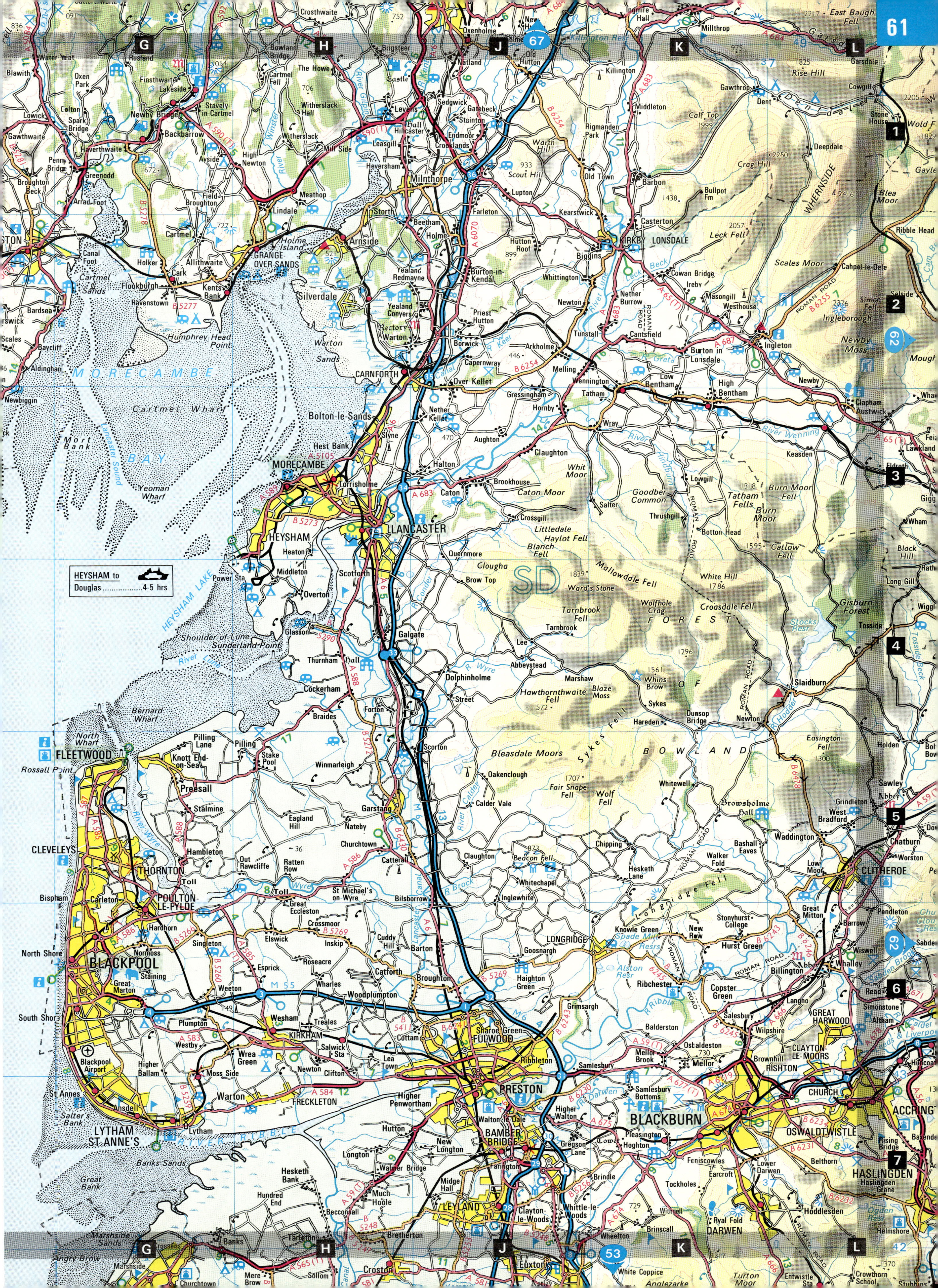
G
H
J
K
L
1
2
3
4
5
6
7
MORECAMBE BAY
HEYSHAM to Douglas 4-5 hrs
SD
Kendal
Milnthorpe
Arnside
Silverdale
Grange-over-Sands
Kirkby Lonsdale
Ingleton
Carnforth
Morecambe
Heysham
Lancaster
Fleetwood
Cleveleys
Thornton
Poulton-le-Fylde
Blackpool
Lytham St Anne's
Garstang
Preston
Fulwood
Longridge
Clitheroe
Blackburn
Great Harwood
Clayton-le-Moors
Rishton
Church
Oswaldtwistle
Accrington
Haslingden
Darwen
Leyland
Bamber Bridge
Kirkham
Freckleton
Warton
Slaidburn
Forest of Bowland
Bleasdale Moors
Whernside
Sedbergh
Dent
Bentham
Hornby
Caton
Galgate
Cockerham
Pilling
Preesall
Knott End-on-Sea
Bolton-le-Sands
Halton
Whalley
Billington
Ribchester
Chipping
Waddington
Grindleton

THE PENNINES
SKIPTON
SETTLE
CLITHEROE
ACCRINGTON
BURNLEY
NELSON
COLNE
PADIHAM
BARROWFORD
BRIERFIELD
BARNOLDSWICK
EARBY
KEIGHLEY
BINGLEY
SHIPLEY
BAILDON
BRADFORD
HALIFAX
HEBDEN BRIDGE
TODMORDEN
BACUP
RAWTENSTALL
HASLINGDEN
OSWALDTWISTLE
CHURCH
GREAT HARWOOD
ILKLEY
BRIGHOUSE
SOWERBY BRIDGE
CLECKHEATON
LIVERSEDGE
HECKMONDWIKE
WHITWORTH
Mytholmroyd
Grassington
Kettlewell
Malham
Hawes
Aysgarth
Ingleton
Gargrave
Hellifield
Long Preston
Gisburn
Haworth
Oxenhope
Silsden
Steeton
Cowling
Trawden
Whalley
Sabden
Wycoller
LANGSTROTHDALE CHASE
THE FOREST OF TRAWDEN
FOREST OF ROSSENDALE
KEIGHLEY MOOR
Ilkley Moor
Embsay Moor
Heptonstall Moor
Widdale Fell
Dodd Fell
Fountains Fell
Darnbrook Fell
Horton Moor
Grassington Moor
Conistone Moor
Ribblesdale
Wharfedale
Airedale
Littondale
Gisburn Forest
Due to open Late 1988

F
G
H
J
K
69
Helmsley Moor
THE HAMBLETON HILLS
Thirsk
Sowerby
Ripon
Boroughbridge
Aldborough
Knaresborough
Harrogate
Wetherby
Tadcaster
York
Haxby
Easingwold
Helmsley
Kirkbymoorside
Rievaulx Abbey
Byland Abbey
Newburgh Priory
Coxwold
Howardian Hills
Castle Howard
Malton
Marston Moor
Selby
Sherburn in Elmet
Leeds
Pudsey
Horsforth
Morley
Batley
Dewsbury
Wakefield
Castleford
Pontefract
Knottingley
Featherstone
Normanton
Rothwell
Garforth
Kippax
Woodlesford
Collingham
Boston Spa
Bramham
Cawood
Riccall
Snaith
Howden
Bubwith
Dunnington
Fulford
Heslington
Strensall
Stamford Bridge
Pocklington
Skipwith
Camblesforth
Carlton
Drax
Hemingbrough
Barlby
Brayton
Eggborough
Whitley
Gowdall
Rawcliffe
Airmyn
Bedale
Leeming
Aiskew
Kirby Hill
Dishforth
Topcliffe
Baldersby
Sutton-under-Whitestonecliffe
Kilburn
Ampleforth
Gilling East
Hovingham
Slingsby
Coneysthorpe
Terrington
Sheriff Hutton
Sutton-on-the-Forest
Shipton
Nether Poppleton
Upper Poppleton
Rufforth
Long Marston
Tockwith
Bilton
Askham Bryan
Bishopthorpe
Appleton Roebuck
Acaster Selby
Ulleskelf
Church Fenton
South Milford
Monk Fryston
Hambleton
Thorpe Willoughby
Fountains Abbey
Studley Roger
Bishop Monkton
South Stainley
Spofforth
Kirk Deighton
North Rigton
Harewood
Scarcroft
Barwick in Elmet
Aberford
Micklefield
Ledsham
Allerton Bywater
Methley
Ackworth
Darrington
Womersley
Cridling Stubbs
Kellington
Temple Hirst
Hirst Courtney
Chapel Haddlesey
Birkin
Burton Salmon
Fairburn
Brotherton
Ferrybridge
Ossett
Horbury
Kirkhamgate
Altofts
Stanley
Outwood
Lofthouse Gate
Tingley
Drighlington
Gildersome
Birstall
Beeston
Headingley
Chapel Allerton
Roundhay
Temple Newsam
Swillington
Great Preston
Oulton
Middleton
Ardsley East
Leeds & Bradford Airport
Eccup
Arthington
Bardsey
Thorner
Scholes
Leathley
Stainburn
Huby
Weeton
Pannal
Follifoot
Beaver Dyke Resr
Scargill Res
Beckwithshaw
Killinghall
Hampsthwaite
Birstwith
Burnt Yates
Ripley
Nidd
Brearton
Farnham
Scotton
Scriven
Goldsborough
Flaxby
Whixley
Green Hammerton
Kirk Hammerton
Cattal
Hunsingore
Nun Monkton
Moor Monkton
Beningbrough
Newton-on-Ouse
Linton-on-Ouse
Aldwark
Tholthorpe
Flawith
Alne
Tollerton
Youlton
Myton-on-Swale
Brafferton
Helperby
Raskelf
Crayke
Brandsby
Stillington
Marton Abbey
Farlington
Whenby
Skewsby
Stearsby
Bulmer
Welburn
Whitwell-on-the-Hill
Foston
Thornton-le-Clay
Flaxton
West Lilling
Huby
Wigginton
Towthorpe
Earswick
Huntington
New Earswick
Rawcliffe
Skelton
Stockton on the Forest
Warthill
Gate Helmsley
Holtby
Murton
Kexby
Elvington
Wheldrake
Thorganby
Escrick
Deighton
Naburn
Stillingfleet
Kelfield
Crockey Hill
Acaster Malbis
Copmanthorpe
Colton
Bolton Percy
Ryther
Saxton
Barkston
Towton
Stutton
Clifford
Newton Kyme
Thorp Arch
Walton
Healaugh
Wighill
Askham Richard
Bilbrough
Catterton
Hutton Wandesley
Angram
Hessay
Knapton
Bickerton
Sicklinghall
Kirkby Overblow
Little Ribston
North Deighton
Staveley
Arkendale
Ferrensby
Minskip
Roecliffe
Langthorpe
Marton
Grafton
Lower Dunsforth
Upper Dunsforth
Ouseburn
Thorpe Underwood
Newby Hall
Skelton
Littlethorpe
Bishop Thornton
Markington
Sawley
Risplith
Aldfield
Grantley
Winksley
Sharow
Copt Hewick
Hutton Conyers
Norton-le-Clay
Cundall
Rainton
Asenby
Dalton
Catton
Skipton-on-Swale
Carlton Miniott
Sandhutton
Kirby Wiske
Pickhill
Ainderby Quernhow
Sinderby
Kirklington
Nosterfield
Thornborough
West Tanfield
Sutton Howgrave
Wath
Melmerby
North Stainley
Mickley
Azerley
Sutton Grange
Galphay
Laverton
Grewelthorpe
Well
Snape
Thornton Watlass
Burrill
Cowling
Thirn
Exelby
Theakston
Burneston
Carthorpe
Londonderry
Gatenby
Maunby
Newby Wiske
South Otterington
Thornton-le-Moor
Borrowby
Knayton
Upsall
North Kilvington
South Kilvington
Felixkirk
Boltby
Kirby Knowle
Thirlby
Old Byland
Cold Kirby
Scawton
Oldstead
Wass
Bagby
Thirkleby
Carlton Husthwaite
Husthwaite
Oulston
Yearsley
Thormanby
Sessay
Hutton Sessay
Cross Lanes
Ampleforth College
Thorpe Hall
Cawton
Coulton
Scackleton
Stonegrave
Nunnington
Muscoates
Harome
Wombleton
Beadlam
Nawton
Pockley
Sproxton
Great Edstone
Keldholme
Gillamoor
Fadmoor
Carlton
Spaunton
Hutton-le-Hole
Appleton-le-Street
Barton-le-Street
Butterwick
Salton
Ness
Oswaldkirk
Kirby Mills
Ganthorpe
Barton-le-Willows
Harton
Bossall
Claxton
Sand Hutton
Upper Helmsley
Low Catton
High Catton
Wilberfoss
Newton upon Derwent
Sutton upon Derwent
East Cottingwith
Ellerton
Aughton
North Duffield
South Duffield
Breighton
Wressle
Newsholme
Barmby on the Marsh
Asselby
Newland
Long Drax
Barlow
Burn
West Haddlesey
Gateforth
Lumby
Hillam
Newthorpe
Biggin
Little Fenton
Wistow
Osgodby
Cliffe
Lund
Hensall
Great Heck
Pollington
Balne
Whitley Bridge
East Hardwick
Sharlston
Warmfield
Whitwood
New Fryston
Rawcliffe Bridge
55
58
64
River Ouse
River Nidd
River Swale
River Wharfe
River Derwent
River Aire
River Foss
River Rye
River Riccal
River Seven
River Dove
Cod Beck
Cock Beck
Selby Canal
ROMAN ROAD
DERE STREET
SE
ISVRIVM
EBVRACVM

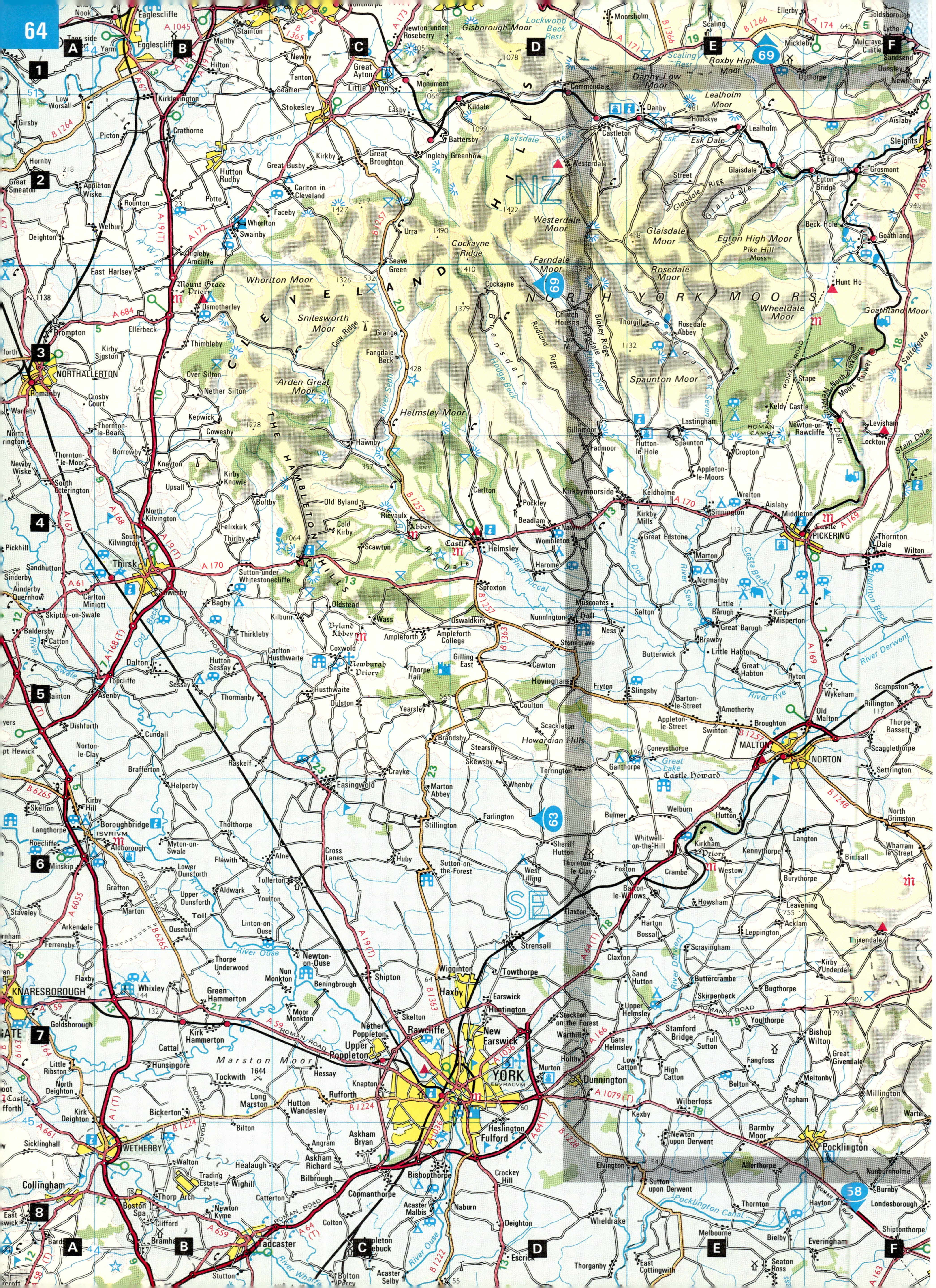
A
B
C
D
E
F
1
2
3
4
5
6
7
8
NZ
SE
Eaglescliffe
Yarm
Maltby
Hilton
Stainton
Newby
Tanton
Seamer
Stokesley
Great Ayton
Little Ayton
Newton under Roseberry
Monument
Gisborough Moor
Lockwood Beck Resr
Moorsholm
Scaling
Scaling Resr
Roxby High Moor
Ellerby
Mickleby
Lythe
Sandsend
Dunsley
Newholm
Ugthorpe
Danby Low Moor
Commondale
Lealholm Moor
Danby
Houlskye
Castleton
Esk Dale
Lealholm
Aislaby
Sleights
Egton
Grosmont
Egton Bridge
Glaisdale
Glaisdale Moor
Egton High Moor
Pike Hill Moss
Beck Hole
Goathland
Westerdale
Westerdale Moor
Kildale
Battersby
Easby
Ingleby Greenhow
Baysdale
Kirkleavington
Crathorne
Picton
Hornby
Great Smeaton
Appleton Wiske
Rounton
Welbury
Deighton
Hutton Rudby
Potto
Great Busby
Kirkby
Great Broughton
Carlton in Cleveland
Faceby
Whorlton
Swainby
Ingleby Arncliffe
East Harlsey
Mount Grace Priory
Osmotherley
Whorlton Moor
Snilesworth Moor
Cow Ridge
Grange
Urra
Seave Green
Cockayne Ridge
Farndale Moor
Cockayne
CLEVELAND HILLS
NORTH YORK MOORS
Church Houses
Low Mill
Rudland Rigg
Blakey Ridge
Rosedale Moor
Rosedale Abbey
Thorgill
Wheeldale Moor
Hunt Ho
Goathland Moor
Stape
North Yorkshire Moors Railway
Newton Dale
Levisham
Lockton
Keldy Castle
Newton-on-Rawcliffe
Cropton
Spaunton Moor
Lastingham
Spaunton
Hutton-le-Hole
Appleton-le-Moors
Gillamoor
Fadmoor
Brompton
Kirby Sigston
Ellerbeck
Thimbleby
Over Silton
Nether Silton
NORTHALLERTON
Romanby
Crosby Court
Warlaby
Thornton-le-Beans
Thornton-le-Moor
South Otterington
Newby Wiske
Borrowby
Kepwick
Cowesby
Knayton
Upsall
Kirby Knowle
Arden Great Moor
Fangdale Beck
Hawnby
Helmsley Moor
Bilsdale
THE HAMBLETON HILLS
Old Byland
Cold Kirby
Boltby
Felixkirk
Thirlby
North Kilvington
South Kilvington
Thirsk
Sowerby
Sutton-under-Whitestonecliffe
Scawton
Rievaulx
Rievaulx Abbey
Helmsley
Carlton
Pockley
Beadlam
Nawton
Wombleton
Harome
Kirkbymoorside
Keldholme
Kirkby Mills
Sinnington
Wrelton
Aislaby
Middleton
PICKERING
Thornton Dale
Wilton
Great Edstone
Marton
Normanby
Salton
Little Barugh
Great Barugh
Kirby Misperton
Brawby
Little Habton
Great Habton
Butterwick
Ryton
Slingsby
Barton-le-Street
Appleton-le-Street
Amotherby
Swinton
Broughton
MALTON
NORTON
Old Malton
Wykeham
Scampston
Rillington
Thorpe Bassett
Scagglethorpe
Settrington
Sandhutton
Sinderby
Ainderby Quernhow
Carlton Miniott
Skipton-on-Swale
Baldersby
Catton
Topcliffe
Dalton
Asenby
Dishforth
Norton-le-Clay
Cundall
Brafferton
Helperby
Kirby Hill
Boroughbridge
Aldborough
Langthorpe
Roecliffe
Minskip
Skelton
Bagby
Kilburn
Thirkleby
Byland Abbey
Wass
Ampleforth
Ampleforth College
Oldstead
Coxwold
Newburgh Priory
Carlton Husthwaite
Hutton Sessay
Sessay
Thormanby
Husthwaite
Oulston
Thorpe Hall
Gilling East
Oswaldkirk
Sproxton
Nunnington
Nunnington Hall
Stonegrave
Muscoates
Ness
Cawton
Hovingham
Fryton
Coulton
Yearsley
Brandsby
Scackleton
Howardian Hills
Stearsby
Skewsby
Terrington
Coneysthorpe
Ganthorpe
Castle Howard
Great Lake
Whenby
Crayke
Raskelf
Easingwold
Marton Abbey
Stillington
Farlington
Sheriff Hutton
Bulmer
Welburn
Whitwell-on-the-Hill
Kirkham Priory
Westow
Hutton
Kennythorpe
Langton
Birdsall
Wharram le Street
North Grimston
Tholthorpe
Myton-on-Swale
Flawith
Alne
Cross Lanes
Huby
Sutton-on-the-Forest
West Lilling
Thornton le Clay
Foston
Barton-le-Willows
Crambe
Burythorpe
Leavening
Acklam
Howsham
Harton
Bossall
Flaxton
Strensall
Claxton
Scrayingham
Leppington
Thixendale
Kirby Underdale
Buttercrambe
Bugthorpe
Tollerton
Youlton
Lower Dunsforth
Upper Dunsforth
Aldwark
Grafton
Marton
Staveley
Arkendale
Ferrensby
Linton-on-Ouse
Ouseburn
Thorpe Underwood
Newton-on-Ouse
Nun Monkton
Beningbrough
Shipton
Wigginton
Haxby
Towthorpe
Earswick
Huntington
Sand Hutton
Upper Helmsley
Skirpenbeck
Youlthorpe
Bishop Wilton
Stockton on the Forest
Warthill
Gate Helmsley
Stamford Bridge
Full Sutton
Fangfoss
Great Givendale
Flaxby
Whixley
Green Hammerton
Kirk Hammerton
Moor Monkton
Skelton
Nether Poppleton
Upper Poppleton
Rawcliffe
New Earswick
KNARESBOROUGH
Goldsborough
Cattal
Hunsingore
Marston Moor
Tockwith
Hessay
Long Marston
Knapton
Rufforth
YORK
Murton
Holtby
Dunnington
Low Catton
High Catton
Bolton
Meltonby
Yapham
Millington
Little Ribston
North Deighton
Kirk Deighton
Bickerton
Hutton Wandesley
Bilton
Askham Bryan
Askham Richard
Angram
Healaugh
Heslington
Fulford
Kexby
Wilberfoss
Newton upon Derwent
Barmby Moor
Pocklington
Warter
WETHERBY
Sicklinghall
Walton
Trading Estate
Wighill
Bilbrough
Catterton
Copmanthorpe
Bishopthorpe
Acaster Malbis
Naburn
Crockey Hill
Elvington
Sutton upon Derwent
Pocklington Canal
Allerthorpe
Thornton
Hayton
Nunburnholme
Burnby
Londesborough
Shiptonthorpe
Collingham
Boston Spa
Thorp Arch
Clifford
Bramham
Newton Kyme
Tadcaster
Colton
Appleton Roebuck
Acaster Selby
Bolton Percy
Deighton
Escrick
Wheldrake
Thorganby
East Cottingwith
Melbourne
Bielby
Everingham
Seaton Ross
Stutton
River Ouse
River Wharfe
River Derwent
River Rye
River Swale
River Nidd
River Seven
River Dove
Costa Beck
Cod Beck
Bardsey
East Keswick
Wetherby
Skipton-on-Swale
Stokesley
Ainderby
69
63
58
A 19(T)
A 172
A 684
A 167
A 168(T)
A 170
A 171
A 173
A 174
A 169
A 64(T)
A 166
A 1079
A 59
A 61
A 6055
A 659
A 661
A 163
B 1257
B 1363
B 1224
B 1228
B 1248
B 1266
B 1366
B 6265
B 1264

G
H
J
K
L
53
51
45
1
2
3
4
5
6
7
8
Saltwick Bay
Abbey
WHITBY
Ruswarp
R Esk
Stainsacre
Sneaton
Hawsker
Ugglebarnby
Ness Point or
North Cheek
Sneatonthorpe
Raw
Fylingthorpe
Robin Hood's Bay
B 14 6
Old Peak or
South Cheek
Ravenscar
A 171
Fylingdales Moor
959
Burn Howe
Rigg
16
Staintondale
Harwood Dale
Cloughton
Newlands
Cloughton Wyke
Cloughton
Burniston
Cromer Point
656
Silpho
Langdale
End
Toll
SCALBY
A 165
Scalby Ness
Rocks
Hackness
Everley
Castle
812
SCARBOROUGH
Black Rocks
A 170
B 1261
Hutton
Buscel
Ayton
Eastfield
Osgodby
Cayton Bay
Sawdon
Irton
Allerston
Ebberston
Wykeham
Seamer
Cayton
The Wyke
A 170
Snainton
Brompton
Due to open
Early 1988
Lebberston
Gristhorpe
A 1039
R Hertford
FILEY
Flixton
Yedingham
THE CARRS
Willerby
Staxton
Folkton
A 1039
6
Muston
Filey Bay
B 1258
Ganton
Holiday Camp
Sherburn
Hunmanby
Knapton
A 64 (T)
East Heslerton
Reighton Sands
15
West
Heslerton
B 1249
Fordon
Reighton
Speeton
Crab Rocks
Wintringham
Wold
Newton
B 1229
Foxholes
Burton
Fleming
Bempton
Danes Dyke
Holiday
Camp
Place Newton
A 165
FLAMBOROUGH HEAD
Willy
Howe
Grindale
Butterwick
Flamborough
B 1259
Helperthorpe
Weaverthorpe
Thwing
Gypsey Race
654
B 1255
West
Lutton
Boynton
Sewerby
B 1253
Duggleby
15
Rudston
B 1254
BRIDLINGTON
Kirby Grindalythe
Duggleby
Howe
Langtoft
ROMAN ROAD
A 166
B 1253
THE WOLDS
Carnaby
Haisthorpe
Hilderthorpe
B 1248
Sledmere
573
Kilham
BRIDLINGTON BAY
Hall
B 1251
365
79
Burton Agnes
A 165
Burdale
510
B 1252
Harpham
Ruston Parva
12
TA
Fraisthorpe
Fimber
B 1249
Lowthorpe
9
Gransmoor
Barmston
Garton-on-
the-Wolds
A 166
Nafferton
Fridaythorpe
Wetwang
B 1248
A 166
Kelk Beck
Lissett
GREAT DRIFFIELD
Great Kelk
B 1242
Ulrome
Wansford
Huggate
Foston
on the
Wolds
B 1249
Tibthorpe
Castle
Skipsea
510
Kirkburn
Old Howe
Skerne
A 164
Canal
Beeford
B 1249
A 163
Brigham
16
56
North
Frodingham
North Dalton
Hutton
Cranswick
Dunnington
Bainton
6
B 1246
Rotsea
Hempholme
Bewholme
Atwick
Middleton-on-
the-Wolds
Watton
480
HORNSEA
Watton Beck
15
Lund
Kilnwick
Burshill
Beswick
163
Lockington
River Hull
Brandesburton
Seaton
13
Holme on
the Wolds
B 1244
Hornsea
Mere
59
Aike
6
Rolston
Scorborough
Canal
Sigglesthorne
South Dalton
Leven
Catwick
Mappleton
Arram
Goxhill
Etton
Goodmanham
Routh
Leconfield
A 1035
Great Hatfield
B 1242
Rise
Long Riston
A 165
Cherry Burton
10
B 1248
Tickton
A 1079 (T)
Market
Withernwick
NORTH SEA

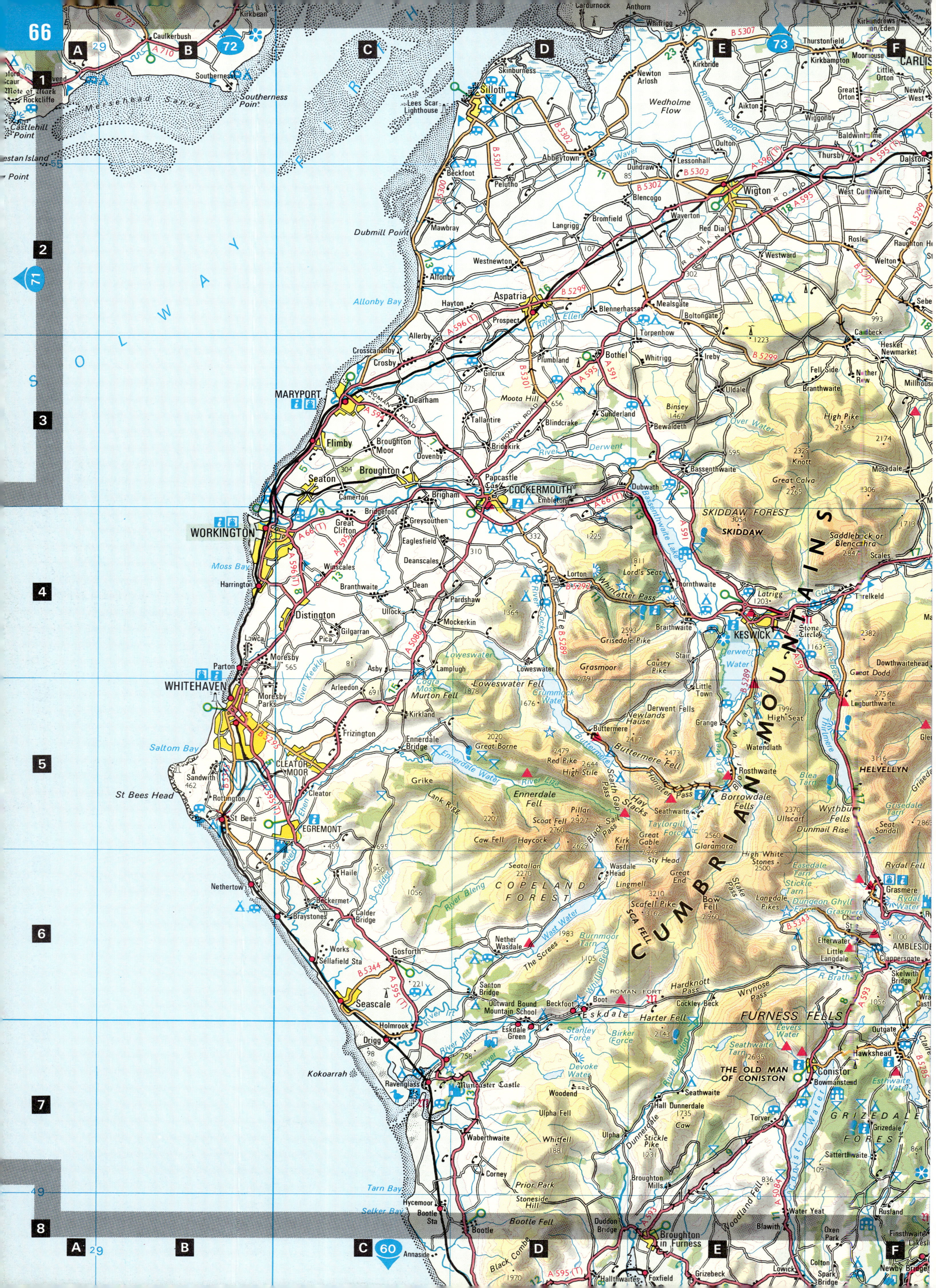

SOLWAY FIRTH
Mersehead Sands
Southerness Point
Caulkerbush
Southerness
Rockcliffe
Hestan Island
Skinburness
Silloth
Lees Scar Lighthouse
Newton Arlosh
Wedholme Flow
Kirkbride
Abbeytown
Beckfoot
Pelutho
Dubmill Point
Mawbray
Westnewton
Allonby
Allonby Bay
Aspatria
Hayton
Prospect
Allerby
Crosscanonby
Crosby
Gilcrux
Moota Hill
Plumbland
Bothel
Torpenhow
Blennerhasset
Mealsgate
Boltongate
Wigton
Waverton
Red Dial
Westward
Dalston
Thursby
West Curthwaite
Rosley
Welton
Raughton Head
Caldbeck
Hesket Newmarket
Fell Side
High Pike
Great Calva
Knott
Mosedale
Saddleback or Blencathra
Scales
Threlkeld
SKIDDAW FOREST
SKIDDAW
MARYPORT
Dearham
Flimby
Broughton Moor
Broughton
Dovenby
Tallantire
Bridekirk
Papcastle
COCKERMOUTH
Embleton
Brigham
Seaton
Camerton
Bridgefoot
Great Clifton
Greysouthen
Eaglesfield
Deanscales
Dean
WORKINGTON
Moss Bay
Harrington
Winscales
Branthwaite
Ullock
Pardshaw
Mockerkin
Distington
Gilgarran
Pica
Lowca
Parton
Moresby
WHITEHAVEN
Moresby Parks
Arlecdon
Asby
Lamplugh
Loweswater
Loweswater Fell
Crummock Water
Murton Fell
Kirkland
Frizington
Ennerdale Bridge
Ennerdale Water
Great Borne
Red Pike
High Stile
Buttermere
Buttermere Fell
Honister Pass
Saltom Bay
St Bees Head
Sandwith
CLEATOR MOOR
Cleator
Rottington
St Bees
EGREMONT
Grike
Lank Rigg
Ennerdale Fell
Pillar
Scoat Fell
Haycock
Caw Fell
Seatallan
COPELAND FOREST
Haile
Nethertown
Beckermet
Braystones
Calder Bridge
Sellafield Sta
Works
Gosforth
Seascale
Santon Bridge
Outward Bound Mountain School
Holmrook
Drigg
Kokoarrah
Ravenglass
Muncaster Castle
Eskdale Green
Beckfoot
Boot
Eskdale
Harter Fell
Hardknott Pass
ROMAN FORT
Cockley Beck
Wrynose Pass
FURNESS FELLS
Stanley Force
Birker Force
Devoke Water
Woodend
Ulpha Fell
Ulpha
Waberthwaite
Whitfell
Corney
Prior Park
Stoneside Hill
Tarn Bay
Selker Bay
Hycemoor
Bootle Sta
Bootle
Bootle Fell
Black Combe
Annaside
Duddon Bridge
Broughton in Furness
Foxfield
Grizebeck
Hallthwaites
Lowick
Dunnerdale
Hall Dunnerdale
Seathwaite
Stickle Pike
Broughton Mills
Woodland Fell
Blawith
Water Yeat
Torver
THE OLD MAN OF CONISTON
Seathwaite Tarn
Levers Water
Coniston
Bowmanstead
Coniston Water
GRIZEDALE FOREST
Grizedale
Satterthwaite
Hawkshead
Esthwaite Water
Outgate
Rusland
Finsthwaite
Oxen Park
Colton
Spark Bridge
Newby Bridge
Nether Wasdale
The Screes
Wast Water
Burnmoor Tarn
Wasdale Head
Lingmell
Scafell Pike
Scafell
Great End
Great Gable
Sty Head
Kirk Fell
Glaramara
Bow Fell
Esk Pike
Langdale Pikes
Stickle Tarn
Dungeon Ghyll
Elterwater
Little Langdale
Skelwith Bridge
Clappersgate
AMBLESIDE
Rydal Fell
Rydal Water
Grasmere
Easedale Tarn
High White Stones
Ullscarf
Seatoller
Seathwaite
Rosthwaite
Borrowdale Fells
Watendlath
Grange
Derwent Fells
Newlands Hause
Causey Pike
Grisedale Pike
Grasmoor
Whinlatter Pass
Lorton
Lord's Seat
Thornthwaite
Braithwaite
Stair
Little Town
Derwent Water
KESWICK
Latrigg
High Seat
Blea Tarn
Thirlmere
Legburthwaite
Great Dodd
HELVELLYN
Wythburn Fells
Dunmail Raise
Seat Sandal
CUMBRIAN MOUNTAINS
Dubwath
Bassenthwaite Lake
Bassenthwaite
Uldale
Ireby
Whitrigg
Binsey
Bewaldeth
Sunderland
Blindcrake
Over Water
Kirkbampton
Thurstonfield
Baldwinholme
Great Orton
Wiggonby
Aikton
Dulton
Lessonhall
Dundraw
Blencogo
Bromfield
Langrigg
Anthorn
Cardurnock
Whitrigg
Kirkbean
CARLISLE

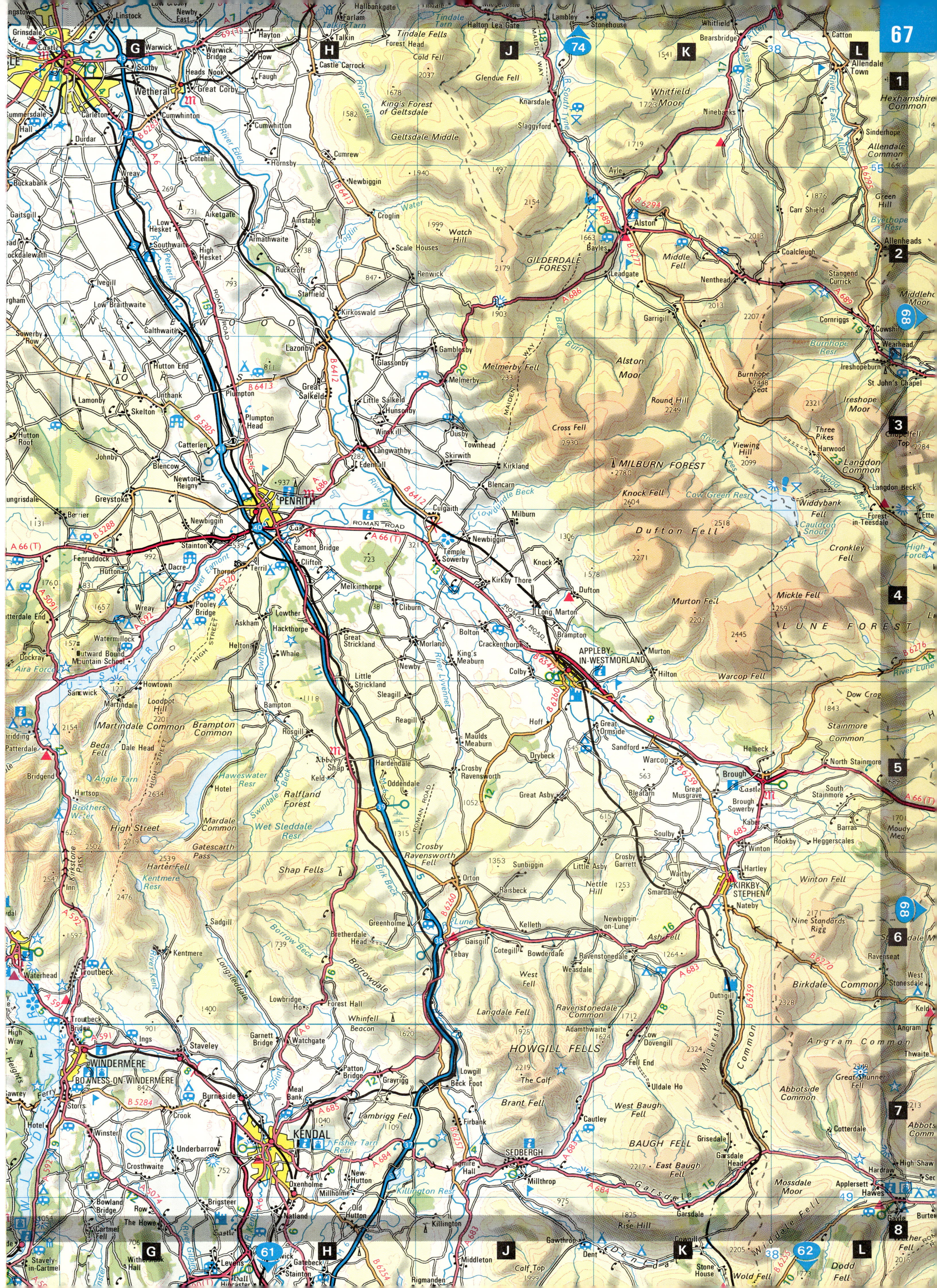
G
H
J
K
L
1
2
3
4
5
6
7
8
74
68
69
61
62
Linstock
Warwick
Hayton
Talkin
Castle Carrock
Scotby
Wetheral
Great Corby
Carleton
Cumwhinton
Cumwhitton
Hornsby
Cumrew
Newbiggin
Croglin
Cotehill
Wreay
Aiketgate
Low Hesket
High Hesket
Southwaite
Armathwaite
Ainstable
Ruckcroft
Staffield
Kirkoswald
Renwick
Scale Houses
Watch Hill
Tindale Fells
Forest Head
Cold Fell
Glendue Fell
King's Forest of Geltsdale
Geltsdale Middle
Knarsdale
Slaggyford
Halton Lea Gate
Lambley
Stonehouse
Whitfield
Bearsbridge
Whitfield Moor
Ninebanks
Catton
Allendale Town
Hexhamshire Common
Sinderhope
Allendale Common
Ayle
Alston
Bayles
Leadgate
Garrigill
Middle Fell
Nenthead
Carr Shield
Coalcleugh
Allenheads
Stangend Currick
Cornriggs
Cowshill
Wearhead
Ireshopeburn
St John's Chapel
Burnhope Resr
GILDERDALE FOREST
Alston Moor
Melmerby Fell
Melmerby
Gamblesby
Glassonby
Lazonby
Great Salkeld
Little Salkeld
Hunsonby
Winskill
Ousby
Townhead
Skirwith
Langwathby
Edenhall
Kirkland
Blencarn
Cross Fell
MILBURN FOREST
Round Hill
Burnhope Seat
Ireshope Moor
Viewing Hill
Three Pikes
Harwood
Langdon Common
Langdon Beck
Knock Fell
Cow Green Resr
Widdybank Fell
Cauldron Snout
Forest in-Teesdale
Dufton Fell
Cronkley Fell
Milburn
Culgarth
Newbiggin
Temple Sowerby
Kirkby Thore
Knock
Dufton
Long Marton
Brampton
Murton Fell
Mickle Fell
LUNE FOREST
APPLEBY-IN-WESTMORLAND
Murton
Hilton
Warcop Fell
Dow Crag
Stainmore Common
Helbeck
North Stainmore
South Stainmore
Brough
Church Brough
Brough Sowerby
Kaber
Great Musgrave
Warcop
Bleatarn
Sandford
Great Ormside
Hoff
Drybeck
Colby
Crackenthorpe
Bolton
King's Meaburn
Morland
Cliburn
Great Strickland
Little Strickland
Newby
Sleagill
Reagill
Maulds Meaburn
Crosby Ravensworth
Great Asby
Soulby
Winton
Hartley
KIRKBY STEPHEN
Nateby
Winton Fell
Nine Standards Rigg
Birkdale Common
Ravenseat
Angram Common
Abbotside Common
Great Shunner Fell
Cotterdale
Hardraw
Hawes
Appersett
Mossdale Moor
Garsdale Head
Grisedale
Garsdale
Rise Hill
Dent
Dentdale
Cowgill
Stone House
Widdale Fell
Wold Fell
Dodd Fell
Mallerstang Common
Outhgill
Ash Fell
Newbiggin-on-Lune
Ravenstonedale
Ravenstonedale Common
Weasdale
Bowderdale
West Fell
Langdale Fell
HOWGILL FELLS
Adamthwaite
Low Dovengill
Fell End
Uldale Ho
The Calf
Brant Fell
West Baugh Fell
BAUGH FELL
East Baugh Fell
Cautley
SEDBERGH
Millthrop
Killington Resr
Killington
Gawthrop
Middleton
Calf Top
Rigmanden
Firbank
Lowgill
Beck Foot
Tebay
Gaisgill
Orton
Raisbeck
Sunbiggin
Little Asby
Nettle Hill
Crosby Garrett
Waitby
Smardale
Kelleth
Cotegill
Greenholme
Crosby Ravensworth Fell
Shap Fells
Bretherdale Head
Borrowdale
Whinfell Beacon
Lowbridge Ho
Forest Hall
Grayrigg
Patton Bridge
Lambrigg Fell
Meal Bank
KENDAL
Fisher Tarn Resr
Oxenholme
New Hutton
Old Hutton
Natland
Sedgwick
Stainton
Gatebeck
Levens
Milnthorpe
Burneside
Garnett Bridge
Watchgate
Staveley
Ings
WINDERMERE
BOWNESS-ON-WINDERMERE
Crook
Underbarrow
Crosthwaite
Brigsteer
Winster
Storrs
Bowland Bridge
Cartmel Fell
The Howe
Witherslack Hall
Stavely in-Cartmel
Troutbeck
Kentmere
Kentmere Resr
Sadgill
High Street
Harter Fell
Gatescarth Pass
Mardale Common
Haweswater Resr
Wet Sleddale Resr
Ralfland Forest
Shap
Keld
Hardendale
Oddendale
Rosgill
Bampton
Brampton Common
Martindale Common
Martindale
Loadpot Hill
Howtown
Sandwick
Aira Force
Dockray
Watermillock
Pooley Bridge
Askham
Helton
Whale
Hackthorpe
Lowther
Melkinthorpe
Clifton
Eamont Bridge
Thorpe
Dacre
Stainton
Penruddock
Hutton
Newbiggin
Greystoke
Berrier
Blencow
Newton Reigny
Catterlen
Johnby
PENRITH
Plumpton
Plumpton Head
Unthank
Skelton
Lamonby
Hutton End
Calthwaite
Ivegill
Low Braithwaite
Sowerby Row
Hutton Roof
Buckabank
Durdar
Gaitsgill
Fenruddock
Beda Fell
Dale Head
Angle Tarn
Patterdale
Hartsop
Brothers Water
Bridgend
Kirkstone Pass
Waterhead
High Wray
Hotel
Sawrey
Ferry
Hutton End
INGLEWOOD FOREST
River Eden
River Eamont
River Lowther
ULLSWATER
WINDERMERE
River Lune
River Kent
River South Tyne
River West Allen
River East Allen
River Nent
River Tees
Crowdundle Beck
Birk Beck
Borrow Beck
Longsleddale
River Sprint
Black Burn
MAIDEN WAY
ROMAN ROAD
HIGH STREET
A 66 (T)
A 6
A 686
A 689
A 685
A 683
A 684
A 591
A 592
A 5074
B 6413
B 6412
B 5288
B 5305
B 6320
B 6542
B 6260
B 6261
B 6259
B 6276
B 6270
B 6294
B 5284
B 6263
B 6254
B 6255

A
B
C
D
E
1
2
3
4
5
6
7
8
74
75
67
62
THE PENNINES
NY
SD
Allendale Town
Hexhamshire Common
Allendale Common
Allenheads
Slaley Forest
Blanchland Moor
Blanchland
Derwent Resr
Edmundbyers
Muggleswick
Muggleswick Common
Hisehope Resr
Smiddy Shaw Resr
Waskerley Resr
Hunstanworth
Nookton Fell
Bolt's Law
Stanhope Common
Collier Law
Tunstall Resr
Wolsingham Park Moor
Crawleyside
Stanhope
Frosterley
Wolsingham
Tow Law
Middlehope Moor
Rookhope
Cowshill
Wearhead
Westgate
Eastgate
Ireshopeburn
St John's Chapel
Daddry Shield
Burnhope Resr
Killhope Cross
Ireshope Moor
Chapelfell Top
Westernhope Moor
Snowhope Hill
Bollihope Common
Hill End
White Kirkley
Langdon Common
Langdon Beck
Newbiggin Common
Middleton Common
Pawlaw Pike
Pikeston Fell
St John's Hall
Forest in-Teesdale
Ettersgill
High Force
Holwick
Newbiggin
Middleton-in-Teesdale
Eggleston Common
Hamsterley Forest
Woodland Fell
Woodland
Eggleston
Cronkley Fell
Lune Moor
Bowbank
Mickleton
Thringarth
Grassholme
Grassholme Resr
Romaldkirk
Hunderthwaite
Selset Resr
Kelton
Hunderthwaite Moor
Hury Resr
Blackton Resr
Balderhead Resr
Clove Lodge
Cotherstone Moor
Cotherstone
Lartington
Kinninvie
Barnard Castle
Startforth
Eggleston Abbey
Boldron
Bowes
Deep Dale
Stainmore
South Stainmore
Bowes Moor
Stainmore Forest
Sleightholme
Sleightholme Moor
Scargill High Moor
The Stang
Barningham Moor
Scargill
Barningham
Brignall
Greta Bridge
Kexwith Moor
Hurst Moor
Hurst
Tan Hill
Arkengarthdale Moor
Stonesdale Moor
Rogan's Seat
Great Pinseat
Whaw
Langthwaite
Ravenseat
West Stonesdale
Keld
Melbecks Moor
Angram
Thwaite
Muker
Gunnerside
Feetham
Low Row
Healaugh
Reeth
Fremington
Grinton
Kearton
Marrick
Skelton
Marske
Crackpot
Gibbon Hill
Swaledale
Abbotside Common
Askrigg Common
East Bolton Moor
Redmire Moor
Castle Bolton
Redmire
High Shaw
Hardraw
Sedbusk
Askrigg
Newbiggin
Woodhall
Carperby
Hawes
Appersett
Gayle
Burtersett
Bainbridge
Wensleydale
Aysgarth
Thornton Rust
Countersett
Semer Water
Marsett
Stalling Busk
Thoralby
West Burton
Newbiggin
Penhill
Swinithwaite
Bolton Hall
Wensley
Preston-under-Scar
Leyburn
Harmby
Spennithorne
Middleham
Coverham
Agglethorpe
Melmerby
Carlton
Caldbergh
Cotterdale
Great Shunner Fell
Whitley Chapel
Dalton
Slaley
Healey
Whittonstall
Newlands
Minsteracres
Kiln Pit Hill
Ebchester
Shotley Bridge
Consett
Carterway Heads
Castleside
Healeyfield
Knitsley
Leadgate
Iveston
Dipton
Hobson
Burnopfield
Tanfield
Tantobie
Stanley
South Moor
Annfield Plain
Craghead
Maiden Law
Lanchester
Burnhope
Butsfield
Satley
Cornsay
Quebec
Esh
Esh Winning
Waterhouses
East Hedleyhope
Sunniside
Stanley Crook
Billy Row
Crook
Thornley
Oakenshaw
Willington
Bearpark
Langley Park
New Brancepeth
Brancepeth
Brandon
Ushaw Moor
Durham
Chester-le-Street
Sacriston
Witton Gilbert
Framwellgate Moor
Pelton
Grange Villa
Waldridge
Edmondsley
Plawsworth
Kimblesworth
Great Lumley
Washington
Birtley
Ouston
Beamish
Kibblesworth
Lamesley
Rowlands Gill
Hamsterley
Chopwell
Sunderland Bridge
Croxdale
Tudhoe
Hett
Spennymoor
Byers Green
Newfield
Hunwick
Fir Tree
Howden-le-Wear
Witton-le-Wear
High Grange
Escomb
Bedburn
Hamsterley
Toft Hill
Witton Park
Etherley
Bishop Auckland
St Helen Auckland
West Auckland
Shildon
Coundon
Coundon Grange
Middlestone Moor
Kirk Merrington
Middridge
Newton Aycliffe
Chilton
Ferryhill
Bradbury
Rushyford
Redworth
Heighington
Evenwood
Butterknowle
Morley
Copley
Cockfield
Raby Castle
Wackerfield
Hilton
Ingleton
Staindrop
Killerby
Langton
Headlam
Bolam
Denton
Walworth
Staintion
Cleatlam
Little Newsham
Winston
Gainford
Piercebridge
High Coniscliffe
Low Coniscliffe
Darlington
Whorlton
Ovington
Caldwell
Hutton Magna
Forcett
Eppleby
Stanwick St John
Aldbrough St John
Manfield
Cleasby
Stapleton
Hurworth-on-Tees
Croft-on-Tees
Dalton-on-Tees
West Layton
East Layton
Newsham
Dalton
Gayles
Kirby Hill
Whashton
Ravensworth
Hartforth
Gilling West
Melsonby
Barton
Scotch Corner
Middleton Tyas
Moulton
North Cowton
East Cowton
Skeeby
Richmond
Uckerby
Scorton
Brompton-on-Swale
Colburn
Catterick Bridge
Catterick
Bolton-on-Swale
Whitwell
Catterick Garrison
Hudswell
Downholme
Hipswell Moor
Scotton
Tunstall
Bellerby
Barden
East Hauxwell
Hunton
Arrathorne
Hackforth
Constable Burton
Finghall
Newton-le-Willows
Patrick Brompton
Langthorne
Crakehall
Bedale
Aiskew
Leeming Bar
Leeming
Thornton Steward
Thornton Resr
Jervaulx Abbey
Thornton Watlass
Thirn
Cowling
Great Langton
Kirby Fleetham
Scruton
Exelby
Theakston
Burneston
Eryholme
Great Burdon
Brafferton
Coatham Mundeville
Great Stainton
Sadberge
River Tees
River Wear
River Swale
River Greta
River Lune
River Cover
River Ure
Roman Road
Dere Street
A 66 (T)
A 67
A 68
A 688
A 689
A 690
A 691
A 692
A 693
A 684
A 6108
A 1 (M)
A 1 (T)
A 167
B 6278
B 6279
B 6282
B 6274
B 6275
B 6270
B 6306
B 6296
B 6299
B 6301
B 6302
B 6310
B 6277
B 6276

NORTH SEA
SUNDERLAND
South Hylton
Hendon
New Silksworth
Silksworth
Ryhope
HOUGHTON LE-SPRING
Seaton
SEAHAM
HETTON LE-HOLE
Murton
Dalton-le-Dale
Cold Hesleden
Easington Lane
Hawthorn
Beacon Point
South Hetton
Haswell
Easington Colliery
Littletown
Horden Point
Horden
Dene Mouth
Shotton Colliery
Ludworth
Thornley
PETERLEE
Shotton
Castle Eden
Blackhall Colliery
Blackhall Rocks
Hesleden
Wheatley Hill
Wingate
Quarrington Hill
Station Town
Hutton Henry
Sheraton
Hart
Kelloe
Trimdon Colliery
Trimdon Grange
Hurworth Burn Resr
Trimdon
Fishburn
Crookfoot Resr
Elwick
HARTLEPOOL
Hartlepool Bay
Dalton Piercy
Owton Manor
Sedgefield
Seaton Carew
Tees Bay
Greatham
Graythorp
Thorpe Larches
Wolviston
Newton Bewley
West Scar
Salt Scar
The Flashes
Coatham
REDCAR
Works
BILLINGHAM
Cowpen Bewley
Oil Refinery
Stillington
Thorpe Thewles
Dormanstown
MARSKE-BY-THE-SEA
Kirkleatham
Whitton
Carlton
Redmarshall
Norton
Bishopton
Toll
Grangetown
SALTBURN-BY-THE-SEA
New Marske
Upleatham
Lazenby
Wilton
BROTTON
Skinningrove
LOFTUS
Boulby
Staithes
Little Stainton
STOCKTON-ON-TEES
MIDDLESBROUGH
Eston
SKELTON
Easington
Port Mulgrave
Hinderwell
Acklam
Sadberge
Elton
Thornaby-on-Tees
Ormesby
GUISBOROUGH
Priory
Boosbeck
Lingdale
Stanghow
Liverton
Roxby
Runswick Bay
Kettleness
Longnewton
Urlay Nook
Marton
Eaglescliffe
Nunthorpe
Moorsholm
Ellerby
Goldsborough
Lythe
Mulgrave Castle
Sandsend
Dunsley
Newholm
Dinsdale Sta
Tees-side Airport
Yarm
Egglescliffe
Stainton
Maltby
Newton under Roseberry
Gisborough Moor
Lockwood Beck Resr
Scaling
Scaling Resr
Roxby High Moor
Mickleby
Ugthorpe
Newby
Hilton
Great Ayton
Tanton
Seamer
Little Ayton
Monument
Commondale
Danby Low Moor
Danby
Lealholm Moor
Houlsyke
Lealholm
Aislaby
Sleights
Kirklevington
Low Worsall
Stokesley
Easby
Kildale
Castleton
Esk Dale
Girsby
Picton
Crathorne
River Leven
Battersby
Ingleby Greenhow
Baysdale Beck
Westerdale
Street
Glaisdale
Glaisdale Rigg
Egton
Egton Bridge
Hornby
Great Smeaton
Appleton Wiske
Hutton Rudby
Great Busby
Kirkby
Great Broughton
Carlton in Cleveland
Rounton
Potto
Faceby
Whorlton
Swainby
Urra
Westerdale Moor
Glaisdale Moor
Egton High Moor
Pike Hill Moss
Beck Hole
Goathland
Welbury
Deighton
Ingleby Arncliffe
Cockayne Ridge
Seave Green
Farndale Moor
Rosedale Moor
Hunt Ho
East Harlsey
Mount Grace Priory
Whorlton Moor
CLEVELAND HILLS
NORTH YORK MOORS
Cockayne
Osmotherley
Snilesworth Moor
Church Houses
Low Mill
Rosedale Abbey
Wheeldale Moor
Brompton
Ellerbeck
Thimbleby
Cow Ridge
Grange
Bransdale
Rudland Rigg
Farndale
Blakey Ridge
Thorgill
Goathland Moor
Yafforth
Kirby Sigston
Fangdale Beck
Hodge Beck
NORTHALLERTON
Over Silton
Nether Silton
Arden Great Moor
Spaunton Moor
Stape
Romanby
Crosby Court
Keldy Castle
Newton-on-Rawcliffe
Levisham
Lockton
Warlaby
Kepwick
Helmsley Moor
Gillamoor
Lastingham
Roman Camps
North Otterington
Thornton-le-Beans
Cowesby
Hawnby
Fadmoor
Hutton-le-Hole
Spaunton
Cropton
Borrowby
Knayton
Appleton-le-Moors
Newby Wiske
South Otterington
Thornton-le-Moor
Kirby Knowle
Carlton
Kirkbymoorside
Keldholme
Wrelton
Aislaby
Sinnington
Middleton
North Kilvington
Boltby
Old Byland
Rievaulx
Pockley
Beadlam
Kirby Mills
Kirby
Hambleton Hills
NZ
SE
North Yorkshire Moors Railway
Roman Road
Newton Dale
Rosedale
Rosedale Seven
Cleveland Way

76
77
NORTH CHANNEL
Ailsa Craig
Castle
Matthew's Port
Dowhill
Dipple
Ladybank
Dalquharran Castle
Dailly
Kilkerran
Glenalla Field
Killochan Castle
Maxwellston
Hadyard Hill
Old Dailly
Penkill
Penwhapple Resr
GIRVAN
Saugh Hill
Mull of Miljoan
Dalquhairn
Garleffin Fell
Tairlaw Ring
Linfern Loch
Black Hill
Woodland Bay
Byne Hill
Tormitchell
Auchensoul
Barr
CHANGUE PLANTATION
Eldrick Hill
CARRICK FOREST
Rig of the Shalloch
Kennedy's Pass
Grey Hill
Pinmore
River Stinchar
Merkland
Polmaddie Hill
Cairn Hill
Currarie
Lendalfoot
Knockdaw Hill
Mid Hill
Greensides
Knockinlochie
Tarfessock
Pinwherry
Bellamore
Pindonnan Craigs
Kirriereoch Hill
Bennane Head
Colmonell
Standing
Loch Moan
MERRICK
Knockdolian Castle
Pinwherry Hill
Black Clauchrie
Garwall Hill
Benyellary
BALLANTRAE BAY
Ballantrae
Knockdhu
Glen Tig
Water of Tig
Cosses
Balkissock
Shiel Hill
Barrhill
Laggan
Palgowan
Downan Point
Glenapp Castle
Corwar Ho
GLENTROOL FOREST
Glencaird Hill
Glen Trool Lodge
Loch Trool
Currarie Port
Beneraird
Strawarren Fell
Drumlamford Loch
Glentrool Village
CAIRNRYAN to Larne 2-2½ hrs
STRANRAER to Larne 2¼ hrs Douglas 4 hrs
Carlock Hill
Milljoan Hill
Chirmorrie
Drumlamford Ho
Bargrennan
Penderry Hill
Water of App
Glen App
High Murdonochee
Loch Maberry
Loch Ochiltree
Polbae
Milleur Point
Corsewall Point
Barnhills
Finnarts Bay
Jamieson Point
Mid Moile
Stab Hill
Craig Airie Fell
Loch Derry
Knowe
Standing Stones
Urrall Fell
North Cairn
South Cairn
Knockcoid
Cairn Point
Cairnryan
Penwhirn Resr
Main Water of Luce
Quarter Fell
Dounan Bay
Kirkcolm
Lamb Hill
Cross Water of Luce
Airtfield Fell
Eldrig Fell
Carseriggan
Loch Connell
Ervie
The Wig
Broad Fell
Balmurrie
Challoch
Portobello
LOCH RYAN
Leswalt
Cairnscarrow
New Luce
Loch Heron
NEWTON STEWART
Slouchnawen Bay
Knocknain
Lochnaw Castle
Innermessan
Drumphail
Bught Fell
Loch Ronald
Tarf Water
Benfield
Lochinch Castle
White Loch
Black Loch
Craig Fell
Glenluce
Glenstockadale
STRANRAER
Aird
Castle Kennedy
Carscreugh
Dergoals
Kirkcowan
Barraer Fell
High Moor of Killiemore
Broadsea Bay
Soulseat Loch
Challoch Hill
Dunragit
Abbey
Castle of Park
Derragulas Loch
Knock Moss
Carsegowan
Craigenlee Fell
Lochans
Cairn Pat
Genoch Mains
Torhousemuir Stone Circle
Black Head
Whitefield Loch
Bladnoch
Portpatrick
Bean Hill
Torrs Warren
Milton
Dunskey Castle
Stoneykirk
Luce Sands
Castle Loch
Craignarget Hill
Doon of May
Mochrum Loch
Port of Spittal Bay
Sandhead
Mochrum Fell
Chapel Finian
Barrachan
Whauphill
Cairngarroch Bay
Cairngarroch
Lake Cottage
Money Head
Meikle Float
Elrig
Mote of Druchtag
Float Bay
Hole Stone Bay
Chapel Rossan Bay
Milton Point
Mochrum
Drumtroddan
Ardwell
Ardwell Point
Logan Mains
Balgowan Point
LUCE BAY
Port William
Monreith Mains
Mull of Logan
Port Nessock or Port Logan Bay
Terally Point
Fell of Barhullion
Monreith
Barsalloch Point
Monreith Bay
Port Logan
Cairnywellan Head
Clanyard Bay
Kilstay Bay
Point of Cairndoon
Glasserton
Fell of Carleton
Clanyard
Kirkmaiden
Drummore
Laggantalluch Head
Cailiness Point
Damnaglaur
Port Castle Bay
Crammag Head
Maryport
Maryport Bay
Scares
Cairngaan
Port Kemin
MULL OF GALLOWAY

F
G
H
J
K
L
1
2
3
4
5
6
7
8
CARSPHAIRN
Carsphairn
Windy Standard
Loch Doon
Loch Muck
Dalmellington
Carsphairn Forest
Craigengillan
Corserine
Millfire
Glenhoul
Knowehead
Earlstoun
Moniaive
Kirkland
Penpont
Thornhill
Dunscore
Tynron
Keir Mill
Closeburn
DALMACALLAN FOREST
St John's Town of Dalry
Balmaclellan
NEW GALLOWAY
Kenmure Castle
Clatteringshaws Loch
CAIRN EDWARD FOREST
Mossdale
Corsock
Parton
Crocketford or Ninemile Bar
Milton
Kirkpatrick Durham
Springholm
Crossmichael
Laurieston
LAURIESTON FOREST
Threave Castle
CASTLE DOUGLAS
DALBEATTIE
DALBEATTIE FOREST
GLENGAP FOREST
Ringford
Tongland
KIRKCUDBRIGHT
Twynholm
GATEHOUSE OF FLEET
Anwoth
Cardoness Castle
Creetown
WIGTOWN
Wigtown Sands
Baldoon Sands
Carsluith
Ravenshall Point
Fleet Bay
Islands of Fleet
Sandgreen
Borgue
Knockbrex
Ross
Borness
Borness Point
Kirkcudbright Bay
Little Ross
Balmae
Townhead
Dundrennan
Abbey
Netherlaw
Port Mary
Abbey Head
Orroland
Rascarrel Bay
Auchencairn
Auchencairn Bay
Hestan Island
Balcary Point
Rough Island
Kippford or Scaur
Palnackie
Rockcliffe
Castlehill Point
Mersehead Sands
Garlieston
Eggerness Point
WIGTOWN BAY
WHITHORN
Priory
Portyerrock Bay
Cairn Head
Port Allen
Isle of Whithorn
St Ninian's Chapel
BURROW HEAD
SOLWAY
GALLOWAY
GLENKENS
72
99
29
54

A
B
C
D
E
1
2
3
4
5
6
7
8
Catrine
Sorn
Netherwood
Smallburn
Muirkirk
Kames
Black Hill
Glenbuck
Hareshaw Hill
Carmacoup
Glespin
Douglas
Urit Hill
Auchensaugh Hill
Middle Muir
Crawfordjohn
Roberton Law
Backstane Hill
Dungavel Hill
Startup Hill
Abington
Drake Law
Crawford
Duncangill Head
Limmerhaugh
Airds Moss
Cairn Table
Wardlaw Hill
Glenmuir
Cronberry
Lugar
Auchinleck
Ochiltree
HOLMHEAD
CUMNOCK
Logan
Dalblair
Glenmuirshaw
Glenmuir Water
Common Hill
Cairn Kinney
Duneaton Water
Kirkton
White Hill
Netherthird
Skares
Auchtitench Hill
Mount Stuart
Fingland
Spango Hill
Rake Law
Wellgrain Dod
Elvanfoot
Clyde Law
Kirkland Hill
Kirkland
Wedder Dod
Leadhills
Wanlockhead
Stood Hill
Dun Law
Watermeetings
Green Lowther
Lowther Hill
LOWTHER HILLS
Carsgailoch Hill
Pathhead
Mansfield
New Cumnock
Connel Park
Bankglen
Burnside
Dalgig
River Nith
Kirkconnel
Crawick
SANQUHAR
Ulzieside
Conrig Hill
Auchengruith
Mennock
Thirstane Hill
Winterhope Fell
Dalleagles
The Knipe
Hare Hill
Benty Cowan Hill
Quintin Knowe
Kello Water
Mid Hill
Euchan Water
Cairn Hill
Comb Law
Well Hill
Ballencleuch Law
Enoch Hill
Prickeny Hill
Milray Hill
Craigdarroch
Blackcraig Hill
White Knowes
Craignane
Afton Resr
Blacklorg Hill
Corse Hill
Shiel Hill
Enterkinfoot
Durisdeer
Wedder Law
Gana Hill
Queensberry
Wee Queensberry
Windy Standard
CARSPHAIRN
Brockloch Rig
Polskeoch
Countam
Cairnkinna Hill
Chanlockfoot
Enoch
Drumlanrig Castle
Tibbers Castle
Morton Castle
Carronbridge
Bellybought Hill
Benbrack
Waterhead Hill
Alhang
Ox Hill
Blackcraig Hill
Colt Hill
Lamford Hill
FOREST
Cairnsmore of Carsphairn
Dodd Hill
Beninner
Corlae
Benbrack
Countam
Auchenbrack
Benbuie
Shinnel Water
Scar Water
Burnhead
Penpont
Thornhill
Gatelawbridge
Locherben
Minnygap Height
Garryhorn
Carsphairn
Cairn Avel
Marscalloch Hill
Manquhill Hill
Wether Hill
Dalwhat Water
Bennan
Fort
Tynron
Keir Mill
Closeburn
Keir Hills
Holehouse Hill
Whitefauld Hill
Gairy Craig
Castlemaddy
Thorny Hill
Knowehead
Culmark Hill
Kendoon Loch
Moniaive
Kirkland
Motte of Dinning
Great Hill
FOREST OF AE
Bownmoor Hill
Troston Loch
Glenmaid Moor
Parkgate
Loch Harrow
Forrest Lodge
Glenhoul
Carsfad Loch
Wether Hill
DALMACALLAN FOREST
Bogrie Hill
Dunscore
Dalswinton
Lochinvar
Loch Urr
Earlstoun
Earlstoun Loch
Blackcraig Hill
Drumbuie
Garroch
Glenlee
Bogue
St John's Town of Dalry
Barr Hill
Duncow
Amisfield
Kirkton
Holywood
Tinwald
Locharbriggs
Bennan
Balmaclellan
NEW GALLOWAY
Kenmure Castle
Brockloch Hill
Lincluden College
Terregles
DUMFRIES
Clatteringshaws Loch
Cairnsmore or Black Craig of Dee
Craig
Ironmacannie
Corsock
Shawhead
Cargenbridge
Kingholm Quay
Knocknevis
Lochfoot
Lochrutton Loch
Round Fell
CAIRN EDWARD
Shaw Hill
Crocketford or Ninemile Bar
Auchenreoch Loch
Milton Loch
Milton
Cargen
Mabie
Fell of Fleet
FOREST
Stroan Loch
Mossdale
Drumrash
Parton
Kirkpatrick Durham
Springholm
Lochaber Loch
Beeswing
Loch Arthur
Glen
Bankend
Loch Grannoch
Craiglowrie
Loch Skerrow
Airie Hill
Woodhall Loch
Old Bridge of Urr
Tower
Lotus Hill
Kirkconnell
Sweetheart Abbey
New Abbey
W
A
Y
Lochenbreck Loch
Laurieston
LAURIESTON FOREST
Crossmichael
Clarebrand
Haugh of Urr
Kirkgunzeon
Townhead of Greenlaw
Mote of Urr
Long Fell
Loch Kindar
Clints of Dromore
Big Water of Fleet
White Top of Culreoch
Glentoo Loch
Threave Castle
CASTLE DOUGLAS
Carlingwark Loch
Meikle Hard Hill
Criffel
Overton
Bargatton Loch
Loch Whinyeon
GLENGAP FOREST
Bengray
Loch Mannoch
Bridge of Dee
Rhonehouse or Kelton Hill
DALBEATTIE
Maidenpap
Boreland Hill
Carsethorn
Pibble Hill
Stey Fell
Gelston
DALBEATTIE FOREST
Palnackie
Caulkerbush
Kirkbean
Glen
Ringford
GATEHOUSE OF FLEET
Anwoth
Cairnharrow
NS
NX
78
77
71
A 70
A 76(T)
A 74(T)
A 702
A 712
A 713
A 762
A 75(T)
A 711
A 710
A 701
A 709
B 729
B 7000
B 794
B 795
B 796
B 797
B 740
B 741
B 7040
B 7036
B 705

SELKIRK
HAWICK
MOFFAT
LOCKERBIE
LOCHMABEN
LANGHOLM
ANNAN
Gretna
Longtown
Eastriggs
Brampton
Canonbie
Ettrick
Eskdalemuir
Tushielaw Inn
Newcastleton
Kershopefoot
Stanhope
Tweedsmuir
Talla Resr
Talla Linnfoots
Meggethead
Megget Resr
Tibbie Shiels Inn
St Mary's Loch
Rodono Hotel
Cappercleuch
Dryhope
Mountbenger
Yarrow Feus
Sundhope
Ettrickbridge
Kirkhope
Ashkirk
Bowhill
Newark Castle
Broadmeadows
Philiphaugh
Midlem
Riddell
Lilliesleaf
Clerklands
Hassendean
Horsleyhill
Appletreehall
Burnfoot
Wilton Dean
Roberton
Borthwickbrae
Branxholm Park
Branxholme
Newmill
Broadhaugh
Teviothead
Craik
CRAIK FOREST
ESKDALEMUIR FOREST
CASTLE O'ER FOREST
NEWCASTLETON FOREST
KERSHOPE FOREST
ETTRICK FOREST
ESKDALE
TEVIOT
LIDDESDALE
ANNANDALE
NT
NY
Gallow Hill
Beattock
Auchen Castle
Dumcrieff
Poldean
Marchbankwood
Lochwood
Newton
St Ann's
Johnstonebridge
Dinwoodie Mains
Blackacre
Nethermill
Templand
Corrie Common
Boreland
Bankshill
Waterbeck
Middlebie
Ecclefechan
Eaglesfield
Kirtlebridge
Kirkpatrick-Fleming
Springkell
Chapelknowe
Evertown
Milltown
Rowanburn
Catlowdy
Nook
Netherby
Easton
Hoddom Castle
Carrutherstown
Mouswald
Dalton
Hightae
Torthorwald
Cleughbrae
Clarencefield
Comlongon Castle
Ruthwell
Cummertrees
Powfoot
Blackshaw
Bowness-on-Solway
Port Carlisle
Glasson
Drumburgh
Burgh by Sands
Beaumont
Rockcliffe
Westlinton
Todhills
Blackford
Harker
Houghton
Kingstown
Scaleby
Smithfield
Kirklinton
Hethersgill
Irthington
Crosby-on-Eden
Newby East
Walton
Newtown
Stapleton
Kirkcambeck
Bewcastle
Cardurnock
Anthorn
Whitrigg
Kirkbride
Thurstonfield
Kirkandrews-on-Eden
Moorhouse
Grinsdale
Solway Moss
Sarkfoot Point
Solway Firth
Priestside Bank
Tweed's Well (Source of R. Tweed)
Devil's Beef Tub
Ericstane
Granton Ho
Grey Mare's Tail
Loch Skeen
White Coomb
Hart Fell
Swatte Fell
Saddle Yoke
Bodesbeck Law
Bell Craig
Herman Law
Ettrick Pen
White Shank
Capel Fell
Croftheads Hill
Loch Fell
Ewelairs Hill
Black Hill
Laverhay
Milne Height
Fingland Fell
Black Esk Resr
Dinnings Hill
Castle O'er
The Knock
Hart Fell
Hallhills Loch
Winterhope Resr
Raes Knowes
Glentenmont Height
Calkin Rig
The Shin
Bentpath
Crumpton Hill
Broad Head
Hog Hill
Jamestown
Fow Side
White Hope Edge
Causeway Grain Head
Blaeberry Hill
Ewesless Knowe
Davington
Cross Hill
Black Knowe
Glenkerry
Sauchie Law
Ramseycleuch
Buccleuch
Redfordgreen
Dun Knowe
Crosslee
Ramsey Knowe
Gilmanscleuch
Black Knowe Head
Cavers Hill
Akermoor Loch
Shaws Under Loch
Hellmoor Loch
Alemoor Loch
Drinkstone Hill
Smasha Hill
Borthwickshiel
Redcleuch Edge
Pike Hill
Craik Cross Hill
Dryden Fell
Skelfhill Pen
Cauldcleuch Head
Greatmoor Hill
The Pike
Maiden Paps
Leap Hill
Millstone Edge
Comb Hill
Wisp Hill
Mosspaul Hotel
Tudhope Hill
Geordie's Hill
Din Fell
Hermitage
Castle
Pike Fell
Roan Fell
North Birny Fell
Watch Hill
Black Edge
Tinnis Hill
Bruntshiel Hill
Caulside
Arthur Seat
Ewes
Hog Fell
Earshaw Hill
Burnswark Fort
Howat's Hill
Grange Fell
Culter Fell
Glenwhappen Rig
Glenmuck Height
Broad Law
Dollar Law
Taberon Law
Black Law
Deer Law
Deuchar Law
Fastheugh Hill
Lochcraig Head
Cape Law
Fruid Resr
Craigmaid
Steele Road
Old Castleton
Wilson Pike
Black Knowe
Crossings
Due to open Mid 1988
A 7(T)
A 74(T)
A 75(T)
A 701
A 708
A 709
A 6071
B 709
B 7009
B 723
B 722
B 7068
B 6357
B 6318
79
80
74
66
67
F
G
H
J
K
L
1
2
3
4
5
6
7
8

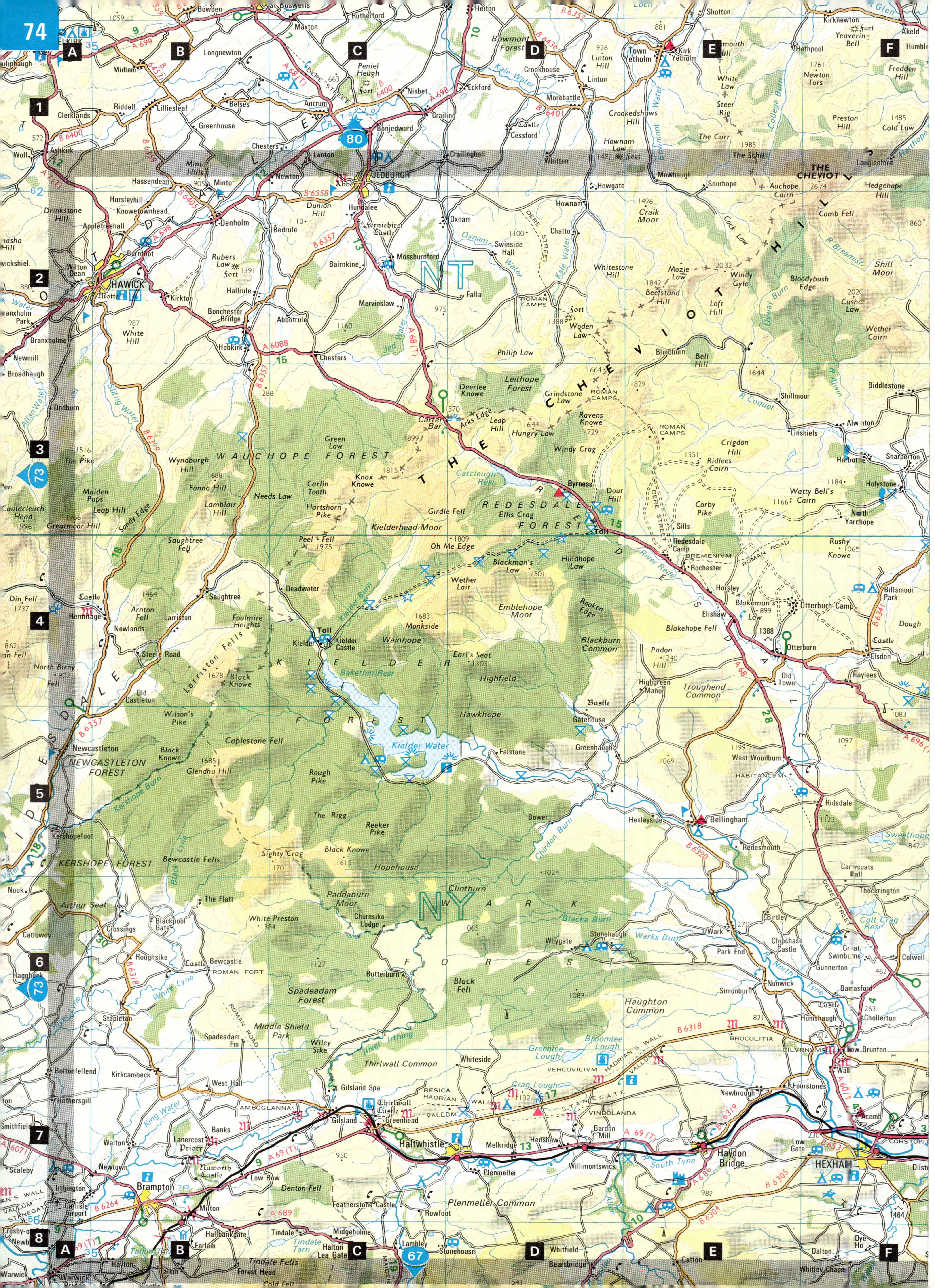
A
B
C
D
E
F
1
2
3
4
5
6
7
8
Maxton
Rutherford
Heiton
Shotton
Kirknewton
Longnewton
Midlem
Peniel Heugh
Bowmont Forest
Crookhouse
Linton
Town Yetholm
Kirk Yetholm
Hethpool
Yeavering Bell
Akeld
Riddell
Clerklands
Lilliesleaf
Belses
Greenhouse
Ancrum
Nisbet
Eckford
Crailing
Morebattle
Bonjedward
Cessford
Crookedshaws Hill
Newton Tors
Preston Hill
Cold Law
Ashkirk
Chesters
Lanton
Minto Hills
Minto
Crailinghall
Whitton
Hownam Law
The Curr
The Schil
JEDBURGH
Hassendean
Newton
Howgate
Mowhaugh
Sourhope
Auchope Cairn
THE CHEVIOT
Langleeford
Hedgehope Hill
Drinkstone Hill
Horsleyhill
Knowetownhead
Dunion Hill
Hundalee
Oxnam
Hownam
Craik Moor
Comb Fell
Appletreehall
Denholm
Bedrule
Ferniehirst Castle
Swinside Hall
Chatto
Wilton Dean
HAWICK
Burnfoot
Rubers Law
Hallrule
Bairnkine
Mossburnford
NT
Falla
Whitestone Hill
Mozie Law
Windy Gyle
Beefstand Hill
Loft Hill
Bloodybush Edge
Shill Moor
Cushat Law
Kirkton
Bonchester Bridge
Abbotrule
Mervinslaw
ROMAN CAMPS
Wether Cairn
White Hill
Hobkirk
Chesters
Jed Water
Philip Law
Woden Law
Blindburn
Bell Hill
Branxholme
Newmill
Broadhaugh
Deerlee Knowe
Leithope Forest
Grindstone Law
ROMAN CAMPS
Shillmoor
Biddlestone
Dodburn
Carter Bar
Arks Edge
Leap Hill
Hungry Law
Ravens Knowe
Windy Crag
ROMAN CAMPS
Linshiels
Alwinton
Crigdon Hill
Harbottle
Sharperton
Holystone
Watty Bell's Cairn
The Pike
Wyndburgh Hill
WAUCHOPE FOREST
Green Law
Fanna Hill
Carlin Tooth
Knox Knowe
THE
Catcleugh Resr
Byrness
Dour Hill
Ridlees Cairn
Maiden Paps
Leap Hill
Lamblair Hill
Needs Law
Hartshorn Pike
Girdle Fell
REDESDALE
Ellis Crag
FOREST
Corby Pike
North Yardhope
Cauldcleuch Head
Greatmoor Hill
Sandy Edge
Kielderhead Moor
Sills
Redesdale Camp
BREMENIVM
ROMAN ROAD
Rushy Knowe
Saughtree Fell
Peel Fell
Oh Me Edge
Blackman's Law
Hindhope Law
Rochester
Deadwater
Saughtree
Wether Lair
Horsley
Billsmoor Park
Din Fell
Hermitage
Arnton Fell
Larriston
Foulmire Heights
Emblehope Moor
Rooken Edge
Elishaw
Blakeman's Law
Otterburn Camp
Newlands
Monkside
Wainhope
Blakehope Fell
Kielder
Kielder Castle
Blackburn Common
Dough Crag
Otterburn
Elsdon
Steele Road
Larriston Fells
Earl's Seat
Padon Hill
North Birny Fell
Black Knowe
Bakethin Resr
Highfield
Old Town
Raylees
Old Castleton
Wilson's Pike
Highgreen Manor
Troughened Common
Hawkhope
Gatehouse
Greenhaugh
Newcastleton
NEWCASTLETON FOREST
Black Knowe
Caplestone Fell
Kielder Water
Falstone
West Woodburn
HABITANCVM
Glendhu Hill
Rough Pike
Kershope Burn
The Rigg
Reeker Pike
Bower
Hesleyside
Bellingham
Ridsdale
Kershopefoot
Sighty Crag
Black Knowe
Redesmouth
Sweethope
KERSHOPE FOREST
Bewcastle Fells
Hopehouse
Carrycoats Hall
Nook
Arthur Seat
The Flatt
Paddaburn Moor
Clintburn
Churnsike Lodge
NY
PARK
Thockrington
White Preston
Birtley
Colt Crag Resr
Catlowdy
Crossings
Blackpool Gate
Whygate
Stonehaugh
Wark
Chipchase Castle
Roughsike
Bewcastle
ROMAN FORT
Butterburn
FORES
Park End
Great Swinburne
Gunnerton
Colwell
Spadeadam Forest
Black Fell
Nunwick
Simonburn
Barrasford
Stapleton
Haughton Common
Humshaugh
Chollerton
Middle Shield Park
Spadeadam Fm
ROMAN ROAD
Wiley Sike
Greenlee Lough
Broomlee Lough
HADRIAN'S WALL
BROCOLITIA
Low Brunton
Boltonfellend
Thirlwall Common
Whiteside
VERCOVICIVM
Kirkcambeck
West Hall
Gilsland Spa
Crag Lough
Newbrough
Fourstones
Hethersgill
CAMBOGLANNA
Gilsland
Thirlwall Castle
Greenhead
VINDOLANDA
STANEGATE
Banks
Haltwhistle
Melkridge
Henshaw
Bardon Mill
Haydon Bridge
Walton
Lanercost Priory
Plenmeller
Willimontswick
South Tyne
HEXHAM
Low Gate
Newtown
Naworth Castle
Low Row
Irthington
Brampton
Denton Fell
Featherstone Castle
Plenmeller Common
Carlisle Airport
Milton
Rowfoot
Tindale
Tindale Tarn
Halton Lea Gate
Lambley
Stonehouse
Whitfield
Hallbankgate
Farlam
Tindale Fells
Forest Head
Cold Fell
Hayton
Talkin
Warwick
Bearsbridge
Catton
Dalton
Dye Ho
Whitley Chapel

NORTH SEA
NEWCASTLE UPON TYNE to
Seasonal
Esbjerg 18½-21 hrs
Bergen 20-27 hrs
Gothenburg 25-27½ hrs
Stavanger 17-20½ hrs
ALNWICK
AMBLE-BY-THE-SEA
Coquet Island
Hauxley Haven
Hadston Carrs
Druridge Bay
Lynemouth
Beacon Point
NEWBIGGIN-BY-THE-SEA
ASHINGTON
MORPETH
BEDLINGTON
BLYTH
Seaton Sluice
St Mary's or Bait Island
WHITLEY BAY
TYNEMOUTH
SOUTH SHIELDS
Marsden Bay
Whitburn
SUNDERLAND
NEWCASTLE UPON TYNE
WALLSEND
JARROW
HEBBURN
GATESHEAD
FELLING
BOLDON
WHICKHAM
BLAYDON
RYTON
NEWBURN
PRUDHOE
Corbridge
Painshawfield
Ponteland
Darras Hall
Cramlington
Seaton Delaval
LONGBENTON
Newcastle Airport
Rothbury
Beadnell
Beadnell Bay
Snook Point
St Mary's or Newton Haven
Embleton Bay
Dunstanburgh Castle
Craster
Howick Haven
Boulmer Haven
Alnmouth
Alnmouth Bay
Warkworth
ROTHBURY FOREST
HARWOOD FOREST
Fontburn Resr
Rothley Lakes
Bolam Lake
Hallington Resr
DERE STREET
VINDOVALA
Thrunton Wood
Hulne Park
Shirlaw Pike
Simonside
Felton
Widdrington
Cresswell
The Scars
Lilburn
Wooler
Chatton
Bamburgh
Due to open Mid 1990
G
H
J
K
L
1
2
3
4
5
6
7
8
44
56
62
68
69
81

82
83
KENNACRAIG to
Port Askaig 1¾ hrs
Port Ellen 2 hrs
Skervuile Lighthouse
Point of Knap
Tarbert
West Tarbert
Kennacraig
Whitehouse
Kilchamaig
Ardpatrick Point
Clachan
Rhunahaorine
Tayinloan
Killean
Beacharr
A' Chleit
Muasdale
Bellochantuy
Glenbarr
Cleongart
Killocraw
Kilchenzie
Machrihanish
Machrihanish Bay
Machrihanish Airport
CAMPBELTOWN
Drumlemble
Knocknaha
Campbeltown Loch
Island Davaar
Kildalloig
Achinhoan Head
Johnston's Point
Polliwilline Bay
Macharioch
Southend
Carskie
Keil Point
Brunerican Bay
Cove Point
Sheep Island
Sanda Island
Black Point
Mull of Kintyre
Rubha Chlachan
Port Mean
Beinn na Lice
Glen Breackerie
Conie Glen
Strone Glen
Earadale Point
Rubha Dùin Bhàin
Rubh'a' Mharaiche
Cnoc Moy
The Slate
Tirfergus Hill
Carradale
Carradale Point
Torrisdale
Torrisdale Castle
Dippen
Saddell
Saddell Bay
Ugadale
Ugadale Point
Black Bay
Ardnacross Bay
Peninver
Glen Lussa
Grogport
Sunadale
Cour
Cour Bay
Crossaig
Claonaig
Claonaig Bay
Skipness
Skipness Point
Glenreasdell Mains
Eascairt Point
GIGHA ISLAND
Sound of Gigha
Ardminish
Cara Island
Mull of Cara
Gigalum Island
Craro Island
KILBRANNAN SOUND
SOUND OF BUTE
Inchmarnock
Ardlamont Point
Ardscalpsie Point
Lochranza
Catacol
Catacol Bay
Cock of Arran
Pirnmill
Imachar
Balliekine
Dougarie Point
Machrie Bay
Auchagallon
Tormore
Drumadoon Point
Drumadoon Bay
Blackwaterfoot
Shiskine
Kilpatrick
Brown Head
Sliddery
Lagg
Kilmory
Torrylinn
Bennan Head
Kildonan
Pladda
Sound of Pladda
Dippen Head
Largybeg Point
Whiting Bay
Kingscross Point
Holy Island
Lamlash
Clauchlands Point
Brodick
Brodick Bay
Merkland Point
Corrie
Sannox
Sannox Bay
GOATFELL
ISLAND OF ARRAN
Beinn Bharrain
Glen Sannox
Millstone Point
Torr Meadhonach
Ailsa Craig
NR

G
H
J
K
L
84
85
78
72
70
71
1
2
3
4
5
6
7
Innellan
Wemyss Bay
Inverkip
Kilmacolm
Bishopton
Bearsden
Clydebank
Port Bannatyne
Rothesay
Toward Point
Skelmorlie
Renfrew
Glasgow Airport
Paisley
Govan
Pollokshields
Linwood
Kilbarchan
Elderslie
Johnstone
Houston
Bridge of Weir
Ranfurly
Lochwinnoch
Largs
Great Cumbrae Island
Millport
Fairlie
Kilbirnie
Beith
Barrhead
Neilston
Newton Mearns
Eaglesham
Little Cumbrae Island
Garroch Head
Farland Head
West Kilbride
Seamill
Dalry
Glengarnock
Dunlop
Stewarton
Kilwinning
Ardrossan
Saltcoats
Stevenston
Horse Isle
Irvine
Irvine Bay
Kilmaurs
Kilmarnock
Hurlford
Galston
Newmilns
Greenholm
Darvel
Riccarton
Dundonald
Barassie
Troon
Symington
Lady Isle
Monkton
Prestwick Airport
Prestwick
New Prestwick
Tarbolton
Mauchline
Catrine
Sorn
Ayr
Annbank
Mossblown
Auchinleck
Cumnock
Heads of Ayr
Alloway
Dalrymple
Drongan
Ochiltree
Culzean Bay
Culzean Castle
Maybole
Maidenhead Bay
Maidens
Turnberry Bay
Turnberry
Kirkoswald
Crossraguel Abbey
Patna
Dalmellington
Bellsbank
Straiton
Kirkmichael
Dailly
Dalquharran Castle
Carsphairn
Firth of Clyde
Cunninghame
Kyle
NS

A
B
C
D
E
1
2
3
4
5
6
7
8
Inchmurrin
Gartocharn
Killearn
Fintry
Carron Valley Resr
Carron Valley Forest
Dunipace
HELENSBURGH
Rhu
Craigendoran
Rosneath Point
ALEXANDRIA
Balloch
Jamestown
Bonhill
Renton
Cardross
Dumbarton Muir
CAMPSIE FELLS
Campsie Hills
Strathblane
Blanefield
Lennoxtown
Milton of Campsie
Kilsyth Hills
KILSYTH
Banton
GREENOCK
PORT GLASGOW
DUMBARTON
KILPATRICK HILLS
MILNGAVIE
BEARSDEN
Old Kilpatrick
Bowling
Duntocher
Kirkintilloch
CUMBERNAULD
Bishopton
ERSKINE
CLYDEBANK
Bishopbriggs
Stepps
Muirhead
Kilmacolm
Bridge of Weir
Houston
RENFREW
Linwood
PAISLEY
GLASGOW
COATBRIDGE
AIRDRIE
Ranfurly
Kilbarchan
Elderslie
JOHNSTONE
Govan
Pollokshields
RUTHERGLEN
Cathcart
Baillieston
Birkenshaw
Viewpark
Holytown
Mistylaw Muir
Lochwinnoch
Howwood
BARRHEAD
Neilston
Nitshill
Pollokshaws
Thornliebank
Newton Mearns
Busby
Clarkston
Cambuslang
Uddingston
Bothwell
Bellshill
MOTHERWELL
Newarthill
Beith
Uplawmoor
EAST KILBRIDE
HAMILTON
Stonefield
Blantyre
Larkhall
Dalry
Eaglesham
Waterfoot
Limekilnburn
Dunlop
STEWARTON
Fenwick
Strathaven
Whitelee Forest
KILWINNING
Kilmaurs
KILMARNOCK
IRVINE
Irvine Bay
Dreghorn
Hurlford
Riccarton
NEWMILNS
Darvel
GREENHOLM
Galston
Dundonald
Symington
TROON
Barassie
Tarbolton
Mauchline
Catrine
Sorn
Kype Muir
PRESTWICK
Monkton
Mossblown
Annbank
AYR
Ochiltree
Auchinleck
Lugar
CUMNOCK
Drongan
Dalrymple
Cronberry
Muirkirk
Glenmuir
Glenmuirshaw
NS

DUNFERMLINE
KINGHORN
BURNTISLAND
Kincardine
Valleyfield
Oakley
Culross
Crossford
Charlestown
Limekilns
Rosyth
INVERKEITHING
Dalgety Bay
Aberdour
Inchkeith
Inchcolm
Oxcars
Inchmickery
Cramond Island
Black Rocks
Leith
EDINBURGH
Portobello
GRANGEMOUTH
BO'NESS
FALKIRK
Stenhousemuir
Larbert
Bonnybridge
Carron
Camelon
Redding
Brightons
Linlithgow Bridge
LINLITHGOW
Philpstoun
Winchburgh
Kirkliston
QUEENSFERRY
North Queensferry
Dalmeny
Cramond
Edinburgh Airport
Ingliston
Ratho
Broxburn
Uphall
Dechmont
BATHGATE
ARMADALE
Whitburn
LIVINGSTON
East Calder
Mid Calder
West Calder
Polbeth
Blackburn
Harthill
Shotts
Fauldhouse
Stane
Breich
Caldercruix
Plains
Salsburgh
Newmains
WISHAW
Carluke
Forth
Carnwath
Carstairs
Carstairs Junction
Ravenstruther
LANARK
New Lanark
Kirkfieldbank
Crossford
Nemphlar
Hazelbank
Auchenheath
Kirkmuirhill
Hawksland
Lesmahagow
Coalburn
Douglas
Douglas Water
Rigside
Symington
Thankerton
BIGGAR
Skirling
Broughton
Libberton
Dolphinton
West Linton
Carlops
Romannobridge
Blyth Bridge
Penicuik
Roslin
Loanhead
Bilston
Straiton
Balerno
Currie
Juniper Green
Colinton
Liberton
Gilmerton
Dalkeith
Carrington
Leadburn
Howgate
Eddleston
Peebles
Glentress
Cardrona
Kirkton Manor
Stobo
Drumelzier
Tweedsmuir
Abington
Crawford
Elvanfoot
Leadhills
Crawfordjohn
Lamington
Wiston
Coulter
Talla Reservoir
Megget Reservoir
Fruid Reservoir
Tibbie Shiels Inn
Ettrick
Ramseycleuch
Tushielaw Inn
Crosslee
PENTLAND HILLS
Tinto
Broad Law
Dollar Law
Culter Fell
Cardrona Forest
Glentress Forest
Union Canal
Firth of Forth
M 8
M 9
M 90
A 70
A 702
A 701
A 72
A 721
A 73
A 74
A 71
A 706
86
87
80
73
1
2
3
4
5
6
7
8
F
G
H
J
K

FIRTH OF FORTH
Inchkeith
Fidra
Eyebroughy
NORTH BERWICK
Craigleith
Bass Rock
Tantallon Castle
St Baldred's Boat
Auldhame
Scoughall
St Baldred's Cradle
Tyne Mouth
Gullane Bay
Dirleton
Gullane
Kingston
Whitekirk
Tyninghame
Aberlady Bay
Aberlady
Gosford Bay
Drem
Athelstaneford
Ballencrieff
East Linton
Preston
DUNBAR
West Barns
Broxburn
Barns Ness
Black Rocks
Leith
EDINBURGH
Portobello
Arthur's Seat
Duddingston
COCKENZIE AND PORT SETON
PRESTONPANS
Seton Ho
Longniddry
Spittal
Garleton Hills
Haddington
Hailes Castle
Traprain
Traprain Law
Luggate Burn
Stenton
Pitcox
Spott
Brunt Hill
Skateraw
Power Station
Thorntonloch
Innerwick
MUSSELBURGH
Inveresk
TRANENT
Wallyford
Macmerry
New Winton
Elvingston
Gladsmuir
Samuelston
Papple
Whitelaw Hill
Garvald
Halls
Cocklaw Hill
Cockburnspath
Oldhamstocks
Craigmillar
Danderhall
Liberton
Millerhill
Elphinstone
Whitecraig
Ormiston
New Town
Bolton
Gilmerton
Carfrae
Dunbar Common
Bransly Hill
Monynut Edge
Heart Law
Ecclaw
Gilmerton
Kaimes
DALKEITH
Cousland
Pencaitland
East Saltoun
West Saltoun
Gifford
Yester Ho
Yester Castle
Danskine
Lammer Law
Hopes Resr
Meikle Says Law
Whiteadder Resr
Spartleton Edge
Cranshaws
Cranshaws Hill
Abbey St Bathans
Ellemford
BONNYRIGG AND LASSWADE
Eskbank
Mayfield
Oxenford Castle
Peaston Bank
Peaston
Gilchriston
Longyester
Roslin
Bilston
Rosewell
Newtongrange
Newlandrig
Pathhead
Humbie
Blegbie
Arniston Engine
Gorebridge
Crichton Castle
Fala Dam
Fala
Carrington
Temple
North Middleton
Middleton
Tynehead
Soutra Mains
Crib Law
LAMMERMUIR
Hunt Law
Cairn
Dye Water
Longformacus
Watch Water Resr
Blythe Edge
Twinlaw Cairns
Edgelaw Resr
Rosebery Resr
Gladhouse Resr
Torfichen Hill
Fala Moor
Gilston
Falahill
Heriot
Hogs Law
Dirrington Great Law
Dirrington Little Law
DUNS
Gavinton
MOORFOOT HILLS
Oxton
Hotel
Forts
Soonhope Burn
Scoured Rig
Polwarth
Fogo
Portmore Loch
Collie Law
Fort
Edgarhope Wood
Hare Faulds
Blythe
Westruther
Halliburton
Hule Moss
Dun Law
Fountainhall
Thirlestane Castle
LAUDER
Thirlestane
Houndslow
Blackhope Scar
Ladyside Height
Torquhan
Lauder Common
Greenlaw
Lugate Water
Killochyett
Stow
Netherr Blainslie
Legerwood
Whitehope Law
Dunslair Heights
Leithen Water
Windlestraw Law
Gordon
Tower
Glentress Forest
Black Law
Great Law
Broch
Bowland
Fans
Hume
Eccles
PEEBLES
Glentress
Priesthope Hill
Caddon Water
William Law
Langshaw
Mellerstain Ho
Sweethope Hill
Stichill
Lee Pen
Blackhaugh
Knowes Hill
Buckholm
Earlston
Nenthorn
Ednam
Cardrona
Walkerburn
GALASHIELS
Fort
Redpath
Smailholm
Hendersyde Park
Cardrona Forest
Traquair House
INNERLEITHEN
Elibank and Traquair Forest
Clovenfords
Meigle Hill
Caddonfoot
Lowood
Gattonside
Abbey
Tower
Sprouston
Wallace's Hill
Traquair
Ashiestiel Hill
Boleside
Abbotsford
MELROSE
Newstead
Bemersyde
Floors Castle
KELSO
Maxwellheugh
Kirkhouse
Minch Moor
Yair Hill Forest
Eildon Hills
Newtown St Boswells
Dryburgh Abbey
Clintmains
Makerstoun
Trows
Glen Ho
Broomy Law
Cauldshiels Hill
St Boswells
Roxburgh
Blakelaw
Blake Muir
Broadmeadows
Lindean
Linglie Hill
Bowden
Rutherford
Heiton
Deuchar Law
Newark Castle
SELKIRK
Maxton
Bowmont Forest
Yarrow
Yarrow Water
Bowhill
Fastheugh Hill
Philiphaugh
Midlem
Longnewton
Peniel Heugh
Crookhouse
Yarrow Feus
Sundhope
Mountbenger
Hotel
Ettrickbridge
Riddell
Lilliesleaf
Belses
Ancrum
Nisbet
Eckford
Morebattle
ETTRICK FOREST
Kirkhope
Clerklands
Greenhouse
Bonjedward
Crailing
Cessford
Black Knowe Head
Cavers Hill
Ashkirk
Chesters
Lanton
Crailinghall
Whitton
Gilmanscleuch
Akermoor Loch
Woll
Minto Hills
Minto
Newton
JEDBURGH
Shaws Under Loch
Hassendean
Horsleyhill
Knowetownhead
Dunion Hill
Hundalee
Hownam
Crosslee
Dun Knowe
Drinkstone Hill
Appletreehall
Denholm
Bedrule
Ferniehirst Castle
Oxnam
Chatto
Hellmoor Loch
Smasha Hill
Burnfoot
Rubers Law
Swinside Hill
Redfordgreen
Alemoor Loch
Roberton
Wilton Dean
Mossburnford
Bairnkine
Ettrick
Ramseycleuch
HAWICK
Hallrule
DERE STREET

G
H
J
K
L
42
1
68
2
3
4
5
6
7
8
62
NORTH SEA
Reed Point
Siccar Point
Wheat Stack
Fast Castle
Telegraph Hill
ST ABB'S HEAD
Lumsdaine
Meikle Black Law
Cross Law
Coldingham Moor
Northfield
St Abbs
Coldingham Bay
Grantshouse
Coldingham
Priory
Buss Craig
EYEMOUTH
Houndwood
Ale Water
Eye Water
Horseley Hill
Reston
Ayton
Castle
Burnmouth
Auchencrow
Lamberton Beach
Preston
Lintlaw
Chirnsidebridge
Edrom
Chirnside
Foulden
Lamberton
Clappers
Halidon Hill
Whiteadder Water
Allanton
Hutton
Paxton
BERWICK-UPON-TWEED
Tweedmouth
Spittal
Redshin Cove
Blackadder
East Ord
Loanend
Fishwick
Whitsome
New Horndean
Horncliffe
Longridge Towers
Scremerston
Cheswick Black Rocks
Cheswick
Swinton
Ladykirk
Norham
Murton
Thornton
Shoresdean
West Allerdean
Shoreswood
Grindon
Simprim
Leitholm
Swintonmill
Felkington
Ancroft
Haydon Dean
Goswick
Haggerston Castle
Beal
Emmanuel Head
HOLY ISLAND
Lindisfarne
Holy Island Sands
Holy Island
Castle
Priory
Castle Point
Guile Point
Berrington
Duddo
Castle Heaton
Bowsden
Barmoor Castle
Lennel
COLDSTREAM
Birgham
Cornhill-on-Tweed
Pallinsburn Ho
Etal
Crookham
Lowick
Kyloe
Kyloe Hills
Fenwick
Buckton
Carham
Wark
Learmouth
Branxton
Ford
Elwick
Ross
FARNE ISLANDS
Staple Sound
Inner Sound
Budle Bay
Hadden
Pressen
Flodden
Kimmerston
Holburn
Detchant
Middleton
Waren Mill
Budle
Bamburgh
Downham
Fenton
Nesbit
Belford
Easington
Spindlestone
Burton
Lempitlaw
Mindrum
Pawston
Kilham
Housedon Hill
Milfield
Doddington
Hazelrigg
Mousen
Bradford
Seahouses
North Sunderland
Hoselaw Loch
Lanton
Coupland
River Till
Horton
Bellshill
Adderstone
Lucker
Elford
Newham Hall
Beadnell
Shotton
Kirknewton
Yeavering Bell
Akeld
Weetwood Hall
Fowberry Tower
Greendykes
Warenton
Warenford
Newham
Swinhoe
West Fleetham
Town Yetholm
Kirk Yetholm
Coldsmouth Hill
Hethpool
Humbleton
Wooler
Chatton
Newstead
Chathill
Beadnell Bay
Linton Hill
White Law
Newton Tors
Fredden Hill
Earle
Haugh Head
Chillingham
Castle
Ellingham
Preston
High Newton-by-the-Sea
Snook Point
Steer Rig
Middleton Hall
Newtown
Hepburn
Brunton
St Mary's or Newton Haven
Crookedshaws Hill
Preston Hill
Cold Law
Middleton
Lilburn Tower
Middle Moor
Brownieside
Christon Bank
Embleton Bay
Hownam Law
The Curr
The Schil
Harthope Burn
Langlee Crags
East Lilburn
Cateran Hill
North Charlton
Embleton
Dunstanburgh Castle
Fort
Langleeford
Ilderton
Roseden
Old Bewick
West Ditchburn
THE CHEVIOT
Howgate
Mowhaugh
Sourhope
Auchope Cairn
Hedgehope Hill
Roddam
Wooperton
Harehope
New Bewick
Eglingham
South Charlton
Rock
Dunston
Craster
Craik Moor
Comb Fell
Dunmoor Hill
Beanley
Rennington
Howick
Cock Law
Brandon
Powburn
Littlehoughton
Howick Haven
Whitestone Hill
Mozie Law
Windy Gyle
Bloodybush Edge
Greensidehill
Ingram
Glanton Pike
Glanton
Titlington
Shawdon Hall
Hulne Park
Longhoughton
Denwick
Boulmer
Beefstand
A1(T)
A1107
A6105
A698
A697
A6112
B6438
B6355
B6437
B6460
B6461
B6470
B6354
B6353
B6525
B6352
B6351
B6348
B6349
B6396
B6342
B6341
B6340
B6346
B6347
B1342
B1341
B1340
B1339
DEVIL'S CAUSEWAY
NU
75

88
76
INNER HEBRIDES
Dubh Artach
Port na Cuilce
Rubh' a' Geodha
Balnahard
Carnan Eoin
Cailleach Uragaig
Kiloran Bay
Port Ceann a' Gharraidh
Beinn Bhreac
Kiloran
Port a' Bhàta
An Rubha
Kilchattan
Scalasaig
Port an Obain
COLONSAY
COLONSAY to Oban 2½ hrs
Loch Staosnaig
Eilean a' Chladaich
Ardskenish
Garvard
Rubha Dubh
Meall an Arbhair
Dubh Eilean
Priory
Eilean Treadhrach
ORONSAY
Eilean Ghaoideamal
Eilean nan Ròn
Caolas Mòr
Ceann Riobha
Glendebadel Bay
Rubha na h-Uamha-sàile
Corpach Bay
Beinn Bhreac
Allt an Tairbh
Shian Bay
Rainberg Mòr
Shian River
Loch Righ Mòr
Cruib
Rubh' an t-Sàilein
J U R A
Rubh' a' Chrois-aoinidh
Loch Tarbert
Cnoc an Ime
Lagg
Rubha a' Chamais
Post Rocks
Rubh' Aird na Sgitheich
RUBH' A' MHAIL
Glen Batrick
Beinn Bhreac
Lundale
Allt na Gile
Loch an Aircill
Loch Lesgamail
Port Doir' a' Chrorain
Rubha Bholsa
Sgarbh Breac
Nave Island
Ardnave Point
Eilean Beag
Gortantaoid Point
Sgairail
Sgarbh Dubh
Loch Smigeadail
Bachlaig
Beinn an Oir
PAPS OF JURA
Bunnahabhainn
Gortantaoid
Giur-bheinn
Loch a' Chnuic Bhric
JURA FOREST
Corran R
An Dùnan
Carraig Bhàn
Ardnave
Killinallan
Beinn Bhreac
Inver Cottage
Gleann Astaile
NR
Leargybreck
Knockrome
Skervuile Lighthouse
An Clachan
Tòn Mhòr
Kilnave
Loch Gruinart
SOUND OF ISLAY
Caol Ila
Port Askaig
Feolin Ferry
Abhainn Gleann Iubharnadeal
Glas Bheinn
Loch na Mile
Lowlandman's Bay
Sanaigmore
Braigo
B 8018
Rubha Lamanais
Keills
Keils
Small Isles
Ballinaby
Carnduncan
Gruinart
B 8017
Craigens
Abhainn Gleann Ulibh
Loch Finlaggan
Castle
Craighouse
Saligo Bay
Loch Gorm
Loch Cam
Ballygrant
Loch Ballygrant
Brat Bheinn
Port na Birlinne
Coul Point
Foreland Ho
Kilmeny
Esknish
A 846
Glas Eilean
Cabrach
Na Cùiltean
Kilchoman
Blackrock
Islay Ho
Dun Bhruichlinn
Beinn Dubh
Jura Ho
Am Fraoch Eilean
Claig Castle
Brosdale Island
Rubha na Tràille
Machir Bay
Cnoc Dubh
Conisby
Carraig Dubh
Daill
Bridgend
Bruichladdich
Kilchiaran
ISLAY
Kilchiaran Bay
Bowmore
Dùn
Rosebridge
McArthur's Head
Rubha Ghlamraidh
Tormisdale
Port Charlotte
Cairn
LOCH INDAAL
Mulindry
Glas Bheinn
1¾ hrs
Gartbreck
Kilennan R
Lossit Point
Lossit
Beinn Tart a' Mhill
Lossit Bay
Nereabolls
RHINNS OF ISLAY
A 847
Laggan
Laggan Point
Port Mòr na Carraig
R Laggan
Duich R
Duich Lots
B 8016
A 846
Beinn Bhàn
Beinn Bheigier
Gleann Leora
Claggain River
Carraig Mhòr
Ardtalla
Claggain Bay
Rubha na Faing
Portnahaven
Port Wemyss
Ellister
Port Gleann na Gaoidh
Orsay
RHINNS POINT
LAGGAN BAY
Beinn Uraraidh
Loch Beinn Uraraidh
Cnoc Mòr na Claigin
Aros Bay
Islay (Port Ellen) Aerodrome
Knockangle Point
Glenegedale
Glenegedale R
Loch Leathann an Sgorra
Kintour
Kildalton Cross
Ardmore
Ardmore Point
Allt nan Airighean
Beinn Sholum
Eilean Craobhach
Eilean a' Chuirn
2¼ hrs
Hotel
Port Alsaig
Kintra
Rubha Mòr
Leorin
Kildalton House
Eilean Bhrìde
Slugaide Glas
Mool
Buidhe
Cornabus
Cragabus
Port Ellen
Lagavulin
Ardbeg
Rubha na Gainmhich
Eilean Imersay
Glenastle
Dùn Mòr Ghil
Laphroaig
Dunyvaig Castle
THE OA
Risabus
Carraig Fhada
Texa
Lower Killeyan
Loch Kinnabus
Inerval
MULL OF OA
Beinn Mhòr
Rubha na Mèise Bàine
Port an Eas
Rubha nan Leacan
Craro Island
Grob

F
G
H
J
K
L
1
2
3
4
5
6
7
8
89
84
76
NM
NN
NS
JURA
SCARBA
LOCHGILPHEAD
Kilmory Castle
Ardrishaig
Tarbert
West Tarbert
Tayvallich
Crinan
Bellanoch
Kilmartin
Cairnbaan
Ford
Furnace
Strachur
Minard
Otter Ferry
Kilfinan
Tighnabruaich
Kames
Portavadie
Millhouse
Port Bannatyne
ROTHESAY
Ascog
Kingarth
Kilchattan Bay
Garroch Head
Inchmarnock
ISLAND OF BUTE
SOUND OF BUTE
KYLES OF BUTE
LOCH FYNE
LOCH STRIVEN
LOCH SWEEN
LOCH CAOLISPORT
KNAPDALE
KINTYRE
Skipness
Claonaig
Lochranza
Catacol
Pirnmill
Imachar
Sannox
Corrie
GOATFELL
Brodick
ISLAND OF ARRAN
Carradale
Grogport
Whitehouse
Kennacraig
Clachan
Tayinloan
Killean
Muasdale
GIGHA
SOUND OF GIGHA
Ardminish
Gigalum Island
Cara Island
Achnamara
Inverneill
Castleton
Port Ann
Glendaruel
Clachan of Glendaruel
Auchenbreck
Colintraive
Inveraray
Inverchaolain
Kilmun
Dunoon
Toward Point
Millstone Point
Cock of Arran
Rubha Creagan Dubha
Merkland Point
Ardlamont Point
Stravanan Bay
Scalpsie Bay
St Ninian's Point
Ettrick Bay
Kildavanan
Ardmarnock Bay
Barmore Island
East Loch Tarbert
West Loch Tarbert
Kilberry Head
Point of Knap
Eilean Mòr
Gulf of Corryvreckan
Summer only

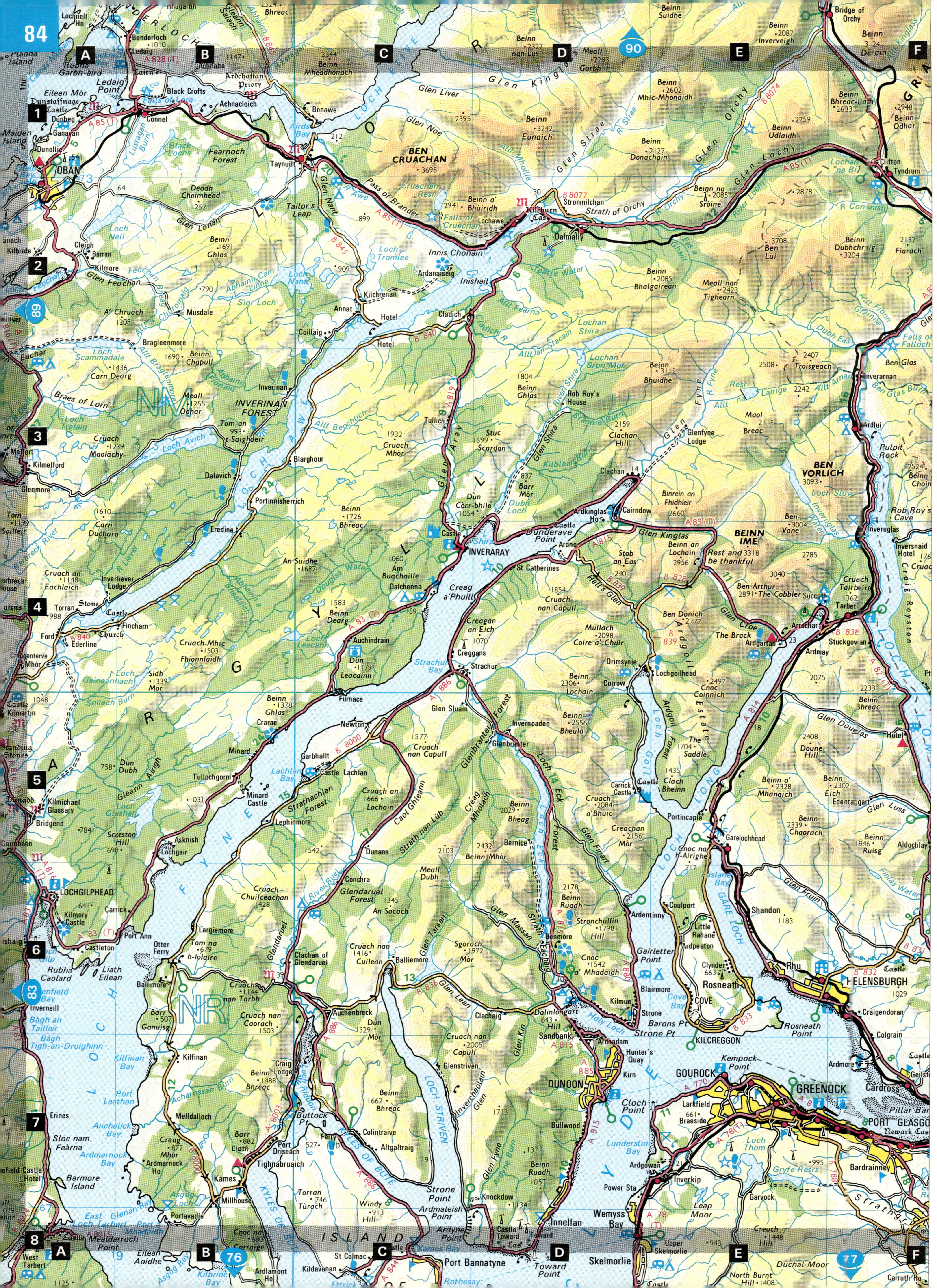

OBAN
INVERARAY
LOCHGILPHEAD
DUNOON
GOUROCK
GREENOCK
HELENSBURGH
PORT GLASGOW
BEN CRUACHAN
BEN VORLICH
BEINN IME
Loch Awe
Loch Fyne
Loch Long
Loch Eck
Loch Striven
Loch Goil
Gare Loch
Holy Loch
Kyles of Bute
Pass of Brander
Strath of Orchy
Glen Kinglas
Glen Orchy
Glen Lochy
Glen Aray
Glen Shira
Glen Fyne
Glen Douglas
Glen Luss
Glen Fruin
Dalmally
Taynuilt
Connel
Kilmore
Kilchrenan
Furnace
Strachur
Tighnabruaich
Port Bannatyne
Innellan
Wemyss Bay
Skelmorlie
Arrochar
Tarbet
Ardlui
Lochgoilhead
Sandbank
Kilmun
Rosneath
Kilcreggan
Ardentinny
Cairndow
Clachan
Otter Ferry
Kilfinan
Kames
Portavadie
Ardrishaig
Minard

Glen Lochay
Loch Tay
Killin
Finlarig
Morenish
Ardeonaig
Ardtalnaig
Lawers
Tullich Hill
Glen Dochart
Strath Fillan
Crianlarich
Lochdochart Ho
Loch Iubhair
Ledcharrie
Auchlyne
Ardchyle
Glen Ogle
Lochearnhead
Loch Earn
Edinample
Ardvorlich
St Fillans
Dundurn
Glen Lednock
Invergeldie
Comrie
Dalginross
Tullybannocher
Ross
Lawers
Ochtertyre
Quoig
Hosh
Lochlane
Culloch
Ballindalloch
Drummond Castle
Tigh-na-Blair
Ochtermuthill
Dalchruin
Glen Artney
Forest of Glenartney
Ben Vorlich
Strathyre Forest
Strathyre
Kingshouse
Balquhidder
Braes of Balquhidder
Loch Voil
Loch Doine
Monachyle
Craigruie
Auchtubh
Inverlochlarig
Ballimore
Ben More
Stob Binnein
Beinn Tulaichean
Ben Ledi
Loch Lubnaig
Ardchullarie More
Anie
Pass of Leny
Leny Ho
Callander
Bochastle
Coilantogle
Loch Venachar
Lendrick
Brig o' Turk
The Trossachs
Ellen's Isle
Loch Katrine
Stronachlachar
Glen Finglas
Glen Finglas Resr
Strath Gartney
Ben Venue
Loch Achray
Loch Drunkie
Achray Forest
Ben Lomond
Kinlochard
Loch Ard
Loch Chon
Loch Arklet
Aberfoyle
Milton
Menteith Hills
Port of Menteith
Lake of Menteith
Priory
Loch Ard Forest
Duchray Water
Gartmore
Gartmore Ho
Dalmary
Flanders Moss
Cardross
Dykehead
River Forth
Goodie Water
Ruskie
Thornhill
Braes of Doune
Burn of Cambus
Deanston
Doune
Lanrick
Drumvaich
Kilbryde Castle
Blair Drummond
Brock
Blairdrummond Moss
Drip Moss
Dunblane
Kippenross
Ashfield
Kinbuck
Sheriff Muir
Bridge of Allan
Blairlogie
Causewayhead
Cambuskenneth
Stirling
St Ninians
Bannockburn
Cambusbarron
Gillies Hill
Chartershall
Touch Hills
Gargunnock
Gargunnock Hills
Kippen
Cauldhame
Arnprior
Buchlyvie
Old Leckie
Carleatheran
Stronend
Cringate Law
Earl's Hill
North Third Resr
Loch Coulter Resr
Cowie
Plean
Torwood
Dunipace
Denny
Larbert
Carron Valley Forest
Carron Valley Resr
Carron Bridge
Cairnoch Hill
Tomtain
High Banton
Banton
Kilsyth
Kilsyth Hills
Dennyloanhead
Head of Muir
Bonnybridge
Longcroft
Castlecary
Cumbernauld
Dullatur
Twechar
Eastfield
Campsie Fells
Lennoxtown
Milton of Campsie
Queenzieburn
Kirkintilloch
Waterside
Condorrat
Mollinsburn
Birdston
Torrance
Balgrochan
Clachan of Campsie
Strathblane
Blanefield
Netherton
Strathblane Hills
Dumgoyne
Killearn
Ballikinrain Castle
Fintry
Loch Walton
Earl's Seat
Balfron
Boquhan
Gartness
Croftamie
Drymen
Buchanan Smithy
Garadhban Forest
Moor Park
Milton of Buchanan
Balmaha
Conic Hill
Inchcailloch
Inchmurrin
Ross Priory
Gartocharn
Arden
Balloch
Balloch Castle
Jamestown
Alexandria
Bonhill
Renton
Dumbarton
Dumbarton Muir
Kilpatrick Hills
Bellsmyre
Duntocher
Old Kilpatrick
Erskine Br Toll
Bishopton
Erskine
Clydebank
Bearsden
Milngavie
Mugdock
Blairskaith
Bardowie
Balmore
Bishopbriggs
Lenzie
Auchinloch
Chryston
Muirhead
Stepps
Garnkirk
Gartcosh
Glenboig
Coatbridge
Airdrie
Caldercruix
Longriggend
Greengairs
Wattston
Riggend
Annathill
Stand
Renfrew
Glasgow Airport
Kilmacolm
Bridge of Weir
Houston
Georgetown
Drumchapel
Yoker
Langbank
Finlaystone
Slamannan
Limerigg
Binniehill
Muckhart
Braco
Braco Castle
Greenloaning
Cromlet
Slymaback
Glenlichorn
Muir of Orchill
Strath Allan
Dumyat
Menstrie
Tullibody
Fallin
Throsk
Crieff

NN
NS
NT
Glen Quaich
Glen Almond
Glen Lednock
Glen Artney
Glen Devon
Glen Eagles
STRATHEARN
STRATHALLAN
OCHIL HILLS
FOREST OF GLENARTNEY
Ardtalnaig
Tullich Hill
Ardeonaig
Amulree
Corrymuckloch
Auchnafree
Meall nam Fuaran
Meall nan Caorach
Meall Reamhar
Meall Tarsuinn
Auchnafree Hill
Ben Chonzie
Loch Lednock Resr
Loch Turret Resr
Invergeldie
Càrn Chois
St Fillans
Dunira
Comrie
Dalginross
Ross
Lawers
Ochtertyre
Crieff
Monzie
Gilmerton
Fowlis Wester
Keillour Forest
Madderty
Methven
Tibbermore
Almondbank
Huntingtower
Perth
Scone Palace
New Scone
Kinnoull Hill
Tarsappie
Craigend
Bridge of Earn
Aberargie
Kintillo
Forteviot
Aberdalgie
Milltown of Aberdalgie
Forgandenny
Dunning
Aberuthven
Auchterarder
Muirton
Kincardine Castle
Gleneagles Hotel
Gleneagles Sta
Gleneagles Ho
Blackford
Braes of Ogilvie
Greenloaning
Braco
Braco Castle
Muir of Orchil
Muthill
Drummond Castle
The Balloch
Kinkell Bridge
Strathallan Cas
Trinity Gask
Findo Gask
Clathy
St David
Kirkton
Chapel
Glendevon
Yetts o' Muckhart
Pool of Muckhart
Rumbling Bridge
Crook of Devon
Carnbo
Drum
Coldrain
Kinross
Loch Leven
Milnathort
Tullibole Castle
Dollar
Castle Campbell
Tillicoultry
Alva
Menstrie
Tullibody
New Sauchie
Cambus
Alloa
Clackmannan
Kennet
Devonside
Coalsnaughton
Forest Mill
Blairingone
Saline
Knock Hill
Powmill
Cleish Hills
Kelty
Lochore
Ballingry
Glencraig
Cowdenbeath
Crossgates
Dunfermline
Oakley
Carnock
Gowkhall
Cairneyhill
Crossford
Limekilns
Charlestown
Rosyth
Inverkeithing
North Queensferry
Queensferry
Dalmeny
Kirkliston
Newbridge
Ratho
Broxburn
Uphall
Livingston
Bathgate
Linlithgow
Philpstoun
Winchburgh
Ecclesmachan
Bo'ness
Grangemouth
Kinneil
Carriden
Blackness
Muirhouses
Falkirk
Camelon
Larbert
Stenhousemuir
Carron
Skinflats
Airth
Kincardine
Tulliallan
Culross
Valleyfield
Torryburn
Low Torry
Dunmore
Throsk
Cowie
Plean
Bannockburn
Stirling
St Ninians
Cambuskenneth
Causewayhead
Bridge of Allan
Blairlogie
Dunblane
Kippenross
Doune
Deanston
Kilbryde Castle
Kinbuck
Ashfield
Sheriff Muir
Blairdenon Hill
Ben Cleuch
Core Hill
Braes of Doune
Drumvaich
Burn of Cambus
Lanrick
Gargunnock
Gargunnock Hills
Touch Hills
Cambusbarron
Gillies Hill
Chartershall
North Third Resr
Carron Valley Forest
Carron Valley Resr
Cringate Law
Cairnoch Hill
Carron Bridge
Dunipace
Denny
Dennyloanhead
Head of Muir
High Bonnybridge
Bonnybridge
Banknock
Longcroft
Kilsyth
Kilsyth Hills
Queenzieburn
Banton
Tomtain
Cumbernauld
Dullatur
Kirkintilloch
Waterside
Condorrat
Luggiebank
Castlecary
Greenhill
Glen Village
Redding
Shieldhill
Reddingmuirhead
California
Maddiston
Whitecross
Standburn
Avonbridge
Slamannan
Binniehill
Limerigg
Armadale
Blackridge
Westrigg
Torphichen
Cairnpapple
Dechmont
Uphall Station
Bonnington
Hermiston
Gogar
Edinburgh Airport
Ingliston
Hopetoun House
Forth Road Bridge
Forth Bridge
Dalgety Bay
St Davids
Inchcolm
Aberdour
Hillend
Fordell
Donibristle
Townhill
Wellwood
Kingseat
Bowershall
Craigluscar Hill
Lassodie
Loch Fitty
Balgonar
Cult Hill
Gairney Bank
Blairadam
Scotlandwell
Balgedie
Kinnesswood
Middleton
Dunecrievie
Glenfarg
Glenfarg Resr
Path of Condie
Rossie Ochill
Balmanno Hill
Dron
Glenfoot
Newton of Balcanquhal
Gateside
Burnside
Dochrie Hill
Middle Rigg
Innerdouny Hill
Common of Dunning
Craig Rossie
Steele's Knowe
Wether Hill
Glendevon Resrs
Commonedge Hill
Lendrick Hill
Seamab Hill
Whitemoss
Busby
Logiealmond Lodge
Harrietfield
Chapelhill
Moneydie
Luncarty
Pitcairngreen
Stanley
Gellyburn
Murthly
Meikleour
Kinclaven
Cargill
Obney Hills
Waterloo
Bankfoot
Airntully
Glenshee
Milton
Dunsinnan
Guildtown
Wolfhill
Kinrossie
Balbeggie
St Martins
Newmill
Colenden
Kinfauns
Glencarse
Moncrieffe Hill
Elcho
Rhynd
Tigh-na-Blair
Dalchruin
Culloch
Ochtermuthill
Torlum
Torlum Wood
Cromlet
Glenlichorn
Slymaback
Ben Clach
Creag Beinn nan Eun
Uamh Bheag
Ben Halton
Tullybannocher
Loch Boltachan
Lochlane
Hosh
Falls of Keltie
Loch Meallbrodden
Cairnie
Braegrum
Keillour
Buchanty
Glenalmond
College
Braes of Doune
Blair Drummond
Leckie
Carleatheran
Earlsburn Resrs
Earl's Hill
Drip Moss
Abbey Craig
Fallin
Torwood
Plean
Longannet Point
Torry Bay
Preston Island
Whitecross
Linlithgow Bridge
Union Canal
Lochcote Resr
Westfield
Avonbridge
Caldercruix
Airdrie
Coatbridge
Plains
Muirhead
Gartcosh
Glenboig
Annathill
Riggend
Greengairs
Wattston
Longriggend
Forrestfield
Blackridge
Seafield
Deans
Livingston Village
East Calder
Kirknewton
Dedridge
Oakbank
Wilkieston
Balerno
Juniper Green

DUNDEE
Tay Road Bridge
Tay Bridge
FIRTH OF TAY
NEWPORT ON TAY
TAYPORT
TENTSMUIR FOREST
BUDDON NESS
CARNOUSTIE
MONIFIETH
Broughty Ferry
Invergowrie
Eden Mouth
ST ANDREWS BAY
ST ANDREWS
NEWBURGH
CUPAR
LADYBANK
FALKLAND
AUCHTERMUCHTY
PITMEDDEN FOREST
GLENROTHES
LESLIE
MARKINCH
Kennoway
LEVEN
METHIL
BUCKHAVEN
East Wemyss
West Wemyss
LARGO BAY
ELIE
EARLSFERRY
ST MONANCE
PITTENWEEM
ANSTRUTHER
KILRENNY
CRAIL
FIFE NESS
Kingsbarns
Isle of May
NORTH SEA
FIRTH OF FORTH
KIRKCALDY
KINGHORN
BURNTISLAND
Inchkeith
Inchmickery
Cramond Island
EDINBURGH
Leith
Portobello
MUSSELBURGH
PRESTONPANS
COCKENZIE AND PORT SETON
Longniddry
Aberlady
Gullane
Dirleton
NORTH BERWICK
Bass Rock
Fidra
HADDINGTON
EAST LINTON
DUNBAR
Tyne Mouth
Dunbar Common
TRANENT
Dalkeith
Pencaitland
Gifford
F
G
H
J
K
1
2
3
4
5
6
7
8
93
80

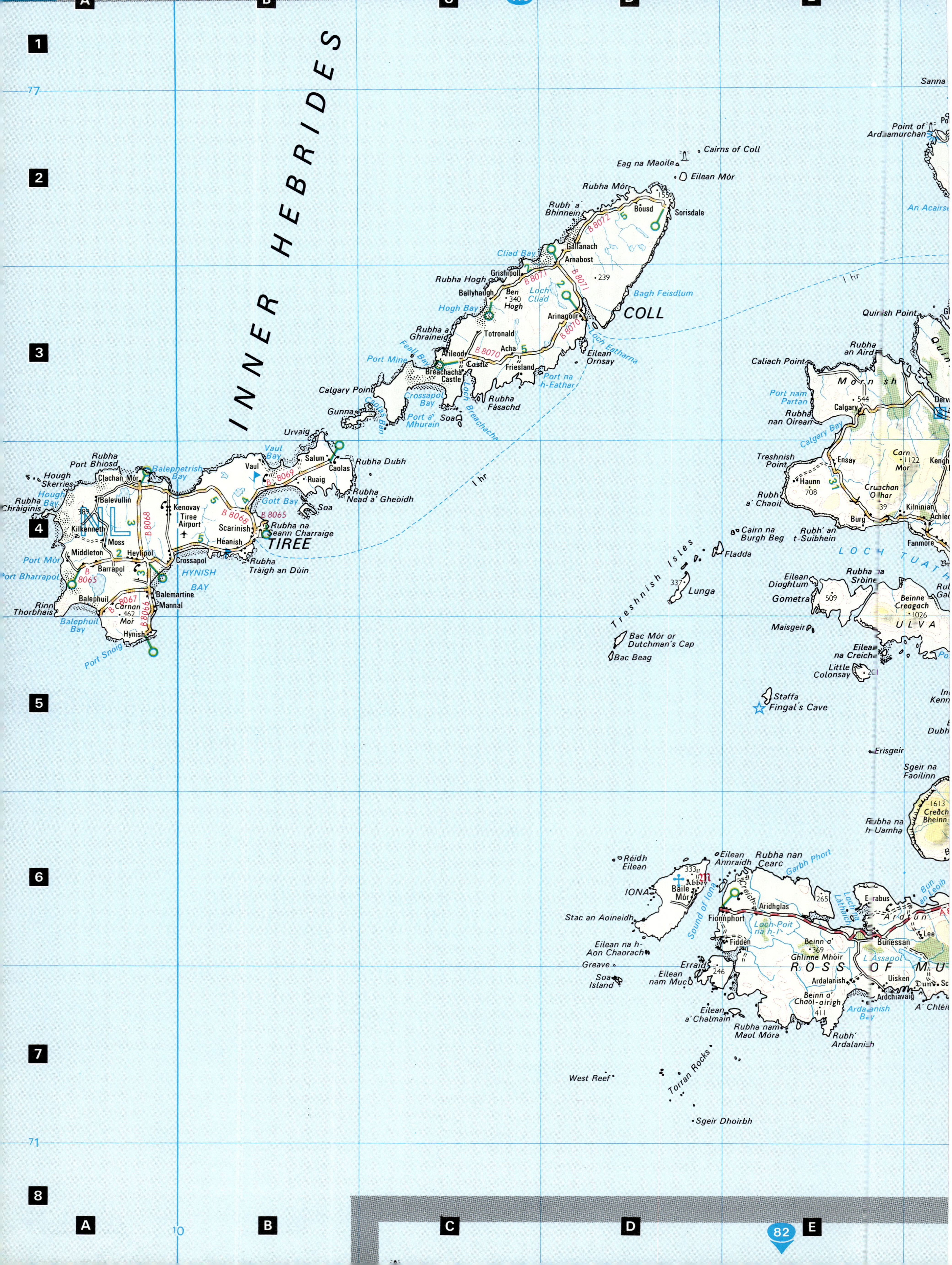
INNER HEBRIDES
COLL
TIREE
IONA
ROSS OF MULL
ULVA
Treshnish Isles
110
82
Cairns of Coll
Eag na Maoile
Eilean Mór
Rubha Mòr
Rubh' a' Bhinnein
Bousd
Sorisdale
Cliad Bay
Gallanach
Arnabost
Grishipoll
Rubha Hogh
Ballyhaugh
Ben Hogh
Loch Cliad
Hogh Bay
Arinagour
Bagh Feisdlum
Rubha a' Ghraineig
Totronald
Acha
Arileod
Feall Bay
Port Mine
Breachacha Castle
Castle
Friesland
Eilean Ornsay
Loch Eatharna
Port na h-Eathar
Calgary Point
Crossapol Bay
Gunna
Port a' Mhurain
Soa
Rubha Fàsachd
Loch Breachacha
Urvaig
Vaul Bay
Salum
Caolas
Rubha Dubh
Vaul
Ruaig
Rubha Nead a' Gheòidh
Rubha Port Bhiosd
Balephetrish Bay
Hough Skerries
Clachan Mòr
Hough Bay
Balevullin
Rubha Chràiginis
Kenovay
Tiree Airport
Gott Bay
Soa
Kilkenneth
Scarinish
Rubha na Seann Charraige
Heanish
Moss
Middleton
Heylipol
Port Mòr
Crossapol
Barrapol
Port Bharrapol
HYNISH BAY
Rubha Tràigh an Dùin
Balemartine
Mannal
Balephuil
Rinn Thorbhais
Carnan Mòr
Balephuil Bay
Hynish
Port Snoig
1 hr
Sanna
Point of Ardnamurchan
An Acairseid
Quinish Point
Rubha an Aird
Caliach Point
Mornish
Port nam Partan
Calgary
Rubha nan Oirean
Calgary Bay
Treshnish Point
Ensay
Carn Mòr
Haunn
Rubh' a' Chaoil
Cruachan Odhar
Kilninian
Burg
Rubh' an t-Suibhein
Fanmore
Cairn na Burgh Beg
Fladda
LOCH TUATH
Eilean Dioghlum
Rubha na Sròine
Gometra
Beinne Creagach
Lunga
Maisgeir
Bac Mór or Dutchman's Cap
Bac Beag
Eilean na Creiche
Little Colonsay
Staffa
Fingal's Cave
Erisgeir
Sgeir na Faoilinn
Rubha na h-Uamha
Creach Bheinn
Réidh Eilean
Eilean Annraidh
Rubha nan Cearc
Garbh Phort
Baile Mòr
Abbey
Sound of Iona
Stac an Aoineidh
Aridhglas
Fionnphort
Fidden
Loch Poit na h-I
Erraid
Beinn a' Ghlinne Mhòir
Bunessan
Loch Assapol
Eilean na h-Aon Chaorach
Greave
Soa Island
Eilean nam Muc
Ardalanish
Uisken
Ardchiavaig
Beinn a' Chaol-airigh
Ardalanish Bay
Eilean a' Chalmain
Rubha nam Maol Móra
Rubh' Ardalanish
West Reef
Torran Rocks
Sgeir Dhoirbh
B 8072
B 8071
B 8070
B 8069
B 8068
B 8065
B 8067
B 8066

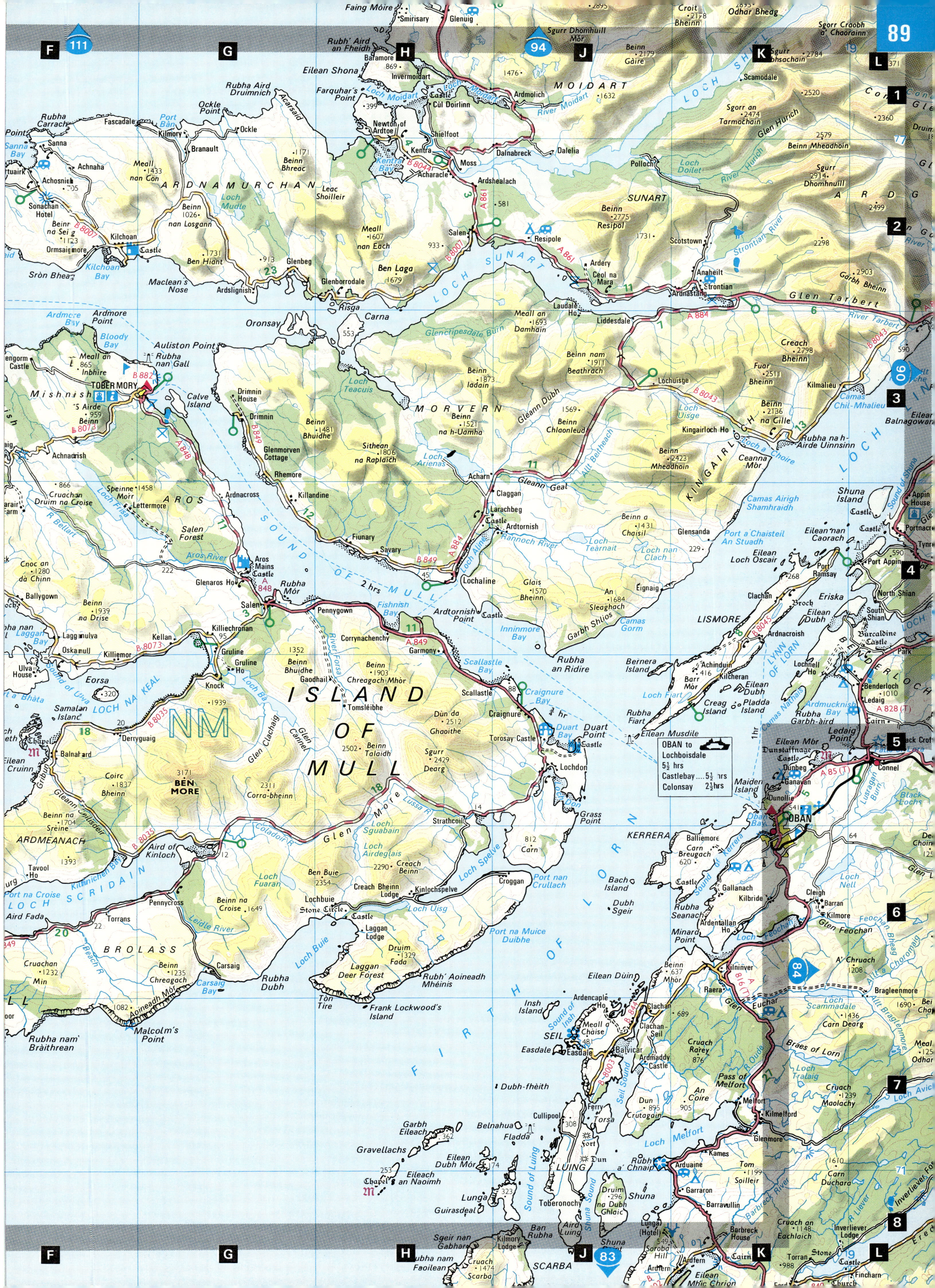
111
94
90
84
83
F
G
H
J
K
L
1
2
3
4
5
6
7
8
Faing Mòire
Smirisary
Glenuig
Sgurr Dhomhnuill Mòr
Odhar Bheag
Sgorr Craobh a' Chaorainn
Rubh' Aird an Fheidh
Eilean Shona
Inverfoidart
MOIDART
Loch Moidart
Castle
Cùl Doirlinn
Ardmolich
River Moidart
Loch Shiel
Scamodale
Sgorr an Tarmachain
Glen Hurich
Beinn Mheadhoin
Sgurr Dhomhnuill
Rubha Aird Druimnich
Ockle Point
Farquhar's Point
Newton of Ardtoe
Kentra
Kentra Bay
Acharacle
Moss
Shielfoot
Dalnabreck
Dalelia
Polloch
Loch Doilet
River Hurich
Rubha Carrach
Fascadale
Port Bàn
Kilmory
Ockle
Sanna
Sanna Bay
Achnaha
Achosnich
Sonachan Hotel
Branault
Beinn Bhreac
Meall nan Con
ARDNAMURCHAN
Leac Shoilleir
Loch Mudle
Ardshealach
Ardnamurchan
Beinn Resipol
SUNART
Strontian River
Beinn na Seilg
Ormsaigmore
Kilchoan
Castle
Kilchoan Bay
Beinn nan Losgann
Ben Hiant
Maclean's Nose
Glenbeg
Glenborrodale
Ardslignish
Meall nan Each
Ben Laga
Salen
Resipole
Ardery
Ceol na Mara
Scotstown
Anaheilt
Strontian
Ardnastang
Glen Tarbert
Garbh Bheinn
Sròn Bheag
LOCH SUNART
Risga
Carna
Oronsay
Ardmore Point
Ardmore Bay
Bloody Bay
Auliston Point
Rubha nan Gall
TOBERMORY
Mishnish
Airde Beinn
Calve Island
Meall an Inbhire
Drimnin House
Drimnin
Loch Teacuis
Glencripesdale Burn
Meall an Damhain
Laudale Ho
Liddesdale
Beinn nam Beathrach
River Tarbert
Creach Bheinn
Fuar Bheinn
Kilmalieu
Beinn Iadain
MORVERN
Gleann Dubh
Beinn na h-Uamha
Beinn Chaorach
Beinn Chlaonleud
Lochuisge
Loch Uisge
Kingairloch Ho
Beinn na Cille
Rubha na h-Airde Uinnsinn
Camas Chil-Mhalieu
Balnagowan
LOCH LINNHE
Achnacraig
Speinne Mòr
Lettermore
AROS
Loch Frisa
Cruachan Druim na Croise
Glenmorven Cottage
Beinn Bhuidhe
Sithean na Raplaich
Loch Arienas
Rhemore
Killundine
Ardnacross
Acharn
Gleann Geal
Allt Beitheach
Beinn Mheadhoin
KINGAIRLOCH
Ceanna Mòr
Shuna Island
Appin House
Portnacroish
Salen Forest
Aros River
R. Bellart
SOUND OF MULL
Fiunary
Savary
Claggan
Larachbeg
Castle
Ardtornish
Rannoch River
Loch Aline
Beinn a Chaisil
Loch Tearnait
Loch nan Clach
Glensanda
Port a Chaisteil
An Stuadh
Eilean Loch Oscair
Eilean nan Caorach
Port Ramsay
Port Appin
Cnoc an Dà Chinn
Ballygown
Aros Mains
Castle
Glenaros Ho
Rubha Mòr
Salen
Lochaline
Glais Bheinn
An Sleaghach
Eignaig
Clachan
Eriska
North Shian
South Shian
Beinn na Drise
Laggan Bay
Laggan Ulva
Oskamull
Killiemor
Kellan
Killiechronan
Pennygown
Fishnish Bay
Corrynachenchy
Garmony
Ardtornish Point
Castle
Inninmore Bay
Garbh Shlios
Camas Gorm
LISMORE
Eilean Dubh
Ardnacroish
Lochnell Ho
Barcaldine Castle
Ulva House
Eorsa
Gruline
Gruline Ho
Knock
Beinn Bhuidhe
Gaodhail
Beinn Chreagach Mhòr
River Forsa
Scallastle Bay
Scallastle
Rubha an Ridire
Bernera Island
Achinduin
Kilcheran
Barr Mòr
Loch Fiart
Creag Island
Eilean Dubh
Pladda Island
LYNN OF LORN
Benderloch
Ledaig
Samalan Island
LOCH NA KEAL
NM
ISLAND OF MULL
Tomsléibhe
Glen Clachaig
Glen Cannel
Dùn da Ghaoithe
Craignure
Craignure Bay
Torosay Castle
Duart Bay
Duart Point
Castle
Rubha Fiart
Eilean Musdile
OBAN to
Lochboisdale
Castlebay
Colonsay
Eilean Mòr
Dunstaffnage Castle
Dunbeg
Ledaig Point
Connel
Derryguaig
Balnahard
Eilean Chruinn
BEN MORE
Coirc Bheinn
Corra-bheinn
Beinn Talaidh
Sgurr Dearg
Lochdon
Grass Point
Maiden Island
Ganavan
Dunollie
OBAN
Black Lochs
Lusragan Burn
Beinn na Sreine
ARDMEANACH
Gleann Seilisdeir
Aird of Kinloch
Glen More
Coladoir R.
Loch Sguabain
Strathcoil
Carn
KERRERA
Balliemore
Carn Breugach
Castle
Gallanach
Kilbride
Tavool Ho
Kinlochbeg
LOCH SCRIDAIN
Pennycross
Beinn na Croise
Loch Fuaran
Loch Airdeglais
Creach Beinn
Ben Buie
Creach Bheinn Lodge
Kinlochspelve
Loch Spelve
Croggan
Port nan Crullach
Bach Island
Dubh Sgeir
Rubha Seanach
Ardentallan Ho
Minard Point
Cleigh
Barran
Kilmore
Glen Feochan
Loch Nell
Port na Croise
Aird Fada
Torrans
BROLASS
Leidle River
Lochbuie
Stone Circle
Castle
Laggan Lodge
Loch Uisg
Druim Fada
Port na Muice Duibhe
Loch Feochan
A' Chruach
Bragleenmore
Loch Scammadale
Carn Dearg
Cruachan Min
Beinn Chreagach Mhòr
Carsaig
Carsaig Bay
Loch Buie
Laggan Deer Forest
Rubh' Aoineadh Mhèinis
Rubha Dubh
Tòn Tìre
Frank Lockwood's Island
Eilean Dùin
Beinn Mhòr
Kilninver
Raera
Glen Euchar
Braes of Lorn
Aoineadh Mòr
Malcolm's Point
Rubha nam Bràithrean
FIRTH OF LORN
Insh Island
Ardencaple Ho
Sound of Insh
SEIL
Easdale
Meall a' Chaise
Balvicar
Clachan
Clachan-Seil
Ardmaddy Castle
Cruach Rarey
Pass of Melfort
Loch Tralaig
Cruach Moolachy
Dubh-fhèith
Sail Sound
Dun
Crutagain
An Coire
Melfort
Kilmelford
Glenmore
Garbh Eileach
Gravellachs
Eilean Dubh Mòr
Eileach an Naoimh
Belnahua
Fladda
Cullipool
LUING
Toberonochy
Sound of Luing
Shuna Sound
Loch Melfort
Rubh' a' Chnaip
Arduaine
Kames
Tom Soilleir
Carn Chuduaro
Barbreck River
Inverliever Forest
Lunga
Guirasdeal
Druim na Dubh Ghlaic
Shuna
Garraron
Barravullin
Barbreck House
Cruach an Eachlaich
Inverliever Lodge
Torran
Stone
Castle
Sgeir nan Gabhar
Kilmory Lodge
Bàn Rubha
Aird Luing
Shuna Island
Lunga (Hotel)
Ardfern
Rubha nam Faoilean
Cruach Scarba
SCARBA
Eilean Mhic Chrion
Craobh Haven
Fincharn
Church

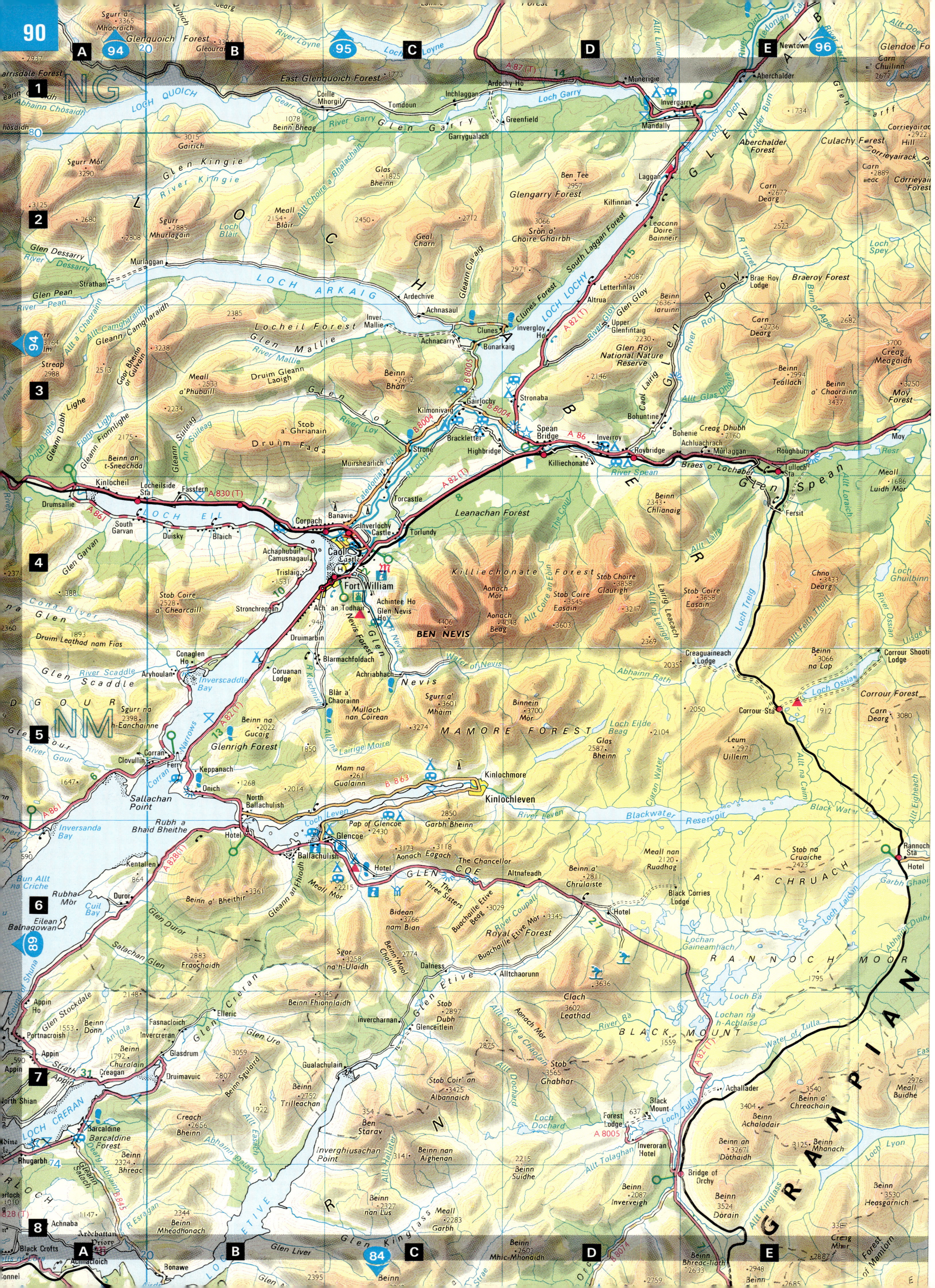
94
95
96
84
89
A
B
C
D
E
1
2
3
4
5
6
7
8
Sgurr a' Mhaoraich
Gleanquoich
Forest
River Loyne
Loch Loyne
East Glenquoich Forest
Coille Mhorgil
Tomdoun
Inchlaggan
Ardochy Ho
Munerigie
Invergarry
Aberchalder
Newtown
Loch Garry
Greenfield
Mandally
LOCH QUOICH
Abhainn Chosaidh
Gearr Garry
River Garry
Glen Garry
Garrygualach
Beinn Bheag
Gairich
Sgurr Mòr
Glen Kingie
River Kingie
Glas Bheinn
Ben Tee
Glengarry Forest
Kilfinnan
Laggan
GLEN ALBYN
Loch Oich
Calder Burn
Aberchalder Forest
Culachy Forest
Corrieyairack Forest
Carn Dearg
Meall Blàir
Sgurr Mhurlagain
Loch Blair
Geal Charn
Sròn a' Choire Ghairbh
South Laggan Forest
Leacann Doire Bainneir
Glen Dessarry
River Dessarry
Murlaggan
Strathan
Glen Pean
River Pean
LOCH ARKAIG
LOCHABER
Ardechive
Achnasaul
Gleann Cia-aig
Clunes Forest
LOCH LOCHY
Letterfinlay
Altrua
Glen Gloy
Beinn Iaruinn
Glen Roy
Brae Roy Lodge
Braeroy Forest
River Roy
Upper Glenfintaig
Glen Roy National Nature Reserve
Locheil Forest
Glen Mallie
River Mallie
Inver Mallie
Achnacarry
Clunes
Bunarkaig
Invergloy Ho
Gleann Camgharaidh
Gleann Fionnlighe
Streap
Meall a' Phubuill
Druim Gleann Laoigh
Glen Loy
River Loy
Beinn Bhàn
Stronaba
Kilmonivaig
Gairlochy
Spean Bridge
Creag Meagaidh
Moy Forest
Beinn Teallach
Beinn a' Chaorainn
Gleann Dubh Lighe
Gleann Suileag
Stob a' Ghrianain
Druim Fada
Brackletter
Highbridge
Strone
Inverroy
Roybridge
Bohenie
Achluachrach
Creag Dhubh
Murlaggan
Roughburn
Tulloch Sta
Glen Spean
River Spean
Braes o' Lochaber
Killiechonate
Beinn an t-Sneachda
Kinlocheil
Locheilside Sta
Fassfern
Muirshearlich
Caledonian Canal
Torcastle
Leanachan Forest
Beinn Chlianaig
Fersit
Meall Luidh Mòr
Drumsallie
LOCH EIL
South Garvan
Duisky
Blaich
Corpach
Banavie
Inverlochy Castle
Torlundy
Glen Garvan
Achaphubuil
Camusnagaul
Caol
Castle
Trislaig
Fort William
Killiechonate Forest
Stob Choire Claurigh
Aonach Mòr
Stob Coire Easain
Stob Coire Easain
Chno Dearg
Loch Treig
Loch Ghuilbinn
Stob Coire a' Chearcaill
Cona River
Glen
Stronchreggan
Ach' an Todhair
Achintee Ho
Glen Nevis Ho
Nevis Forest
Aonach Beag
BEN NEVIS
Lairig Leacach
Druim Leathad nam Fias
Conaglen Ho
Druimarbin
Blarmachfoldach
Glen Nevis
Water of Nevis
Abhainn Rath
Creaguaineach Lodge
Beinn na Lap
Corrour Shooting Lodge
Loch Ossian
Corrour Forest
Glen Scaddle
River Scaddle
Aryhoulan
Inverscaddle Bay
Corunanan Lodge
Achriabhach
Nevis
Blàr a' Chaorainn
Sgurr a' Mhaim
Binnein Mòr
Corrour Sta
Carn Dearg
Sgurr na h-Eanchainne
Mullach nan Coirean
MAMORE FOREST
Loch Eilde Beag
Glas Bheinn
Leum Uilleim
Gleann Gour
River Gour
Corran
Clovullin
Ferry
Corran Narrows
Beinn na Gucaig
Glenrigh Forest
Lairig Moire
Allt na h-Eilde
Keppanach
Onich
North Ballachulish
Sallachan Point
Mam na Gualainn
Kinlochmore
Kinlochleven
Ciaran Water
Blackwater Reservoir
Allt na Cailliche
Rannoch Sta
Inversanda Bay
Rubh a' Bhaid Bheithe
Hotel
Loch Leven
Pap of Glencoe
Garbh Bheinn
River Leven
Glencoe
Ballachulish
Aonach Eagach
The Chancellor
GLEN COE
Altnafeadh
Beinn a' Chrulaiste
Meall nan Ruadhag
Stob na Cruaiche
A' CHRUACH
Kentallen
Rubha Mòr
Duror
Cuil Bay
Beinn a' Bheithir
Meall Mòr
Gleann an Fhiodh
The Three Sisters
Bidean nam Bian
Buachaille Etive Beag
Buachaille Etive Mòr
Royal Forest
River Coupall
Black Corries Lodge
Loch Laidon
Eilean Balnagowan
Glen Duror
Salachan Glen
Fraochaidh
Sgor na h-Ulaidh
Beinn Maol Chaluim
Dalness
Lochan Gaineamhach
RANNOCH MOOR
Alltchaorunn
Glen Etive
Beinn Fhionnlaidh
Glen Creran
Clach Leathad
Loch Ba
Lochan na h-Achlaise
Appin Ho
Glen Stockdale
Beinn Donn
Fasnacloich
Invercreran
Glenure
Elleric
Glen Ure
Inverchaorain
Stob Dubh
Glenceitlein
Aonach Mòr
Coire a' Chaolain
River Ba
BLACK MOUNT
Water of Tulla
Portnacroish
Appin
Strath of Appin
Beinn Churalain
Creagan
Glasdrum
Druimavuic
Beinn Sgulaird
Gualachulain
Beinn Trilleachan
Stob Coir' an Albannaich
Allt Dochard
Stob Ghabhar
Achallader
Beinn a' Chreachain
Beinn Achaladair
Port Shian
LOCH CRERAN
Barcaldine
Barcaldine Forest
Creach Bheinn
Abhainn Teithil
Ben Starav
Loch Dochard
Forest Lodge
Black Mount
Loch Tulla
Inveroran Hotel
Beinn an Dòthaidh
Beinn Mhanach
Loch Lyon
Rhugarbh
Beinn Bhreac
Inverghiusachan Point
Beinn nan Aighenan
Beinn Suidhe
Allt Tolaghan
Bridge of Orchy
Beinn Inverveigh
Beinn Dòrain
GRAMPIAN
Beinn Heasgarnich
LOCH ETIVE
Beinn nan Lus
Meall Garbh
Glen Kinglass
Achnaba
Beinn Mheadhonach
Black Crofts
Ardchattan Priory
Achnacloich
Bonawe
Glen Liver
Beinn Mhic-Mhonaidh
Beinn Bhreac-liath
Creag Mhòr
Forest of Mamlorn
A 82 (T)
A 87 (T)
A 86
A 830 (T)
A 861
A 828 (T)
A 8005
B 8004
B 8005
B 863
B 845

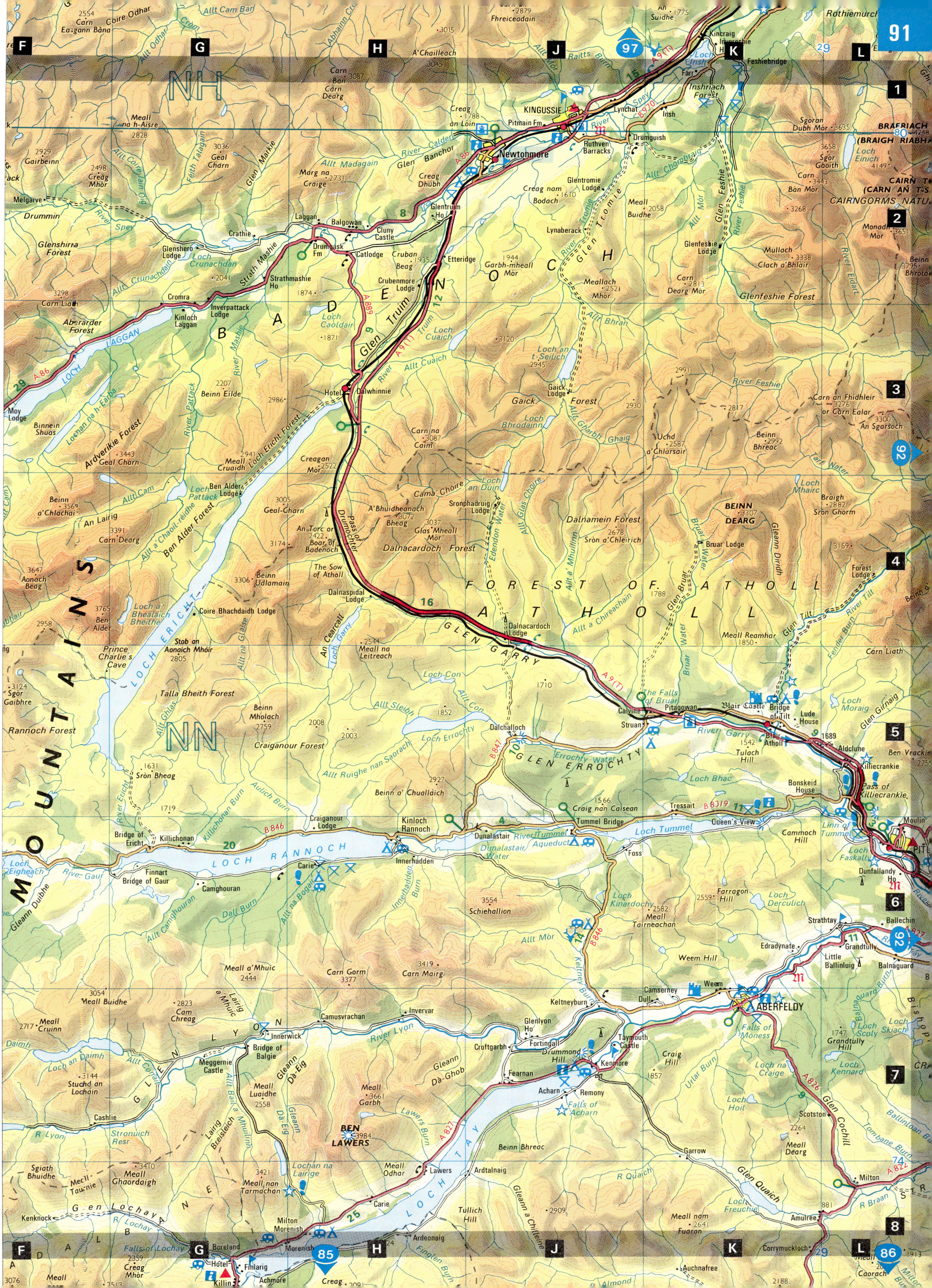

NH
NN
MOUNTAINS
FOREST OF ATHOLL
BADENOCH
KINGUSSIE
Newtonmore
Kincraig
Feshiebridge
Insh
Inshriach Forest
Ruthven Barracks
Drumguish
Glen Banchor
River Calder
Creag Dhubh
River Spey
Laggan
Balgowan
Cluny Castle
Catlodge
Crathie
Drumgask Fm
Cruban Beag
Crubenmore Lodge
Etteridge
Glentruim Ho
Strathmashie Ho
Strath Mashie
Garbh-mheall Mòr
Glentromie Lodge
Creag nam Bodach
Lynaberack
Meallach Mhòr
Meall Buidhe
Glen Feshie
Glenfeshie Lodge
Mullach Clach a' Bhlair
Carn Dearg Mòr
Glenfeshie Forest
Sgoran Dubh Mòr
BRAERIACH (BRAIGH RIABHACH)
Loch Einich
Sgor Gaoith
Carn Bàn Mòr
CAIRNGORMS
Monadh Mòr
Glenshirra Forest
Glenshero Lodge
Loch Crunachdan
Carn Liath
Aberarder Forest
Cromra
Kinloch Laggan
Inverpattack Lodge
LOCH LAGGAN
Loch Caoldair
Glen Truim
Loch Cuaich
Allt Cuaich
Dalwhinnie
Hotel
Loch an t-Seilich
Gaick Lodge
Gaick Forest
Loch Bhrodainn
Allt Gharbh Ghaig
Uchd a' Chlarsair
Beinn Bhreac
River Feshie
Carn an Fhidhleir or Carn Ealar
An Sgarsoch
Tarf Water
Beinn Eilde
Moy Lodge
Binnein Shuas
Ardverikie Forest
Geal Charn
Meall Cruaidh
Loch Ericht Forest
Creagan Mòr
Carn na Caim
Càma' Choire
Loch an Duin
Stronphadruig Lodge
A' Bhuidheanach Bheag
Glas Mheall Mòr
Dalnacardoch Forest
Pass of Drumochter
Dalnamein Forest
Sròn a' Chleirich
Bruar Lodge
BEINN DEARG
Loch Mhairc
Bràigh Sròn Ghorm
Forest Lodge
Beinn a' Chlachair
An Lairig
Carn Dearg
Ben Alder Lodge
Loch Pattack
Ben Alder Forest
Geal-Charn
An Torc or Boar of Badenoch
The Sow of Atholl
Dalnaspidal Lodge
Beinn Udlamain
Aonach Beag
Ben Alder
Loch a' Bhealaich Bheithe
Coire Bhachdaidh Lodge
LOCH ERICHT
Stob an Aonaich Mhòir
Prince Charlie's Cave
Dalnacardoch Lodge
GLEN GARRY
Meall na Leitreach
Loch Garry
Loch Con
Meall Reamhar
Glen Tilt
Carn Liath
Loch Moraig
Glen Girnaig
The Falls of Bruar
Calvine
Struan
Pitagowan
Blair Castle
Blair Atholl
Bridge of Tilt
Lude House
Aldclune
Killiecrankie
Pass of Killiecrankie
Tulach Hill
Loch Bhac
Bonskeid House
Dalchalloch
Loch Errochty
GLEN ERROCHTY
Talla Bheith Forest
Beinn Mholach
Craiganour Forest
Sròn Bheag
Rannoch Forest
Sgor Gaibhre
Beinn a' Chuallaich
Craig nan Caisean
Tummel Bridge
Tressait
Queen's View
Loch Tummel
Cammoch Hill
Loch Faskally
Dunfallandy Ho
Bridge of Ericht
Killichonan
Craiganour Lodge
Kinloch Rannoch
Dunalastair
Dunalastair Water
Foss
LOCH RANNOCH
Loch Eigheach
Bridge of Gaur
Finnart
Camghouran
Carie
Innerhadden
Schiehallion
Loch Kinardochy
Meall Tairneachan
Farragon Hill
Loch Derculich
Strathtay
Ballechin
Grandtully
Little Ballinluig
Balnaguard
Gleann Duibhe
Meall a' Mhuic
Carn Gorm
Carn Mairg
Weem Hill
Weem
ABERFELDY
Camserney
Dull
Keltneyburn
Meall Buidhe
Cam Chreag
Meall Cruinn
Camusvrachan
Innerwick
Invervar
Glenlyon Ho
Fortingall
Drummond Hill
Taymouth Castle
Kenmore
Croftgarbh
Fearnan
Acharn
Remony
Falls of Acharn
Falls of Moness
Grandtully Hill
Loch Kennard
Loch na Craige
Loch Scoly
Loch Skiach
Craig Hill
Urlar Burn
Loch Hoil
Scotstown
Meall Dearg
Glen Cochill
Bridge of Balgie
Meggernie Castle
Gleann Dà-Eig
River Lyon
GLEN LYON
Loch an Daimh
Stuchd an Lochain
Cashlie
Stronuich Resr
Meall Luaidhe
Meall Garbh
BEN LAWERS
Lawers
Meall Odhar
Ardtalnaig
LOCH TAY
Beinn Bhreac
Garrow
Glen Quaich
Loch Freuchie
Amulree
Milton
Sgiath Bhuidhe
Meall Ghaordaigh
Meall nan Tarmachan
Lochan na Lairige
Milton Morenish
Morenish
Carie
Kenknock
Glen Lochay
Falls of Lochay
Boreland
Finlarig
Killin
Achmore
Creag Mhor
Tullich Hill
Ardeonaig
Meall nam Fuaran
Corrymuckloch
Auchnafree
River Almond
F G H J K L
1 2 3 4 5 6 7 8
97 92 85 86

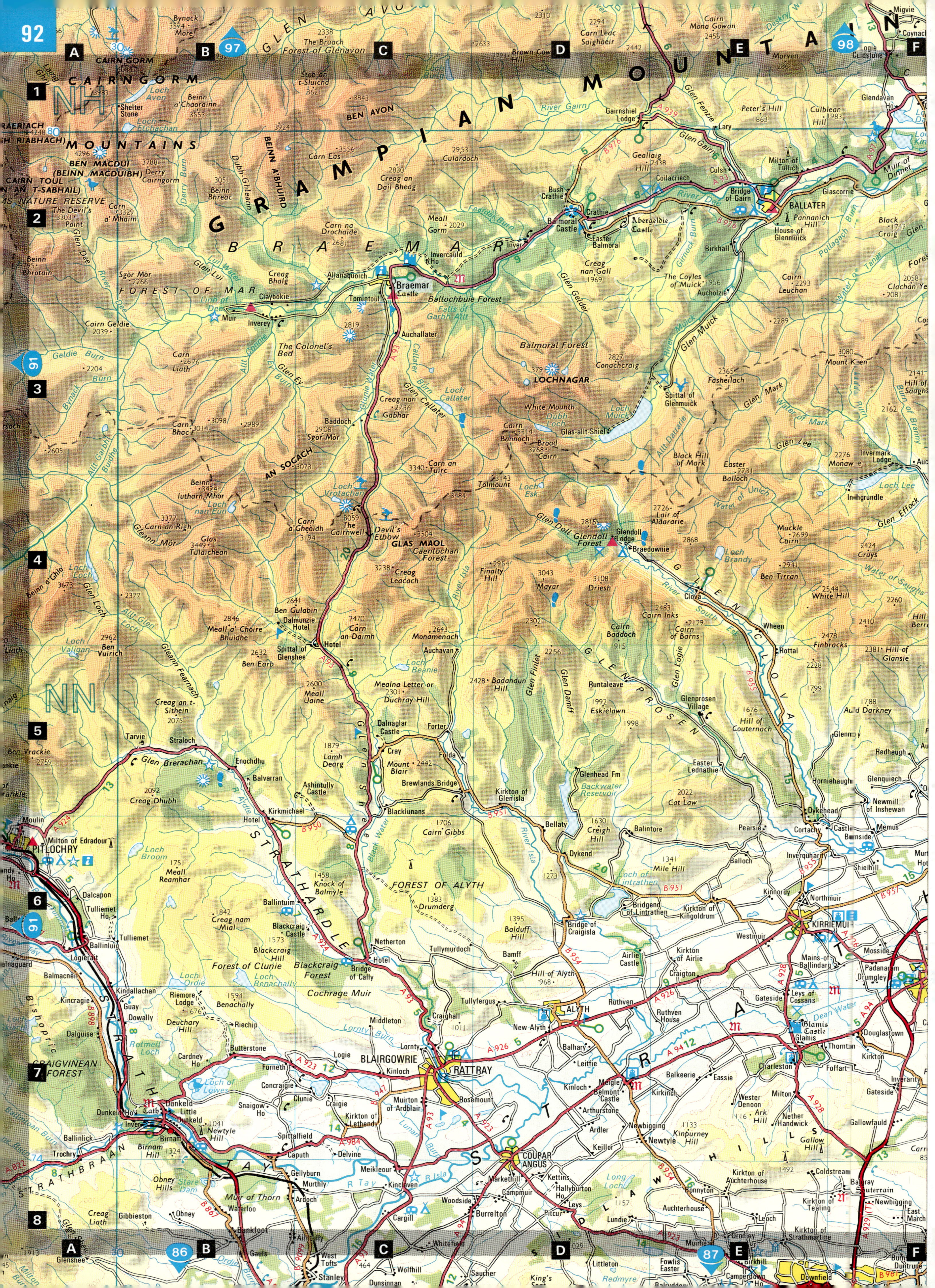

A
B
C
D
E
F
97
98
86
87
91
1
2
3
4
5
6
7
8
NH
NN
CAIRN GORM
CAIRNGORM
MOUNTAINS
BEN MACDUI
(BEINN MACDUIBH)
CAIRN TOUL
(AN T-SABHAIL)
NATURE RESERVE
The Devil's Point
GRAMPIAN MOUNTAINS
BRAEMAR
GLEN AVON
BEN AVON
BEINN A'BHUIRD
Forest of Glenavon
The Bruach
Carn Eas
Culardoch
Creag an Dail Bheag
Beinn a'Chaorainn
Beinn Bhreac
Derry Cairngorm
Carn a' Mhaim
Sgòr Mòr
FOREST OF MAR
Linn of Dee
Inverey
Claybokie
Allanaquoich
Braemar
Braemar Castle
Invercauld
Ballochbuie Forest
Falls of Garbh Allt
Creag Bhalg
Morrone
Cairn Geldie
Geldie Burn
Glen Dee
Glen Lui
Glen Ey
The Colonel's Bed
Carn Liath
Auchallater
Callater Burn
Loch Callater
Glen Callater
Creag nan Gabhar
Carn Bhac
Baddoch
Sgor Mor
AN SOCACH
Beinn Iutharn Mhor
Loch nan Eun
Glas Tulaichean
Gleann Mòr
Carn a' Gheoidh
The Cairnwell
Devil's Elbow
GLAS MAOL
Caenlochan Forest
Creag Leacach
Carn an Tuirc
Tolmount
Loch Esk
Balmoral Forest
LOCHNAGAR
White Mounth
Dubh Loch
Loch Muick
Glas-allt-Shiel
Cairn Bannoch
Broad Cairn
Glen Doll
Glendoll Lodge
Glendoll Forest
Braedownie
Clova
GLEN CLOVA
Mayar
Driesh
Finalty Hill
Spittal of Glenmuick
Glen Muick
Conachcraig
Fasheilach
Black Hill of Mark
Easter Balloch
Glen Mark
Water of Mark
Glen Lee
Muckle Cairn
Ben Tirran
White Hill
Mount Keen
Loch Lee
Invermark Lodge
Inchgrundle
Glen Effock
Balmoral Castle
Crathie
Abergeldie Castle
Easter Balmoral
Creag nan Gall
Glen Gelder
Bush Crathie
Gairnshiel Lodge
Geallaig Hill
Coilacriech
River Dee
Bridge of Gairn
BALLATER
Pannanich Hill
House of Glenmuick
Birkhall
The Coyles of Muick
Glen Gairn
River Gairn
Brown Cow Hill
Carn Leac Saighaeir
Morven
Peter's Hill
Culblean Hill
Glendavan
Logie Coldstone
Migvie
Coynach
Milton of Tullich
Glascorrie
Muir of Dinnet
Black Craig
Cairn Leuchan
Clachan Yell
Lary
Culsh
Tullich
Glen Tanar
Water of Tanar
Cairn Mona Gowan
Fearder Burn
Inver
Meall Gorm
Carn na Drochaide
Bynack More
Beinn Bhrotain
Beinn a' Ghlo
Glen Loch
Loch Loch
Loch Valigan
Ben Vuirich
Meall a' Choire Bhuidhe
Ben Earb
Ben Gulabin
Dalmunzie Hotel
Spittal of Glenshee
Glen Shee
Carn an Daimh
Monamenach
Auchavan
Loch Beanie
Glen Isla
Glen Finlet
Glen Damff
Glen Prosen
Mealna Letter
Dalnaglar Castle
Cray
Mount Blair
Forter
Folda
Brewlands Bridge
Meall Uaine
Lamh Dearg
Ashintully Castle
Kirkmichael
Enochdhu
Balvarran
Glen Brerachan
Straloch
Tarvie
Ben Vrackie
Creag Dhubh
Glenn Fearnach
Creag an t-Sithein
Blacklunans
Kirkton of Glenisla
Glenhead Fm
Backwater Reservoir
Runtaleave
Eskielown
Cairn Baddoch
Cairn of Barns
Cairn Inks
Glenprosen Village
Easter Lednathie
Hill of Couternach
Cat Law
Rottal
Wheen
Finbracks
Hill of Glansie
Auld Darkney
Glenmoy
Redheugh
Glenquiech
Horniehaugh
Newmill of Inshewan
Dykehead
Cortachy
Castle
Burnside
Memus
Pearsie
Balintore
Inverquharity
Shielhill
Balloch
Kinnordy
Northmuir
KIRRIEMUIR
Westmuir
Mains of Ballindarg
Gateside
Leys of Cossans
Glamis Castle
Glamis
Dean Water
Douglastown
Padanaram
Drumgley
Mosside
Thornton
Charleston
Kirkton
Foffart
Invararity
Gateside
Milton
Nether Handwick
Gallowfauld
Gallow Hill
Wester Denoon
Ark Hill
Eassie
Balkeerie
Kirkinch
Kinpurney Hill
Newtyle
Newbigging
Keillor
SIDLAW HILLS
STRATHMORE
Kirkton of Auchterhouse
Auchterhouse
Coldstream
Balgray
Kirkton of Tealing
Newbigging
Kirkton of Strathmartine
Muirhead
Birkhill
Fowlis Easter
Camperdown Ho
Downfield
Pitcur
Lundie
Kettins
Hallyburton Ho
Leys
Campmuir
Burrelton
Woodside
Cargill
Whitefield
Wolfhill
Saucher
King's Seat
Littleton
Redmyre
Dunsinnan
Kinclaven
Meikleour
Markethill
COUPAR ANGUS
Arthurstone
Ardler
Meigle
Belmont Castle
Kinloch
Leitfie
Balhary
Ruthven
Ruthven House
Craigton
Kirkton of Airlie
Airlie Castle
ALYTH
Hill of Alyth
New Alyth
Bamff
Balduff Hill
Bridge of Craigisla
Bridgend of Lintrathen
Kirkton of Kingoldrum
Loch of Lintrathen
Mile Hill
Dykend
Bellaty
Creigh Hill
Cairn Gibbs
FOREST OF ALYTH
Drumderg
Tullymurdoch
Netherton
Bridge of Cally
Hotel
Blackcraig Castle
Blackcraig Hill
Blackcraig Forest
Forest of Clunie
Ballintuim
Knock of Balmyle
STRATHARDLE
Creag nam Mial
Cochrage Muir
Middleton
Lornty Burn
Craighall
Tullyfergus
BLAIRGOWRIE
RATTRAY
Kinloch
Muirton of Ardblair
Rosemount
Lethendy
Kirkton of Lethendy
Craigie
Clunie
Loch of Clunie
Concraigie
Forneth
Butterstone
Riechip
Benachally
Loch Benachally
Loch Ordie
Riemore Lodge
Deuchary Hill
Cardney Ho
Snaigow Ho
Loch of Lowes
Dunkeld
Little Dunkeld
Newtyle Hill
Birnam
Birnam Hill
Inver
Spittalfield
Caputh
Delvine
Gellyburn
Murthly
Ardoch
Mur of Thorn
Waterloo
Bankfoot
Airntully
Gauls
Stanley
West Tofts
Obney Hills
Obney
Gibbieston
Creag Liath
Glenshee
Trochry
Ballinlick
STRATHBRAAN
River Tay
Stare Dam
CRAIGVINEAN FOREST
Bishopric
Dalguise
Guay
Dowally
Kindallachan
Kincraigie
Balmacneil
Logierait
Ballinluig
Tulliemet
Dalcapon
Meall Reamhar
Loch Broom
PITLOCHRY
Milton of Edradour
Moulin
Loch Tummel
River Isla
Lunan Burn
Long Loch
Glen Fenzie
Deskry Water
Glen Tanar
Hill of Saughs
Wester Glansie

NORTH SEA
ABOYNE
BANCHORY
STONEHAVEN
INVERBERVIE
LAURENCEKIRK
BRECHIN
MONTROSE
FORFAR
ARBROATH
CARNOUSTIE
MONIFIETH
Tarland
Lumphanan
Torphins
Echt
Cults
Peterculter
Milltimber
Bieldside
Nigg
Cove Bay
Findon Ness
Portlethen
Portlethen Village
Downies
Newtonhill
Muchalls
Doonie Point
Garron Point
Castle Haven
Dunnottar Castle
Thornyhive Bay
Crawton Bay
Braidon Bay
Todhead Point
Catterline
Kinneff
Little John's Haven
Bervie Bay
Gourdon
Doolie Ness
Johnshaven
Milton Ness
St Cyrus
Lunan Bay
Boddin Point
Scurdie Ness
Ferryden
Long Craig
Lang Craig
Red Head
Prail Castle
Meg's Craig
The Deil's Heid
Auchmithie
Inverkeilor
Letham
Friockheim
Edzell
Fettercairn
Fordoun
Drumlithie
Marykirk
Hillside
Kirkton of Craig
Fasque
Auchenblae
Glen Dye
Glen Esk
Mount Battock
Hill of Fare
Midmar Forest
Durris Forest
Fetteresso Forest
Blackhall Forest
Forest of Birse
Drumtochty Forest
Montreathmont Forest
HOWE OF THE MEARNS
Montrose Basin
River Dee
Water of Feugh
River North Esk
River South Esk
Lunan Water
Bervie Water
Cairn-mon-earn
Clachnaben
Hill of Wirren
West Knock
Mount Battock
Kerloch
Strathfinella Hill
Hill of Trusta
Tornaveen
Drumoak
Crathes
Crathes Castle
Kirkton of Durris
Kirkton of Maryculter
Netherley
Rickarton
Tipperweir
Glenfarquhar Lodge
Arbuthnott
Allardice
Benholm
Lauriston Castle
Logie Pert
Dun
Bridge of Dun
Kinnaird Castle
Farnell
Rossie Moor
Dunninald Castle
Usan
Ethie Castle
Colliston
Carmyllie
Kellie Castle
East Haven
Monikie
Kirkton of Monikie
Barry
Aldbar Castle
Finavon
Tannadice
Careston
Balnamoon
Inchbare
Tigerton
Tulloes
Guthrie
Kinnell
Redford
Arbirlot
G
H
J
K
L
1
2
3
4
5
6
7
8
99
87

A
B
C
D
E
1
2
3
4
5
6
7
8
100
111
89
90
NG
NM
ISLAND OF RAASAY
INNER
SOUND
SOUND OF SLEAT
SOUND OF ARISAIG
SCALPAY
CROWLIN ISLANDS
Longay
Pabay
Guillamon Island
Applecross
Camusteel
Camusterrach
Culduie
Toscaig
Ard-dhubh
Kishorn Island
Loch Kishorn
Lochcarron
Strathcarron
Achintee
Kirkton
Attadale
Stromeferry
Plockton
Achmore
Ardaneaskan
Stromemore
Castle
Duirinish
Drumbuie
Port-an-eorna
Erbusaig
Badicaul
Kyle of Lochalsh
Balmacara
Balmacara Square
Auchtertyre
Nostie
Ardelve
Dornie
Keppoch
Bundalloch
Killilan
LOCH ALSH
Kyleakin
Skye Broadford
Broadford
Harrapool
Breakish
Skulamus
Drochaid Lusa
Sgurr na Coinnich
Glen Arroch
Kylerhea
Bernera
Galltair
Glenelg
Eilanreach
Dùn Grugaig
Ratagan
Ratagan Forest
Shiel Bridge
Morvich
Inverinate
Dorusduain
LOCH DUICH
FIVE SISTERS
Kintail Forest
BEINN FHADA
BEN ATTOW
Glenshiel Forest
The Saddle
Kinlochhourn Forest
Sandaig Islands
Rubha Buidhe
BEINN SGRITHEALL
Arnisdale
Corran
Glen Arnisdale
LOCH HOURN
Barrisdale Bay
Barrisdale Forest
Glen Barrisdale
Kinloch Hourn
Glen Quoich
LOCH QUOICH
Sleat
Kinloch
Isleornsay or Eilean Iarmain
Ornsay
Duisdalemore
Drumfearn
Eilean Heast
Heast
Loch Eishort
Ord
Tokavaig
Tarskavaig
Tarskavaig Point
Achnacloich
Teangue
Saasaig
Ferrindonald
Kilmore
Knock Bay
Kilbeg
Armadale
Armadale Castle
Armadale Bay
Ardvasar
Calligarry
Aird of Sleat
Point of Sleat
Ard Thurinish
Rubha Charn nan Cearc
Geur Rubha
Inverie
Inverie Bay
Scottas
Airor
Sandaig
KNOYDART
Glas Eilean
Rubha Raonuill
LOCH NEVIS
Mallaig
Courteachan
Morar
Beoraidbeg
Sgeir Mhòr
Bracora
Bracorina
Stoul
Tarbet
Swordland
Brinacory
NORTH MORAR
LOCH MORAR
SOUTH MORAR
Glenancross
Traigh Ho
Portnaluchaig
Bunacaimb
Back of Keppoch
Arisaig
ARISAIG
Eilean Ighe
Luinga Bheag
Luinga Mhòr
Rubh' Arisaig
Druimindarroch
Eilean a' Ghaill
Eilean an t-Snidhe
Loch nan Uamh
Ardnish
Lochailort
Inveralort
Peanmeanach
Glen Beasdale
Glenfinnan
Kinlocheil
Locheilside Sta.
Fassfern
Drumsallie
Duisky
LOCH EIL
ROIS-BHEINN
Roshven
Glenuig
Smirisary
Samalaman Island
Eilean Shona
Moidart
Kinlochmoidart
Glen Dessarry
Strathan
Glen Pean
Camusrory
LOCH SHIEL
Glen Garvan
Cona Glen

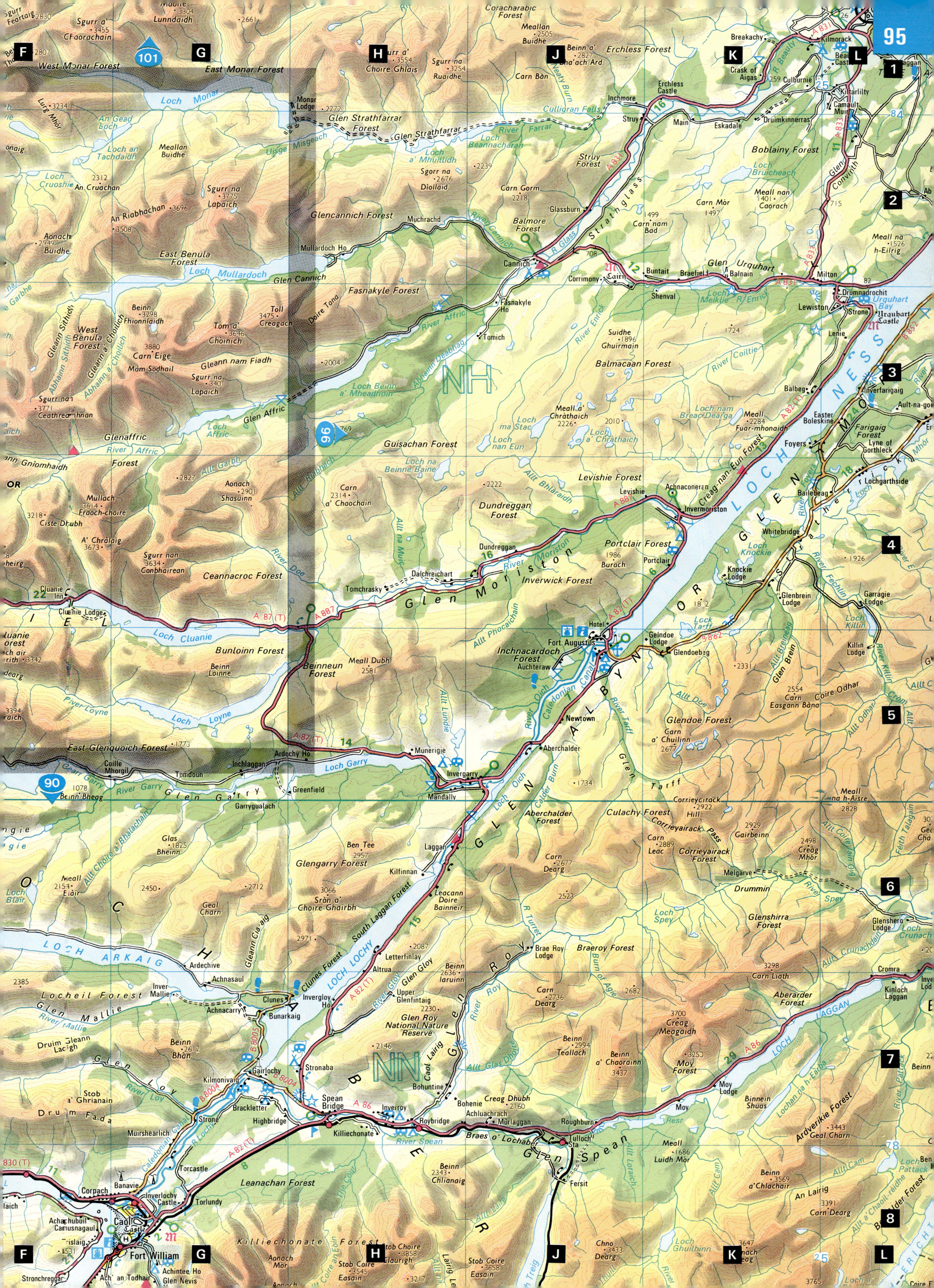
F
G
H
J
K
L
1
2
3
4
5
6
7
8
101
96
90
West Monar Forest
East Monar Forest
Loch Monar
Monar Lodge
Glen Strathfarrar
Glen Strathfarrar Forest
River Farrar
Uisge Misgeach
Loch Beannacharan
Loch a' Mhuillidh
Sgorr na Diollaid
Struy Forest
Struy
Culligran Falls
Inchmore
Erchless Forest
Erchless Castle
Breakachy
Crask of Aigas
Culburnie
Kilmorack
Beauly
Kiltarlity
Camault Muir
Main
Eskadale
Druimkinnerras
Boblainy Forest
Loch Bruicheach
Glen Convinth
Meall nan Caorach
Meall na h-Eilrig
Strathglass
Glassburn
Carn Gorm
Balmore Forest
Carn Mòr
Carn nam Bad
Cannich
Comrie
Corrimony
Cairn
Buntait
Braefield
Balnain
Glen Urquhart
Loch Meikle
R. Enrick
Shenval
Milton
Drumnadrochit
Lewiston
Urquhart Bay
Urquhart Castle
Strone
Lenie
Carn na Coire
Glencannich Forest
Muchrachd
Mullardoch Ho
Loch Mullardoch
Glen Cannich
East Benula Forest
An Riabhachan
An Cruachan
Sgurr na Lapaich
Meallan Buidhe
Loch Cruoshie
Loch an Tachdaidh
An Gead Loch
Lurg Mhòr
Aonach Buidhe
Fasnakyle Forest
Fasnakyle Ho
Tomich
River Affric
Abhainn Deabhag
Toll Creagach
Tom a' Choinich
Beinn Fhionnlaidh
West Benula Forest
Carn Eige
Màm Sodhail
Gleann nam Fiadh
Sgurr na Lapaich
Sgurr nan Ceathreamhnan
Glenaffric
Gleann Sìthidh
Abhainn a' Choilich
Loch Beinn a' Mheadhoin
Loch Affric
Glen Affric
River Affric
NH
Suidhe Ghuirmain
Balmacaan Forest
River Coiltie
Meall a' Chrathaich
Loch ma Stac
Loch nan Eun
Loch a' Chrathaich
Loch nam Breac Dearga
Meall Fuar-mhonaidh
Balbeg
Inverfarigaig
Easter Boleskine
Foyers
Farigaig Forest
Lyne of Gorthleck
Lochgarthside
Loch Ness
Guisachan Forest
Loch na Beinne Bàine
Allt Garbh
Aonach Shasuinn
Carn a' Chaochain
Levishie Forest
Levishie
Achnaconeran
Invermoriston
Creag nan Eun Forest
Dundreggan Forest
Allt Bhlàraidh
Dundreggan
Portclair Forest
Burach
Portclair
Whitebridge
Loch Knockie
Knockie Lodge
Glenbrein Lodge
River Fechlin
Garragie Lodge
Ciste Dhubh
Mullach Fraoch-choire
A' Chràlaig
Sgurr nan Conbhairean
Ceannacroc Forest
Cluanie Inn
Cluanie Lodge
Loch Cluanie
River Doe
Dalchreichart
Tomchrasky
Glen Moriston
River Moriston
Inverwick Forest
Allt na Muic
Allt Phocaichain
Inchnacardoch Forest
Auchteraw
Fort Augustus
Glendoe Lodge
Glendoebeg
Loch Tarff
Loch Killin
Killin Lodge
Glen Brein
Carn Easgann Bàna
Coire Odhar
Bunloinn Forest
Beinn Loinne
Beinneun Forest
Meall Dubh
River Loyne
Loch Loyne
East Glenquoich Forest
Allt Lùndie
Newtown
Aberchalder
Glendoe Forest
Carn a' Chuilinn
Glen Tarff
River Tarff
Caledonian Canal
River Oich
Loch Oich
Ardochy Ho
Coille Mhorgil
Tomdoun
Inchlaggan
Loch Garry
Munerigie
Invergarry
Mandally
Greenfield
Garrygualach
River Garry
Glen Garry
Beinn Bheag
Calder Burn
Aberchalder Forest
Culachy Forest
Corrieyairack Hill
Corrieyairack Pass
Corrieyairack Forest
Carn Leac
Gairbeinn
Creag Mhòr
Meall na h-Aisre
Melgarve
Drummin
River Spey
Glenshero Lodge
Glenshirra Forest
Loch Spey
Glas Bheinn
Allt Choire a' Bhalachain
Ben Tee
Glengarry Forest
Kilfinnan
Laggan
Glen Albyn
Sròn a' Choire Ghairbh
Meall an t-Eidir
Geal Charn
South Laggan Forest
Leacann Doire Bainneir
Carn Dearg
River Turret
Brae Roy Lodge
Braeroy Forest
Burn of Agie
Glen Roy
Carn Liath
Aberarder Forest
Cromra
Kinloch Laggan
Loch Laggan
Loch Arkaig
Ardechive
Achnasaul
Glen Cia-aig
Clunes Forest
Clunes
Loch Lochy
Letterfinlay
Altrua
Glen Gloy
Beinn Iaruinn
Upper Glenfintaig
Glen Roy National Nature Reserve
River Roy
Creag Meagaidh
Moy Forest
Loch Moy Lodge
Binnein Shuas
Beinn Teallach
Beinn a' Chaorainn
Inver Mallie
Achnacarry
Bunarkaig
Invergloy Ho
Locheil Forest
Glen Mallie
River Mallie
Druim Gleann Laoigh
Gleann Loy
Beinn Bhàn
Stob a' Ghrianain
River Loy
Druim Fada
Kilmonivaig
Gairlochy
Stronaba
NN
Caol Lairig
Bohuntine
Allt Glas Dhoire
Bohenie
Creag Dhubh
Achluachrach
Murlaggan
Roughburn
Moy
Spean Bridge
Inveroy
Roybridge
River Spean
Braes o' Lochaber
Tulloch Sta
Glen Spean
Allt Laire
Ardverikie Forest
Geal Charn
Lochan na h-Earba
Meall Luidh Mòr
Beinn a' Chlachair
Loch Pattack
Ben Alder Forest
Brackletter
Strone
Highbridge
Killiechonate
Muirshearlich
Torcastle
Leanachan Forest
Beinn Chlianaig
Fersit
Caledonian Canal
Corpach
Banavie
Inverlochy Castle
Torlundy
Achaphubuil
Camusnagaul
Trislaig
Caol
Fort William
Achintee Ho
Glen Nevis
Ach' an Todhair
Stronchreggan
Killiechonate Forest
Aonach Mòr
Stob Choire Claurigh
Stob Coire Easain
Stob Coire Easain
Lairig Leacach
Chno Dearg
Loch Ghuilbinn
Allt Coire an Eoin
An Lairig
Carn Dearg
Loch Ericht
A 831
A 833
A 82 (T)
A 887
A 86
A 830 (T)
B 862
B 852
B 8004
B 8005

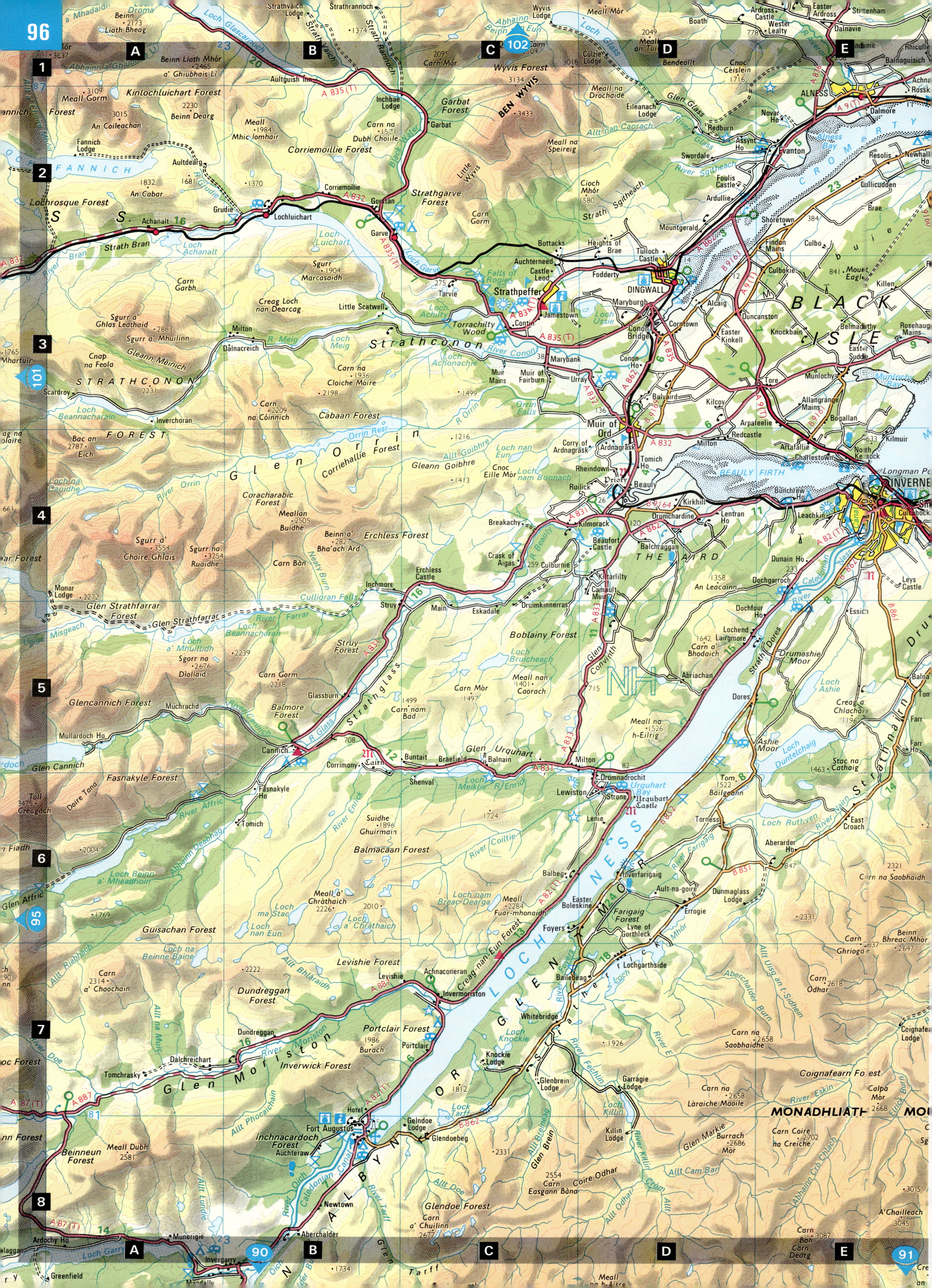
A
B
C
D
E
1
2
3
4
5
6
7
8
102
101
95
90
91
Wyvis Forest
BEN WYVIS
ALNESS
Evanton
Dalmore
CROMARTY
Kinlochluichart Forest
Meall Gorm
An Coileachan
Beinn Dearg
Corriemoillie Forest
Fannich Lodge
FANNICH
Lochrosque Forest
Achanalt
Lochluichart
Garve
Strathgarve Forest
Garbat Forest
Little Wyvis
Strath Bran
Sgurr Marcasaidh
Strathpeffer
Falls of Rogie
Auchterneed
Castle Leod
Fodderty
DINGWALL
Tulloch Castle
Maryburgh
Conon Bridge
Muir of Ord
BLACK ISLE
Culbokie
Munlochy
Tore
Kilcoy
Redcastle
Arpafeelie
Kilmuir
North Kessock
INVERNESS
Culloden
Leys Castle
BEAULY FIRTH
Beauly
Kirkhill
Bunchrew
Lentran
Tomich
Rheindown
Kilmorack
Beaufort Castle
Kiltarlity
THE AIRD
Drumchardine
Dochgarroch
Dunain Ho
Strathconon
STRATHCONON
Milton
Dalnacreich
Scardroy
Invervhoran
Cabaan Forest
FOREST
Glen Orrin
Corriehallie Forest
Gleann Goibhre
Erchless Forest
Erchless Castle
Corachabric Forest
Struy
Inchmore
Eskadale
Culburnie
Glen Strathfarrar
Glen Strathfarrar Forest
Monar Lodge
Boblainy Forest
Glencannich Forest
Muchrachd
Mullardoch Ho
Glen Cannich
Cannich
Strathglass
Corrimony
Glen Urquhart
Balnain
Milton
Drumnadrochit
Urquhart Castle
Lewiston
Strone
Fasnakyle Forest
Fasnakyle Ho
Tomich
Balmacaan Forest
Guisachan Forest
Levishie Forest
Dundreggan Forest
Invermoriston
Portclair Forest
Glen Moriston
Inverwick Forest
Tomchrasky
Dalchreichart
Fort Augustus
Inchnacardoch Forest
Caledonian Canal
Glendoe Forest
Newtown
Aberchalder
Invergarry
Munerigie
LOCH NESS
Foyers
Inverfarigaig
Farigaig Forest
Errogie
Dunmaglass Lodge
Whitebridge
Knockie Lodge
Glenbrein Lodge
Garrogie Lodge
Killin Lodge
Stratherrick
Coignafearn Forest
MONADHLIATH
Dores
Abriachan
Ashie Moor
Loch Duntelchaig
Torness
Aberarder Ho
Loch Ruthven
East Croach
Strathnairn
Greenfield
Glen Affric
Loch Beinn a' Mheadhoin
NH

MORAY FIRTH
BURGHEAD BAY
BURGHEAD
Hopeman
Duffus
Cummingstown
Roseisle
Findhorn
Findhorn Bay
Kinloss
Culbin Forest
FORRES
NAIRN
Auldearn
Cawdor
Croy
Petty
Balloch
Culloden Muir
Ardersier
Fort George
CROMARTY
INVERGORDON
Rosemarkie
FORTROSE
Whiteness Head
Darnaway Forest
Altyre Woods
Dallas
Knockando
Ferness
Dulsie
River Findhorn
Lochindorb
Carn Kitty
Advie
Bridge of Avon
Cromdale
HILLS OF CROMDALE
GRANTOWN-ON-SPEY
Dulnain Bridge
Carrbridge
Boat of Garten
Abernethy Forest
Nethy Bridge
Aviemore
Coylumbridge
Rothiemurchus
Tomatin
Slochd
Tomintoul
Glenlivet
Braes of Abernethy
The Queen's Forest
Glenmore Lodge
Loch Morlich
CAIRN GORM
CAIRNGORM
KINGUSSIE
Feshiebridge
Kincraig
Inshriach Forest
GLEN AVON
BEN AVON
BRAERIACH
Strathspey
Glenferness Ho
Ardclach
Redburn
Beachans
Dava
Mains of Dalvey
Craggan
Kinveachy
Tulloch
Forest Lodge
Glen Kyllachy
Moy
Ruthven
Balvraid
Inverbrough
Findhorn Bridge
Dalmigavie Lodge
Corrievorrie
Carn nan Tri-tighearnan
Beinn Bhuidhe Mhòr
Saddle Hill

A
B
C
D
E
F
1
2
3
4
5
6
7
8
Covesea Skerries
Halliman Skerries
LOSSIEMOUTH
Hopeman
Duffus
Gordonstoun
Boar's Head Rock
SPEY BAY
PORTKNOCKIE
FINDOCHTY
Cullen Bay
Logie Head
CULLEN
Redhythe Point
PORTSOY
Sandend Bay
Sandend
Boyne Bay
Knock Head
Whitehills
BUCKIE
Rathven
Spey Bay
Portgordon
Kingston
Garmouth
Lochhill
Leuchars Ho
Loch Spynie
Palace
Quarrywood
ELGIN
Pittendreich
Miltonduff
Moss of Barmuckity
Urquhart
Lhanbryde
Muir of Lochs
Mosstodloch
Bogmoor
Nether Dallachy
Upper Dallachy
Auchenhalrig
Slackhead
Cairnfield Ho
Drybridge
Bin of Cullen
Lintmill
Milton
Kirktown of Deskford
Fordyce
Durn Hill
Eastern Whyntie
Boyndie
Mid Culbeuchly
Paddockhaugh
Thornhill
Longmorn
Blackhills
Whitewreath
Teindland
Fochabers
Dipple
Urdiquish
Inchberry
Wood of Ordiequish
Whiteash Hill Wood
Broadley
Clochan
Shiel Muir
Craibstone
Berryhillock
Toux
Cornhill
Oldtown of Ord
Auchmore
Aultmore
Deerhill
Millstone Hill
Lurg Hill
Gordonstown
Kellas
Glenlatterach Rest
Pikey Hill
Glen of Rothes
Newlands
Mulben
Hill of Mulderie
Forgie
Broadrashes
Newmill
Grange Crossroads
Cairnie
Davoch of Grange
Edingight Ho
Knock Hill
Sillyearn
Bracobrae
Knock
Drumnagorrach
Farmtown
Finnygaud
ABERCHIRDER
Marnoch
Cunie
Cairn Uish
ROTHES
Carn na Cailliche
Burn of Rothes
Cairn Cattoch
Elchies Forest
Ben Aigan
Arndilly Ho
Dandaleith
Hill of Towie
Towiemore
KEITH
The Balloch
Glen of Coachford
Little Pitlurg
Milltown of Rothiemay
Yonder Bognie
Fourman Hill
Bogniebrae
Inverkeithny
Auchininna
Archiestown
Whiteacen
Ringorm
Craigellachie
Maggieknockater
Kininvie Ho
Drummuir Castle
Upper Knockando
Cardow
Knockando Ho
Knockando
Carron
CHARLESTOWN OF ABERLOUR
Midtown of Buchromb
Balvenie Castle
Daugh of Kinermony
Milltown of Edinvillie
DUFFTOWN
Milltown of Auchindoun
Daugh of Invermarkie
Daugh of Cairnborrow
Cairnborrow
The Bin Forest
HUNTLY
Castle
Lessendrum
Drumblade
Corse
Drumblair
Aucharnie
Balgave
Daugh of Carron
Marypark
Belleheiglash
Bridge of Avon
BEN RINNES
Beatshach
Meikle Conval
Favillar
Auchindoun Castle
The Scalp
Haugh of Glass
River Deveron
Torry
Glen of Artloch Croft
Clashmach Hill
STRATHBOGIE
Bailiesward
Bridgend
Sliach
Bridewells
Denend
Thomastown
Ythanwells
Cairnacay
Glen Rinnes
Glenfiddich Lodge
Glen Fiddich
Glenfiddich Forest
Succoth
Tillathrowie
Tomnaven
Coynachie
Culdrain
Kirkstile
Shanquhar
Backburn
Gartly Moor
Hill of Foudland
Glens of Foudland
Kirkton of Culsalmond
Gartly
Cults
Knockandy Hill
Largie
Picardy Stone
Colp
Wrangham
Drumin
Glenlivet
Shenval
Corryhabbie Hill
Bridgend
Inverharroch
Black Water
Tap o' North
Milton of Noth
Milton of Lesmore
Leith Hall
Kennethmont
Ardlair
Aulton
Insch
Ardoyne
Carn Liath
Cairn Muldonich
Carn an t-Suidhe
Blackwater Forest
Round Hill
Aldunie
Mount of Haddoch
Cabrach
Wheedlemont
Rhynie
Cottown
Clatt
Duncanstone
Leslie
Kirkton
Carn Daimh
Tomnavoulin
Knockandhu
Cairnbrallan
Hill of Three Stones
The Buck
Craig Castle
Church
Clova
Knockespock Ho
Suie Hill
Coldwells Croft
CORREEN HILLS
Mucklestown
Auchleven
Licklyhead Castle
Bennachie
Clashnoir
Braes of Glenlivet
Chapeltown
River Livet
Lumsden
Lord Arthur's Hill
Tullynessle
Keig
Castle Forbes
Gl'nton
Milton
Carn Mor
LADDER HILLS
Glenbuchat Lodge
Badenyon
Creag an Eunan
Moss Hill
Ladylea Hill
Belnacraig
Rinmore
Kildrummy
Mossat
Bridge of Alford
Montgarrie
HOWE OF ALFORD
Alford
Auchintoul
Rorandle
Cairn William
Pitfichie Forest
Pitmunie
Blairnamarrow
Carn Liath
The Socach
Kirkton of Glenbuchat
Castle
Breagach Hill
Bellabeg
Strathdon
Glenkindie
Towie
Sinnahard
Milltown of Towie
Asloun
Muir of Fowlis
Little Lynturk
Whitehouse
Kirkton of Tough
Tillyfour
Black Hill
Ordhead
Lecht Road
Carn Ealasaid
Craig Veann
Geal Charn
Inverernan Ho
Candacraig Ho
Heugh-head
Boultenstone
Frosty Hill
Milton of Cushnie
Leochel Cushnie
Corrennie Moor
Corrennie Forest
Cock Bridge
Delnadamph Lodge
Corgarff
Colnabaichin
Tornahaish
Craig of Bunzeach
Hillockhead
Pressendye
Craigievar Castle
Kintocher
Benaquhallie
Bankhead
Milton of Corsindae
The House of Corse
East Davoch
Migvie
Coynach
Douneside
Craskins
Craigfleich
Tornaveen
Comers
Findrack Ho
Drumlassie
Carn Leac Saighdeir
Cairn Mona Gowan
Brown Cow Hill
Morven
Logie Coldstone
Tarland
Culsh
Souterrain
Lumphanan
Stone Circle
Peel
Coull
Milton of Auchlossan
Mortlich
Auchlossan
Torphins
Milton of Campfield
Mid Beltie
HILL
Loch Builg
River Gairn
Gairnshiel Lodge
Peter's Hill
Culblean Hill
Scar Hill
Glendavan
Ordie
Loch Davan
CROMAR
MOUNTAINS
NJ
River Spey
River Isla
River Avon
River Don
River Bogie
Kirkney Water
Water of Buchat
Water of Nochty
Deskry Water
Ernan Water
Conglass Water
A 96
A 95
A 941
A 920
A 97
A 944
A 980
A 939
A 98
B 9040
B 9015
B 9013
B 9014
B 9103
B 9102
B 9009
B 9008
B 9115
B 9117
B 9022
B 9018
B 9016
B 9002
B 9119
B 9025
B 9023
B 9001
B 9094
B 992
B 973
97
92
81
32

G
H
J
K
L
1
2
3
4
5
6
7
8
NORTH SEA
B U C H A N
F O R M A R T I N E
Troup Head
Pennan Head
Quarry Head
ROSEHEARTY
Sandhaven
KINNAIRD HEAD
FRASERBURGH
Fraserburgh Bay
Cairnbulg Point
Inverallochy
Cairnbulg Castle
St Combs
Inzie Head
Head of Garness
Gamrie Bay
MACDUFF
Crovie
Gardenstown
Pennan
Silverford
Greenskares
Dubford
Cushnie
Longmanhill
New Aberdour
Windyheads Hill
Woodhead
Boyndlie
Peathill
Percyhorner
Mid Ardlaw
Pitblae
Memsie
Cairn
Rathen
Cairness
Crimonmogate
Loch of Strathbeg
Crimond
RATTRAY HEAD
Blackhill
Gas Terminal
St Fergus
Scotstown Head
Kirkton Head
Kirkton
Rora
Newseat
St Fergus Moss
Rora Moss
Backfolds
Hythie
Denhead
Leys
New Leeds
Strichen
Mormond Hill
Adziel
North Ugie Water
Fetterangus
Forest of Deer
Toux
Mintlaw
Old Deer
Abbey
South Ugie Water
Stuartfield
Longside
Inverquhomery
Flushing
Torterston
Inverugie
PETERHEAD
Peterhead Bay
Sandford Bay
Buchan Ness
Boddam
Sandfordhill
Longhaven
Hilhead of Cocklaw
Little Dens
Nether Kinmundy
Millbreck
Clola
Kinknockie
Smallburn
Moss of Cruden
Stoneygate
Teucan
Coldwells
North Haven
Hatton
Twa Havens
Cruden Bay
Bay of Cruden
Chapel Hill
Whinnyfold
Auchleuchries
Bogbrae
Toll of Birness
Birness
Leask
Artrochie
Auchmacoy
Collieston
St Catherine's Dub
Hackley Head or Forvie Ness
Sands of Forvie
Newburgh
Newburgh Bar
Foveran
Drums
Delfrigs
Eventide Home
Balmedie
Belhelvie
Potterton
Blackdog
Blackdog Rock
Denmore
Bridge of Don
ABERDEEN
Old Aberdeen
Girdle Ness
Nigg Bay
Greg Ness
Kincorth
Nigg
Souter Head
Cove Bay
Charlestown
Cults
Bieldside
Milltimber
Peterculter
Craigton
Hardgate
Drum Castle
West Cullerley
Blacktop
Easter Ord
Westhill
Kingswells
Elrick
Kirkton of Skene
Mastrick
Mannofield
Northfield
Bucksburn
Bankhead
Craibstone
Stoneywood
DYCE
Aberdeen (Dyce) Airport
Parkhill
Corby Loch
Mundurno
Kinmundy
Drumligair
Whitecairns
Causeyend
Craigie
Middlemuir
Beauty Hill
Straloch
Newmachar
Cairnpark
Hatton of Fintray
Cothall
Wester Fintray
Balbithan
Kinmuck
Middleton
Tyrebagger Hill
Overton
Blackburn
Clinterty
Brimmond Hill
East Auchronie
Skene
Loch of Skene
Garlogie
Echt
Barmekin Hill
South Kirkton
Redhill
Landerberry
Midmar Forest
OF FARE
Dunecht
Old Kinnernie
Corsindae
Sauchen
Cluny Castle
Castle Fraser
Lauchintilly
Lyne of Skene
Achath
Leylodge
Craigearn
Kemnay
KINTORE
Cottown
Clovenstone
Grantlodge
Blairdaff
Bograxie
Gallow Hill
Burnhervie
Aquhythie
Kinkell Church
Keith Hall
Port Elphinstone
INVERURIE
Chapel of Garioch
Stone Circle
Manar
Balhalgardy
Harlaw
Pitcaple
Inveramsay
Maiden Stone
Pittodrie
Bennachie Forest
Westhall
Whiteford
Durno
Old Rayne
Kirkton of Rayne
Tocher
Newseat
Rothmaise
Meikle Wartle
Daviot
Stone Circle
Mounie Castle
Fingask
Barra Castle
Hill of Barra
OLDMELDRUM
Meldrum Ho
Cairnbrogie
Mill of Kingoodie
Kirkton of Bourtie
Hattoncrook
Nether Crimond
Hillbrae
Whiterashes
Afflleck
Tillykerrie
Tillygreig
Udny Station
Udny Green
Pettymuick
Culterculllen
Minnes
Tillycorthie
Ardo Ho
Pitmedden
Pitmedden Ho
Tolquhon Castle
Auquhorthies
Craigdam
Tarves
Ythsie
Kinharrachie
Esslemont
ELLON
Kirkton of Logie-Buchan
Meikle Tarty
Tipperty
Tarty Burn
Foveran Burn
Wedderlaie
Inverebrie
Ythanbank
Hilton
Broomfield
Earlsford
Balgove
Barthol Chapel
Haddo Ho
Methlick
Collynie
Crofts of Haddo
Ardo
Skelmonae
Quilquox
Drumwhindle
Inkhorn
Milton Coldwells
Hill of Dudwick
Arthrath
Muirtack
Backhill
Fortrie
Annochie
Skelmuir
Kinnadie
Crichie
Auchnagatt
Barrack
Knaven
Nethermuir
Dumuir
Backhill of Clackriach
Maud
Culsh
New Deer
Fedderate
Ironside
Bonnykelly
Oldwhat
Hill of Corsegight
Balthangie
Garmond
New Byth
New Pitsligo
Glasslaw
Craigmaud
Ladysford
Hillhead of Auchentumb
Hill of Fishrie
Netherbrae
Minnonie
Keilhill
Gorrachie
Milltown of Craigston
Craigston Castle
Fintry
Muiryfold
Delgatie Castle
TURRIFF
Kirkton
Colp
Darra
Hatton Castle
Birkenhills
Crofts of Inverthernie
Kirkton of Auchterless
Steinmanhill
Gourdas
Tifty
Monkshill
Lethenty
Cottown
Millbrex
Deer's Hill
Murtack
Howe of Teuchar
Cuminestown
Slacks of Cairnbanno
Cairnbanno Ho
Brownhill
Cairnorrie
Hill of Skilmafilly
Woodhead
Fyvie
Petty
Rothienorman
Crombie
St Katherines
Folla Rule
Core Hill
Cross of Jackston
Jackstown
Rothiebrisbane
Gordonstown
Badenscoth
Backhill
Little Water
Ythan
River Ythan
Duff Ho
Montcoffer Ho
Mundren
Forglen Ho
ABERDEEN to
Stromness ... 8 hrs
Lerwick ... 14 hrs
A 90
A 92
A 96(T)
A 947
A 948
A 950
A 952(T)
A 975
A 981
A 98(T)
A 920
A 944
A 93
B 9000
B 9005
B 9031
B 9032
B 9033
B 9093
B 9170
B 9029
B 9030
B 9105
B 9119
B 9125
B 9001
B 9002
B 977
B 979
B 993
B 999
41
87
81
93

ULLAPOOL to Stornoway 3½ hrs
NB
NG
Eilean Mullagrach
Isle Ristol
Glas-leac Mòr
Tanera Beg
Summer Isles
Tanera Mòr
Glas-leac Beag
Eilean Dubh
Horse Island
Priest Island
Bottle Island
Càrn nan Sgeir
Greenstone Point
Cailleach Head
Leac Dhonn
Rubha Beag
Rubha Mòr
Opinan
Mellon Udrigle
Stattic Point
Scoraig
Carnach
Gob a' Chuaille
Slaggan Bay
Gruinard Island
GRUINARD BAY
Badluarach
Mungasdale
Gruinard House
Inchina
Dundonnell
Badcaul
Sròn a' Gheodha Dhubh
Eilean Furadh Mòr
Rubha Reigh
Rubha nan Sasan
Cove
Achgarve
Beinn Dearg Mhòr
Mellon Charles
Ormiscaig
Bualnaluib
Tighnafiline
Aultbea
Laide
Sand
Coast
Little Gruinard
Sròn na Cléite
An Cuaidh
Loch an Draing
Isle of Ewe
Drumchork
Beinn Dearg Bad Chailleach
Melvaig
Aultgrishan
LOCH EWE
Midtown
Inverasdale
Brae
Tournaig
Seana Chamas
Peterburn
Cnoc Breac
Naast
Rubha' Ard na Bà
Loch Fada
Aird Dhubh
Beinn a' Chàisgein Beag
Meall na Mèine
Bad Bog
Fisherfield Forest
Beinn Dearg Mòr
North Erradale
Port Erradale
Rubha Bàn
Big Sand
Poolewe
Londubh
Loch Kernsary
Beinn a' Chàisgein Mòr
Longa Island
Caolas Beag
Lonemore
Strath
Loch Tollaidh
Dubh Loch
Fionn Loch
Gairloch
An Ard
Charlestown
LOCH GAIRLOCH
Meall an Doirein
Beinn Airigh Charr
Port Henderson
Sròn na Carra
Eilean Horrisdale
Badachro
Opinan
Kerrysdale
Shieldaig
Eilean Ruairidh Mòr
Beinn Làir
Mullach Mhic Fh
Letterewe Forest
Letterewe
South Erradale
Loch Clàir
River Kerry
Loch Bad an Sgalaig
Eilean Subhainn
Talladale
LOCH MAREE
Redpoint
Dubh Loch
Loch Garbhaig
SLIOCH
Sgeir Ghlas
Flowerdale Forest
Meall na h-Uamha
Sgeir na Trian
Craig River
Shieldaig Forest
Beinn Bhreac
Loch na h-Oidhche
Beinn an Eòin
Strath Lungard
Glen Grudie
Meall Ghiubhais
Rhu Noa
Anancaun
Kinlochewe
Rubha na Fearn
LOCH TORRIDON
Lower Diabaig
Loch Diabaigas Airde
Beinn Alligin
Ruadh-stac Mòr
BEINN EIGHE
Beinn Eighe National Nature Reserve
WESTER ROSS
Fearnmore
Fearnbeg
Rubha Chuaig
Arinacrinachd
Cuaig
Kenmore
Ardheslaig
Alligin Shuas
Rechullin
Torridon Forest
LIATHACH
Glen Torridon
Torridon Ho
Inveralligin
UPPER LOCH TORRIDON
Torridon
Annat
River Torridon
Sgurr Dubh
Coulin Lodge
Loch Clàir
Loch Coulin
Coulin Forest
Kalnakill
Lonbain
Loch Shieldaig
Shieldaig
Balgy
Ben-damph Forest
Beinn Damh
Maol Chean-dearg
Beinn Liath Mhòr
Sgorr Ruadh
Loch Gaineamhach
Cròic-bheinn
Ben Shieldaig
Loch Damh
Loch an Eòin
Achnashellach Sta
Lair
Glenshieldaig Forest
Loch Lundie
Beinn a' Chlachain
Applecross Forest
Balnacra
Loch Dùghaill
Achnashellach Forest
Applecross Ho
Loch Coultrie
Coulags
River Carron
Applecross
Beinn Bhan
Sgurr a' Gharaidh
Glas Bheinn
New Kelso
Strathcarron
Achintee
Creag a' Chaorainn
Camusteel
Camusterrach
Sgurr a' Ghaorachain
Meall Gorm
Kirkton
Ard-dhubh
Culduie
Ardarroch
Lochcarron
Attadale
Càrn Geuradainn
Bendronaig
ISLAND OF RONA
Loch a' Bhràige
Port an Fhearainn
Eilean Garbh
Eilean Tigh
Garbh Eilean
Caol Rona
Loch a' Sguirr
An Caol
Eilean Fladday
Torran
Loch Arnish
Arnish
Brochel
Screapadal
ISLAND OF RAASAY
SOUND OF RAASAY
Manish Point
Holm Island
Prince Charles's Cave
Bearreraig Bay
Leac Tressirnish
Rubha nam Brathairean
Culnaknock
Valtos
Rock
Glame
Balmeanach
Balachuirn
Dùn Caan
Holoman Bay
Ard na Claise Mòire
INNER SOUND
A 832
A 896
B 8021
B 8056
B 8057
111
94

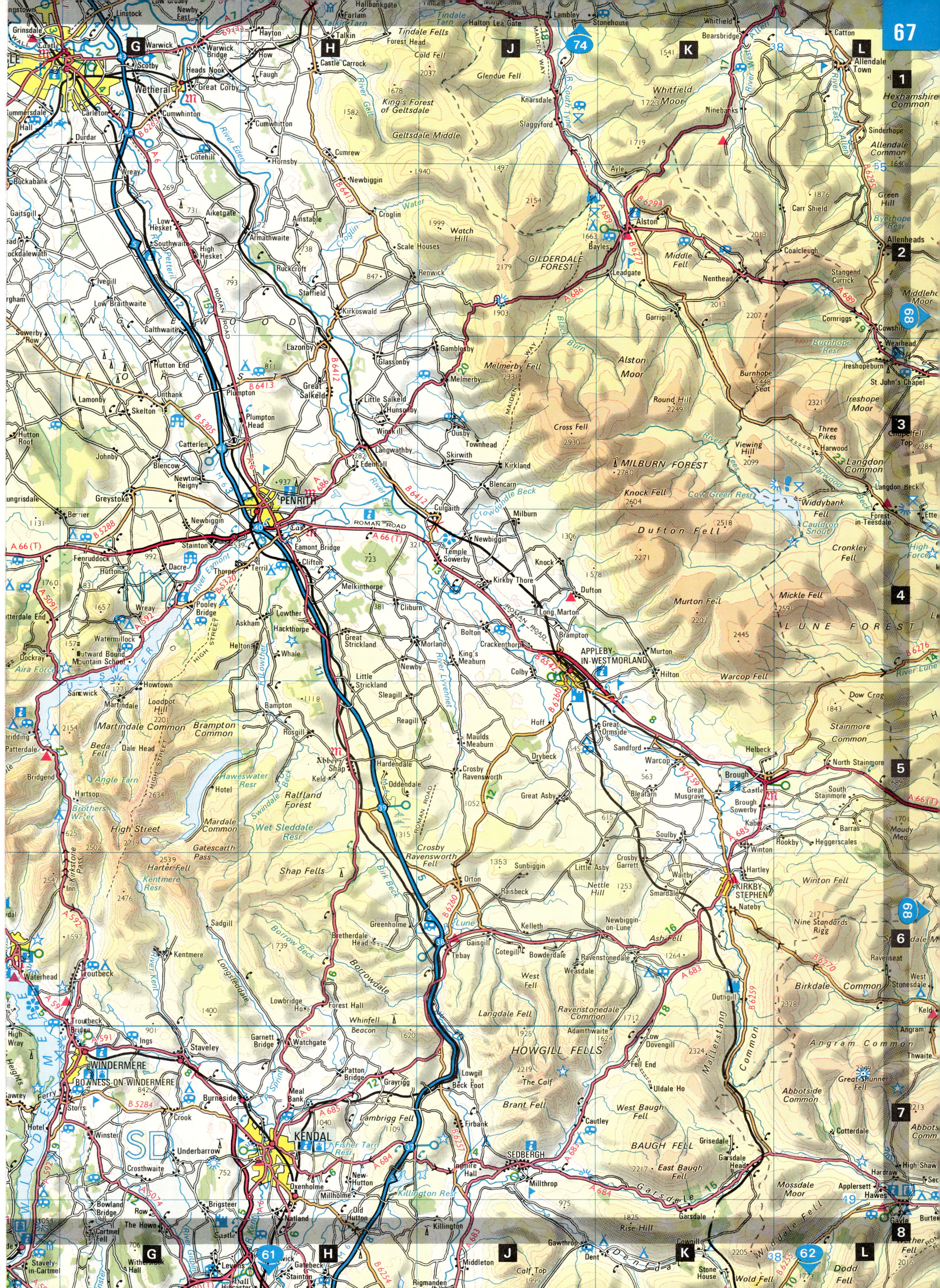

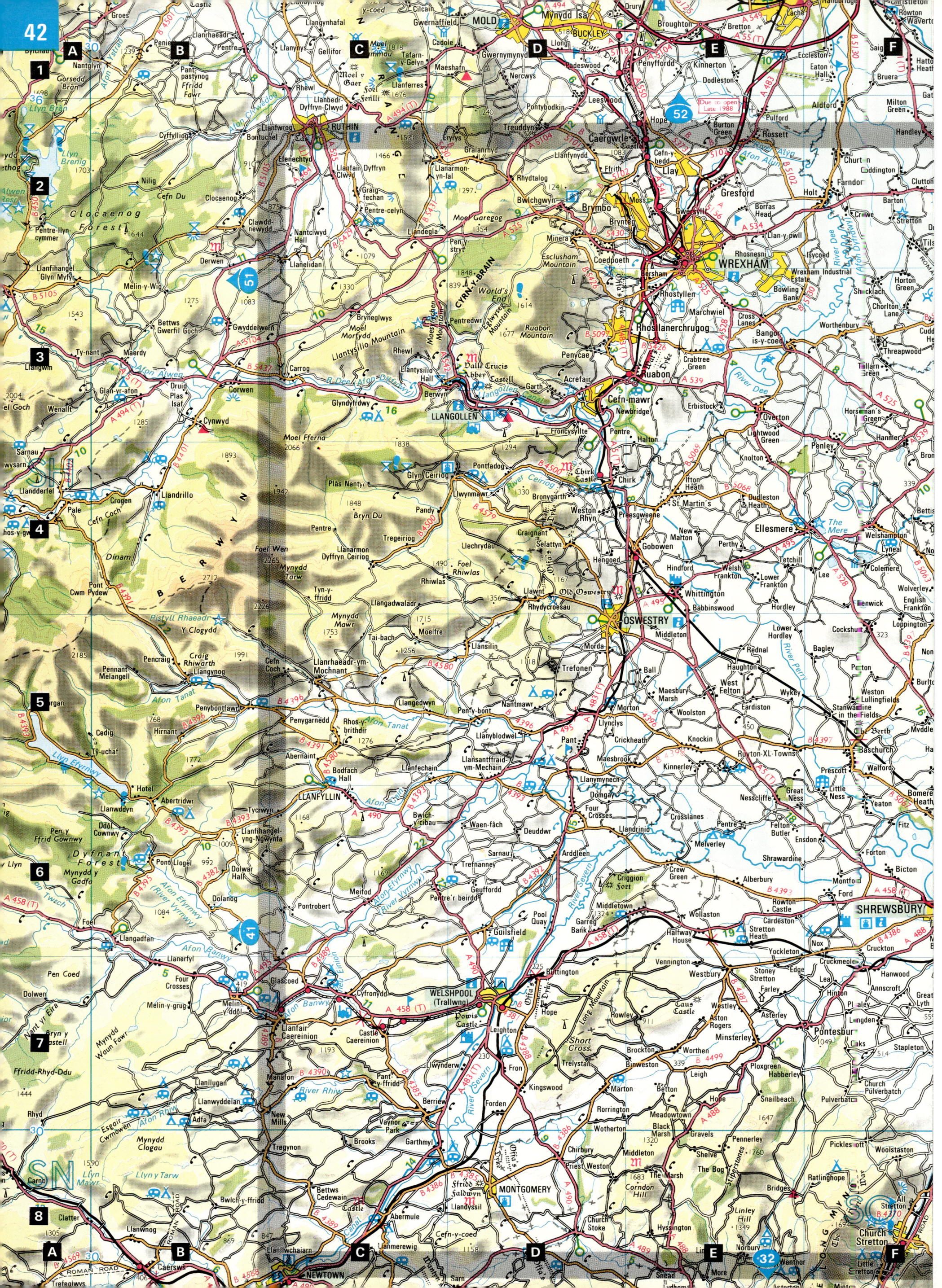

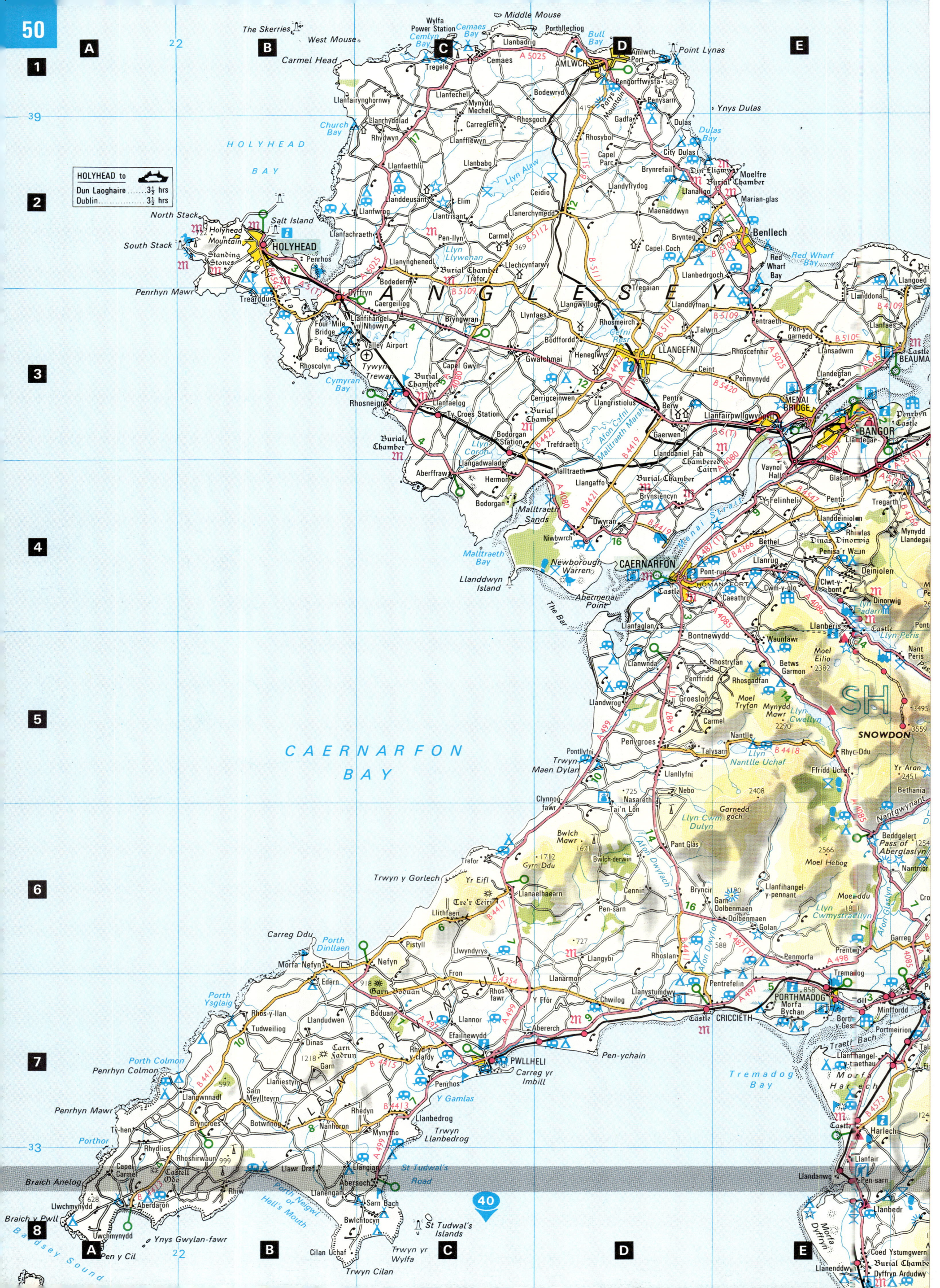

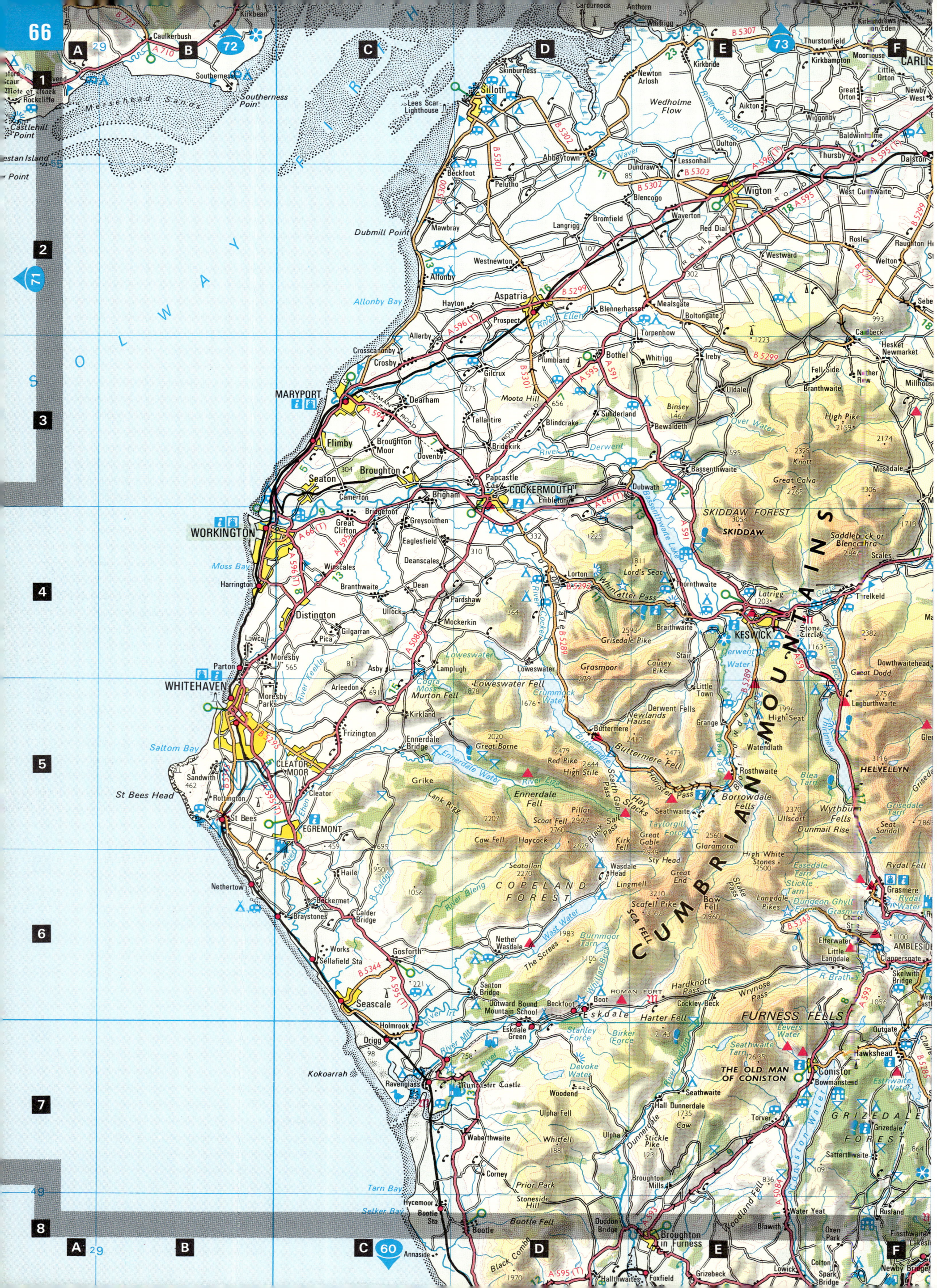

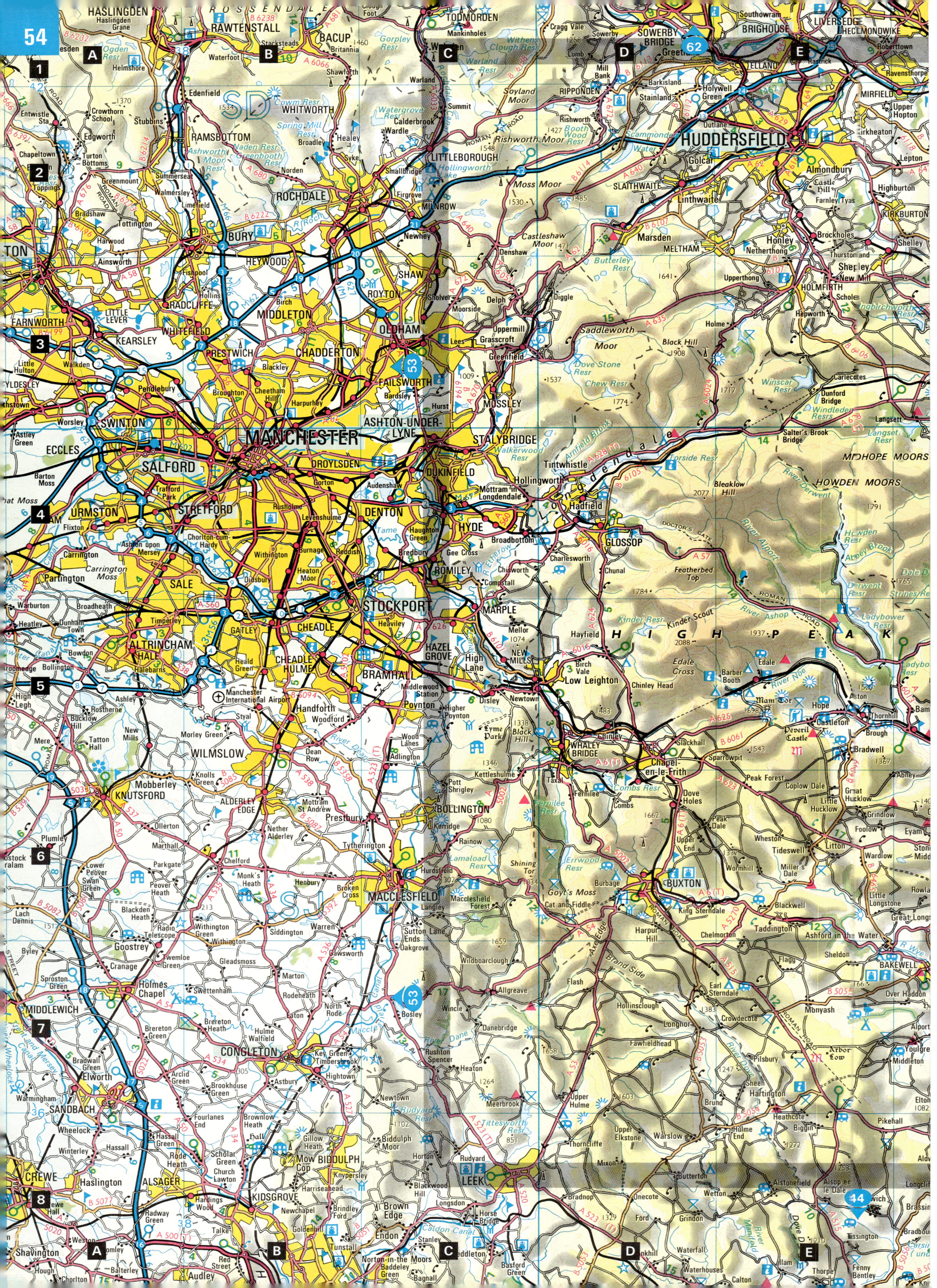

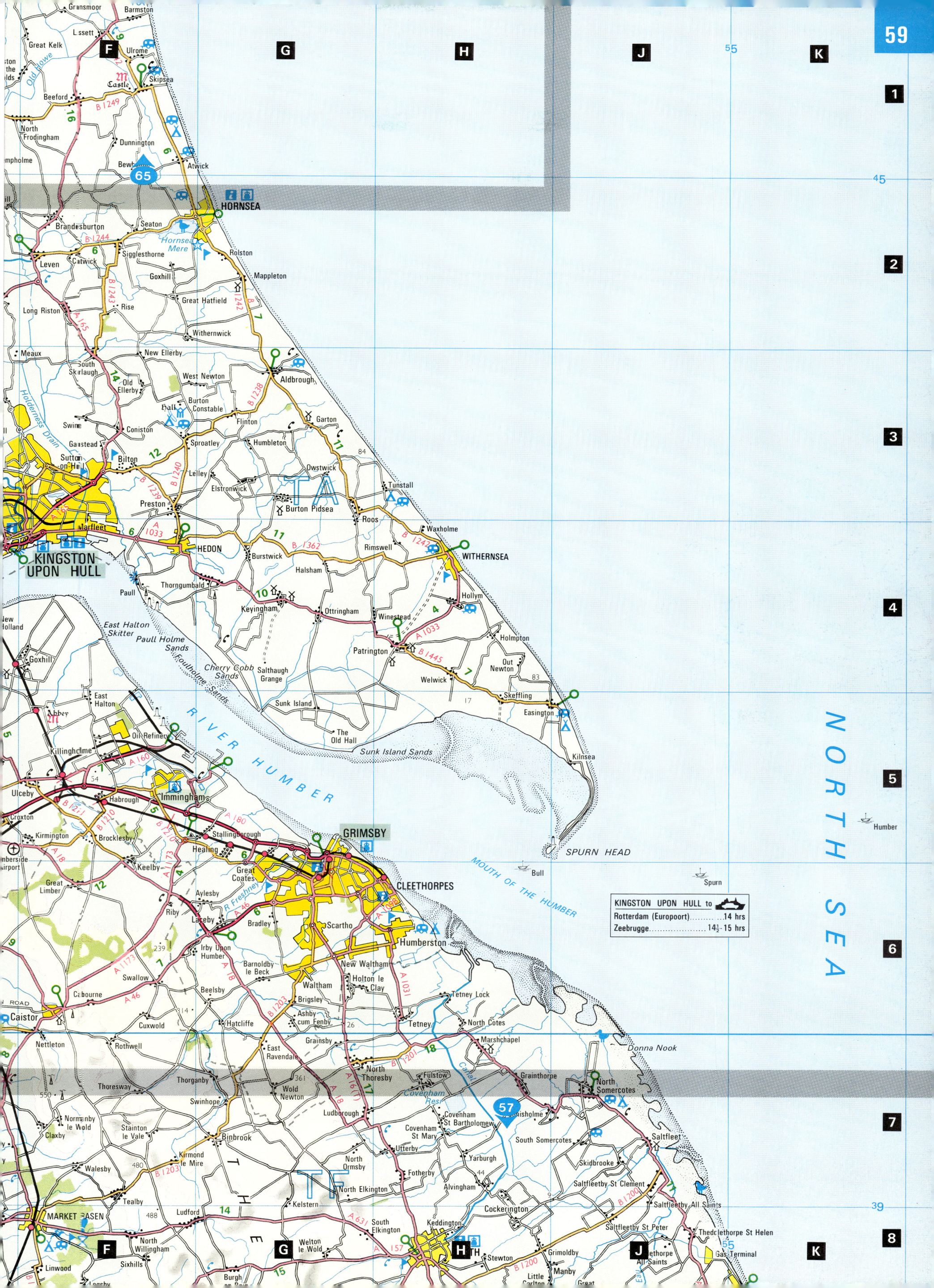

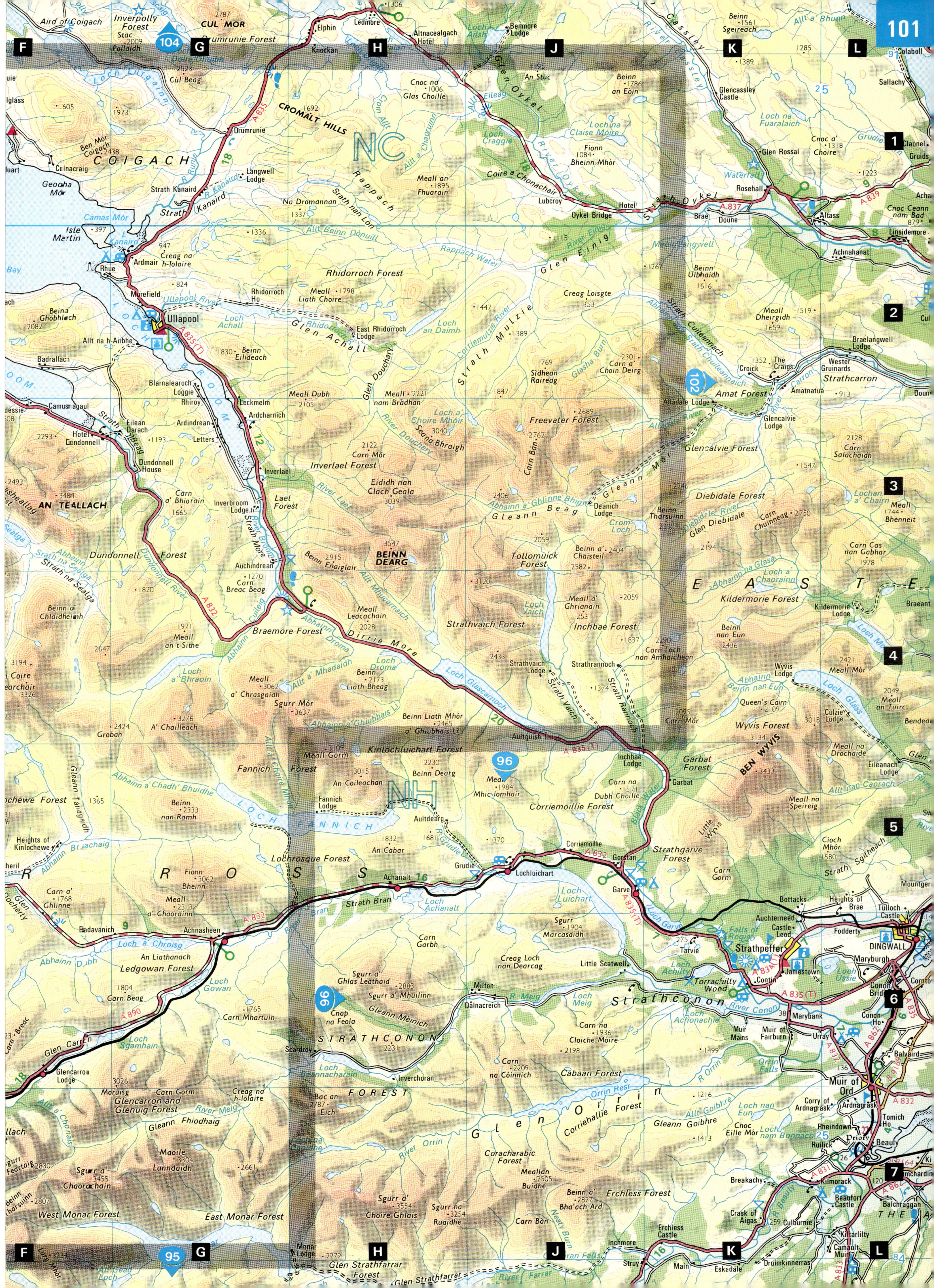
F
G
H
J
K
L
104
102
96
95
1
2
3
4
5
6
7
NC
NH
CUL MOR
Drumrunie Forest
Inverpolly Forest
Stac Pollaidh
Aird of Coigach
Elphin
Knockan
Ledmore
Altnacealgach Hotel
Benmore Lodge
Glen Oykel
An Stùc
Beinn an Eòin
Beinn Sgeireach
Glencassley Castle
Sallachy
Cul Beag
Loch Lurgainn
CROMALT HILLS
Drumrunie
Cnoc na Glas Choille
Loch Ailsh
Loch na Claise Moire
Fionn Bheinn Mhòr
Loch Craggie
River Oykel
Coire a' Chonachair
Lubcroy
Oykel Bridge
Hotel
Strath Oykel
Glen Rossal
Rosehall
Altass
Brae
Doune
Cnoc a' Choire
Loch na Fuaralaich
Claonel
Gruids
Cnoc Ceann nam Bad
Linsidemore
Achnahanat
COIGACH
Ben Mòr Coigach
Culnacraig
Geodha Mòr
Strath Kanaird
Langwell Lodge
Rappach
Na Dromannan
Strath nan Lòn
Meall an Fhuarain
Camas Mòr
Isle Martin
Allt Beinn Donuill
Rappach Water
River Einig
Glen Einig
Meoir Langwell
Ardmair
Creag na h-Iolaire
Rhue
Rhidorroch Forest
Rhidorroch Ho
Meall Liath Choire
Morefield
Ullapool
Ullapool River
Loch Achall
Glen Achall
East Rhidorroch Lodge
Loch an Daimh
Strath Mulzie
Corriemulzie River
Creag Loisgte
Beinn Ulbhaidh
Meall Dheirgidh
Strath Cuileannach
Braelangwell Lodge
Wester Gruinards
Strathcarron
Croick
The Craigs
Amatnatua
Amat Forest
Beinn Ghobhlach
Allt na h-Airbhe
Badrallach
LOCH BROOM
Beinn Eilideach
Blarnalearoch
Loggie
Rhiroy
Leckmelm
Meall Dubh
Glen Douchary
Sidhean Raireag
Carn a' Choin Deirg
Glasha Burn
Alladale Lodge
Glencalvie Lodge
Glencalvie Forest
Alladale River
Camusnagaul
Strath Beag
Eilean Darach
Hotel Dundonnell
Ardindrean
Ardcharnich
Letters
Meall nam Bradhan
Loch a' Choire Mhòir
Seana Bhraigh
Freevater Forest
Carn Bàn
Gleann Mòr
Carn Salachaidh
Dundonnell House
Inverlael
Carn Mòr
Inverlael Forest
Eididh nan Clach Geala
Diebidale Forest
Glen Diebidale
Carn Chuinneag
Lochan a' Chairn
Meall Bhenneit
AN TEALLACH
Carn a' Bhiorain
Inverbroom Lodge
Strath More
Lael Forest
River Lael
Abhainn a' Ghlinne Bhig
Gleann Beag
Deanich Lodge
Crom Loch
Beinn Tharsuinn
Carn Cas nan Gabhar
Dundonnell
Forest
Strath na Sealga
BEINN DEARG
Beinn Enaiglair
Auchindrean
Tollomuick Forest
Beinn a' Chaisteil
Abhainn na Glasa
Loch a' Chaorainn
EASTER ROSS
Kildermorie Forest
Kildermorie Lodge
Braeantra
Carn Breac Beag
Allt Mhucarnaich
Beinn a' Chlaidheimh
Loch Vaich
Meall a' Ghrianain
Inchbae Forest
Strathvaich Forest
Beinn nan Eun
Meall Mòr
Loch Morie
Braemore Forest
Meall Ledcachain
Meall an t-Sithe
Dirrie More
Abhainn Droma
Loch Droma
Carn Loch nan Amhaichean
Strathrannoch
Strathvaich Lodge
Strath Vaich
Wyvis Lodge
Abhainn Beinn nan Eun
Loch Glass
Meall an Tuirc
Culzie Lodge
Bendeallt
Coire Mhearchair
Loch a' Bhraoin
Meall a' Chrasgaidh
Allt a' Mhadaidh
Beinn Liath Bheag
Loch Glascarnoch
Queen's Cairn
Wyvis Forest
BEN WYVIS
Groban
A' Chailleach
Sgurr Mòr
Beinn Liath Mhòr a' Ghiubhais Li
Aultguish Inn
Carn Mòr
Abhainn a' Ghiubhais Li
Meall Gorm
Kintochluichart Forest
Inchbae Lodge
Garbat Forest
Garbat
Meall na Drochaide
Eileanach Lodge
Allt nan Caorach
Fannich Forest
An Coileachan
Beinn Dearg
Meall Mhic Iomhair
Black Water
Carn na Dubh Choille
Corriemoillie Forest
Gleann Tanagaidh
Kinlochewe Forest
Beinn nan Ramh
LOCH FANNICH
Fannich Lodge
Aultdearg
Little Wyvis
Meall na Speireig
Heights of Kinlochewe
Abhainn Bruachaig
Lochrosque Forest
An Cabar
Grudie
Corriemoillie
Strathgarve Forest
Carn Gorm
Cioch Mhòr
Strath Sgitheach
ROSS
Fionn Bheinn
Achanalt
Lochluichart
Gorstan
Garve
Bottacks
Heights of Brae
Tulloch Castle
Mountgerald
Carn a' Ghlinne
Meall a' Ghrianain
Strath Bran
Loch Achanalt
Loch Luichart
Achnasheen
Badavanich
Glen Docherty
Loch a' Chroisg
An Liathanach
Ledgowan Forest
Carn Garbh
Sgurr Marcasaidh
Auchterneed
Castle Leod
Tarvie
Falls of Rogie
Strathpeffer
Fodderty
DINGWALL
Maryburgh
Jamestown
Contin
Loch Ussie
Conon Bridge
Creag Loch nan Dearcag
Little Scatwell
Loch Achilty
Torrachilty Wood
Carn Beag
Loch Gowan
Sgurr a' Ghlas Leathaid
Sgurr a' Mhuilinn
Gleann Meinich
Milton
Dalnacreich
River Meig
Loch Meig
Strathconon
River Conon
Loch Achonachie
Marybank
Conon Ho
Carn Mhartuin
Cnap na Feola
STRATHCONON
Carn na Cloiche Moire
Muir Mains
Muir of Fairburn
Urray
Glen Carron
Loch Sgamhain
Scardroy
Orrin Falls
Balvaird
Glencarron Lodge
Loch Beannacharain
Invercharon
Carn na Còinnich
Cabaan Forest
River Orrin
Muir of Ord
Ardnagrask
Maoruisg
Glencarron and Glenuig Forest
Carn Gorm
Creag na h-Iolaire
River Meig
FOREST
Bac an Eich
Orrin Reservoir
Glen Orrin
Corriehallie Forest
Allt Goibhre
Gleann Goibhre
Corry of Ardnagrask
Tomich Ho
Gleann Fhiodhaig
Loch na Caoidhe
Loch nan Eun
Cnoc Eille Mòr
Loch nam Bonnach
Rheindown
Ruilick
Beauly
Priory
Maoile Lunndaidh
River Orrin
Coracharabic Forest
Sgurr a' Chaorachain
Meallan Buidhe
Beinn a' Bha'ach Ard
Erchless Forest
Breakachy
Kilmorack
Beaufort Castle
Balchraggan
Crask of Aigas
Culburnie
Kiltarlity
Camault Muir
Sgurr a' Choire Ghlais
Sgurr na Ruaidhe
Carn Bàn
Erchless Castle
Inchmore
Struy
Main
Eskadale
Druimkinnerras
West Monar Forest
East Monar Forest
Monar Lodge
Glen Strathfarrar Forest
Glen Strathfarrar
River Farrar
A 835
A 835(T)
A 832
A 837
A 839
A 890
A 831
A 862
A 833
A 834

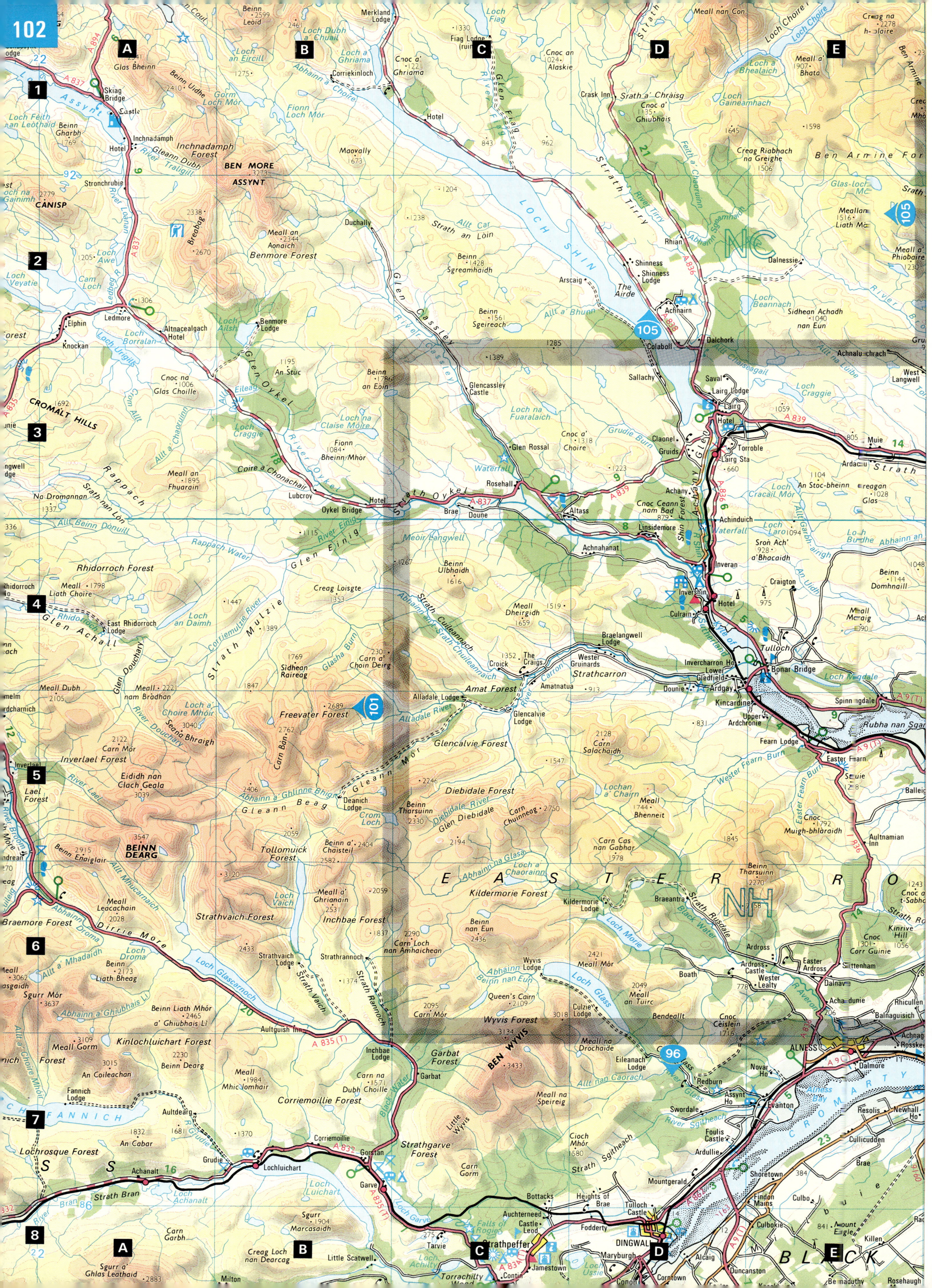
A
B
C
D
E
1
2
3
4
5
6
7
8
Glas Bheinn
Skiag Bridge
Loch Assynt
Inchnadamph
Inchnadamph Forest
BEN MORE ASSYNT
CANISP
Stronchrubie
Loch Awe
Ledmore
Elphin
Knockan
Altnacealgach Hotel
Loch Borralan
CROMALT HILLS
Rappach
Strath nan Lon
Benmore Forest
Benmore Lodge
Loch Ailsh
Glen Oykel
River Oykel
Oykel Bridge
Lubcroy
Rhidorroch Forest
East Rhidorroch Lodge
Glen Achall
Strath Mulzie
Corriemulzie River
Glen Douchary
Freevater Forest
Seana Bhraigh
Inverlael Forest
BEINN DEARG
Gleann Beag
Deanich Lodge
Tollomuick Forest
Strathvaich Forest
Inchbae Forest
Loch Glascarnoch
Aultguish Inn
Strathvaich Lodge
Kinlochluichart Forest
Corriemoillie Forest
Loch Fannich
Lochrosque Forest
Achanalt
Strath Bran
Lochluichart
Loch Luichart
Gorstan
Garve
Strathgarve Forest
Little Wyvis
BEN WYVIS
Garbat Forest
Inchbae Lodge
Strathpeffer
Contin
Jamestown
Torrachilty Wood
Tarvie
Castle Leod
Auchterneed
Dingwall
Maryburgh
Conon Bridge
Tulloch Castle
Fodderty
Heights of Brae
Strath Sgitheach
Mountgerald
Ardullie
Foulis Castle
Evanton
Swordale
Alness
Novar Ho
Redburn
Cromarty Firth
Culbokie
Findon Mains
Black Isle
Wyvis Forest
Wyvis Lodge
Loch Glass
Kildermorie Forest
Kildermorie Lodge
Loch Morie
Ardross
Ardross Castle
Boath
EASTER ROSS
NH
NC
Diebidale Forest
Glencalvie Forest
Glencalvie Lodge
Alladale Lodge
Amat Forest
Strathcarron
Croick
Ardgay
Bonar Bridge
Kincardine
Dornoch Firth
Spinningdale
Fearn Lodge
Culrain
Invershin
Inveran
Achnahanat
Linsidemore
Shin Forest
Rosehall
Altass
Glen Rossal
Glencassley Castle
Glen Cassley
Loch Shin
Lairg
Lairg Sta
Saval
Sallachy
Colaboll
Achnairn
Shinness
Strath Tirry
Rhian
Crask Inn
Strath a' Chraisg
Merkland Lodge
Corriekinloch
Loch a' Ghriama
Fiag Lodge
Glen Fiag
Dalnessie
Achnaluachrach
Muie
Torroble
Achany
Achinduich
Craigton
Tulloch
Loch Buidhe
Meall nan Con
Ben Armine Forest
Strath Rusdale
Saltburn
Balnaguisich
Rosskeen
Achnagarron
Dalmore
Newhall
Resolis
Cullicudden
Braemore Forest
Loch Droma
Dirrie More
Fannich Lodge
Aultdearg
Grudie
Corriemoillie
Milton
Little Scatwell
Creag Loch nan Dearcag
Sgurr Marcasaidh
Carn Garbh
Sgurr a' Ghlas Leathaid
Carn Gorm
Cioch Mhor
Meall na Speireig
Eileanach Lodge
Culzie Lodge
Meall Mor
Bendeallt
Cnoc Ceislein
Queen's Cairn
Carn Loch nan Amhaichean
Beinn nan Eun
Carn Chuinneag
Carn Salachaidh
Meall Bhenneit
Carn Cas nan Gabhar
Beinn Tharsuinn
Cnoc Muigh-bhlaraidh
Aultnamain Inn
Beinn Ulbhaidh
Meall Dheirgidh
Braelangwell Lodge
Westerr Gruinards
Invercharron Ho
Gledfield
Creag Loisgte
Sidhean Rairéag
Carn a' Choin Deirg
Meall Dubh
Meall nam Bradhan
Eididh nan Clach Geala
Carn Ban
Meall Leacachain
Beinn Enaiglair
Beinn Liath Mhor a' Ghiubhais Li
Meall Gorm
An Coileachan
Sgurr Mor
Beinn Liath Bheag
Meall Mhic Iomhair
Carn na Dubh Choille
Beinn a' Chaisteil
Meall a' Ghrianain
Little Wyvis
Loch Vaich
Strath Vaich
Strathrannoch
Strath Rannoch
Beinn an Eoin
An Stuc
Fionn Bheinn Mhor
Coire a' Chonachair
Meall an Fhuarain
Loch Craggie
Loch na Claise Moire
Cnoc na Glas Choille
Loch Urigill
Cam Loch
Loch Veyatie
Meall an Aonaich
Breabag
Loch na Gainimh
Fionn Loch Mor
Gorm Loch Mor
Maovally
Duchally
Strath an Loin
Beinn Sgreamhaidh
Beinn Sgeireach
Arscaig
The Airde
Cnoc a' Choire
Cnoc Ceann nam Bad
Loch Craggie
An Stoc-bheinn
Sron Ach' a' Bhacaidh
Beinn Domhnaill
Creag Riabhach na Greighe
Meall a' Bhata
Mealan Liath Mor
Meall a' Phiobaire
Sidhean Achadh nan Eun
Loch Beannach
Carn Mor
Beinn Deorg

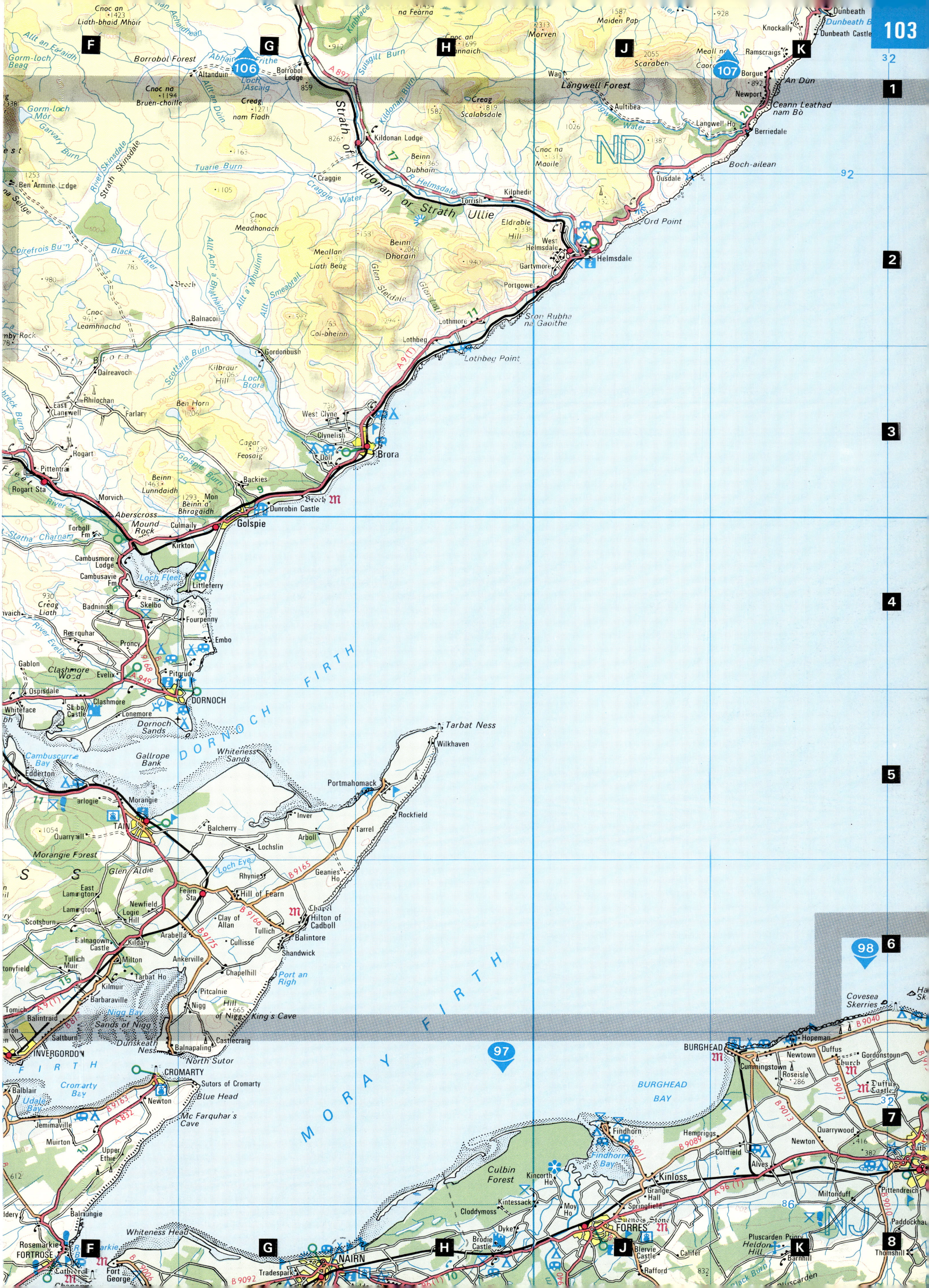

F
G
H
J
K
1
2
3
4
5
6
7
8
ND
NJ
106
107
98
97
Dunbeath
Dunbeath Castle
Knockally
Ramscraigs
Borgue
Newport
An Dùn
Ceann Leathad nam Bò
Berriedale
Langwell Forest
Langwell Ho
Boch-ailean
Ousdale
Ord Point
Morven
Maiden Pap
Scaraben
Creag Scalabsdale
Cnoc na Maoile
Kildonan Lodge
Strath of Kildonan or Strath Ullie
Helmsdale
Kilphedir
Torrish
Eldrable Hill
West Helmsdale
Gartymore
Portgower
Lothmore
Lothbeg
Lothbeg Point
Beinn Dhorain
Beinn Dubhain
Craggie
Craggie Water
Borrobol Forest
Borrobol Lodge
Altanduin
Loch Ascaig
Cnoc na Bruen-choille
Creag nam Fiadh
Liath-bhaid Mhòr
Gorm-loch Beag
Gorm-loch Mòr
Garvary Burn
Tuarie Burn
Ben Armine Lodge
Strath Skinsdale
River Skinsdale
Black Water
Cnoc Meadhonach
Meallan Liath Beag
Coir-bheinn
Broch
Balnacoil
Cnoc Leamhnachd
Gordonbush
Strath Brora
Dalreavoch
Rhilochan
East Langwell
Farlary
Ben Horn
Kilbraur Hill
Loch Brora
Scotarie Burn
West Clyne
Clynelish
Doll
Brora
Cagar Feosaig
Golspie Burn
Backies
Beinn Lunndaidh
Mon Beinn a' Bhragaidh
Hogart
Pittentrail
Rogart Sta
River Fleet
Morvich
Aberscross
Mound Rock
Culmaily
Golspie
Dunrobin Castle
Broch
Kirkton
Cambusmore Lodge
Cambusavie Fm
Loch Fleet
Littleferry
Skelbo
Badninish
Creag Liath
Fourpenny
Embo
Proncy
River Evelix
Clashmore Wood
Evelix
Pitgrudy
DORNOCH
Dornoch Sands
Clashmore
Lonemore
Skibo Castle
Ospisdale
Gablon
Whiteface
DORNOCH FIRTH
Gallrope Bank
Whiteness Sands
Cambuscurrie Bay
Edderton
Tarbat Ness
Wilkhaven
Portmahomack
Rockfield
Inver
Tarrel
Arboll
Balcherry
Lochslin
Morangie
Morangie Forest
TAIN
Glen Aldie
Loch Eye
Rhynie
Geanies Ho
Hill of Fearn
Fearn Sta
Chapel
Hilton of Cadboll
Balintore
Shandwick
Clay of Allan
Tullich
Cullisse
Arabella
Newfield
Logie Hill
East Lamington
Lamington
Scotsburn
Kildary
Balnagown Castle
Milton
Ankerville
Chapelhill
Port an Righ
Tulloch Muir
Tarbat Ho
Kilmuir
Barbaraville
Pitcalnie
Nigg
Hill of Nigg
King's Cave
Nigg Bay
Sands of Nigg
Dunskeath Ness
Balnapaling
Castlecraig
North Sutor
Balintraid
Saltburn
INVERGORDON
CROMARTY
Cromarty Bay
Udale Bay
Balblair
Newton
Sutors of Cromarty
Blue Head
Mc Farquhar's Cave
Jemimaville
Muirton
Upper Ethie
Balmungie
Rosemarkie
FORTROSE
Fort George
Whiteness Head
MORAY FIRTH
NAIRN
Tradespark
Brodie Castle
Dyke
Culbin Forest
Kintessack
Cloddymoss
Kincorth Ho
Moy Ho
FORRES
Sueno's Stone
Blervie Castle
Califer
Rafford
Springfield
Grange Hall
Kinloss
Findhorn
Findhorn Bay
BURGHEAD BAY
BURGHEAD
Cummingstown
Hopeman
Newtown
Duffus
Roseisle
Gordonstoun
Duffus Castle
Hempriggs
Coltfield
Alves
Newton
Quarrywood
Miltonduff
Pittendreich
Paddockhaugh
Pluscarden Priory
Heldon Hill
Barnhill
Thornhill
Covesea Skerries
Ord Point
Sron Rubha na Gaoithe
A 897
A 9(T)
A 949
B 9168
B 9165
B 9166
B 9175
A 832
B 9163
A 96(T)
B 9089
B 9011
B 9013
B 9012
B 9040
B 9092

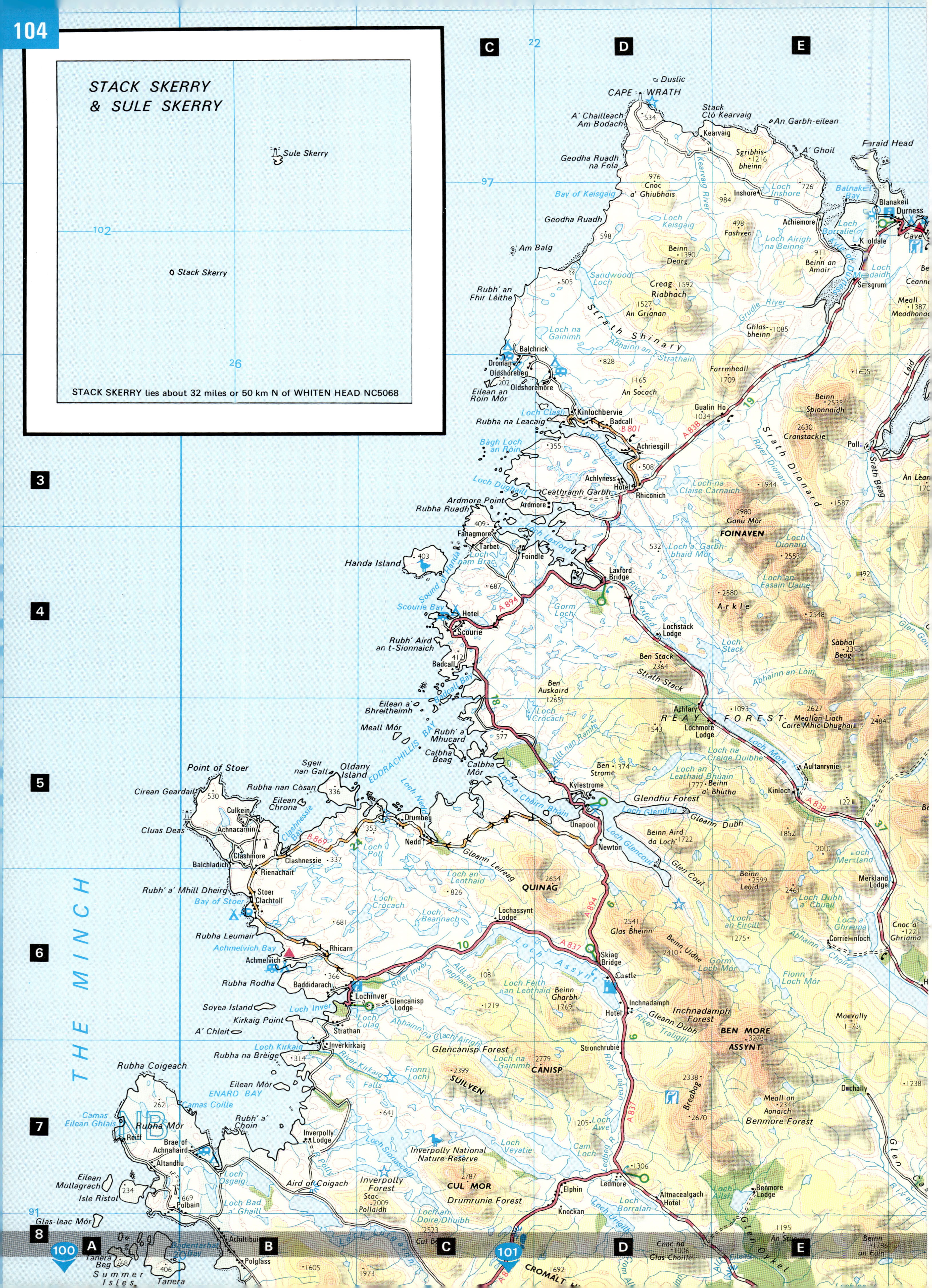
STACK SKERRY
& SULE SKERRY
Sule Skerry
Stack Skerry
STACK SKERRY lies about 32 miles or 50 km N of WHITEN HEAD NC5068
C
D
E
3
4
5
6
7
8
A
B
100
101
Duslic
CAPE WRATH
A' Chailleach
Am Bodach
Stack Clò Kearvaig
An Garbh-eilean
Kearvaig
A' Ghoil
Faraid Head
Geodha Ruadh na Fola
Sgribhis-bheinn
Bay of Keisgaig
Cnoc a' Ghiubhais
Inshore
Loch Inshore
Balnakeil Bay
Balnakeil
Durness
Cave
Geodha Ruadh
Loch Keisgaig
Fashven
Achiemore
Loch Borralie
Loch Airigh na Beinne
Am Balg
Beinn Dearg
Beinn an Amair
Sangomore
Smoo
Rubh' an Fhir Leithe
Sandwood Loch
Creag Riabhach
An Grianan
Grudie River
Loch na Gainimh
Strath Shinary
Ghlas-bheinn
Meall Meadhonach
Balchrick
Droman
Oldshorebeg
Oldshoremore
Eilean an Ròin Mòr
An Socach
Farrmheall
Beinn Spionnaidh
Loch Clash
Kinlochbervie
Badcall
Rubha na Leacaig
Gualin Ho
Cranstackie
Bàgh Loch an Ròin
Achriesgill
Strath Dionard
Loch Dughaill
Achlyness
Rhiconich
Loch na Claise Carnaich
Ceathramh Garbh
Ardmore Point
Rubha Ruadh
Ardmore
Ganu Mòr
FOINAVEN
Fanagmore
Tarbet
Foindle
Loch Dionard
Loch Laxford
Handa Island
Laxford Bridge
Loch a' Garbh-bhaid Mòr
Loch an Easain Uaine
Sound of Handa
Scourie Bay
Scourie
Hotel
Gorm Loch
Arkle
Rubh' Aird an t-Sionnaich
Lochstack Lodge
Sabhal Beag
Badcall
Badcall Bay
Ben Stack
Strath Stack
Loch Stack
Abhainn an Loin
Eilean a' Bhreitheimh
Ben Auskaird
Loch Crocach
Achfary
REAY FOREST
Meallan Liath Coire Mhic Dhughail
Meall Mòr
Rubh' a' Mhucard
Calbha Beag
Calbha Mòr
EDDRACHILLIS BAY
Lochmore Lodge
Loch na Creige Duibhe
Loch More
Point of Stoer
Sgeir nan Gall
Oldany Island
Ben Strome
Loch an Leathaid Bhuain
Beinn a' Bhùtha
Aultanrynie
Kinloch
Cirean Geardail
Rubha nan Còsan
Eilean Chrona
Kylestrome
Glendhu Forest
Loch a' Chàirn Bhàin
Loch Glendhu
Gleann Dubh
Culkein
Achnacarnin
Clashnessie Bay
Drumbeg
Unapool
Cluas Deas
Nedd
Beinn Aird da Loch
Clashmore
Clashnessie
Loch Poll
Gleann Leireag
Newton
Loch Glencoul
Glen Coul
Balchladich
Rienachait
Loch an Leothaid
QUINAG
Beinn Leòid
Loch Dubh a' Chuail
Merkland Lodge
Rubh' a' Mhill Dhìrg
Stoer
Clachtoll
Bay of Stoer
Loch Crocach
Loch Beannach
Lochassynt Lodge
Glas Bheinn
Beinn Uidhe
Loch an Eircill
Corriekinloch
Cnoc a' Ghriama
Rubha Leumair
Achmelvich Bay
Achmelvich
Rhicarn
Loch Assynt
Skiag Bridge
Castle
Gorm Loch Mòr
Rubha Rodha
Baddidarach
River Inver
Allt an Tiaghaich
Loch Fèith an Leothaid
Beinn Gharbh
Inchnadamph
Fionn Loch Mòr
Lochinver
Glencanisp Lodge
Inchnadamph Forest
Gleann Dubh
Soyea Island
Loch Inver
Loch Culag
Hotel
BEN MORE ASSYNT
Maovally
Kirkaig Point
A' Chleit
Strathan
Abhainn na Clach Airigh
River Traligill
Rubha na Brèige
Loch Kirkaig
Inverkirkaig
Glencanisp Forest
Loch na Gainimh
Stronchrubie
Rubha Coigeach
River Kirkaig
Fionn Loch
CANISP
Breabag
Duchally
THE MINCH
Eilean Mòr
ENARD BAY
Falls
SUILVEN
Meall an Aonaich
Camas Eilean Ghlais
Camas Coille
Rubha Mòr
Rubh' a' Choin
Loch Awe
River Loanan
Benmore Forest
Reiff
Brae of Achnahaird
Inverpolly Lodge
Loch Sionascaig
Loch Veyatie
Cam Loch
Altandhu
Inverpolly National Nature Reserve
Ledbeg River
Eilean Mullagrach
Isle Ristol
Loch Osgaig
Aird of Coigach
Inverpolly Forest
CUL MOR
Elphin
Ledmore
Altnacealgach Hotel
Loch Ailsh
Benmore Lodge
Polbain
Loch Bad a' Ghaill
Stac Pollaidh
Drumrunie Forest
Knockan
Loch Borralan
Glas-leac Mòr
Loch Doire Dhubh
Achiltibuie
Loch Lurgainn
Cùl Beag
Cnoc na Glas Choille
River Oykel
Glen Oykel
Tanera Beg
Badentarbat Bay
Summer Isles
Polglass
Tanera
CROMALT HILLS
Beinn an Eoin
A 838
A 894
A 837
B 869
B 801

ATLANTIC OCEAN
F
G
H
J
K
L
1
2
3
4
5
6
7
8
STRATHY POINT
Whiten Head or An Ceann Geal
Rubha Thormaid
Eilean nan Ròn
Neave or Coomb Island
Rabbit Islands
Farr Point
Kirtomy Point
Ardmore Point
Strathy Bay
Red Point
Melvich Bay
Portskerra
Bighouse
Sandside
Drum Hollistan
Golval
Melvich
Baligill
Strathy
Brawl
Aultiphurst
Totegan
Lednagullin
Armadale
Kirtomy
Swordly
Farr
Bettyhill
Achina
Invernaver
Leckfurin
Borgie
Torrisdale
Skerray
Skullomie
Coldbackie
Tongue
Tongue Bay
Talmine
Midfield
Ben Hutig
A' MHOINE
A 838
Moine Ho
Hope
Heilam
Lochside
Portnancon
Eriboll
Ben Arnaboll
Achuvoldrach
Ribigill
Kinloch Lodge
Cnoc Craggie
Beinn Bhreac
Loch Craggie
Borgie Forest
A 836
Skelpick
B 871
Strathnaver
River Naver
Rhifail
Beinn Rifa-gil
Skail
Syre
Dalvina Lodge
Naver Forest
Loch Naver
B 873
BEN HOPE
Cashel Dhu
Loch na Seilg
Meallan Liath
Ben Loyal
Loch an Dherue
Loch Loyal
Loch Loyal Lodge
Cnoc nan Cùilean
Beinn Stumanadh
Loch Haluim
Loch Coulside
NC
Alltnacaillich
Strathmore River
Allnabad
Gobernuisgach Lodge
Loch Meadie
Loch Eileanach
Pole Hill
Mudale
Altnaharra
Hotel
Klibreck
BEN KLIBRECK
Meall nan Con
Loch Choire Forest
Loch Choire Lodge
Creag na h-Iolaire
Ben Armine
Meall an Fhuarain
Strath Vagastie
Loch a' Ghorm-choire
Loch Fiag
Fiag Lodge (ruin)
Cnoc an Alaskie
Crask Inn
Strath a' Chràisg
Cnoc a' Ghiubhais
Strath Tirry
River Tirry
Loch Shin
Rhian
Shinness
Shinness Lodge
The Airde
Achnairn
Dalchork
Colaboll
Sallachy
Saval
Lairg Lodge
Lairg
Arscaig
Strath an Loin
Beinn Sgreamhaidh
Beinn Sgeireach
Dalnessie
Creag Riabhach na Greighe
Ben Armine Forest
Glas-loch Mòr
Meallan Liath Mòr
Meall a' Phiobaire
Sidhean Achadh nan Eun
Grumby Rock
Achnaluachrach
West Langwell
Strath Brora
Dalreavoch
Rhilochan
Ben Armine Lodge
Strath na Seilge
Loch Beannach
Loch Rimsdale
Loch nan Clàr
Loch Truderscaig
Badanloch Forest
Badanloch Lodge
Loch Badanloch
Loch an Alltan Fhearna
Loch Achnamoine
Strath Beg
Kinbrace
Achentoul
Lochside
Achentoul Forest
Knockfin Heights
Ben Griam Mòr
Ben Griam Beg
Loch Druim a' Chliabhain
Cnoc nan Tri-chlach
Loch Croacach
Forsinard
Hotel
Forsinain Fm
Slethill Hill
Strath Halladale
Halladale River
Trantlemore
Dalhalvaig
Craigtown
Achiemore
Cnoc Badaireach na Gaoithe
Strathy Forest
Bowside Lodge
Beinn Ruadh
Loch Strathy
Loch nam Breac
Cnoc Coire na Fearna
Borrobol Forest
Borrobol Lodge
Altanduin
Cnoc nan Bruen-choille
Creag nam Fladh
A 897
Strath of Kildonan
Kildonan Lodge
Craggie
Cnoc Meadhonach
Meallan Liath Beag
Col-bheinn
Balnacoil
Broch
Strath Skinsdale
Black Water
Cnoc Leamhnachd
Gordonbush
Kilbraur Hill
Ben Horn
Beinn Dubhain
Lothbeg
A 836
A 838
106
103
102
28
21
17
29
97
91

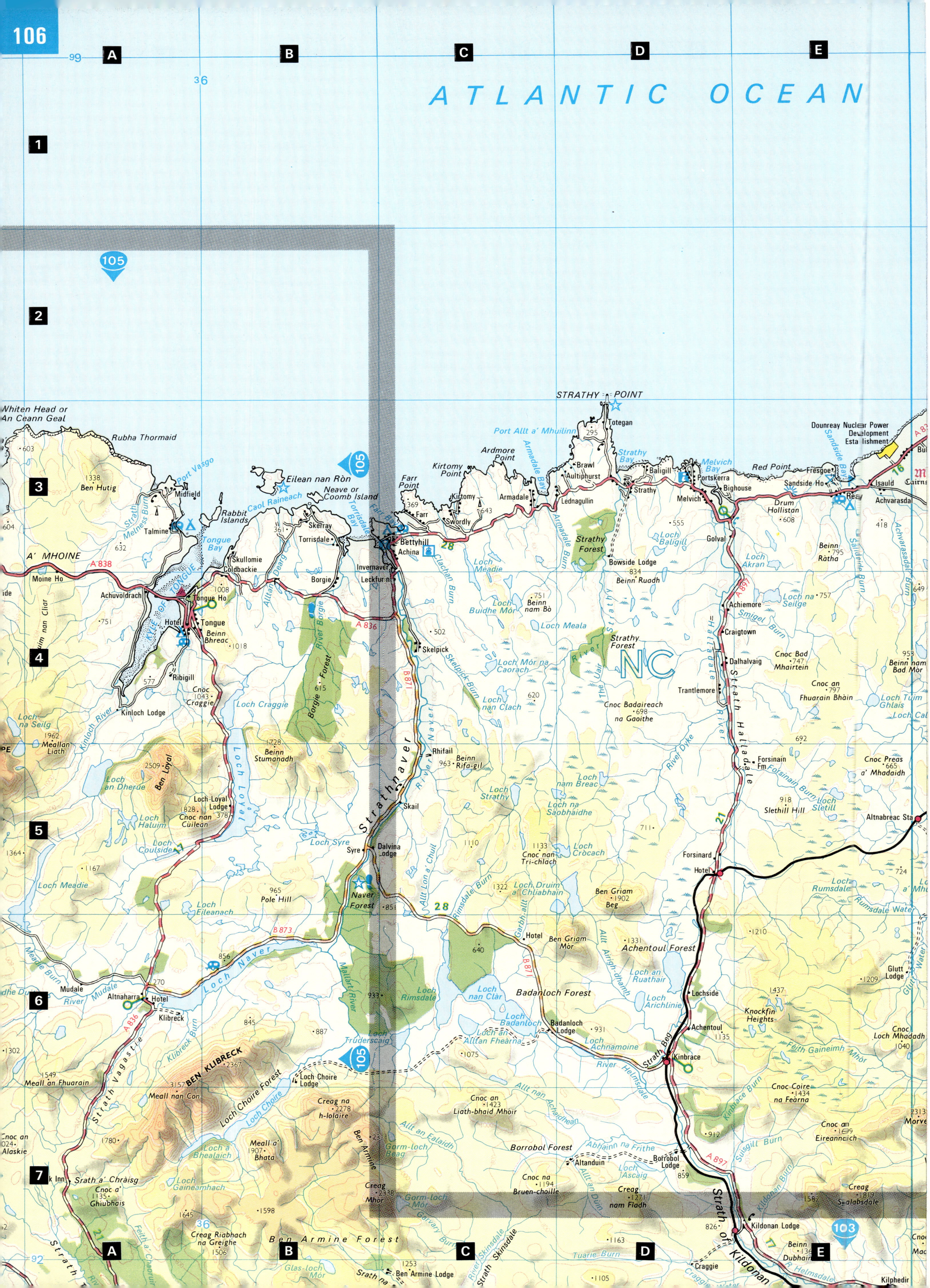
ATLANTIC OCEAN
STRATHY POINT
Whiten Head or An Ceann Geal
Rubha Thormaid
Ben Hutig
Port Vasgo
Midfield
Talmine
Melness Burn
Rabbit Islands
Tongue Bay
A' MHOINE
A 838
Moine Ho
Achuvoldrach
Kyle of Tongue
Tongue Ho
Tongue
Hotel
Beinn Bhreac
Ribigill
Cnoc Craggie
Kinloch Lodge
Kinloch River
Loch na Seilg
Meallan Liath
Ben Loyal
Loch an Dherue
Loch Loyal Lodge
Cnoc nan Cùilean
Loch Hakel
Loch Coulside
Loch Meadie
Loch Eileanach
Mudale
Altnaharra
Hotel
Klibreck
Mudale River
Meall an Fhuarain
Strath Vagastie
BEN KLIBRECK
Meall nan Con
Loch Choire Forest
Loch Choire Lodge
Loch a' Bhealaich
Loch Gaineamhach
Meall a' Bhata
Creag na h-Iolaire
Ben Armine
Creag Mhor
Ben Armine Forest
Creag Riabhach na Greighe
Strath a' Chràisg
Cnoc a' Ghiubhais
Ben Armine Lodge
Caol Raineach
Eilean nan Ròn
Neave or Coomb Island
Skullomie
Coldbackie
Skerray
Torrisdale
Torrisdale Bay
Borgie
Borgie Forest
River Borgie
Loch Craggie
Loch Loyal
Beinn Stumanadh
Loch Syre
Syre
Dalvina Lodge
Naver Forest
Pole Hill
B 873
Loch Naver
Mallart River
Loch Rimsdale
Loch Truderscaig
Strathnaver
River Naver
Skail
Rhifail
Beinn Rifa-gil
Skelpick
Skelpick Burn
Invernaver
Leckfurin
A 836
B 871
Bettyhill
Achina
Farr Bay
Farr Point
Farr
Kirtomy Point
Kirtomy
Swordly
Ardmore Point
Armadale
Armadale Bay
Armadale Burn
Port Allt a' Mhuilinn
Totegan
Brawl
Aultiphurst
Lednagullin
Strathy Bay
Strathy
Baligill
Strathy Forest
Bowside Lodge
Beinn Ruadh
Loch Meadie
Loch Buidhe Mòr
Beinn nam Bò
Loch Meala
Loch Mòr na Caorach
Loch nan Clach
Cnoc Badaireach na Gaoithe
Loch nam Breac
Loch na Saobhaidhe
Loch Strathy
Loch Crocach
Cnoc nan Tri-chlach
Allt Lon a' Chuil
Rimsdale Burn
Loch Druim a' Chliabhain
Ben Griam Beg
Hotel
Ben Griam Mor
Achentoul Forest
Badanloch Forest
Loch nan Clàr
Loch Badanloch
Badanloch Lodge
Loch an Alltan Fhearna
Loch Achnamoine
River Helmsdale
Loch an Ruathair
Loch Arichlinie
Cnoc an Liath-bhaid Mhòir
Allt an Eilaidh
Borrobol Forest
Altanduin
Borrobol Lodge
Cnoc na Bruen-choille
Creag nam Fladh
Loch Ascaig
Gorm-loch Mòr
Gorm-loch Beag
Strath Skinsdale
Tuarie Burn
Strathy Forest
NC
River Strathy
Trantlemore
Dalhalvaig
Craigtown
Achiemore
Golval
Melvich
Melvich Bay
Portskerra
Bighouse
Halladale River
Strath Halladale
Forsinain Fm
Forsinard
Hotel
Lochside
Achentoul
Kinbrace
Strath Beg
A 897
Strath of Kildonan
Kildonan Lodge
Kinbrace Burn
Suisgill Burn
Knockfin Heights
Cnoc Coire na Fearna
Cnoc an Eireannaich
Creag Scalabsdale
Beinn Dubhain
Helmsdale
Red Point
Fresgoe
Sandside Ho
Sandside Bay
Dounreay Nuclear Power Development Establishment
Reay
Isauld
Achvarasdal
Drum Hollistan
Loch Akran
Beinn Ràtha
Loch na Seilge
Cnoc Bad Mhairtein
Cnoc an Fhuarain Bhàin
Loch Tuim Ghlais
Cnoc Preas a' Mhadaidh
Slethill Hill
Loch Sletill
Forsinain Burn
Altnabreac Sta
Loch Rumsdale
Rumsdale Water
Glutt Lodge
Cnoc Loch Mhadadh
Forth Gaineimh Mhòr
Kilphedir
Craggie

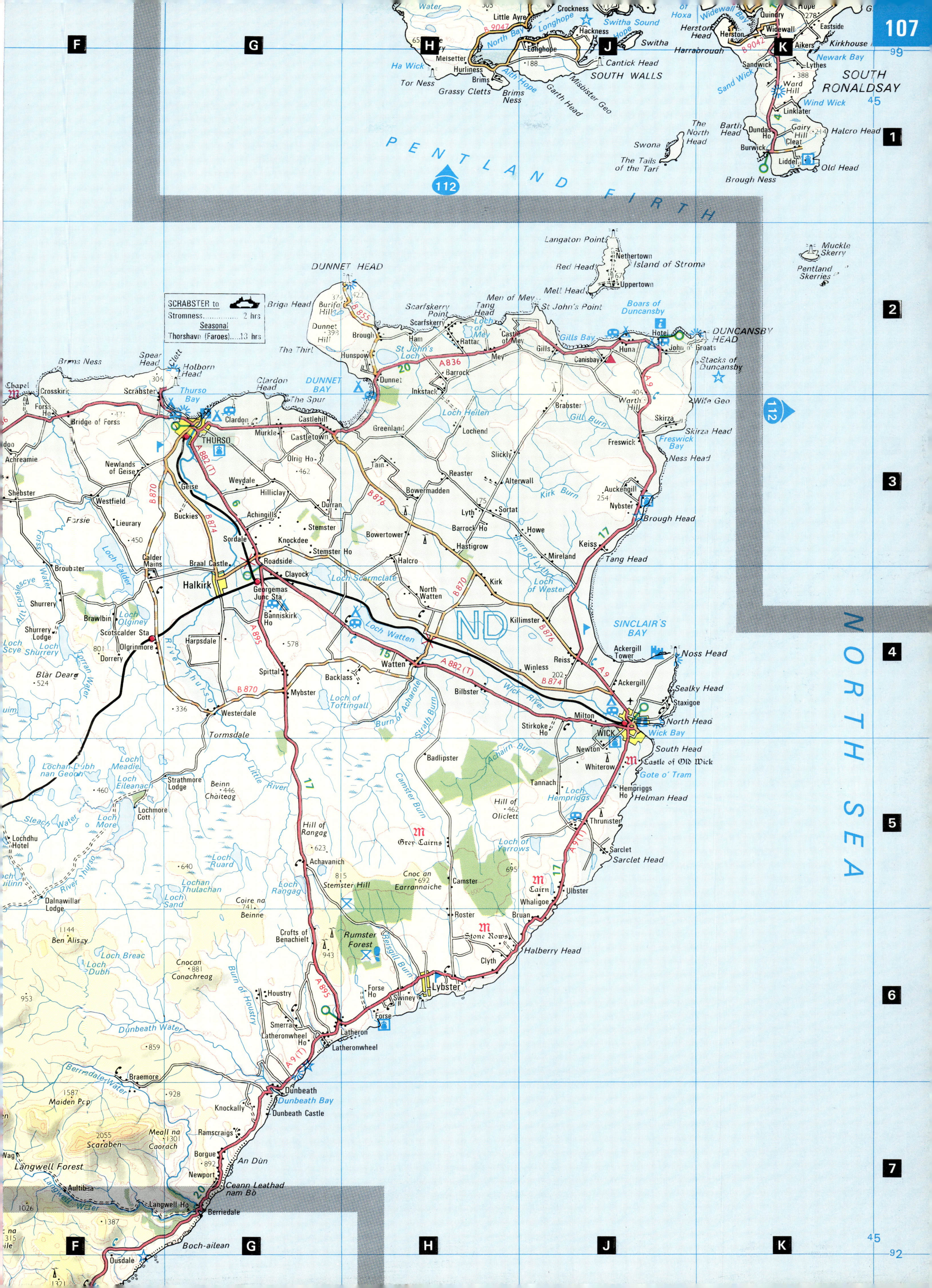
F
G
H
J
K
1
2
3
4
5
6
7
99
45
92
PENTLAND FIRTH
112
NORTH SEA
ND
Little Ayre
Crockness
Switha Sound
Hoxa
Widewall Bay
Quindry
Hope
Herston Head
Herston
Widewall
Eastside
Hackness
Switha
Harrabrough
Aikers
Kirkhouse
Meiselter
Longhope
North Bay
Cantick Head
Newark Bay
Ha Wick
Hurliness
SOUTH WALLS
Sandwick
Lythes
Brims
Aith Hope
SOUTH RONALDSAY
Tor Ness
Grassy Cletts
Brims Ness
Garth Head
Misbister Geo
Sand Wick
Ward Hill
Wind Wick
Linklater
The North Head
Barth Head
Dundas Ho
Halcro Head
Swona
Gairy Hill
Cleat
The Tails of the Tarf
Burwick
Liddel
Old Head
Brough Ness
Langaton Point
Nethertown
Island of Stroma
Muckle Skerry
Red Head
Uppertown
Pentland Skerries
Mell Head
DUNNET HEAD
Men of Mey
St John's Point
Boars of Duncansby
SCRABSTER to
Stromness 2 hrs
Seasonal
Thorshavn (Faroes) 13 hrs
Briga Head
Burifa Hill
Scarfskerry Point
Tang Head
Scarfskerry
Loch of Mey
Castle of Mey
Gills Bay
Hotel
DUNCANSBY HEAD
Dunnet Hill
Brough
Ham
Rattar
Mey
Gills
Canisbay
Huna
John o' Groats
The Thirl
Hunspow
St John's Loch
A 836
Stacks of Duncansby
Brims Ness
Spear Head
Clett
Holborn Head
Barrock
Dunnet
Inkstack
Clardon Head
DUNNET BAY
Scrabster
Thurso Bay
The Spur
Brabster
Worth Hill
Wife Geo
Chapel
Crosskirk
Forss Ho
Bridge of Forss
Clardon
Castlehill
Loch Heilen
Gill Burn
Skirza
Murkle
Greenland
Skirza Head
THURSO
Castletown
Lochend
Freswick
Freswick Bay
Achreamie
Newlands of Geise
Olrig Ho
Slickly
Ness Head
Tain
Reaster
Shebster
Westfield
Geise
Weydale
Hilliclay
Bowermadden
Alterwall
Auckengill
Nybster
Forsie
Lieurary
Buckies
Achingills
Durran
Lyth
Sortat
Kirk Burn
Brough Head
Stemster
Barrock Ho
Howe
Sordale
Knockdee
Bowertower
Hastigrow
Mireland
Keiss
Broubster
Calder Mains
Stemster Ho
Tang Head
Loch Calder
Braal Castle
Roadside
Halcro
Burn of Lyth
Halkirk
Clayock
Loch Scarmclate
Kirk
Loch of Wester
Georgemas Junc Sta
North Watten
Shurrery
Banniskirk Ho
Brawlbin
Loch Olginey
Killimster
SINCLAIR'S BAY
Shurrery Lodge
Scotscalder Sta
Loch Watten
Olgrinmore
Harpsdale
Ackergill Tower
Noss Head
Loch Scye
Loch Shurrery
Dorrery
Watten
Reiss
Spittal
Winless
Blàr Dearg
Backlass
Bilbster
Ackergill
Sealky Head
Mybster
Loch of Toftingall
Wick River
Staxigoe
Westerdale
Milton
North Head
Stirkoke Ho
WICK
Wick Bay
Tormsdale
Newton
South Head
Whiterow
Castle of Old Wick
Gote o' Tram
Loch Meadie
Strathmore Lodge
Badlipster
Loch Eileanach
Beinn Chaiteag
Little River
Tannach
Hempriggs Ho
Loch Hempriggs
Helman Head
Lochmore Cott
Hill of Oliclett
Loch More
Thrumster
Lochdhu Hotel
Hill of Rangag
Grey Cairns
Loch of Yarrows
Sarclet
Sarclet Head
Achavanich
Loch Ruard
Stemster Hill
Cnoc an Earrannaiche
Camster
Cairn
Ulbster
Lochan Thulachan
Loch Rangag
Whaligoe
Dalnawillar Lodge
Loch Sand
Coire na Beinne
Roster
Bruan
Ben Alisky
Crofts of Benachielt
Rumster Forest
Stone Rows
Halberry Head
Loch Breac
Loch Dubh
Cnocan Conachreag
Clyth
Burn of Houstry
Forse Ho
Lybster
Swiney
Houstry
Forse
Dunbeath Water
Smerral
Latheronwheel Ho
Latheron
Latheronwheel
Berriedale Water
Braemore
Dunbeath
Maiden Pap
Dunbeath Bay
Knockally
Dunbeath Castle
Scaraben
Meall na Caorach
Ramscraigs
Langwell Forest
Borgue
An Dùn
Newport
Aultibea
Ceann Leathad nam Bò
Langwell Water
Langwell Ho
Berriedale
Boch-ailean
Ousdale
A 882 (T)
B 874
B 870
B 876
B 855
A 895
A 9
A 9 (T)
B 9042
B 9047

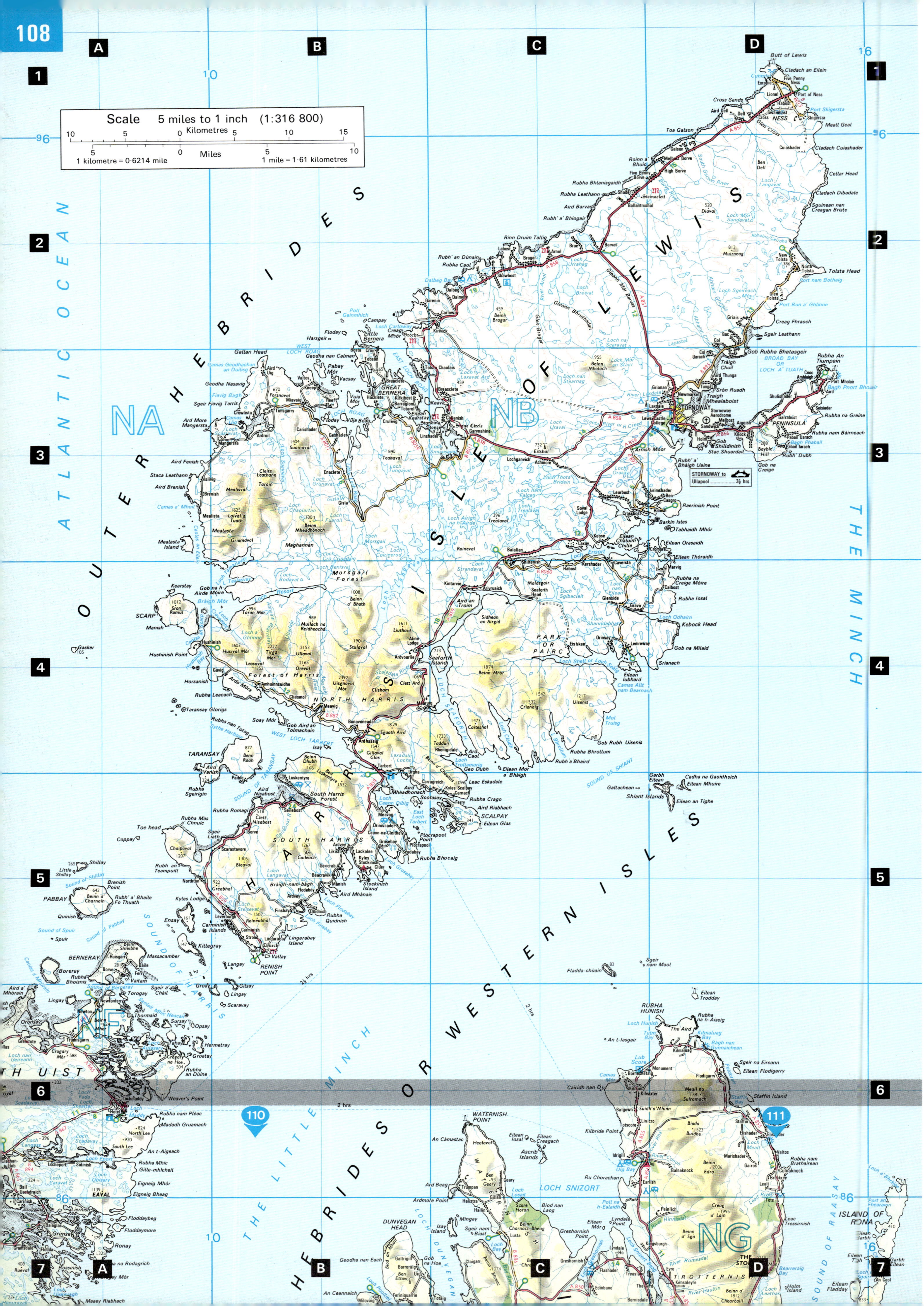
A
B
C
D
1
2
3
4
5
6
7
10
16
96
86
Scale 5 miles to 1 inch (1:316 800)
Kilometres
10 5 0 5 10 15
Miles
5 0 5 10
1 kilometre = 0·6214 mile
1 mile = 1·61 kilometres
ATLANTIC OCEAN
OUTER HEBRIDES
ISLE OF LEWIS
HARRIS
HEBRIDES OR WESTERN ISLES
THE MINCH
THE LITTLE MINCH
NA
NB
NF
NG
Butt of Lewis
Cladach an Eilein
Port of Ness
Cross Sands
NESS
Meall Geal
Toa Galson
Cladach Cuiashader
Cellar Head
Cladach Dibadale
Rubha Bhlanisgaidh
Rubha Leathann
Aird Barvas
Rubh' a' Bhiogair
Rinn Druim Tallig
Rubh' an Dùnain
Rubha Caol
Tolsta Head
Dalbeg Bay
Barvas
Carloway
Great Bernera
Little Bernera
Gallan Head
Pabay Mòr
Vacsay
Uig
Timsgarry
Geodha Nasavig
Ard More Mangersta
Mangersta
Sgeir Rubha Bhatasgeir
Broad Bay or Loch a' Tuath
Rubha An Tiumpain
STORNOWAY
Stornoway Aerodrome
EYE PENINSULA
Rubha na Greine
Rubha nam Bàirneach
Stac Shuardail
Arnish Moor
STORNOWAY to Ullapool 3½ hrs
Aird Fenish
Staca Leathann
Aird Brenish
Camas a' Mhoil
Mealasta Island
Mealisval
Loch Erisort
Eilean Thòraidh
Kebock Head
Gob na Milaid
Srianach
Balallan
Kintarvie
PARK OR PAIRC
Eishken
Morsgail Forest
Kearstay
SCARP
Gasker
Hushinish Point
Taransay Glorigs
Forest of Harris
NORTH HARRIS
Tarbert
Clisham
Beinn Mhòr
Uisenis
Rubha Bhrollum
Gob Rubh Uisenis
Sound of Shiant
Garbh Eilean
Shiant Islands
Eilean an Tighe
TARANSAY
West Loch Tarbert
Sound of Taransay
South Harris Forest
SOUTH HARRIS
SCALPAY
Eilean Glas
Rubha Bhocaig
Toe Head
Coppay
Shillay
Pabbay
Sound of Pabbay
Killegray
Ensay
Rodel
RENISH POINT
BERNERAY
Boreray
Sound of Harris
Fladda-chùain
Sgeir nam Maol
RUBHA HUNISH
Eilean Trodday
Staffin Island
Sgeir na Eireann
Eilean Flodigarry
WATERNISH POINT
Ascrib Islands
Loch Snizort
Uig Bay
Trotternish
Eilean Mhuire
DUNVEGAN HEAD
Loch Dunvegan
Ardmore Point
ISLAND OF RONA
Sound of Raasay
Lochmaddy
EAVAL
Madadh Gruamach
Weaver's Point
110
111
2 hrs

COLL
ISLAND OF MULL
NM
BEN MORE
ROSS OF MULL
IONA
Staffa
Fingal's Cave
ULVA
MORVERN
LISMORE
KERRERA
OBAN
FIRTH OF LORN
ATLANTIC OCEAN
Scale 5 miles to 1 inch (1:316 800)
Kilometres
10 5 0 5 10 15
Miles
5 0 5 10
1 kilometre = 0·6214 mile
1 mile = 1·61 kilometres
COLONSAY
ORONSAY
JURA
PAPS OF JURA
SOUND OF JURA
NR
ISLAY
RHINNS POINT
MULL OF OA
GIGHA ISLAND
SOUND OF GIGHA
KNAPDALE
LOCHGILPHEAD
Loch Fyne
Tobermory
Craignure
Port Ellen
Bowmore
Port Askaig
Tarbert
Crinan
Scarba
Garvellachs
Luing
Seil
Shuna
Islay
Kintyre
West Loch Tarbert
Loch Sween
Loch Tarbert
Loch na Keal
Loch Scridain
Loch Buie
Sound of Mull
Kingairloch
A B C D E
1 2 3 4 5 6
74 64 12 18

108
88
A
B
C
D
E
1
2
3
4
5
6
NORTH UIST
BENBECULA
SOUTH UIST
BARRA
ERISKAY
VATERSAY
MINGULAY
Barra Head
CANNA
COLL
ATLANTIC OCEAN
OUTER HEBRIDES
HEBRIDES OR WESTERN ISLES
INNER HEBRIDES
SEA OF THE HEBRIDES
THE LITTLE MINCH
SOUND OF MONACH
Monach Islands
Lochmaddy
Lochboisdale
Castlebay
HECLA
BEINN MHOR
EAVAL
DUNVEGAN HEAD
WATERNISH POINT
HEALABHAL MHOR
HEALABHAL BHEAG
IDRIGILL POINT
LOCH BRACADALE
LOCH SNIZORT
Dunvegan
NF
NL
LOCHMADDY to Tarbert 2¼ hrs
LOCHBOISDALE to Oban 6 hrs
CASTLEBAY to Oban 6 hrs
UIG to Tarbert 2 hrs
COLL to Tiree 1 hr
Scale 5 miles to 1 inch (1:316 800)
Kilometres
Miles
1 kilometre = 0·6214 mile
1 mile = 1·61 kilometres

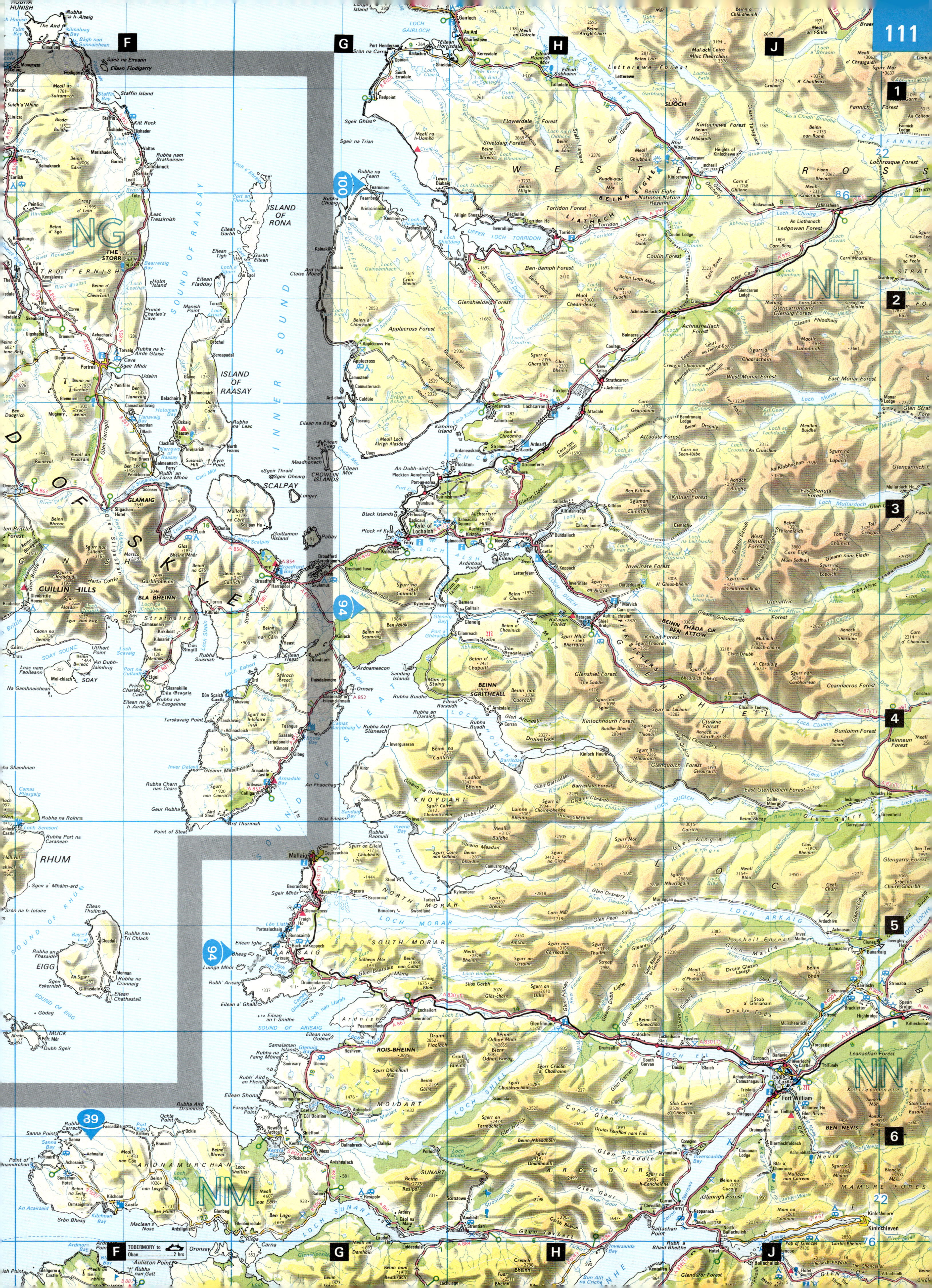
NG
NH
NM
NN
ISLAND OF RAASAY
SKYE
RHUM
EIGG
KNOYDART
ARDNAMURCHAN
MOIDART
SUNART
ARDGOUR
Fort William
Mallaig
Kyle of Lochalsh
Portree
BEN NEVIS
TOBERMORY to Oban 2 hrs

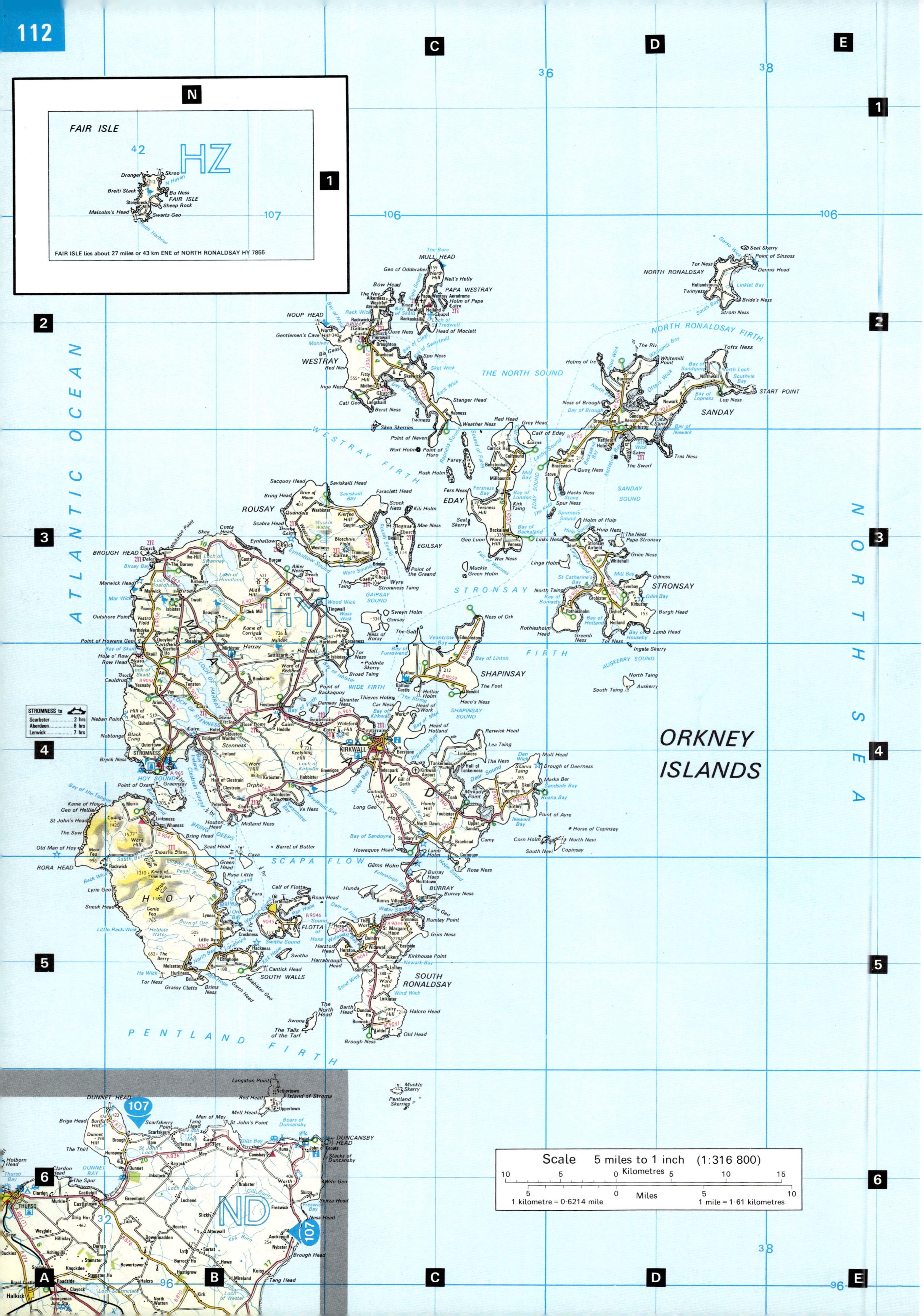
FAIR ISLE
HZ
FAIR ISLE lies about 27 miles or 43 km ENE of NORTH RONALDSAY HY 7855
ATLANTIC OCEAN
NORTH SEA
ORKNEY ISLANDS
HY
ND
WESTRAY
PAPA WESTRAY
NORTH RONALDSAY
SANDAY
EDAY
ROUSAY
EGILSAY
STRONSAY
SHAPINSAY
MAINLAND
KIRKWALL
STROMNESS
HOY
FLOTTA
BURRAY
SOUTH RONALDSAY
SOUTH WALLS
SCAPA FLOW
PENTLAND FIRTH
WESTRAY FIRTH
STRONSAY FIRTH
NORTH RONALDSAY FIRTH
THE NORTH SOUND
SANDAY SOUND
WIDE FIRTH
HOY SOUND
THURSO
DUNNET HEAD
DUNCANSBY HEAD
STRONMESS to
Scarbster 2 hrs
Aberdeen 8 hrs
Lerwick 7 hrs
Scale 5 miles to 1 inch (1:316 800)
Kilometres
Miles
1 kilometre = 0·6214 mile
1 mile = 1·61 kilometres
A B C D E
1 2 3 4 5 6

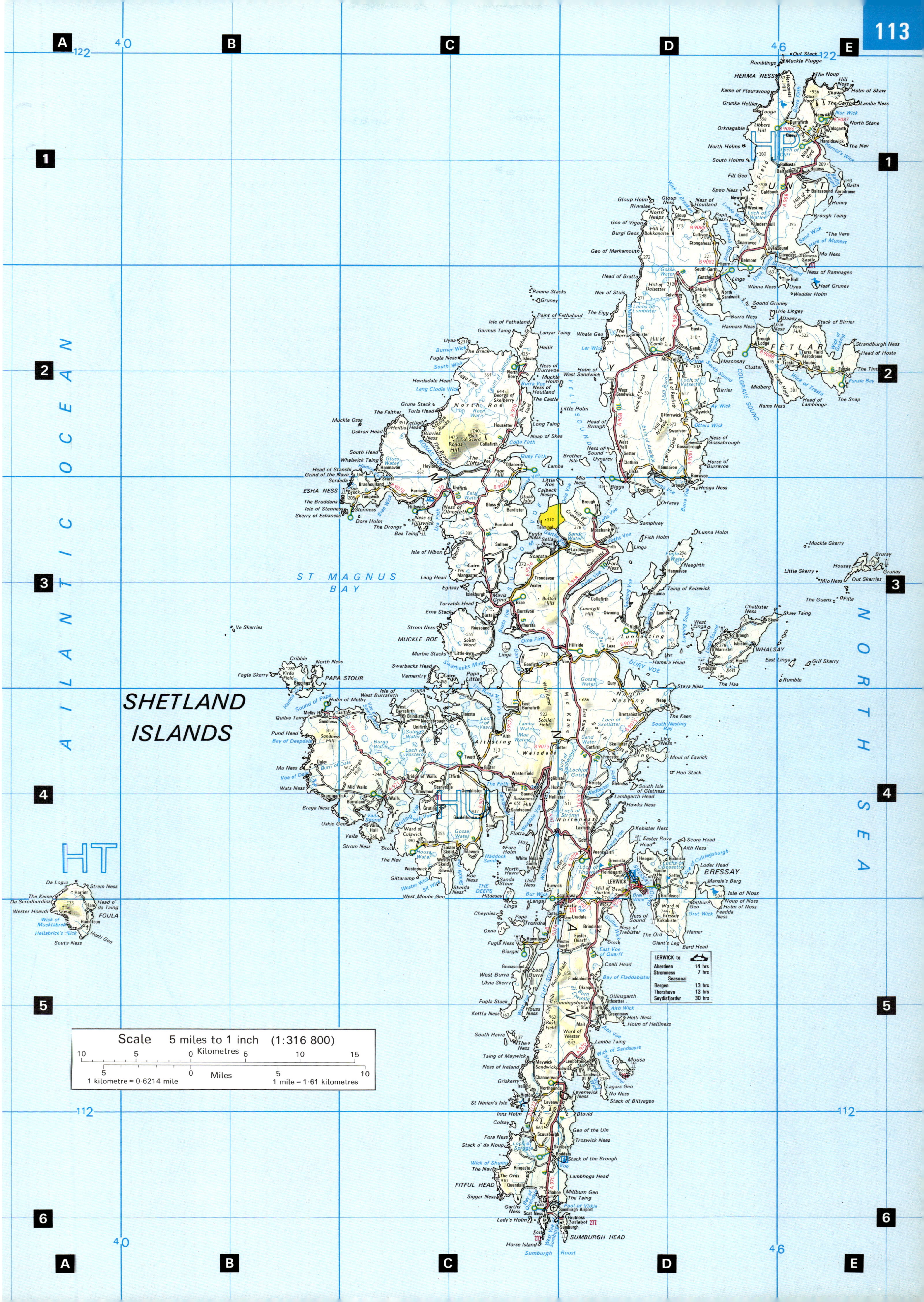
SHETLAND ISLANDS
ATLANTIC OCEAN
NORTH SEA
ST MAGNUS BAY
HP
HU
HT
UNST
YELL
FETLAR
BRESSAY
FOULA
PAPA STOUR
WHALSAY
MUCKLE ROE
ESHA NESS
HERMA NESS
SUMBURGH HEAD
FITFUL HEAD
LERWICK
YELL SOUND
COLGRAVE SOUND
Scale 5 miles to 1 inch (1:316 800)
Kilometres
Miles
1 kilometre = 0·6214 mile
1 mile = 1·61 kilometres
LERWICK to
Aberdeen 14 hrs
Stromness 7 hrs
Seasonal
Bergen 13 hrs
Thorshavn 13 hrs
Seydisfjordur 30 hrs

BIRMINGHAM

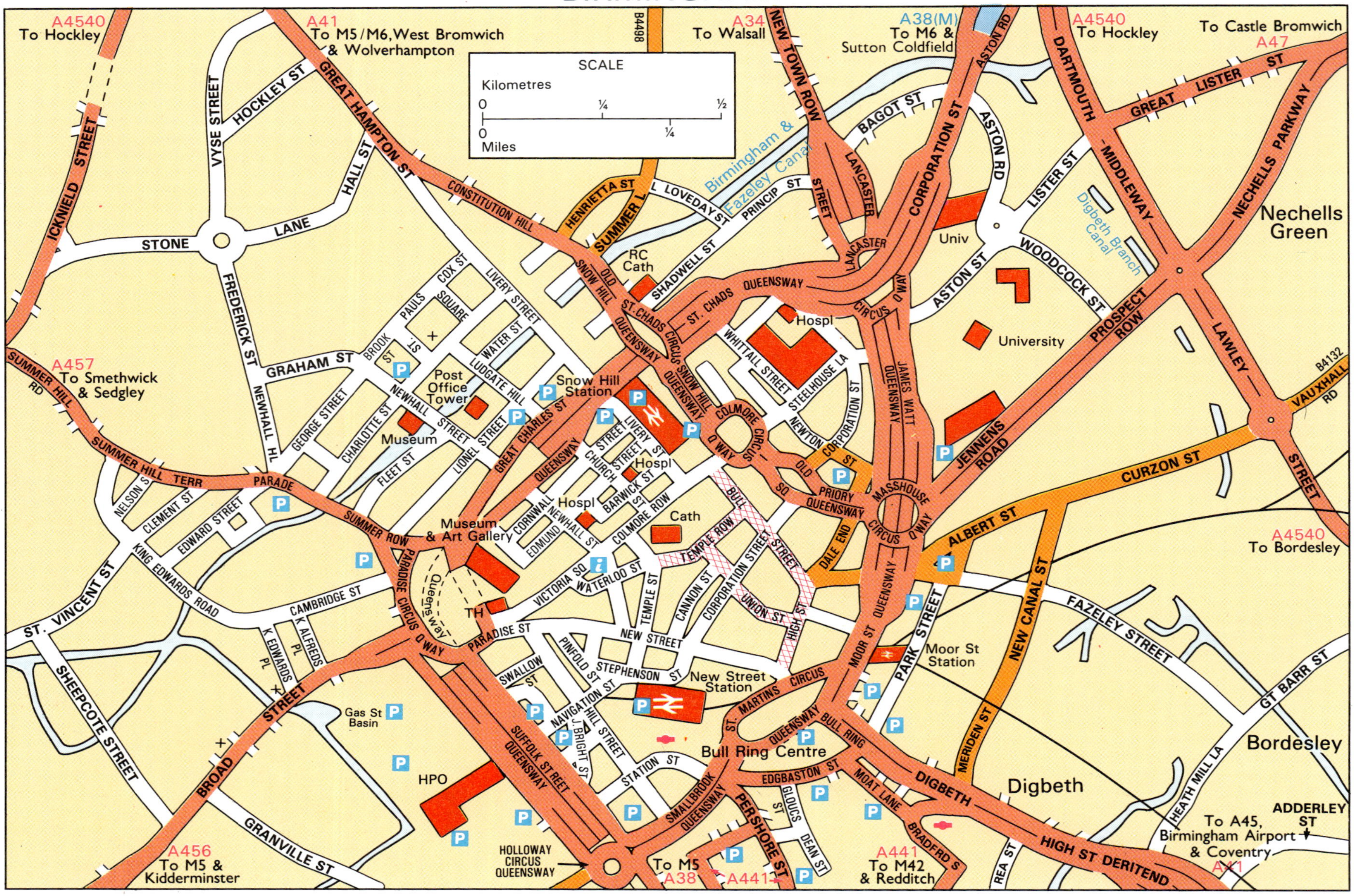

BRISTOL

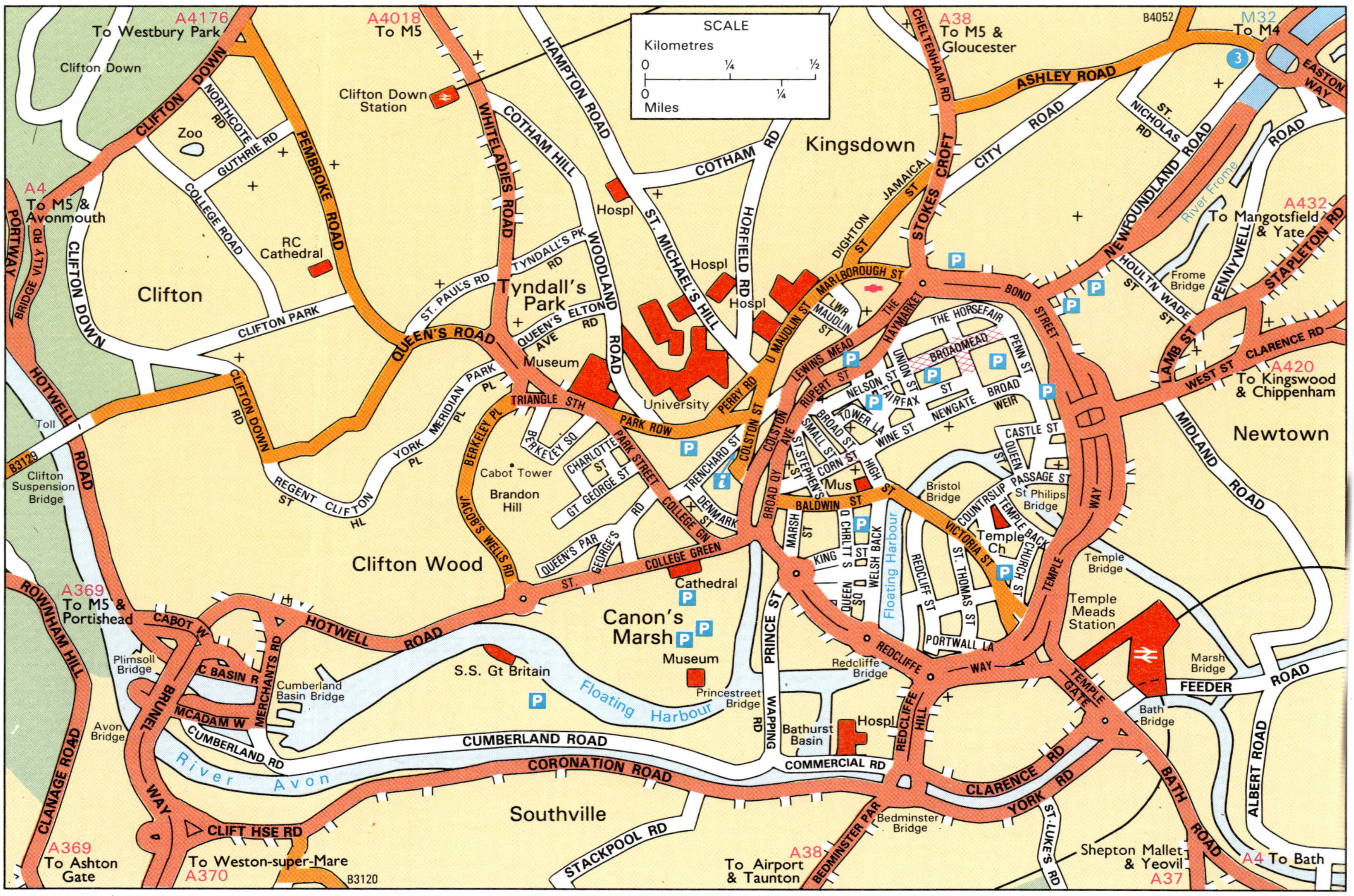

CARDIFF

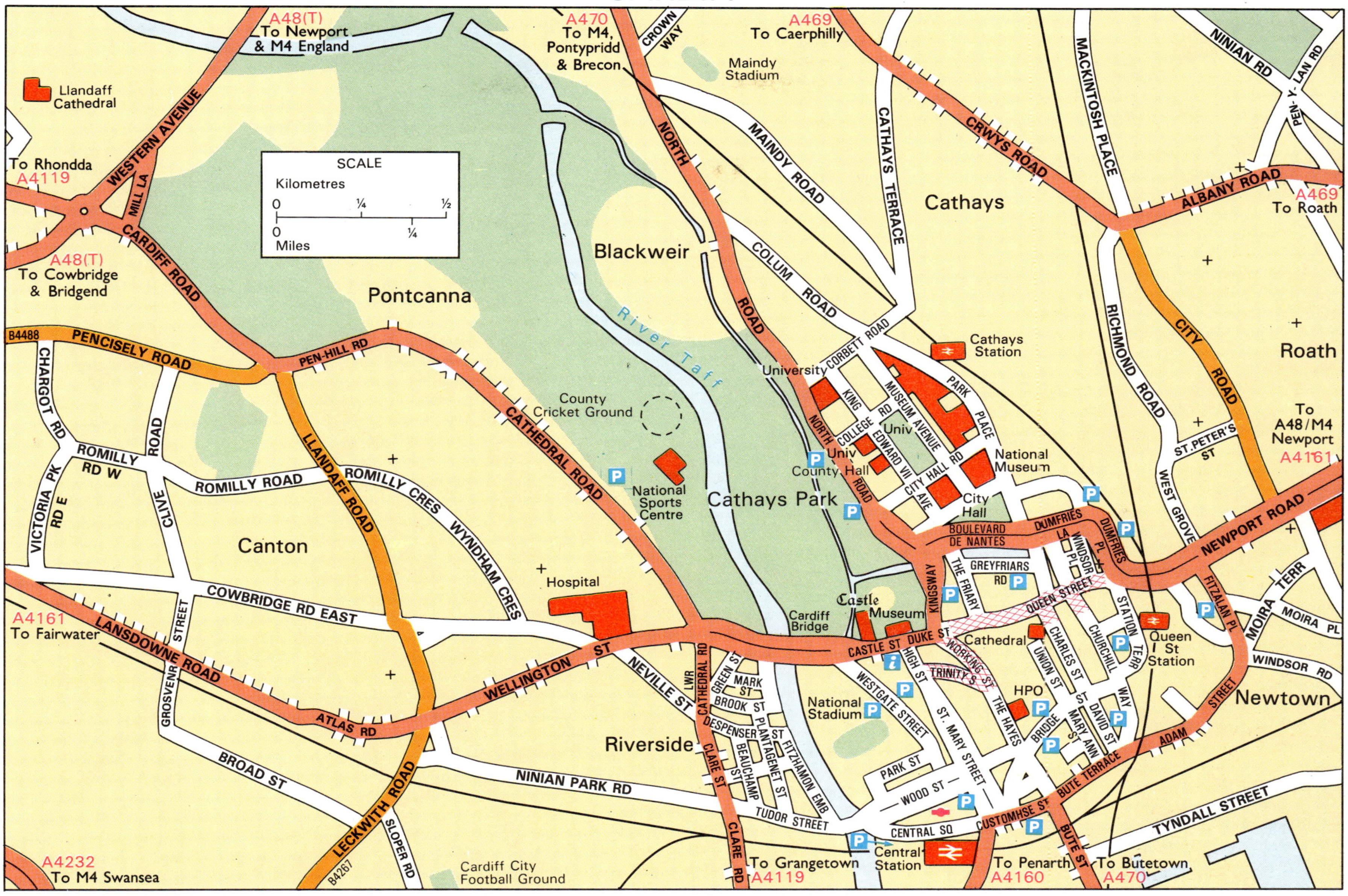
A48(T)
To Newport
& M4 England
A470
To M4,
Pontypridd
& Brecon
A469
To Caerphilly
Maindy
Stadium
Llandaff
Cathedral
To Rhondda
A4119
WESTERN AVENUE
MILL LA
CARDIFF ROAD
SCALE
Kilometres
0
¼
½
0
¼
Miles
A48(T)
To Cowbridge
& Bridgend
Pontcanna
Blackweir
CROWN WAY
NORTH ROAD
MAINDY ROAD
COLUM ROAD
CATHAYS TERRACE
MACKINTOSH PLACE
CRWYS ROAD
NINIAN RD
PEN-Y-LAN RD
ALBANY ROAD
A469
To Roath
Cathays
Roath
CITY ROAD
RICHMOND ROAD
B4488
PENCISELY ROAD
PEN-HILL RD
CHARGOT RD
River Taff
County
Cricket Ground
CATHEDRAL ROAD
University
CORBETT ROAD
Cathays
Station
KING
PARK PLACE
MUSEUM AVENUE
NORTH ROAD
COLLEGE RD
EDWARD VII
Univ
County Hall
CITY HALL RD
National
Museum
City
Hall
To
A48/M4
Newport
A4161
ST. PETER'S ST
WEST GROVE
NEWPORT ROAD
Canton
ROMILLY RD W
ROMILLY ROAD
ROMILLY CRES
CLIVE ROAD
VICTORIA PK RD E
LLANDAFF ROAD
WYNDHAM CRES
National
Sports
Centre
Cathays Park
BOULEVARD DE NANTES
DUMFRIES PL
WINDSOR PL
LA
GREYFRIARS RD
THE FRIARY
KINGSWAY
QUEEN STREET
STATION TERR
CHURCHILL WAY
Queen
St
Station
FITZALAN PL
MOIRA TERR
MOIRA PL
WINDSOR RD
Newtown
Hospital
COWBRIDGE RD EAST
A4161
To Fairwater
LANSDOWNE ROAD
GROSVENOR STREET
Castle
Museum
Cardiff
Bridge
CASTLE ST
DUKE ST
WORKING ST
Cathedral
UNION ST
CHARLES ST
HIGH ST
TRINITY ST
HPO
WELLINGTON ST
NEVILLE ST
CATHEDRAL RD
GREEN ST
MARK ST
BROOK ST
WESTGATE STREET
National
Stadium
ST. MARY STREET
THE HAYES
BRIDGE ST
MARY ANN ST
DAVID ST
ADAM ST
ATLAS RD
Riverside
DESPENSER ST
BEAUCHAMP
PLANTAGENET ST
FITZHAMON EMB
PARK ST
WOOD ST
BUTE TERRACE
BROAD ST
NINIAN PARK RD
TUDOR STREET
CLARE ST
CLARE RD
CENTRAL SQ
CUSTOMHSE ST
TYNDALL STREET
LECKWITH ROAD
SLOPER RD
B4267
A4232
To M4 Swansea
Cardiff City
Football Ground
To Grangetown
A4119
Central
Station
To Penarth
A4160
BUTE ST
To Butetown
A470

EDINBURGH

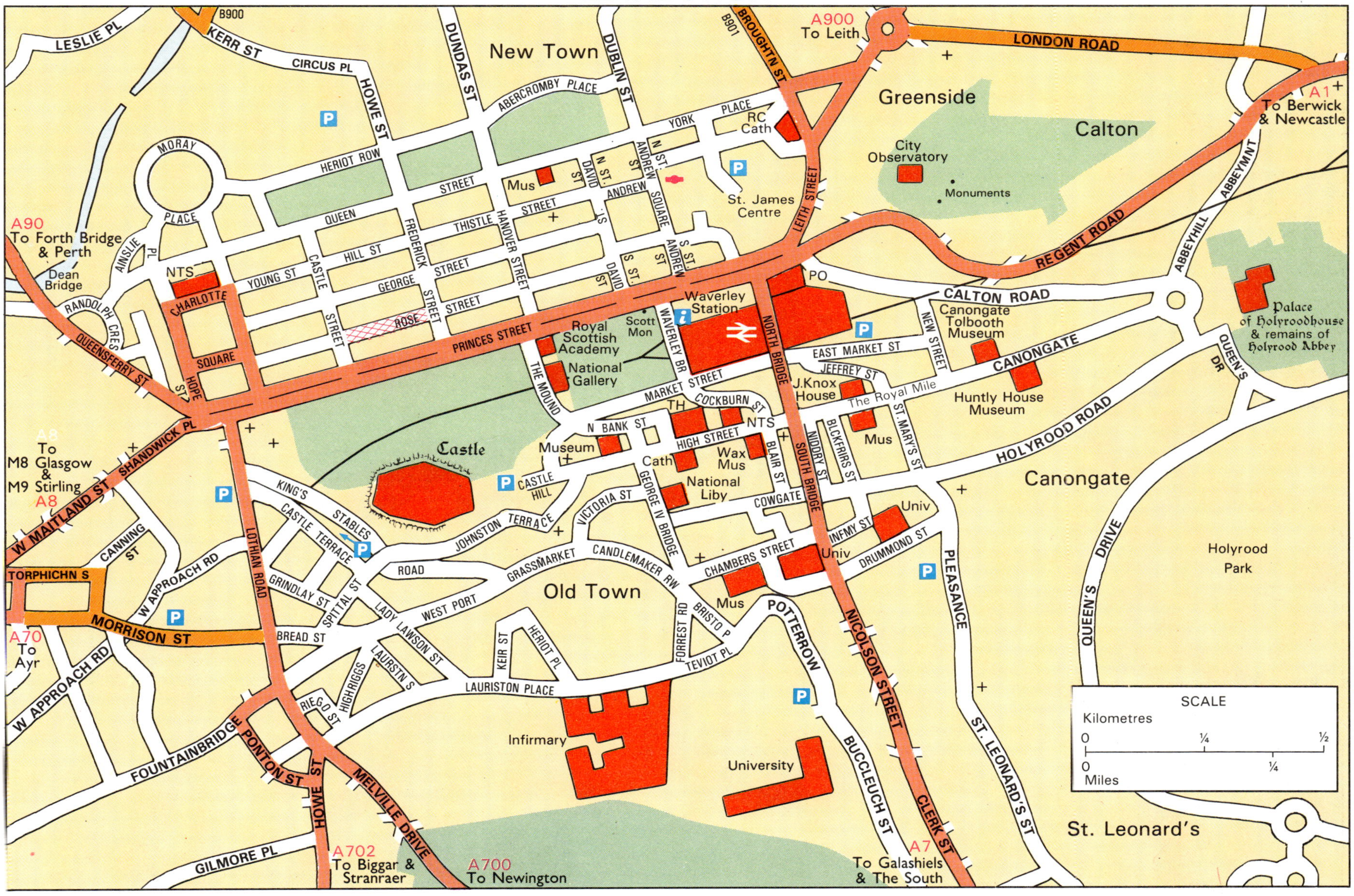
B900
LESLIE PL
KERR ST
CIRCUS PL
HOWE ST
DUNDAS ST
New Town
DUBLIN ST
B901
BROUGHTN ST
A900
To Leith
LONDON ROAD
Greenside
Calton
A1
To Berwick
& Newcastle
ABERCROMBY PLACE
YORK PLACE
RC
Cath
MORAY PLACE
HERIOT ROW
N ST. ANDREW ST
N ST. DAVID ST
N ST. ANDREW ST
Mus
ANDREW SQUARE
St. James
Centre
LEITH STREET
City
Observatory
Monuments
ABBEYMNT
ABBEYHILL
A90
To Forth Bridge
& Perth
Dean
Bridge
AINSLIE PL
QUEEN STREET
THISTLE ST
FREDERICK STREET
HANOVER STREET
S ST. DAVID ST
S ST. ANDREW ST
REGENT ROAD
NTS
YOUNG ST
HILL ST
CASTLE STREET
GEORGE STREET
ROSE STREET
PO
CALTON ROAD
Palace
of Holyroodhouse
& remains of
Holyrood Abbey
RANDOLPH CRES
QUEENSFERRY ST
CHARLOTTE SQUARE
HOPE ST
PRINCES STREET
Royal
Scottish
Academy
Scott
Mon
Waverley
Station
WAVERLEY BR
NORTH BRIDGE
EAST MARKET ST
NEW STREET
Canongate
Tolbooth
Museum
CANONGATE
QUEEN'S DR
National
Gallery
THE MOUND
MARKET STREET
JEFFREY ST
J.Knox
House
The Royal Mile
ST. MARY'S ST
Huntly House
Museum
A8
To
M8 Glasgow
&
M9 Stirling
A8
SHANDWICK PL
W MAITLAND ST
N BANK ST
TH
COCKBURN ST
NTS
HIGH STREET
Museum
Castle
Cath
Wax
Mus
BLAIR ST
NIDDRY ST
SOUTH BRIDGE
BLACKFRIARS ST
Mus
HOLYROOD ROAD
Canongate
CASTLE HILL
KING'S STABLES
CASTLE TERRACE
JOHNSTON TERRACE
VICTORIA ST
GEORGE IV BRIDGE
National
Liby
COWGATE
INFMY ST
Univ
DRUMMOND ST
CANNING ST
TORPHICHN S
W APPROACH RD
GRINDLAY ST
SPITTAL ST
ROAD
LADY LAWSON ST
GRASSMARKET
WEST PORT
CANDLEMAKER RW
CHAMBERS STREET
Univ
Old Town
Mus
BRISTO P
POTTERROW
NICOLSON STREET
PLEASANCE
QUEEN'S DRIVE
Holyrood
Park
LOTHIAN ROAD
MORRISON ST
BREAD ST
A70
To
Ayr
W APPROACH RD
LAURISTN S
HIGHRIGGS
KEIR ST
HERIOT PL
FORREST RD
TEVIOT PL
LAURISTON PLACE
RIEGO ST
Infirmary
University
BUCCLEUCH ST
ST. LEONARD'S ST
SCALE
Kilometres
0
¼
½
0
¼
Miles
FOUNTAINBRIDGE
PONTON ST
HOME ST
MELVILLE DRIVE
GILMORE PL
A702
To Biggar &
Stranraer
A700
To Newington
A7
To Galashiels
& The South
CLERK ST
St. Leonard's

GLASGOW

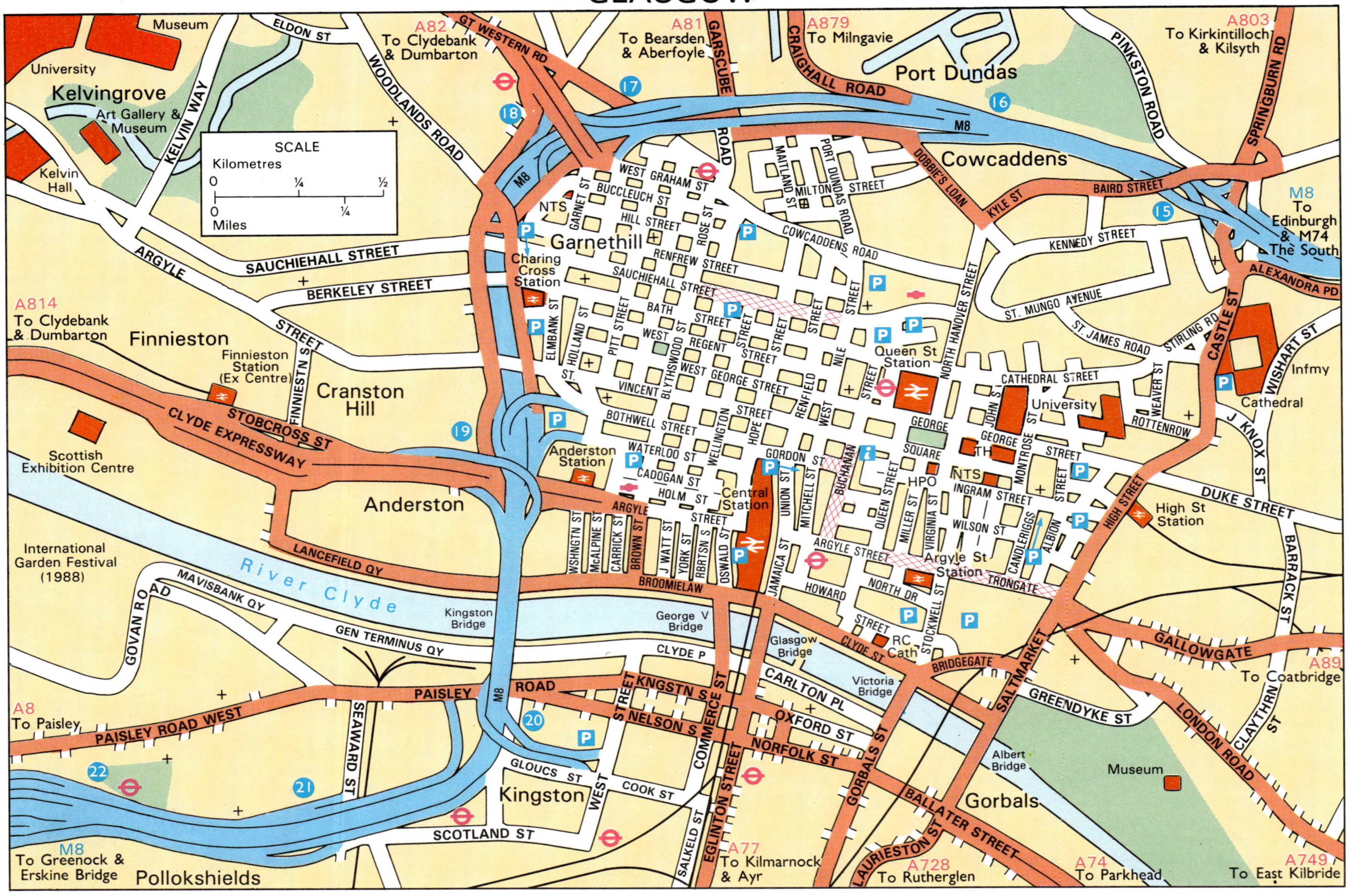
Museum
University
Kelvingrove
Art Gallery & Museum
Kelvin Hall
KELVIN WAY
ELDON ST
WOODLANDS ROAD
A82
To Clydebank & Dumbarton
GT WESTERN RD
A81
To Bearsden & Aberfoyle
GARSCUBE ROAD
A879
To Milngavie
CRAIGHALL ROAD
Port Dundas
A803
To Kirkintilloch & Kilsyth
PINKSTON ROAD
SPRINGBURN RD
Cowcaddens
M8
To Edinburgh & M74 The South
SCALE
Kilometres
0
¼
½
0
¼
Miles
WEST GRAHAM ST
BUCCLEUCH ST
HILL STREET
NTS
Garnethill
Charing Cross Station
RENFREW STREET
SAUCHIEHALL STREET
SAUCHIEHALL STREET
BERKELEY STREET
ARGYLE
A814
To Clydebank & Dumbarton
Finnieston
Finnieston Station (Ex Centre)
Cranston Hill
STOBCROSS ST
CLYDE EXPRESSWAY
Scottish Exhibition Centre
Anderston
Anderston Station
International Garden Festival (1988)
LANCEFIELD QY
River Clyde
MAVISBANK QY
GEN TERMINUS QY
GOVAN ROAD
Kingston Bridge
George V Bridge
BROOMIELAW
Central Station
COWCADDENS ROAD
DOBBIE'S LOAN
KYLE ST
BAIRD STREET
KENNEDY STREET
ST. MUNGO AVENUE
ST. JAMES ROAD
NORTH HANOVER STREET
Queen St Station
ALEXANDRA PD
CASTLE ST
WISHART ST
Infmy
Cathedral
CATHEDRAL STREET
University
ROTTENROW
GEORGE SQUARE
GEORGE STREET
HPO
NTS
INGRAM STREET
DUKE STREET
High St Station
HIGH STREET
ARGYLE STREET
Argyle St Station
TRONGATE
BUCHANAN
GORDON ST
BOTHWELL STREET
WATERLOO ST
CADOGAN ST
VINCENT ST
HOLM ST
HOWARD
RC Cath
CLYDE ST
STOCKWELL ST
Glasgow Bridge
Victoria Bridge
BRIDGEGATE
SALTMARKET
GALLOWGATE
BARRACK ST
A89
To Coatbridge
GREENDYKE ST
LONDON ROAD
CLAYTHRN ST
Museum
Albert Bridge
Gorbals
BALLATER STREET
GORBALS ST
CARLTON PL
OXFORD ST
NORFOLK ST
NELSON S
KINGSTN ST
COMMERCE ST
EGLINTON STREET
SALKELD ST
PAISLEY ROAD
PAISLEY ROAD WEST
A8
To Paisley
SEAWARD ST
Kingston
GLOUCS ST
COOK ST
SCOTLAND ST
WEST STREET
M8
To Greenock & Erskine Bridge
Pollokshields
A77
To Kilmarnock & Ayr
A728
To Rutherglen
LAURIESTON ST
A74
To Parkhead
A749
To East Kilbride

LEEDS

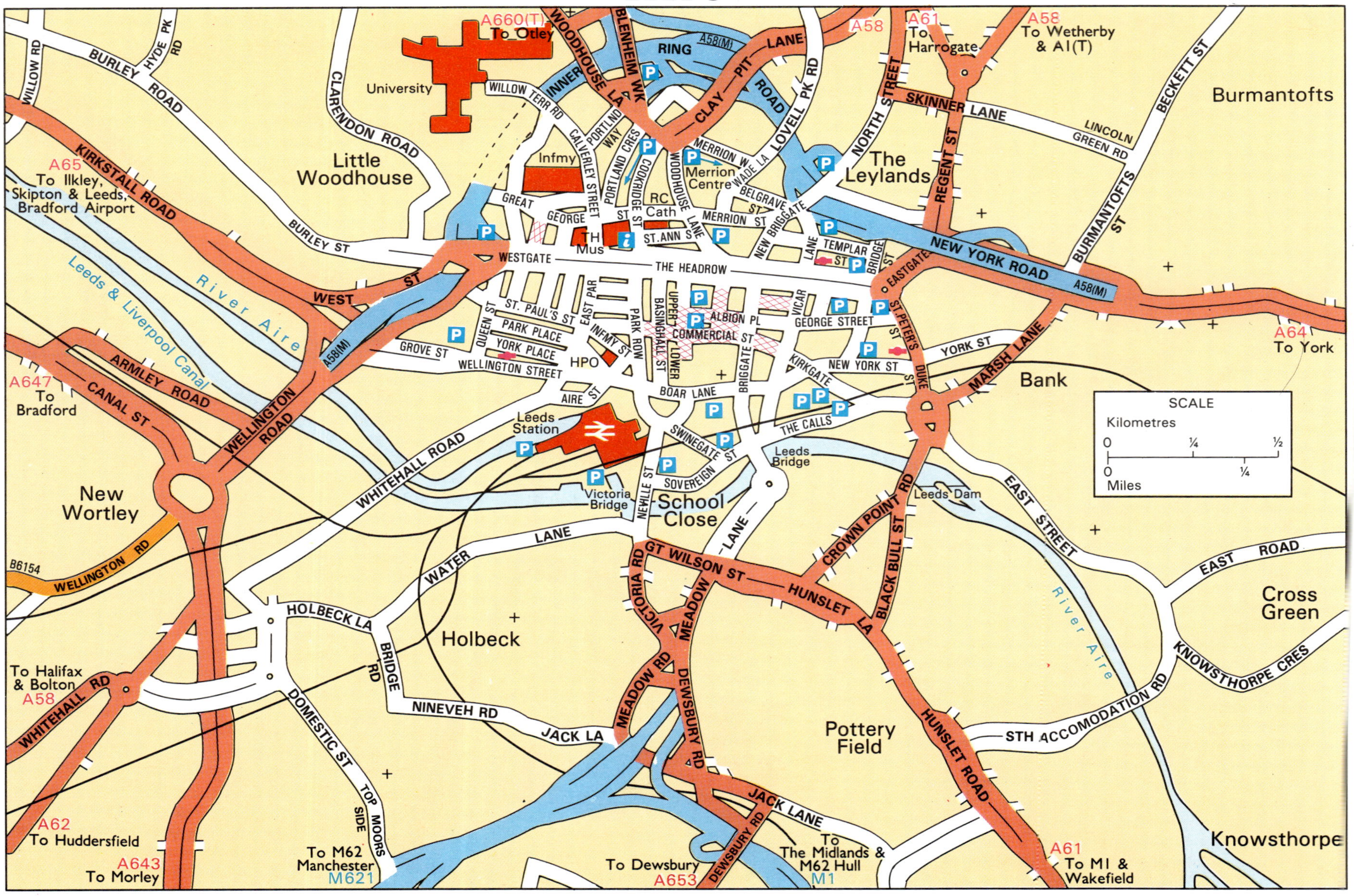
A660(T)
To Otley
University
WILLOW RD
BURLEY RD
HYDE PK RD
CLARENDON ROAD
Little Woodhouse
A65
To Ilkley, Skipton & Leeds, Bradford Airport
KIRKSTALL ROAD
BURLEY ST
WOODHOUSE LA
BLENHEIM WK
INNER RING
A58(M)
CLAY PIT LANE
LOVELL PK RD
A58
A61
To Harrogate
A58
To Wetherby & A1(T)
NORTH STREET
SKINNER LANE
LINCOLN GREEN RD
BECKETT ST
Burmantofts
The Leylands
REGENT ST
BURMANTOFTS ST
NEW YORK ROAD
Infmy
Merrion Centre
MERRION ST
GREAT GEORGE ST
TH
Mus
RC Cath
ST. ANN'S ST
WESTGATE
THE HEADROW
EASTGATE
TEMPLAR ST
WEST ST
River Aire
Leeds & Liverpool Canal
A647
To Bradford
ARMLEY ROAD
CANAL ST
WELLINGTON ROAD
WELLINGTON STREET
GROVE ST
PARK PLACE
YORK PLACE
HPO
ST. PAUL'S ST
BOAR LANE
ALBION PL
COMMERCIAL ST
GEORGE STREET
KIRKGATE
NEW YORK ST
YORK ST
MARSH LANE
Bank
A64
To York
SCALE
Kilometres
0
¼
½
0
¼
Miles
Leeds Station
THE CALLS
Leeds Bridge
Leeds Dam
SWINEGATE
SOVEREIGN ST
Victoria Bridge
School Close
NEVILLE ST
WHITEHALL ROAD
New Wortley
B6154
WELLINGTON RD
WATER LANE
HOLBECK LA
Holbeck
BRIDGE RD
NINEVEH RD
JACK LA
DOMESTIC ST
GT WILSON ST
CROWN POINT RD
HUNSLET LA
BLACK BULL ST
EAST STREET
EAST ROAD
Cross Green
KNOWSTHORPE CRES
River Aire
VICTORIA RD
MEADOW LA
MEADOW RD
DEWSBURY RD
Pottery Field
HUNSLET ROAD
STH ACCOMODATION RD
To Halifax & Bolton
A58
WHITEHALL RD
A62
To Huddersfield
A643
To Morley
To M62 Manchester
M621
TOP MOORS SIDE
To Dewsbury
A653
JACK LANE
To The Midlands & M62 Hull
M1
A61
To M1 & Wakefield
Knowsthorpe

LIVERPOOL

A5036 To Crosby
A565 To Southport
A5038 To Litherland
A59 To Ormskirk & Preston
A580 To Manchester
A5049 To West Derby
Elm Park
Kingsway (Road Tunnel)
WATERLOO ROAD
GREAT HOWARD ST
PALL MALL
VAUXHALL ROAD
SCOTLAND ROAD
ST. ANNE STREET
FOX ST
EVERTON BROW
VILLAGE ST
SHAW STREET
EVERTON ROAD
WEST DERBY RD
SHEIL ROAD
BRUNSWICK RD
LOW HILL
LEEDS STREET
ADDISON S
BYROM ST
CHRISTIAN S
GT CROSSHALL ST
ISLINGTON
KENSINGTON
A57 To Prescot & St. Helens
CHURCHILL WAY
Mus
Art Gallery
PRESCOT ST
HOLT RD
W. BROWN ST
LONDON ROAD
PEMBROKE PLACE
Infirmary
HALL LANE
BATH STREET
KING EDWARD ST
EAST ST
OLD HALL ST
BIXTETH ST
TITHEBARN ST
DALE STREET
Moorfields Station
Hall
LD NELSON ST
Lime St Station
Infmy
WEST DERBY ST
Univ
EDGE LANE
A5047 To M62 Manchester & St. Helens
NEW QUAY
Ferries to Ireland & Isle of Man
CHAPEL ST
TH
WATER STREET
VICTORIA ST
WHITECHAPEL
ROE ST
COPPERAS HILL
HPO
RUSSELL ST
BROWNLOW HILL
CROWN ST
IRVINE ST
DURNING ROAD
Ferry (Foot)
Royal Liver Building
THE STRAND
NICHOLAS PL
CASTLE ST
MAYHEW S
St. John's Centre
CHURCH ST
LORD STREET
BRUNSWICK ST
JAMES ST
James St Station
SCHOOL LA
PARADISE ST
RANELAGH ST
Central Station
RC Cathedral
Univ
WAVERTREE ROAD
B5178
Queensway (Mersey Tunnel)
Museum
STRAND STREET
HANOVER STREET
RENSHAW ST
MOUNT PLEASANT
GRINFIELD ST
OXFORD ST
Edge Hill
OVERBURY ST
TUNNEL ROAD
Edge Hill Station
Ferry (Foot)
River Mersey
WAPPING
DUKE ST
BERRY ST
RODNEY STREET
HARDMAN ST
Hospl
Univ
GROVE STREET
PARK LA
UPPER DUKE ST
Hospl
CATHARINE ST
Hospl
ST JAMES STREET
GT GEORGE ST
ST JAMES RD
HOPE ST
Cath
SMITHDOWN ROAD
SCALE
Kilometres
0 ¼ ½
0 ¼
Miles
CHALONER ST
UPPER PARLIAMENT STREET
PRINCES RD
MULGRAVE ST
Princes Park
KINGSLEY RD
LODGE LANE
A562 To Widnes & Runcorn
A5036 To Toxteth
PARLIAMENT ST
A561 To Widnes
B5175
B5174
B5173
B5182
B5186

MANCHESTER

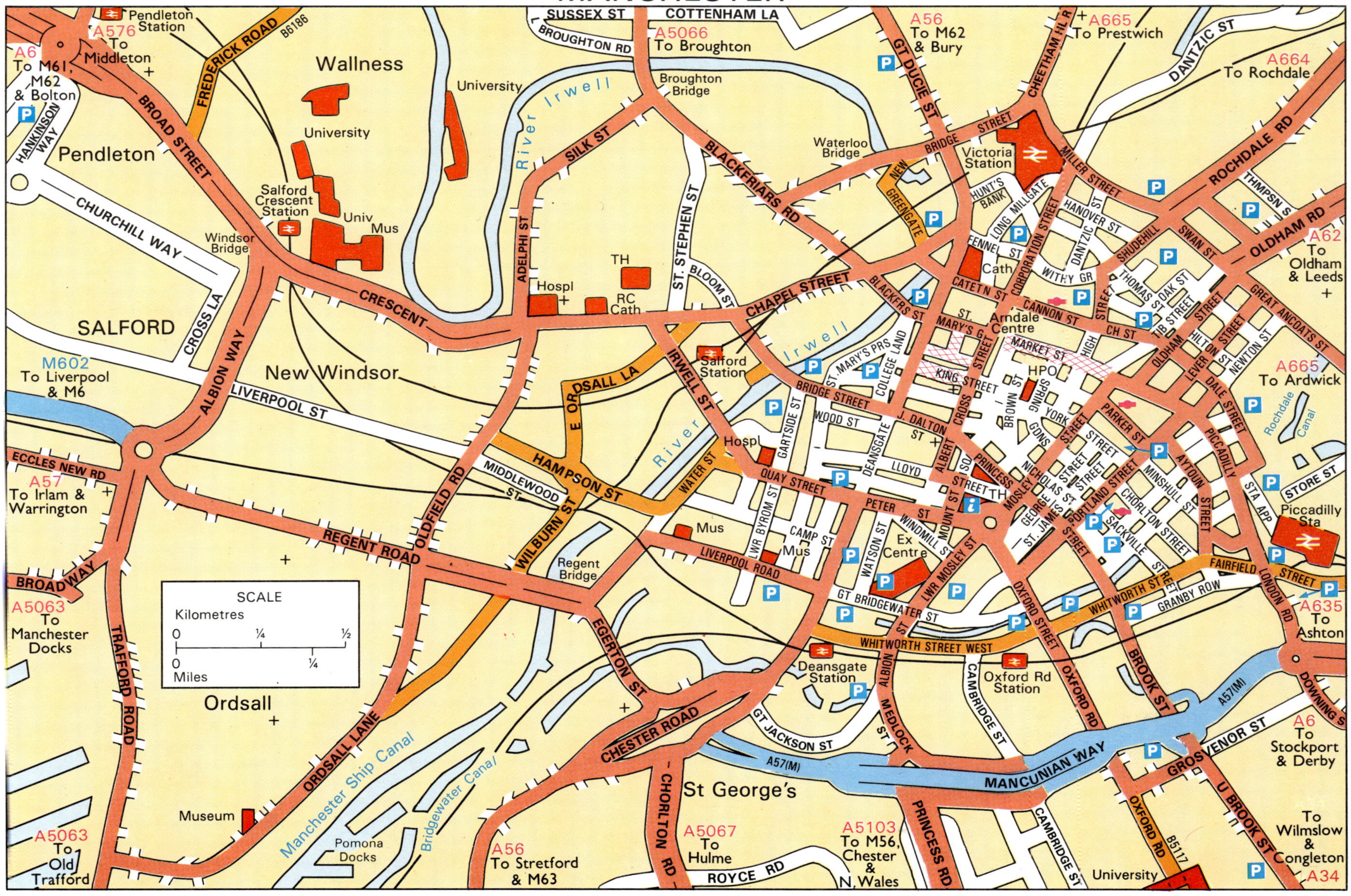

Pendleton Station
A576 To Middleton
FREDERICK ROAD
B6186
SUSSEX ST
COTTENHAM LA
BROUGHTON RD
A5066 To Broughton
A56 To M62 & Bury
A665 To Prestwich
DANTZIC ST
A664 To Rochdale
A6 To M61, M62 & Bolton
Wallness
University
River Irwell
Broughton Bridge
HANKINSON WAY
Pendleton
BROAD STREET
University
SILK ST
BLACKFRIARS RD
GT DUCIE ST
CHEETHAM H RD
Victoria Station
BRIDGE STREET
MILLER STREET
ROCHDALE RD
Waterloo Bridge
CHURCHILL WAY
Salford Crescent Station
Univ
Mus
Windsor Bridge
ADELPHI ST
ST. STEPHEN ST
GREENGATE
CORPORATION STREET
SWAN ST
OLDHAM RD
A62 To Oldham & Leeds
Cath
TH
Hospl
RC Cath
BLOOM ST
CHAPEL STREET
BLACKFRIARS ST
CANNON ST
SALFORD
CROSS LA
CRESCENT
Arndale Centre
MARKET ST
HPO
GREAT ANCOATS ST
M602 To Liverpool & M6
New Windsor
Salford Station
ST. MARY'S PRS
KING STREET
A665 To Ardwick
ALBION WAY
LIVERPOOL ST
ORDSALL LA
IRWELL ST
BRIDGE STREET
DEANSGATE
PICCADILLY
ECCLES NEW RD
A57 To Irlam & Warrington
MIDDLEWOOD ST
HAMPSON ST
River Irwell
Hospl
QUAY STREET
LLOYD ST
PRINCESS ST
PORTLAND STREET
Rochdale Canal
STORE ST
Piccadilly Sta
REGENT ROAD
OLDFIELD RD
WILBURN ST
WATER ST
Mus
LWR BYROM ST
CAMP ST
Mus
PETER ST
Ex Centre
WATSON ST
LIVERPOOL ROAD
GT BRIDGEWATER ST
WHITWORTH ST
FAIRFIELD STREET
LONDON RD
A635 To Ashton
BROADWAY
A5063 To Manchester Docks
Regent Bridge
SCALE
Kilometres
0 ¼ ½
0 ¼
Miles
EGERTON ST
WHITWORTH STREET WEST
GRANBY ROW
TRAFFORD ROAD
Deansgate Station
ALBION ST
CAMBRIDGE ST
Oxford Rd Station
OXFORD ROAD
BROOK ST
A57(M)
Ordsall
ORDSALL LANE
CHESTER ROAD
GT JACKSON ST
MEDLOCK ST
MANCUNIAN WAY
GROSVENOR ST
DOWNING ST
A6 To Stockport & Derby
Manchester Ship Canal
Bridgewater Canal
CHORLTON RD
St George's
U BROOK ST
Museum
Pomona Docks
A5063 To Old Trafford
A56 To Stretford & M63
A5067 To Hulme
ROYCE RD
A5103 To M56, Chester & N. Wales
PRINCESS RD
CAMBRIDGE ST
University
OXFORD RD
B5117
To Wilmslow & Congleton A34

Lord's Cricket Ground
London Mosque
Boat Houses
Boating Lake
The Holme
Open Air Theatre
Queen Mary's Gardens
Inner Circle
New Lodge
Regent's College
Park Square Gardens
Crescent Gardens
Park Crescent
Madame Tussaud's & Planetarium
Royal Academy of Music
London Clinic and Nursing Home
British Telecom Tower
Lisson Green
Lisson Grove
Marylebone Station
Paddington Green Children's Hosp
Florence Nightingale Hospital
Samaritan Hosp for Women
Western Ophthalmic Hosp
Paddington Basin
St Mary's Hospital
Synagogue
Marylebone Road
Edgware Road
Wallace Collection
Wigmore Hall
Marylebone
Oxford Street
London Palladium
Marble Arch
Tyburn Tree (site of)
Speakers Corner
Cumberland Gate
Bayswater Road
North Carriage Drive
Victoria Gate
Westbourne Gate
Marlborough Gate
The Fountains
Buck Hill Walk
North Ride
The Long Water
Nursery
New Lodge
Ranger's Lodge
Serpentine Lodge
Serpentine Bridge
Hyde Park
The Serpentine
Boat Houses
Restaurant
Rotten Row
Bowling Greens
Prince of Wales Gate
Albert Gate
Broad Walk
Park Lane
Grosvenor Square
Mayfair
Museum of Mankind
Royal Academy of Arts
Piccadilly
Green Park
Lancaster House
Clarence House
Constitution Hill
Queen Victoria Memorial
Buckingham Palace
Queens Gallery
The Royal Mews
Wellington Barracks
Hyde Park Corner
Apsley House
Knightsbridge
Albert Memorial
Alexandra Gate
Imperial College
Science Museum
Geo Mus
British Museum (Natural History)
Victoria & Albert Museum
The Oratory
Brompton
Brompton Road
Sloane Street
Belgrave Square
Grosvenor Place
Westminster Cathedral
Victoria Station

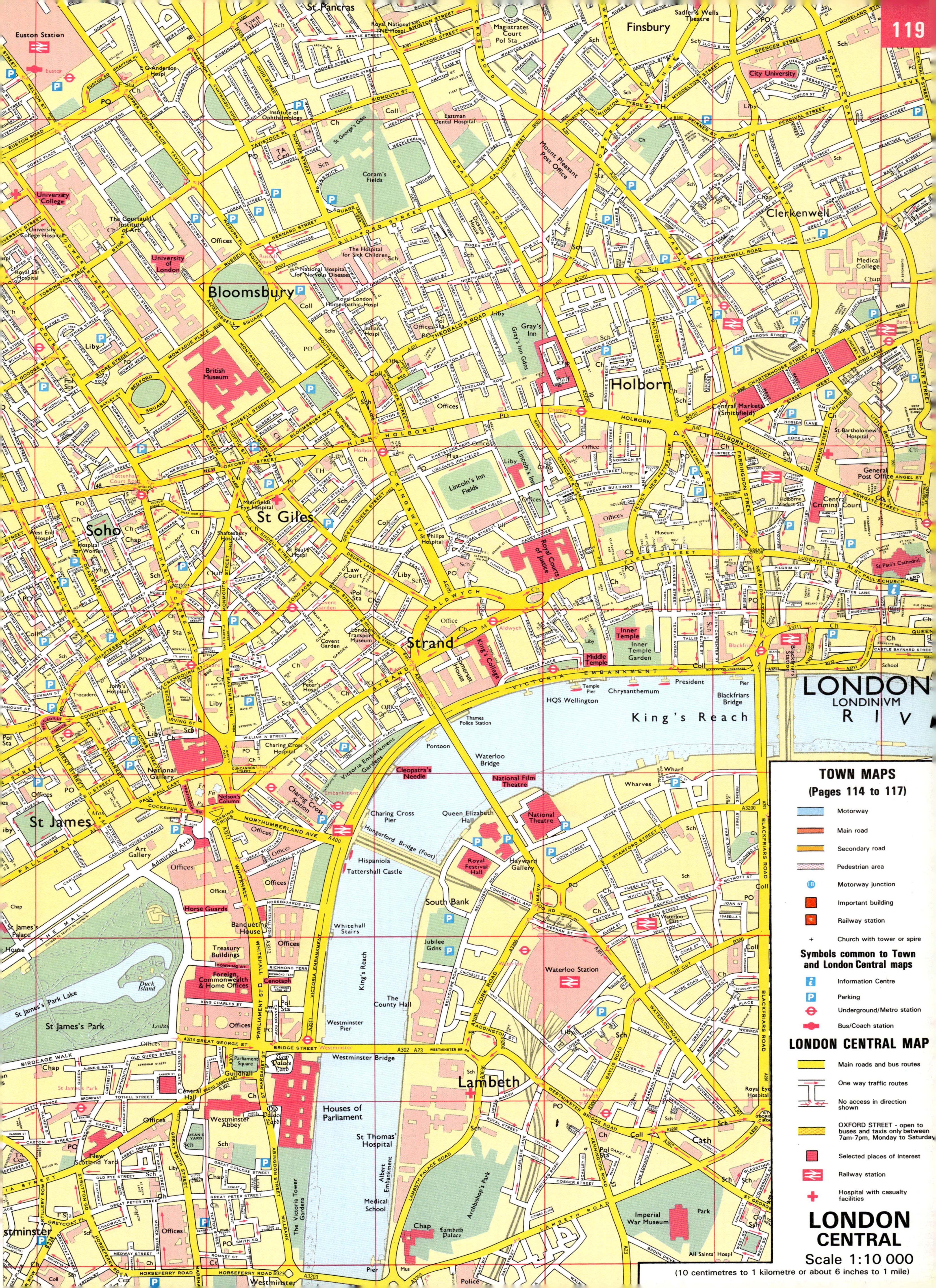
Bloomsbury
Holborn
Clerkenwell
Finsbury
Soho
St Giles
Strand
St James
Lambeth
St Pancras
Euston Station
University College
University of London
British Museum
Coram's Fields
Gray's Inn
Lincoln's Inn Fields
Royal Courts of Justice
Inner Temple
Middle Temple
King's College
Somerset House
Central Markets (Smithfield)
St Bartholomew's Hospital
Central Criminal Court
General Post Office
St Paul's Cathedral
Blackfriars Bridge
Blackfriars Station
King's Reach
Waterloo Bridge
Cleopatra's Needle
National Film Theatre
National Theatre
Queen Elizabeth Hall
Royal Festival Hall
Hayward Gallery
Hungerford Bridge (Foot)
Charing Cross Station
Nelson's Column
National Gallery
Admiralty Arch
Horse Guards
Banqueting House
Treasury Buildings
Foreign, Commonwealth & Home Offices
Cenotaph
St James's Park
St James's Park Lake
Duck Island
Westminster Bridge
Houses of Parliament
Westminster Abbey
New Scotland Yard
The County Hall
Waterloo Station
St Thomas' Hospital
Lambeth Palace
Archbishop's Park
Imperial War Museum
Victoria Embankment
Albert Embankment
Mount Pleasant Post Office
City University
LONDON
LONDINIVM
TOWN MAPS
(Pages 114 to 117)
Motorway
Main road
Secondary road
Pedestrian area
Motorway junction
Important building
Railway station
Church with tower or spire
Symbols common to Town and London Central maps
Information Centre
Parking
Underground/Metro station
Bus/Coach station
LONDON CENTRAL MAP
Main roads and bus routes
One way traffic routes
No access in direction shown
OXFORD STREET - open to buses and taxis only between 7am-7pm, Monday to Saturday
Selected places of interest
Railway station
Hospital with casualty facilities
LONDON CENTRAL
Scale 1:10 000
(10 centimetres to 1 kilometre or about 6 inches to 1 mile)

M25 and Routes into London

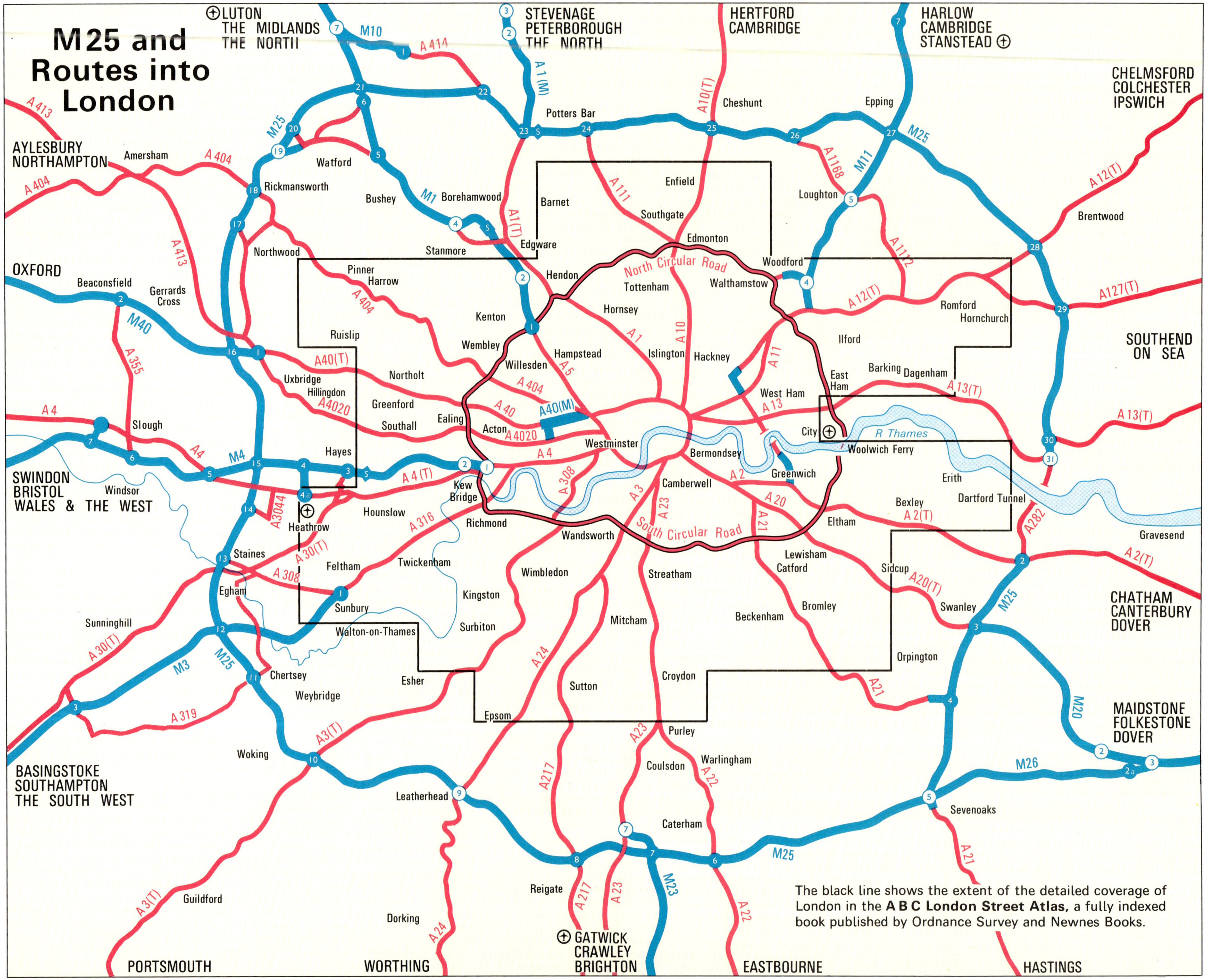

The black line shows the extent of the detailed coverage of London in the **A B C London Street Atlas**, a fully indexed book published by Ordnance Survey and Newnes Books.

Index

How to use this Index

For each entry the Atlas page number is listed and an alphanumeric map reference is given to the grid square in which the name appears. For example:

Barnstaple 7 F2

Barnstaple will be found on page 7, square F2.

County Names showing abbreviations used in this Index

England

Avon	Avon
Bedfordshire	Beds
Berkshire	Berks
Buckinghamshire	Bucks
Cambridgeshire	Cambs
Cheshire	Ches
Cleveland	Cleve
Cornwall	Corn
Cumbria	Cumbr
Derbyshire	Derby
Devon	Devon
Dorset	Dorset
Durham	Durham
East Sussex	E. Susx
Essex	Essex
Gloucestershire	Glos
Greater London	G. Lon
Greater Manchester	G. Man
Hampshire	Hants
Hereford & Worcester	H. & W
Hertfordshire	Herts
Humberside	Humbs
Isle of Wight	I. of W
Kent	Kent
Lancashire	Lancs
Leicester	Leic
Lincolnshire	Lincs
Merseyside	Mers
Norfolk	Norf
North Yorkshire	N. Yks
Northamptonshire	Northnts
Northumberland	Northum
Nottinghamshire	Notts
Oxfordshire	Oxon
Shropshire	Shrops
Somerset	Somer
South Yorkshire	S. Yks
Staffordshire	Staffs
Suffolk	Suff
Surrey	Surrey
Tyne and Wear	T. & W
Warwickshire	Warw
West Midlands	W.Mids
West Sussex	W. Susx
West Yorkshire	W. Yks
Wiltshire	Wilts

Wales

Clwyd	Clwyd
Dyfed	Dyfed
Gwent	Gwent
Gwynedd	Gwyn
Mid Glamorgan	M. Glam
Powys	Powys
South Glamorgan	S. Glam
West Glamorgan	W. Glam

Other Areas

Isle of Man	I. of M
Isles of Scilly	I. Scilly

Region & Island Area Names Scotland

Regions

Region	Abbreviation
Borders	Border
Central	Central
Dumfries & Galloway	D. & G
Fife	Fife
Grampian	Grampn
Highland	Highl
Lothian	Lothn
Strathclyde	Strath
Tayside	Tays

Island Areas

Orkney	Orkney
Shetland	Shetld
Western Isles	W. Isles

The National Grid

The blue grid lines which appear on the Atlas map pages are from the Ordnance Survey National Grid. The National Grid is a reference system which breaks the country down into squares to enable a unique reference to be given to a place or feature. This reference will always be the same no matter which Ordnance Survey map product is used. The squares which form the basic grid cover an area of 100 kilometres by 100 kilometres and are identified by letters; eg SU, TQ. These squares are subdivided by grid lines each carrying a reference number. The numbering sequence runs East and North from the South West corner of the country.

Grid lines on the Atlas map pages appear at 10 kilometre intervals. The 100 kilometre lines are shown in a darker blue. Those grid lines which fall at the top, bottom and outside edge of each page of Atlas mapping also carry their reference numbers (eg $^{2}4$) printed in blue. The larger number is the reference of the actual grid line, the smaller that of the preceding 100 kilometre grid line. The letters which identify each 100 kilometre square appear on the Atlas mapping also printed in blue.

A leaflet on the National Grid referencing system is available from Information and Enquiries, Ordnance Survey, Romsey Road, Maybush, Southampton SO9 4DH.

A

B

Baltasound 113 E1
Balterley 43 J2
Balthangie 99 H3
Baltonsborough 9 G1
Balvaird 96 D3
Balvicar 89 J7
Balvraid 97 G5
Bamber Bridge 61 J7
Bamburgh 81 K6
Bamford 54 F5
Bampton, Cumbr 67 H5
Bampton, Oxon 31 H7
Bampton, Devon 7 K3
Banavie 90 C4
Banbury 35 F6
Banchory 93 J2
Banchory-Devenick 93 L1
Bancyfelin 27 H6
Bancyffordd 27 J4
Banff 99 F2
Bangor 50 E3
Bangor-is-y-coed 42 E3
Banham 38 F2
Bank 10 D4
Bankend, Strath 79 F6
Bankend, D. & G 72 F7
Bankfoot 86 D1
Bankglen 72 A2
Bankhead, Grampn 98 F8
Bankhead, Grampn 99 H7
Bank Newton 62 C4
Banknock 85 K7
Banks, Cumbr 74 B7
Banks, Lancs 52 E1
Bankshill 73 G5
Bank Street 33 F4
Banningham 49 H5
Bannister Green 22 C1
Bannockburn 85 L5
Banstead 21 G8
Bantham 5 G7
Banton 85 K7
Banwell 16 E5
Bapchild 14 F2
Bapton 18 A7
Baramore 89 H1
Barassie 77 J4
Barbaraville 103 F6
Barber Booth 54 E5
Barbon 61 K1
Barbridge 53 H7
Barbrook 7 H1
Barby 35 G3
Barcaldine 90 A7
Barcheston 34 D7
Barcombe 13 H5
Barcombe Cross 13 H5
Barden 68 D7
Bardfield Saling 38 B8
Bardney 57 F7
Bardowie 85 H7
Bardrainney 84 F7
Bardsea 61 G2
Bardsey 63 G5
Bardsley 53 L3
Bardwell 38 E3
Barewood 29 H2
Barford, Warw 34 D4
Barford, Norf 49 G7
Barford St Martin 10 B1
Barford St Michael 31 J4
Barfrestone 15 J3
Bargoed 25 K5
Bargrennan 70 E4
Barham, Cambs 36 E3
Barham, Suff 39 G5
Barham, Kent 15 J3
Bar Hill 37 G4
Barholm 46 D6
Barkby 45 H7
Barkestone-le-Vale 45 J4
Barkham 20 B7
Barking, Suff 38 F5
Barking, G. Lon 21 J5
Barkingside 21 J5
Barkisland 54 D2
Barkston, Lincs 46 C3
Barkston, N. Yks 63 H6
Barkway 37 G7
Barkwith 57 F5
Barlaston 43 K4
Barlavington 12 C5
Barlborough 55 H6
Barlby 63 K6
Barlestone 44 F7
Barley, Lancs 62 B5
Barley, Herts 37 G7
Barleythorpe 46 B7
Barling 22 F5
Barlow, Derby 55 G6
Barlow, T. & W 75 H7
Barlow, N. Yks 63 K7
Barmby Moor 64 E8
Barmby on the Marsh 58 A4
Barmer 48 D4
Barmouth 40 F2
Barmpton 68 F5
Barmston 65 J7
Barnack 46 D7
Barnacle 34 E2
Barnard Castle 68 C5
Barnard Gate 31 J6
Barnardiston 38 C6
Barnburgh 55 H3
Barnby 39 K2
Barnby Dun 55 K3
Barnby in the Willows 46 B2
Barnby Moor 55 K5
Barnes 21 G6
Barnet 21 G4
Barnetby le Wold 58 E6
Barney 48 E4
Barnham, W. Susx 12 C6
Barnham, Suff 38 D3
Barnham Broom 49 F7
Barnhead 93 H6
Barnhill 97 K3
Barningham, Durham 68 C5
Barningham, Suff 38 E3
Barnoldby le Beck 59 G6
Barnoldswick 62 B5
Barns Green 12 E4
Barnsley, Glos 30 E7
Barnsley, S. Yks 55 G3
Barnstaple 7 F2
Barnston, Essex 22 C2
Barnston, Mers 52 D5
Barnt Green 34 B3
Barnton 53 H6
Barnwell 36 D2
Barnwood 30 C6
Barony, The 112 B3
Barr 70 D2
Barrachan 70 E7
Barrack 99 H4
Barraglom 108 B3
Barrahormid 83 G3
Barran 84 A2
Barrapol 88 A4
Barras, Grampn 93 K3
Barras, Cumbr 67 L5
Barrasford 74 F6
Barravullin 89 K8
Barregarrow 60 Q3
Barrhead 77 L2
Barrhill 70 D3
Barrington, Somer 8 E3
Barrington, Cambs 37 G6
Barripper 2 E6
Barrmill 77 J2
Barrock 107 H2
Barrow, Leic 46 B6
Barrow, Suff 38 C4
Barrow, Shrops 43 H7
Barrow, Lancs 61 L6
Barroway Drove 47 J7
Barrowby 46 B4
Barrowden 46 C7
Barrowford 62 B6
Barrow Gurney 17 G4
Barrow-in-Furness 60 E3
Barrow Street 9 K1
Barrow upon Humber 58 E4
Barrow upon Soar 45 G6
Barrow upon Trent 44 E5
Barry, Tays 87 J1
Barry, S. Glam 25 K8
Barry Island 25 K8
Barsby 45 H6
Barsham 39 J2
Barston 34 D3
Bartestree 29 K3
Bartford St Martin 10 B1
Barthol Chapel 99 H5
Barthomley 43 J2
Bartley 10 E3
Bartlow 37 J6
Barton, Warw 34 C5
Barton, N. Yks 68 E6
Barton, Ches 42 F2
Barton, Glos 30 F5
Barton, Cambs 37 H5
Barton, Lancs 61 J6
Barton, Devon 5 K5
Barton Bendish 48 C7
Barton Hartshorn 35 H7
Barton in Fabis 45 G4
Barton in the Beans 44 E7
Barton-le-Clay 36 D7
Barton-le-Street 64 E5
Barton-le-Willows 64 E6
Barton Mills 38 C3
Barton Moss 53 J4
Barton on Sea 10 D5
Barton-on-the-Heath 31 G4
Barton Seagrave 36 B3
Barton Stacey 18 F6
Barton St David 9 G1
Barton Turf 49 J5
Barton-under-Needwood 44 C6
Barton-upon-Humber 58 E4
Barvas 108 C2
Barway 37 J3
Barwell 45 F8
Barwick 9 G3
Barwick in Elmet 63 H6
Baschurch 42 F5
Bascote 34 F4
Basford Green 44 A2
Bashall Eaves 61 L5
Bashley 10 D5
Basildon 22 D5
Basingstoke 19 H5
Baslow 55 F6
Bason Bridge 16 E6
Bassaleg 16 D2
Bassenthwaite 66 E3
Bassett 10 F3
Bassingbourn 37 G6
Bassingfield 45 H4
Bassingham 56 D8
Bassingthorpe 46 C5
Basta 113 D2
Baston 46 E6
Bastwick 49 K6
Batchcott 32 D3
Batcombe, Dorset 9 H4
Batcombe, Somer 17 H7
Bate Heath 53 H6
Bath 17 J4
Bathampton 17 J4
Bathealton 8 B2
Batheaston 17 J4
Bathford 17 J4
Bathgate 79 G3
Bathley 56 B8
Bathpool 4 C4
Batley 63 F7
Batsford 31 F4
Battersby 69 H6
Battersea 21 G6
Battisford 38 F5
Battisford Tye 38 F5
Battle, Powys 25 J1
Battle, E. Susx 13 L5
Battlefield 43 G6
Battlesbridge 22 D4
Battlesden 36 C8
Battleton 7 K3
Battramsley 10 E5
Bauds of Cullen 98 D2
Baughurst 19 G4
Baulking 18 E1
Baumber 57 G6
Baunton 30 E7
Baverstock 10 B1
Bawburgh 49 G7
Bawdeswell 48 F5
Bawdrip 16 E7
Bawdsey 39 J6
Bawtry 55 K4
Baxenden 62 A7
Baxterley 44 D8
Baycliff 61 F2
Baydon 18 D3
Bayford 21 H3
Bayham Abbey 13 K3
Bayhead 110 B1
Bayles 67 K2
Baylham 39 G5
Bayston Hill 43 F7
Bayton 33 F3
Beachampton 35 J7
Beachamwell 48 C7
Beachans 97 J4
Beachborough 15 H5
Beachley 17 G1
Beacon 8 C4
Beacon End 22 F1
Beacon's Bottom 20 B4
Beaconsfield 20 D4
Beacravik 108 B5
Beadlam 63 K1
Beadnell 81 L7
Beaford 7 F4
Beal, Northum 81 J5
Beal, N. Yks 63 J7
Beaminster 9 F4
Beamish 68 E1
Beamsley 62 D4
Bean 14 B1
Beanacre 17 L4
Beanley 75 G2
Beaquoy 112 B3
Beare Green 12 E2
Bearley 34 C4
Bearpark 68 E2
Bearsbridge 67 K1
Bearsden 85 H7
Bearsted 14 D3
Bearwood 10 B5
Beattock 73 F3
Beauchamp Roding 21 K2
Beauchief 55 G5
Beaufort 25 K3
Beaulieu 10 E4
Beauly 96 D4
Beaumaris 50 F3
Beaumont, Essex 39 G8
Beaumont, Cumbr 73 J8
Beausale 34 D3
Beaworthy 4 E2
Beazley End 38 C8
Bebington 52 E5
Bebside 75 J5
Beccles 39 K1
Becconsall 61 H7
Beckbury 43 J7
Beckenham 21 H7
Beckermet 66 C6
Beckermonds 62 B1
Beckfoot, Cumbr 66 C2
Beckfoot, Cumbr 66 D6
Beck Foot, Cumbr 67 J7
Beckford 30 D4
Beckhampton 18 B4
Beck Hole 64 F2
Beckingham, Lincs 46 B2
Beckingham, Notts 56 B4
Beckington 17 K5
Beckley, E. Susx 14 E6
Beckley, Oxon 31 K6
Beck Row 38 B3
Beck Side 60 F1
Beckton 21 J5
Beckwithshaw 63 F4
Becontree 21 J5
Bedale 63 F1
Bedburn 68 D3
Beddau 25 J6
Beddgelert 50 E6
Beddingham 13 H6
Beddington 21 H7
Bedfield 39 H4
Bedford 36 D6
Bedhampton 11 J4
Bedingfield 39 G4
Bedlington 75 J5
Bedlinog 25 J4
Bedmond 20 E3
Bednall 44 A6
Bedrule 74 C2
Bedstone 32 C3
Bedwas 25 K6
Bedworth 34 E2
Beeby 45 H7
Beech, Hants 19 H7
Beech, Staffs 43 K4
Beech Hill 20 A7
Beechingstoke 18 B5
Beedon 19 F3
Beeford 65 J7
Beeley 55 F7
Beelsby 59 G6
Beenham 19 G4
Beer 8 D6
Beer Crocombe 8 E2
Beer Hackett 9 G3
Beesands 5 J7
Beesby 57 J5
Beeson 5 J7
Beeston, Beds 36 E6
Beeston, Norf 48 E6
Beeston, W. Yks 63 F6
Beeston, Notts 45 G4
Beeston, Ches 53 G8
Beeston Regis 49 G3
Beeswing 72 D7
Beetham 61 H2
Beetley 48 E6
Begbroke 31 J6
Begelly 26 F7
Beguildy 32 A3
Beighton, S. Yks 55 H5
Beighton, Norf 49 J7
Beith 77 J2
Bekesbourne 15 J3
Belaugh 49 H6
Belbroughton 33 J3
Belchamp Otten 38 D6
Belchamp St Paul 38 C6
Belchamp Walter 38 D6
Belchford 57 G6
Belford 81 K6
Belhelvie 99 J7
Bellabeg 98 C7
Bellanoch 83 G2
Bell Busk 62 C4
Belleau 57 J6
Belleheiglash 98 A5
Bellerby 68 D7
Bellever 5 G4
Bellingdon 20 D3
Bellingham 74 E5
Belloch 76 B4
Bellochantuy 76 B4
Bellsbank 77 K7
Bellshill, Strath 78 E4
Bellshill, Northum 81 K6
Bellsmyre 85 G7
Bellspool 79 J6
Bellsquarry 79 H3
Bells Yew Green 13 K3
Belmesthorpe 46 D6
Belmont 53 H2
Belnacraig 98 C7
Belowda 3 H3
Belper 44 E3
Belsay 75 H6
Belses 80 D7
Belsford 5 H6
Belstead 39 G6
Belston 77 J5
Belstone 5 G2
Belthorn 61 L7
Beltinge 15 J2
Beltoft 58 C6
Belton, Humbs 58 B6
Belton, Leic 46 B7
Belton, Lincs 46 C4
Belton, Leic 45 F5
Belton, Norf 49 K7
Belvedere 21 J6
Belvoir 46 B4
Bembridge 11 H6
Bemersyde 80 D6
Bempton 65 J5
Benacre 39 L2
Benbuie 72 C4
Ben Casgro 108 D3
Benderloch 89 L5
Benenden 14 E5
Bengate 49 J5
Benholm 93 K5
Beningbrough 63 J4
Benington, Lincs 47 G3
Benington, Herts 21 H1
Benllech 50 E2
Benmore, Strath 84 D6
Benmore, Central 85 G2
Bennacott 4 C2
Bennan 76 E5
Bennecarrigan 76 E5
Benniworth 57 G5
Benover 13 L2
Benson 20 A4
Benthall 43 H7
Bentham 30 D6
Benthoul 99 H8
Bentley, H. & W 34 A4
Bentley, Warw 44 D8
Bentley, Humbs 58 E3
Bentley, S. Yks 55 J3
Bentley, Hants 19 J6
Bentley Heath 34 C3
Bentpath 73 J4
Bentworth 19 H6
Benwick 37 G1
Beoley 34 B4
Beoraidbeg 94 B6
Bepton 11 K3
Berden 37 H8
Berea 26 B5
Bere Alston 4 E5
Bere Ferrers 4 E5
Berepper 2 E7
Bere Regis 9 K5
Bergh Apton 49 J7
Berinsfield 31 K8
Berkeley 30 A8
Berkhamsted 20 D3
Berkley 17 K6
Berkswell 34 D3
Bermondsey 21 H6
Bernera 94 D3
Bernisdale 111 F2
Berrick Salome 20 A4
Berriedale 103 K1
Berrier 67 G4
Berriew 42 C7
Berrington, Shrops 43 G7
Berrington, Northum 81 J5
Berrow 16 D5
Berrow Green 33 G5
Berry Hill 29 K6
Berryhillock 98 E2
Berrynarbor 7 F1
Berry Pomeroy 5 J5
Bersham 42 E3
Bersted 12 C6
Berwick 13 J6
Berwick Bassett 18 C3
Berwick Hill 75 H6
Berwick St James 18 B7
Berwick St John 10 A2
Berwick St Leonard 9 L1
Berwick-upon-Tweed 81 H4
Berwyn 42 C3
Besford 33 J6
Bessacarr 55 K3
Bessels Leigh 31 J7
Bessingham 49 G4
Besthorpe, Notts 56 C7
Besthorpe, Norf 48 F8
Beswick 65 H8
Betchworth 12 F1
Bethania, Dyfed 40 E7
Bethania, Gwyn 51 F5
Bethel 50 E4
Bethersden 14 F4
Bethesda, Dyfed 26 E6
Bethesda, Gwyn 51 F4
Bethlehem 24 E2
Bethnal Green 21 H5
Betley 43 J3
Betsham 14 C1
Betteshanger 15 K3
Bettiscombe 8 E5
Bettisfield 42 F4
Betton, Shrops 42 E7
Betton, Shrops 43 H4
Bettws, Gwent 16 D1
Bettws, Gwent 29 G6
Bettws Cedewain 42 C8
Bettws Gwerfil Goch 51 K6
Bettws Newydd 29 H7
Bettyhill 105 J2
Betws, M. Glam 25 H6
Betws, Dyfed 27 L6
Betws Bledrws 27 K2
Betws Garmon 50 E5
Betws Ifan 27 H3
Betws-y-Coed 51 G5
Betws-yn-Rhos 51 J3
Beulah, Powys 28 D2
Beulah, Dyfed 27 G3
Bevendean 13 G6
Bevercotes 56 A6
Beverley 58 E3
Beverston 30 C8
Bevington 30 A8
Bewaldeth 66 E3
Bewcastle 74 B6
Bewdley 33 G3
Bewerley 62 E3
Bewholme 65 J7
Bexhill 13 L6
Bexley 21 J6
Bexwell 48 B7
Beyton 38 E4
Bibury 30 F7
Bicester 31 K5
Bickenhall 8 D3
Bickenhill 34 C2
Bicker 47 F4
Bickerstaffe 52 F3
Bickerton, Ches 43 G2
Bickerton, N. Yks 63 H4
Bickington, Devon 7 F2
Bickington, Devon 5 H4
Bickleigh, Devon 5 F5
Bickleigh, Devon 7 K5
Bickleton 6 F2
Bickley 21 J7
Bickley Moss 43 G3
Bicknacre 22 D3
Bicknoller 16 C7
Bicknor 14 E3
Bickton 10 C3
Bicton, Shrops 32 B2
Bicton, Shrops 42 F6
Bidborough 13 J2
Biddenden 14 E5
Biddenham 36 D5
Biddestone 17 K3
Biddisham 16 E5
Biddlesden 35 H6
Biddlestone 74 F3
Biddulph 53 K8
Biddulph Moor 53 L8
Bideford 6 E3
Bidford-on-Avon 34 C5
Bielby 58 B2
Bielby 58 B2
Bieldside 93 K1
Bierley 11 G7
Bierton 20 C2
Bigbury 5 G7
Bigbury-on-Sea 5 G7
Bigby 58 E6
Biggar, Cumbr 60 E3
Biggar, Strath 79 H6
Biggin, Derby 44 D3
Biggin, Derby 54 E8
Biggin, N. Yks 63 J6
Biggings 113 B3
Biggin Hill 21 J8
Biggins 61 K2
Biggleswade 36 E6
Bighton 11 H1
Bignor 12 C5
Big Sand 100 C4
Bigton 113 C5
Bilberry 3 J3
Bilborough 45 G3
Bilbrook 43 K7
Bilbrough 63 J5
Bilbster 107 H4
Bildeston 38 E6
Billericay 22 C4
Billesdon 45 J7
Billesley 34 C5
Billing 35 K4
Billingborough 46 E4
Billinge 53 G3
Billingford, Norf 48 F5
Billingford, Suff 39 G3
Billingham 69 G4
Billinghay 46 E2
Billingley 55 H3
Billingshurst 12 D4
Billingsley 33 G2
Billington, Beds 20 D1
Billington, Lancs 61 L6
Billockby 49 K6
Billy Row 68 D3
Bilsborrow 61 J5
Bilsby 57 J6
Bilsington 15 G5
Bilsthorpe 55 K7
Bilston, W. Mids 44 A8
Bilston, Lothn 79 K3
Bilstone 44 E7
Bilting 15 G4
Bilton, Warw 35 F3
Bilton, Humbs 59 F3
Bilton, N. Yks 63 H4
Bilton, Northum 75 J2
Bimbister 112 B4
Binbrook 57 G4
Bincombe 9 H6
Binegar 17 H6
Binfield 20 C6
Binfield Heath 20 B6
Bingfield 75 F6
Bingham 45 J4
Bingley 62 E6
Binham 48 E4
Binley, W. Mids 34 E3
Binley, Hants 18 F5
Binley Woods 34 E3
Binniehill 86 B7
Binstead 11 G5
Binsted 19 J6
Binton 34 C5
Bintree 48 F5
Binweston 42 E7
Birch, Essex 22 F2
Birch, G. Man 53 K3
Bircham Newton 48 C4
Bircham Tofts 48 C4
Birchanger 21 K1
Bircher 32 D4
Birch Green 22 F2
Birchgrove 24 F5
Birchington 15 J2
Birchover 55 F7
Birch Vale 54 D5
Birchwood 56 D7
Bircotes 55 K4
Birdbrook 38 C6
Birdham 11 K4
Birdingbury 34 F4
Birdlip 30 D6
Birdsall 64 F6
Birdsgreen 33 G2
Birdston 85 J7
Birdwell 55 G3
Birdwood 30 B6
Birgham 81 F6
Birkdale 52 E2
Birkenhead 52 E5
Birkenhills 99 G4
Birkenshaw, Strath 78 D3
Birkenshaw, W. Yks 62 F7
Birkhall 92 E2
Birkhill 87 G1
Birkin 63 J7
Birley 29 J2
Birling, Kent 14 C2
Birling, Northum 75 J3
Birlingham 33 J6
Birmingham 34 B2
Birnam 92 B7
Birness 99 J5
Birse 93 G2
Birsemore 93 G2
Birstall 45 G7
Birstall Smithies 63 F7
Birstwith 63 F4
Birtley, H. & W 32 C4
Birtley, T. & W 68 E1
Birtley, Northum 74 E6
Birts Street 33 G7
Bisbrooke 46 B8
Bishampton 34 A5
Bishop Auckland 68 E4
Bishopbriggs 85 J7
Bishop Burton 58 D3
Bishop Middleham 68 F3
Bishop Monkton 63 G3
Bishop Norton 56 D4
Bishopsbourne 15 H3
Bisops Cannings 18 B4
Bishop's Castle 32 C2
Bishop's Caundle 9 H3
Bishop's Cleeve 30 D5
Bishop's Frome 33 F6
Bishop's Itchington 34 E5
Bishops Lydeard 8 C2
Bishop's Nympton 7 H3
Bishop's Offley 43 J5
Bishop's Stortford 21 J1
Bishop's Sutton 11 H1
Bishop's Tachbrook 34 E4
Bishop's Tawton 7 F2
Bishopsteignton 5 K4
Bishopstoke 11 F3
Bishopston 24 D6
Bishopstone, Wilts 10 B2
Bishopstone, Bucks 20 C2
Bishopstone, Wilts 18 D2
Bishopstone, E. Susx 13 H6
Bishopstone, H. & W 29 J3
Bishop Sutton 17 G5
Bishop's Waltham 11 G3
Bishopswood, Somer 8 D3
Bishop's Wood, Staffs 43 K7
Bishopsworth 17 G4
Bishop Thornton 63 F3
Bishopthorpe 63 J5
Bishopton, Durham 69 F4
Bishopton, Strath 85 G7
Bishop Wilton 64 E7
Bishton 16 E2
Bisley, Glos 30 D7
Bisley, Surrey 20 D8
Bispham 61 G5
Bissoe 2 F5
Bisterne Close 10 D4
Bitchfield 46 C5
Bittadon 7 F1
Bittaford 5 G6
Bittering 48 E6
Bitterley 32 E3
Bitterne 11 F3
Bitteswell 35 G2
Bitton 17 H4
Bix 20 B5
Bixter 113 C4
Blaby 45 G8
Blackacre 73 F4
Blackadder 81 G4
Blackawton 5 J6
Blackborough 8 B4
Blackborough End 48 B6
Black Bourton 31 G7
Blackboys 13 J4
Blackbrook 43 J4
Blackburn, Lothn 79 G3
Blackburn, Grampn 99 H7
Blackburn, Lancs 61 K7
Black Callerton 75 H7
Black Clauchrie 70 D3
Black Crofts 84 B1
Blackden Heath 53 J6
Black Dog, Devon 7 J5
Blackdog, Grampn 99 J7
Blackfield 11 F4
Blackford, Tays 86 B4
Blackford, Somer 16 F6
Blackford, Somer 9 H2
Blackford, Cumbr 73 J7
Blackfordby 44 E6
Blackgang 11 F7
Blackhall 93 H2
Blackhall Rocks 69 G3
Blackham 13 J3
Blackhaugh 80 C6
Blackheath, W. Mids 34 A2
Blackheath, Essex 23 G1
Blackhill 99 K3
Blackland 18 B4
Blackley 53 K3
Blacklunans 92 C5
Black Marsh 42 E8
Blackmill 25 H6
Blackmoor 11 J1
Blackmore 22 C3
Blackmore End 38 C7
Black Mount 90 D7
Blackness 86 D7
Blacknest 19 J6
Black Notley 22 D1
Blacko 62 B5
Black Pill 24 E5
Blackpool 61 G6
Blackpool Gate 74 B6
Blackridge 79 F3
Blackrock, Strath 82 C5
Blackrock, Gwent 25 L3
Black Rocks 65 H4
Blackrod 53 H2
Blackshaw 73 F7
Blackstone 12 F5
Blackthorn 20 A2
Blackthorpe 38 E4
Blacktoft 58 C4
Blacktop 99 H8
Black Torrington 6 E5
Blackwater, Hants 20 C7
Blackwater, Corn 2 F5
Blackwater, I. of W 11 G6
Blackwaterfoot 76 D5
Blackwell, H. & W 34 A3
Blackwell, Derby 54 E6
Blackwood, Strath 78 E5
Blackwood, Gwent 25 K5
Blackwood Hill 43 L2
Blacon 52 E7
Bladerstone 61 K6
Bladnoch 71 F6
Bladon 31 J6
Blaenannerch 27 G3
Blaenau Ffestiniog 51 G6
Blaenavon 29 G7
Blaendyryn 28 D4
Blaenffos 27 F4
Blaengarw 25 H5
Blaengwrach 25 G4
Blaengwynfi 25 G5
Blaenpennal 41 F7
Blaenplwyf 40 E6
Blaenporth 27 G3
Blaenrhondda 25 H5
Blaenwaun 27 G5
Blagdon, Avon 17 G5
Blagdon, Devon 5 J5
Blagdon Hill 8 D3
Blaich 90 B4
Blaina 25 L4
Blair Atholl 91 K5
Blairdaff 99 F7
Blairgowrie 92 C7
Blairhall 86 D6
Blairingone 86 C5
Blairlogie 85 L5
Blairmore 84 D6
Blairskaith 85 H7
Blaisdon 30 B6
Blakebrook 33 H3
Blakedown 33 H3
Blakelaw 80 F6
Blakemere 29 H3
Blakeney, Glos 30 A7
Blakeney, Norf 48 F3
Blakenhall, Ches 43 J3
Blakenhall, W. Mids 43 L8
Blakeshall 33 H2
Blakesley 35 H5
Blanchland 68 B1
Blandford Camp 9 L4
Blandford Forum 9 K4
Blandford St Mary 9 K4
Bland Hill 62 F4
Blanefield 85 H7
Blankney 56 E7
Blarmachfoldach 90 B5
Blarnalearoch 101 G2
Blashford 10 C4
Blaston 45 K8
Blatherwycke 46 C8
Blawith 61 F1
Blaxhall 39 J5
Blaxton 55 K3
Blaydon 75 H7
Bleadon 16 E5
Blean 15 H2
Bleasby 45 J3
Bleatarn 67 K5
Blebocraigs 87 H3
Bleddfa 32 B4
Bledington 31 G5
Bledlow 20 B3
Bledlow Ridge 20 B4
Blegbie 80 C3
Blencarn 67 J3
Blencogo 66 D2
Blencow 67 G3
Blendworth 11 J3
Blennerhasset 66 D2
Bletchingdon 31 K6
Bletchingley 13 G1
Bletchley, Bucks 36 B7
Bletchley, Shrops 43 H4
Bletherston 26 E5
Bletsoe 36 D5
Blewbury 19 G2
Blickling 49 G5
Blidworth 45 G2
Blindcrake 66 D3
Blindley Heath 13 G2
Blisland 3 K2
Blissford 10 C3
Bliss Gate 33 G3
Blisworth 35 J5
Blockley 31 F4
Blofield 49 J7
Blo' Norton 38 F3
Blore 44 C3
Bloxham 31 J4
Bloxwich 44 B7
Bloxworth 9 K5
Blubberhouses 62 E4
Blue Anchor 16 B6
Blue Bell Hill 14 D2
Blundellsands 52 E4
Blundeston 49 L8
Blunham 36 E5
Blunsdon St Andrew 18 C2
Bluntisham 37 G3
Blyborough 56 D4
Blyford 39 K3
Blymhill 43 K6
Blyth, Northum 75 K5
Blyth, Notts 55 K5
Blyth Bridge 79 J5
Blythburgh 39 K3
Blythe 80 D5
Blythe Bridge 44 A3
Blyton 56 C4
Boarhills 87 J3
Boarhunt 11 H4
Boarshead 13 J3
Boarstall 20 A2
Boasley Cross 4 F2
Boath 102 D6
Boat of Garten 97 H7
Bobbing 14 E2
Bobbington 33 H1
Bocaddon 3 K4
Bockhampton 9 J5
Bocking 22 D1
Bocking Churchstreet 38 C8
Boddam, Shetld 113 C6
Boddam, Grampn 99 L4
Boddington 30 C5
Bodedern 50 C2
Bodelwyddan 51 K3
Bodenham, Wilts 10 C2
Bodenham, H. & W 29 K2
Bodewryd 50 D1
Bodfari 51 K3
Bodffordd 50 D3
Bodham 49 G3
Bodiam 14 D6
Bodicote 35 F7
Bodieve 3 H2
Bodle Street Green 13 K5
Bodmin 3 J3
Bodney 48 D8
Boduan 50 C7
Bogallan 96 E3
Bogbrae 99 K5
Bogend 77 J4
Boghall 79 G3
Bogmoor 98 C2
Bogniebrae 98 E4
Bognor Regis 12 C7
Bograxie 99 G7
Bog, The 42 E8
Bogton 98 F3
Bohenie 90 D3
Bohortha 3 G6
Bohuntine 90 D3
Boisdale 110 B4
Bojewyan 2 B6
Bolam 68 D4
Bold Heath 53 G5
Boldon 75 K7
Boldre 10 E5
Boldron 68 C5
Bole 56 C5
Bolehill 44 D2
Boleside 80 C6
Bolham 7 K4
Bolham Water 8 C3
Bolingey 2 F4
Bollington, Ches 53 J5
Bollington, Ches 53 L6
Bolney 12 F4
Bolnhurst 36 D5
Bolsover 55 H6
Bolsterstone 55 F4
Bolstone 29 K4
Boltby 63 H1
Bolton, Humbs 64 E7
Bolton, Northum 75 H2
Bolton, G. Man 53 J3
Bolton, Cumbr 67 J4
Bolton, Lothn 87 J7
Bolton Abbey 62 D4
Bolton-by-Bowland 62 A5
Boltonfellend 73 K7
Boltongate 66 E2
Bolton-le-Sands 61 H3
Bolton-on-Swale 68 E7
Bolton Percy 63 J5
Bolton upon Dearne 55 H3
Bolventor 3 K2
Bomere Heath 43 F6
Bonar Bridge 102 E4
Bonawe 84 C1
Bonby 58 E5
Boncath 27 G4
Bonchester Bridge 74 B2
Bondleigh 7 G5
Bonehill 44 C7
Bo'Ness 86 C6
Bonhill 85 F7
Boningale 43 K7
Bonjedward 80 E7
Bonkle 79 F4
Bonnington, Northnts 86 E8
Bonnington, Kent 15 G5
Bonnybridge 85 L6
Bonnykelly 99 H3
Bonnyrigg and Lasswade 80 B3
Bonnyton 93 H6
Bonsall 55 F8
Bont 29 H6
Bontddu 41 F2
Bont Dolgadfan 41 H3
Bont-goch or Elerch 41 F5
Bontnewydd, Gwyn 50 D4
Bont-newydd, Clwyd 51 K3
Bontuchel 51 K5
Bonvilston 25 J7
Booker 20 C4
Booley 43 G5
Boosbeck 69 J5
Boot 66 D6
Boothby Graffoe 56 D8
Boothby Pagnell 46 C4
Boothstown 53 J3
Booth Wood Reservoir 60 E1
Bootle, Cumbr 60 E1
Bootle, Mers 52 E4
Boraston 33 F3
Borcheston 45 F7
Borden 14 E2
Bordley 62 C3
Bordon Camp 11 J1
Boreham, Essex 22 D3
Boreham, Wilts 17 K6
Boreham Street 13 K5
Borehamwood 21 F4
Boreland 73 G4
Borgie 105 H3
Borgue, Highld 107 G7
Borgue, D. & G 71 H7
Borley 38 D6
Bornesketaig 108 C6
Borness 71 H7
Boroughbridge 63 G3
Borough Green 13 K1
Borras Head 42 E2
Borreraig 110 D2
Borrowby 69 G8
Borrowwash 44 F4
Borth 40 F4
Borthwickbrae 73 K2
Borthwickshiels 73 K2
Borth-y-Gest 50 E7
Borve, W. Isles 108 A5
Borve, W. Isles 110 B4
Borve, W. Isles 108 B5
Borve, Highld 111 F2
Borwick 61 J2
Bosavern 2 B6
Bosbury 33 F6
Boscastle 4 A2
Boscombe, Dorset 10 C5
Boscombe, Wilts 18 D7
Bosham 11 K4
Bosherton 26 D8
Boskednan 2 C6
Bosley 53 L7
Bossall 64 E6
Bossiney 4 A3
Bossingham 15 H4
Bosta 108 B2
Bostock Green 53 H7
Boston 47 G3
Boston Spa 63 H5
Boswinger 3 H5
Botallack 2 B6
Botany Bay 21 G4
Botesdale 38 F3
Bothal 75 J5
Bothamsall 55 K6
Bothel 66 D3
Bothenhampton 9 F5
Bothwell 78 E4
Botley, Bucks 20 D3
Botley, Hants 11 G3
Botley, Oxon 31 J7
Botolphs 12 E6
Bottacks 96 C2
Bottesford, Humbs 58 C6
Bottesford, Leic 45 K4
Bottisham 37 J4
Bottomcraig 87 G2
Botusfleming 4 E5
Botwnnog 50 B7
Boughrood 28 F4
Boughspring 29 K8
Boughton, Norf 48 B7
Boughton, Northnts 35 J4
Boughton, Notts 55 K7
Boughton Aluph 15 G4
Boughton Lees 15 G4
Boughton Malherbe 14 E4
Boughton Monchelsea 13 L1
Boughton Street 15 G3
Boulby 69 K5
Bouldon 32 E2
Boulmer 75 J2
Boulston 26 D6
Boultham 56 D7
Bourn 37 G5
Bourne 46 D5
Bourne End, Bucks 20 C5
Bourne End, Beds 36 C6
Bourne End, Herts 20 E3
Bournemouth 10 B5
Bournes Green 30 D7
Bourneville 34 B2
Bournheath 34 A3
Bournmoor 68 F1
Bourton, Oxon 18 D2
Bourton, Avon 16 E4
Bourton, Shrops 43 G8
Bourton, Dorset 9 J1
Bourton on Dunsmore 34 F3
Bourton-on-the-Hill 31 F4
Bourton-on-the-Water 31 F5
Boveney 20 D6
Boverton 25 H8
Bovey Tracey 5 J4
Bovingdon 20 E3
Bovington Camp 9 K6
Bow, G. Lon 21 H5
Bow, Devon 7 H5
Bowbank 68 B4
Bow Brickhill 36 C7
Bowburn 68 F3
Bowcombe 11 F6
Bowd 8 C5
Bowden, Border 80 D6
Bowden, Devon 5 J7
Bowden Hill 18 A4
Bowderdale 67 J6
Bowdon 53 J5
Bowerchalke 10 B2
Bowermadden 107 H3
Bowers Gifford 22 D5
Bowershall 86 D5
Bowertower 107 H3
Bowes 68 B5
Bowhill 80 C7
Bowland 80 C5
Bowland Bridge 67 G8
Bowley 29 K2
Bowlhead Green 12 C3
Bowling 85 G7
Bowling Bank 42 E3
Bowling Green 33 H5
Bowmanstead 66 F7
Bowmore 82 C6
Bowness-on-Solway 73 H7
Bowness-on-Windermere 67 G7
Bowsden 81 H5
Bow Street 41 F5
Bowthorpe 49 G7
Box, Glos 30 C7
Box, Wilts 17 K4
Boxbush 30 B6
Boxford, Suff 38 E6
Boxford, Berks 18 F3
Boxgrove 11 L4
Boxley 14 D3
Boxted, Suff 38 D5
Boxted, Essex 38 E7
Boxworth 37 G4
Boylestone 44 C4
Boyndie 98 F2
Boyndlie 99 J2
Boynton 65 J6
Boyton, Wilts 18 A7
Boyton, Corn 4 D2
Boyton, Suff 39 J6
Bozeat 36 C5
Braaid 60 Q4
Brabling Green 39 H4
Brabourne 15 H4
Brabourne Lees 15 G4
Brabster 107 J3
Bracadale 110 E3
Braceborough 46 D6
Bracebridge Heath 56 D7
Braceby 46 D4
Bracewell 62 B5
Brackenfield 55 G8
Brackletter 90 C3
Brackley, Strath 76 C3
Brackley, Northnts 35 G7
Bracknell 20 C7
Braco 85 L4
Bracobrae 98 E3
Bracon Ash 49 G8
Bracora 94 C6
Bracorina 94 C6
Bradbourne 44 D2
Bradbury 68 F4
Bradda 60 N4
Bradden 35 H6
Braddock 3 K3
Bradenham, Bucks 20 C4
Bradenham, Norf 48 E7
Bradenstoke 18 B3
Bradfield, Essex 39 G7
Bradfield, Berks 19 H3
Bradfield, Norf 49 H4
Bradfield Combust 38 D5
Bradfield Green 53 H8
Bradfield St Clare 38 E5
Bradfield St George 38 E4
Bradford, Devon 6 E5
Bradford, W. Yks 62 E6
Bradford, Northum 81 K6
Bradford Abbas 9 G3
Bradford Leigh 17 K4
Bradford-on-Avon 17 K4
Bradford-on-Tone 8 C2
Bradford Peverell 9 H5
Brading 11 K6
Bradley, Derby 44 D3
Bradley, Humbs 59 G6
Bradley, Hants 19 H6
Bradley, Staffs 43 K6
Bradley Green 34 A4
Bradley in the Moors 44 B3
Bradmore 45 G4
Bradninch 7 K5
Bradnop 44 B2
Bradpole 9 F5
Bradshaw 53 J2
Bradstone 4 D3
Bradwall Green 53 J7
Bradwell, Bucks 36 B7
Bradwell, Essex 22 E1
Bradwell, Derby 54 E5
Bradwell, Norf 49 L7
Bradwell Grove 31 G7
Bradwell-on-Sea 23 G3
Bradwell Waterside 23 F3
Bradworthy 6 D4
Brae, Shetld 113 C3
Brae, Highld 100 D3
Brae, Highld 96 E2
Braeantra 102 D6
Braefield 96 C5
Braehead, Orkney 112 C2
Braehead, Strath 79 F6
Braehead, D. & G 71 F6
Braehead, Strath 79 G4
Braehead, Tays 93 H6
Braehoulland 113 C3
Braemar 92 C2
Braemore 107 F6
Brae of Achnahaird 104 B7
Brae Roy Lodge 90 E2
Braeside 84 E7
Braes, The 111 F3
Braeswick 112 D3
Brafferton, Durham 68 E4
Brafferton, N. Yks 63 H2
Brafield-on-the-Green 36 B5
Bragar 108 C2
Bragbury End 21 G1
Braides 61 H4
Braidley 62 D1
Braidwood 79 G3
Brailes 34 E7
Brailsford 44 D3
Braintree 22 D1
Braiseworth 39 G3
Braishfield 10 E2
Braithwaite 66 C4
Braithwell 55 J4
Bramber 12 E5
Bramcote 45 G4
Bramdean 11 H2
Bramerton 49 H7
Bramfield, Herts 21 G2
Bramfield, Suff 39 K3
Bramford 36 E6
Bramhall 53 K5
Bramhan 63 H5
Bramhope 63 F5
Bramley, Hants 20 A8
Bramley, Surrey 12 D2
Bramley, S. Yks 55 H4
Brampford Speke 5 K2
Brampton, Cumbr 74 B7
Brampton, Lincs 56 C6
Brampton, Cambs 36 F3
Brampton, S. Yks 55 H3
Brampton, Norf 49 H5
Brampton, Cumbr 67 J4
Brampton, Suff 39 K2
Brampton Abbots 29 L5
Brampton Ash 35 J2
Brampton Bryan 32 C3
Bramshall 44 B4
Bramshaw 10 D3
Bramshott 11 K1
Brancaster 48 C3
Brancepeth 68 E3
Branchill 97 J3
Branderburgh 98 B1
Brandesburton 59 F2
Brandeston 39 H4
Brandiston 49 G5
Brandon, Suff 38 C2
Brandon, Lincs 46 C3
Brandon, Durham 68 E3
Brandon, Warw 34 F3
Brandon, Northum 75 G2
Brandon Bank 38 B2
Brandon Creek 38 B1
Brandon Parva 48 F7
Brandsby 63 J2
Brane 2 C7
Bran End 38 B8
Branksome 10 B5
Branksome Park 10 B5
Branscombe 8 C6
Bransford 33 G5
Bransgore 10 C5
Branston, Leic 46 B5
Branston, Staffs 44 D5
Branston, Lincs 56 E7
Branstone 11 H6
Brant Broughton 46 C2
Brantham 36 E7
Branthwaite 66 E3
Brantingham 58 D4
Branton, Northum 75 G2
Branton, S. Yks 55 K3
Branxholme 73 K2
Branxholm Park 73 K2
Branxton 81 G6
Brassington 44 D2
Brasted 13 H1
Brasted Chart 13 H1
Brathens 93 H2
Bratoft 57 J7
Brattleby 56 D5
Bratton 17 L5
Bratton Clovelly 4 E2
Bratton Fleming 7 G2
Bratton Seymour 9 H2
Braughing 37 G8
Braunston, Leic 46 B7
Braunston, Northnts 35 G4
Braunstone 45 G7
Braunton 6 E2
Brawby 64 E5
Brawl 106 D3
Brawlbin 107 F4
Bray 20 D6
Braybrooke 35 J2
Brayford 7 G2
Bray Shop 4 D4
Braystones 66 C6
Brayton 63 K6
Brazacott 4 C2
Breachwood Green 21 F1
Breaclete 108 B3
Breadstone 30 B7
Breage 2 E7
Breakachy 96 C4
Breakish 94 B3
Bream 29 L7
Breamore 10 C3
Brean 16 D5
Brearton 63 G3
Breasclete 108 C3
Breaston 45 F4
Brechfa 27 K4
Brechin 93 G5
Breckles 48 E8
Breckrey 111 F1
Brecon 25 J2
Bredbury 53 L4
Brede 14 E7
Bredenbury 29 L2
Bredfield 39 H5
Bredgar 14 E2
Bredhurst 14 D2
Bredon 33 J7
Bredon's Norton 33 J7
Bredwardine 29 H3
Breedon on the Hill 44 F5
Breich 79 G3
Breighton 58 B3
Bremhil 18 A3
Brenchley 13 K2
Brendon 7 H1
Brenish 108 A3
Brent Eleigh 38 E6
Brentford 21 F6
Brent Knoll 16 E5
Brent Pelham 37 H7
Brentwood 22 B4
Brenzett 15 G6
Brereton 44 B6
Brereton Green 53 J7
Brereton Heath 53 K7
Bressingham 39 F2
Bretby 44 D5
Bretford 34 F3
Bretforton 34 B6
Bretherdale Head 67 H6
Bretherton 53 F1
Brettabister 113 D4
Brettenham, Norf 38 E2
Brettenham, Suff 38 E5
Bretton 52 E7
Brevig 110 B5
Brewham 9 J1
Brewlands Bridge 92 C5
Brewood 43 K7
Briantspuddle 9 K5
Brickendon 21 H3
Bricket Wood 21 F3
Bricklehampton 34 A6
Bride 60 R1
Bridekirk 66 D3
Bridell 26 F3
Bridestowe 4 F3
Brideswell 98 E5
Bridford 5 J3
Bridge 15 H3
Bridge End 46 E4
Bridgefoot 66 C4
Bridge Green 37 H7
Bridgemary 11 G4
Bridgend, Strath 84 A5
Bridgend, Strath 82 C5
Bridgend, Grampn 98 C5
Bridgend, Lothn 86 D7
Bridgend, Grampn 98 E5
Bridgend, Cumbr 67 F5
Bridgend, Fife 87 G3
Bridgend, Tays 93 G5
Bridgend, M. Glam 25 H6
Bridgend, Strath 85 J7
Bridgend of Lintrathen 92 D6
Bridge of Alford 98 E7
Bridge of Allan 85 K5
Bridge of Avon 98 A5
Bridge of Brown 97 K6
Bridge of Cally 92 C6
Bridge of Canny 93 H2
Bridge of Dee 72 C7
Bridge of Don 99 J7
Bridge of Dun 93 H6
Bridge of Dye 93 H3
Bridge of Earn 86 E3
Bridge of Feugh 93 J2
Bridge of Forss 107 F3
Bridge of Gairn 92 E2
Bridge of Muchalls 93 K2
Bridge of Orchy 90 D8
Bridge of Walls 113 C4
Bridge of Weir 77 J1
Bridgerule 6 C5
Bridges 42 E8
Bridge Sollers 29 J3
Bridge Street 38 D6
Bridgetown 7 K2
Bridge Trafford 52 F6
Bridge Yate 17 H3
Bridgham 38 E2
Bridgnorth 43 J8
Bridgtown 44 A7
Bridgwater 8 E1
Bridlington 65 J6
Bridport 9 F5
Bridstow 29 K5
Brierfield 62 B6
Brierley, S. Yks 55 H2
Brierley, H. & W 29 J2
Brierley, Glos 29 L6
Brierley Hill 33 J2
Brigg 58 E6
Brigham, Cumbr 66 C3
Brigham, Humbs 65 H7
Brighouse 62 E7
Brighstone 10 F6
Brightgate 55 F8
Brighthampton 31 H7
Brightling 13 K4
Brightlingsea 23 G2
Brighton, E. Susx 13 G6
Brighton, Corn 3 H4
Brightons 86 C7
Brightwalton 18 F3
Brightwell 39 H6
Brightwell Baldwin 20 A4
Brightwell-cum-Sotwell 19 G1
Brignall 68 C5
Brig o' Turk 85 H4
Brigsley 59 G6
Brigsteer 67 G8
Brigstock 36 C2
Brill 20 A2
Brilley 29 G3
Brimfield 32 E4
Brimington 55 H6
Brimpsfield 30 D6
Brimpton 19 G4
Brims 112 B5
Brinacory 94 C6
Brind 58 B3
Brindister, Shetld 113 C4
Brindister, Shetld 113 D5
Brindle 61 J7
Brindley Ford 43 K2
Brineton 43 K6
Bringhurst 36 B1
Brington 36 D3
Brinian 112 C3
Briningham 48 F4
Brinkhill 57 H6
Brinkley 38 B5
Brinklow 34 F3
Brinkworth 18 B2
Brinscall 61 K7
Brinsley 45 F3
Brinsop 29 J3
Brinsworth 55 H4
Brinton 48 F4
Brisley 48 E5
Brislington 17 H3
Bristol 17 G3
Briston 48 F4
Britannia 62 B7
Britford 10 C2
Brithdir 41 G2
Briton Ferry 25 F5
Britwell Salome 20 A4
Brixham 5 K6
Brixton, Devon 5 F6
Brixton, G. Lon 21 H6
Brixton Deverill 17 K7
Brixworth 35 J3
Brize Norton 31 H7
Broad Blunsde 18 C1
Broadbottom 54 C4
Broadbridge 11 K4
Broadbridge Heath 12 E3
Broad Campden 34 C7
Broad Chalke 10 B2
Broadclyst 5 K2
Broadford 111 G3
Broad Green 33 G5
Broad Haven 26 C6
Broad Heath, H. & W 33 F4
Broadheath, H. & W 33 H5
Broadheath, G. Man 53 J5
Broadhembury 8 C4
Broadhempston 5 J5
Broad Hill 38 A3
Broad Hinton 18 C3
Broad Laying 18 F4
Broadley, Grampn 98 C2
Broadley, Lancs 53 K2
Broadley Common 21 J3
Broad Marston 34 C6
Broadmayne 9 J6
Broadmeadows 80 C6
Broadmere 19 H6
Broad Oak, E. Susx 14 E6
Broadoak, Dorset 9 F5
Broad Oak, Kent 15 H2
Broad Oak, Dorset 9 J3
Broad Oak, H. & W 29 J5
Broad Oak, E. Susx 13 K4
Broadrashes 98 D3
Broadstairs 15 K2
Broadstone, Dorset 10 B5
Broadstone, Shrops 32 E2
Broad Street 14 E3
Broad Town 18 B3
Broadwas 33 G5
Broadwater 12 E6
Broadway, H. & W 34 C7
Broadway, Somer 8 E3
Broadwell, Warw 35 F4
Broadwell, Glos 31 G5
Broadwell, Oxon 31 G7
Broadwell, Glos 29 K6
Broadwey 9 H6
Broadwindsor 9 F4
Broadwoodkelly 7 G5
Broadwoodwidger 4 E3
Brobury 29 H3
Brochel 111 F2
Brockbridge 11 H3
Brockdish 39 H3
Brockenhurst 10 D4
Brocketsbrae 79 F6
Brockford Street 39 G4
Brockhall 35 H4
Brockham 12 E2
Brockhampton, Glos 30 E5
Brockhampton, H. & W 29 K4
Brockholes 54 E2
Brocklesby 59 F5
Brockley 17 F4
Brockley Green 38 D5
Brockton, Shrops 32 C2
Brockton, Shrops 42 E7
Brockton, Shrops 43 G8
Brockton, Shrops 43 J7
Brockweir 29 K7
Brockwood Park 11 H2
Brockworth 30 C6
Brocton 44 A6
Brodick 76 F4
Brodsworth 55 J3
Brogborough 36 C7
Brokenborough 17 L2
Broken Cross, Ches 53 H6
Broken Cross, Ches 53 K6
Bromborough 52 E5
Bromcote 34 F2
Brome 39 G3
Brome Street 39 G3
Bromeswell 39 J5
Bromfield, Cumbr 66 D2
Bromfield, Shrops 32 D3
Bromham, Wilts 18 A4
Bromham, Beds 36 D5
Bromley 21 J7
Bromley Green 15 G5
Brompton, Kent 14 D2
Brompton, N. Yks 69 F7
Brompton, N. Yks 65 G4
Brompton-on-Swale 68 E7
Brompton Ralph 8 B1
Brompton Regis 7 K2
Bromsash 30 A5
Bromsgrove 34 A3
Bromyard 33 F5
Bromyard Downs 33 F5
Bronaber 51 G7
Bronington 43 F4
Bronllys 25 K1
Bronnant 41 F7
Bronwydd Arms 27 J5
Bronygarth 42 D4
Brook, Surrey 12 C3
Brook, Hants 10 D3
Brook, Hants 10 E2
Brook, I. of W 10 E6
Brook, Kent 15 G4
Brooke, Leic 46 B7
Brooke, Norf 49 H8
Brookfield 77 K1
Brookhouse 61 J3
Brookhouse Green 53 K7
Brookland 15 F6
Brookmans Park 21 G3
Brooks 42 C8
Brook Street 21 K4
Brookthorpe 30 C6
Brookwood 12 C1
Broom, Warw 34 B5
Broom, Bes 36 E6
Broome, Shrops 32 D2
Broome, Norf 39 J1
Broome, H. & W 33 J3
Broomedge 53 J5
Broome Park 75 H2
Broomer's Corner 12 E4
Broomfield, Somer 8 D1
Broomfield, Essex 22 D2
Broomfield, Kent 14 E3
Broomfield, Kent 15 H2
Broomfield, Grampn 99 J5
Broomfleet 58 C4
Broom Hill, Dorset 10 B4
Broomhill, Northum 75 J3
Brora 103 H3
Broseley 43 H7
Brothertoft 47 F3
Brotherton 63 H7
Brotton 69 J5
Broubster 107 F3
Brough, Notts 56 C8
Brough, Shetld 113 D3
Brough, Shetld 113 D3
Brough, Shetld 113 D3
Brough, Humbs 58 D4
Brough, Derby 54 E5
Brough, Highld 107 H2
Brough, Cumbr 67 K5
Broughall 43 G3
Brough Sowerby 67 K5
Broughton, Northnts 36 B3
Broughton, Bucks 36 B6
Broughton, Orkney 112 C2
Broughton, Cumbr 66 C3
Broughton, N. Yks 62 C4
Broughton, Humbs 58 D6
Broughton, Hants 10 E1
Broughton, N. Yks 64 E5
Broughton, Clwyd 52 E7
Broughton, Cambs 37 F3
Broughton, Oxon 34 F7
Broughton, M. Glam 25 H7
Broughton, Lancs 61 J6
Broughton, Border 79 J6
Broughton, G. Man 53 K3
Broughton Astley 35 G1
Broughton Beck 61 F1
Broughton Gifford 17 K4
Broughton Hackett 33 J5
Broughton in Furness 60 F1
Broughton Mills 66 E7
Broughton Moor 66 C3
Broughton Poggs 31 G7
Broughtown 112 D2
Broughty Ferry 87 H1
Browland 113 C4
Brown Candover 19 G7
Brown Edge 43 L2
Brownhill, Grampn 99 H4
Brownhill, Lancs 61 K6
Brownhills 44 B7
Brownieside 81 K7
Brownlow Heath 53 K7
Brownston 5 G6
Brow Top 61 J4
Broxbourne 21 H3
Broxburn, Lothn 86 D7
Broxburn, Lothn 80 E2
Broxted 37 J8
Broxwood 29 H2
Bruan 107 J6
Brue 108 C2
Bruera 52 F7
Bruern Abbey 31 G5
Bruernish 110 B4
Bruichladdich 82 B5
Bruisyard 39 J4
Brumby 58 C6
Brund 54 E7
Brundall 49 J7
Brundish 39 H4
Brundish Street 39 H3
Brunton, Fife 87 G2
Brunton, Northum 81 L7
Brushford, Devon 7 G5
Brushford, Somer 7 K3
Bruton 9 H1
Bryanston 9 K4
Brydekirk 73 G6
Brymbo 42 D2
Bryn, Shrops 32 B2
Bryn, G. Man 53 G3
Bryn, W. Glam 25 G5
Brynamman 24 F3
Brynberian 26 F4
Bryncae 25 H6
Bryncethin 25 H6
Bryncir 50 D6
Bryn-coch 25 F5
Bryncroes 50 B7
Bryncrug 40 F3
Bryneglwys 42 C3
Brynford 52 C6
Bryn Gates 53 G3
Bryngwran 50 C3
Bryngwyn, Powys 29 F3
Bryngwyn, Gwent 29 H7
Bryn-henllan 26 E4
Brynhoffnant 27 H2
Brynithel 25 L4
Brynmawr 25 K3
Brynmenyn 25 H6
Brynna 25 H6
Brynrefail 50 D2
Brynsadler 25 J6
Brynsiencyn 50 D4
Brynteg, Gwyn 50 D2
Brynteg, Clwyd 42 E2
Bryn, The 29 H7
Bryn-y-maen 51 H3
Bualintur 111 F3
Bualnaluib 100 D2
Bubbenhall 34 E3
Bubwith 58 B3
Buccleuch 73 J2
Buchlyvie 85 H5
Buckabank 67 F2
Buckden, N. Yks 62 C2
Buckden, Cambs 36 E4
Buckenham 49 J7
Buckerell 8 C4
Buckfast 5 H5
Buckfastleigh 5 H5
Buckhaven 87 G5
Buckholm 80 C6
Buckhorn Weston 9 J2
Buckhurst Hill 21 J4
Buckie 98 D2
Buckies 107 G3
Buckingham 35 H7
Buckland, Glos 34 B7
Buckland, Bucks 20 C2
Buckland, Surrey 12 F1
Buckland, Herts 37 G7
Buckland, Devon 5 G7
Buckland, Oxon 31 H8
Buckland, Kent 15 K4
Buckland Brewer 6 E3
Buckland Common 20 D3
Buckland Dinham 17 J5
Buckland Filleigh 6 E5
Buckland in the Moor 5 H4
Buckland Monachorum 4 E5
Buckland Newton 9 H4
Buckland St Mary 8 D3
Bucklebury 19 G3
Bucklers Hard 10 F4
Bucklesham 39 H6
Buckley 52 D7
Bucklow Hill 53 J5
Buckminster 46 B5
Bucknall, Lincs 57 F7
Bucknall, Staffs 43 L3
Bucknell, Shrops 32 C3
Bucknell, Oxon 31 K5
Bucksburn 99 H8
Buck's Cross 6 D3
Bucks Green 12 D3
Bucks Hill 20 E3
Bucks Horn Oak 19 K6
Buck's Mills 6 D3
Buckton, H. & W 32 C3
Buckton, Northum 81 J6
Buckworth 36 E3
Budbrooke 34 D4
Budby 55 K6
Bude 6 C5
Budlake 5 K1
Budle 81 K6
Budleigh Salterton 8 B6
Budock Water 3 F6
Buerton 43 H3
Bugbrooke 35 H5
Bugle 3 J4
Bugthorpe 64 E7
Builth Road 28 E2
Builth Wells 28 E2
Bulby 46 D5
Buldoo 106 E3
Bulford 18 C6
Bulford Camp 18 C6
Bulkeley 43 G2
Bulkington, Wilts 18 A5
Bulkington, Warw 34 E2
Bulkworthy 6 D4
Bulley 30 B6
Bullwood 84 D7
Bulmer, Essex 38 D6
Bulmer, N. Yks 64 D6
Bulmer Tye 38 D7
Bulphan 22 C5
Bulverhythe 14 D8
Bulwell 45 G3
Bulwick 46 C8
Bumble's Green 21 J3
Bunacaimb 94 B7
Bunarkaig 90 C3
Bunavoneadar 108 B4
Bunbury 53 G8
Bundalloch 94 D3
Bunessan 88 E6
Bungay 39 J2
Bunnahabhainn 82 D4
Bunny 45 G5
Buntait 96 C5
Buntingford 37 G8
Bunwell 49 G8
Burbage, Wilts 18 D4
Burbage, Derby 54 D6
Burbage, Leic 35 F1
Burcombe 10 B1
Burcot 31 K8
Burdale 65 F6
Bures 38 E7
Burford, Shrops 32 E4
Burford, Oxon 31 G6
Burg 88 E4
Burgess Hill 13 G5
Burgh 39 H5
Burgh by Sands 73 J8
Burgh Castle 49 K7
Burghclere 19 F4
Burghead 97 K2
Burghfield 20 A7
Burghfield Common 20 A7
Burghfield Hill 20 A7
Burgh Heath 12 F1
Burghill 29 J3
Burgh le Marsh 57 K7
Burgh next Aylsham 49 H5
Burgh on Bain 57 G5
Burgh St Margaret or Fleggburgh 49 K6
Burgh St Peter 49 K8
Burghwallis 55 J2
Burham 14 D2
Buriton 11 J2
Burland 43 H2
Burlawn 3 H2
Burlescombe 8 B3
Burleston 9 J5
Burley, Leic 46 B6
Burley, Hants 10 D4
Burleydam 43 H3
Burley Gate 29 K3
Burley in Wharfedale 62 E5
Burley Street 10 D4
Burlingjobb 29 G2
Burlton 42 F5
Burmarsh 15 H5
Burmington 34 D7
Burn 63 J7
Burnage 53 K4
Burnaston 44 D4
Burnby 58 C2
Burneside 67 H7
Burness 112 D2
Burneston 63 G1
Burnett 17 H4
Burnfoot, Border 74 B2
Burnfoot, Border 73 K2
Burnham, Berks 20 D5
Burnham, Humbs 58 E5
Burnham Deepdale 48 D3
Burnham Green 21 G2
Burnham Market 48 D3

Burnham Norton 48 D3
Burnham-on-Crouch 22 F4
Burnham-on-Sea 16 E6
Burnham Overy Staithe 48 D3
Burnham Overy Town 48 D3
Burnham Thorpe 48 D3
Burnhead 72 D4
Burnhervie 99 G7
Burnhill Green 43 J7
Burnhope 68 D2
Burnhouse 77 J2
Burniston 65 H3
Burnley 62 B6
Burnmouth 81 H3
Burn of Cambus 85 K4
Burnopfield 68 D1
Burnsall 62 D3
Burnside, Strath 72 A2
Burnside, Shetld 113 C3
Burnside, Lothn 86 D7
Burnside, Fife 86 E4
Burnside of Duntrune 87 H1
Burntisland 87 F6
Burntwood 44 B7
Burnt Yates 63 F3
Burpham, Surrey 12 D1
Burpham, W. Susx 12 D6
Burradon 75 J6
Burrafirth 113 E1
Burras 2 E6
Burravoe, Shetld 113 C3
Burravoe, Shetld 113 D3
Burray Village 112 C5
Burrelton 92 D8
Burridge 11 G3
Burrill 63 F1
Burringham 58 C6
Burrington, H. & W. 32 D3
Burrington, Avon 17 F5
Burrington, Devon 7 G4
Burrough Green 38 B5
Burrough on the Hill 45 J6
Burrow Bridge 8 E1
Burrowhill 20 D7
Burry Green 24 C5
Burry Port 24 C4
Burscough 52 F2
Burscough Bridge 52 F2
Bursea 58 C3
Burshill 65 H8
Bursledon 11 F4
Burslem 43 K3
Burstall 39 F6
Burstock 9 F4
Burston, Norf 39 G2
Burston, Staffs 43 L4
Burstow 13 G2
Burstwick 59 G4
Burtersett 68 A8
Burtle 16 F6
Burton, Dorset 10 C5
Burton, Somer 16 C6
Burton, Lincs 56 D6
Burton, Dyfed 26 D7
Burton, Ches 52 E6
Burton, Ches 53 G7
Burton, Wilts 17 K3
Burton, Northum 81 K6
Burton Agnes 65 J6
Burton Bradstock 9 F6
Burton Constable 59 F3
Burton Fleming 65 H5
Burton Green, Warw 34 D3
Burton Green, Clwyd 52 E8
Burton Hastings 34 F2
Burton-in-Kendal 61 J2
Burton in Lonsdale 61 K2
Burton Joyce 45 H3
Burton Latimer 36 C3
Burton Lazars 45 J6
Burton-le-Coggles 46 C5
Burton Leonard 63 G3
Burton on the Wolds 45 G5
Burton Overy 45 H8
Burton Pedwardine 46 E3
Burton Pidsea 59 G3
Burton Salmon 63 H7
Burton upon Stather 58 C5
Burton upon Trent 44 D5
Burton Wood 53 G4
Burwardsley 43 G2
Burwarton 33 F2
Burwash 13 K4
Burwash Common 13 K4
Burwell, Lincs 57 H6
Burwell, Cambs 37 J4
Bury, W. Susx 12 D5
Bury, Cambs 37 F2
Bury, G. Man 53 K2
Bury, Somer 7 K3
Bury Green 21 J1
Bury St Edmunds 38 D4
Burythorpe 64 E6
Busbridge 12 C2
Busby 78 C4
Buscot 31 G8
Bushbury 43 L7
Bush Crathie 92 D2
Bushey 21 F4
Bushey Heath 21 F4
Bush Green 39 H2
Bushley 30 C4
Bushton 18 B3
Busta 113 C3
Butcher's Pasture 22 C1
Butcombe 17 G4
Butleigh 9 G1
Butleigh Wootton 9 G1
Butlers Marston 34 E5
Butley 39 J5
Butsfield 68 D2
Buttercrambe 64 E7
Butterknowle 68 D4
Butterleigh 7 K5
Buttermere, Cumbr 66 D5
Buttermere, Wilts 18 E4
Buttershaw 62 E7
Butterstone 92 B7
Butterton 44 B2
Butterwick, Humbs 58 C6
Butterwick, N. Yks 64 E5
Butterwick, Lincs 47 G3
Butterwick, N. Yks 65 G5
Butt Green 43 H2
Buttington 42 D7
Buttonoak 33 G3
Buxhall 38 F5
Buxted 13 H4
Buxton, Derby 54 D6
Buxton, Norf 49 H5
Bwlch 25 K2
Bwlch-derwin 50 D6
Bwlchgwyn 42 D2
Bwlch-Llan 27 K2
Bwlchtocyn 40 C1
Bwlch-y-cibau 42 C6
Bwlch-y-fadfa 27 J3
Bwlch-y-ffridd 41 K4
Bwlchygroes 27 G4
Bwlch-y-sarnau 41 K6
Byers Green 68 E3
Byfield 35 G5
Byfleet 20 E7
Byford 29 H3
Bygrave 37 F7
Byker 75 J7
Bylchau 51 J4
Byley 53 J7
Byrness, Northum 74 D3
Byrness, Northum 67 K6
Bythorn 36 D3
Byton 32 C4
Byworth 12 C4

C

Cabourne 59 F6
Cabrach 98 C6
Cadbury 7 K5
Cadbury Barton 7 G4
Cadder 85 J7
Caddington 20 E2
Caddonfoot 80 C6
Cadeby, Leic 44 F7
Cadeby, S. Yks 55 J3
Cadeleigh 7 K5
Cade Street 13 K4
Cadgwith 2 F8
Cadishead 53 J4
Cadle 24 E5
Cadley 18 D4
Cadmore End 20 B4
Cadnam 10 D3
Cadney 58 E6
Cadole 52 D7
Caeathro 50 E4
Caehopkin 25 G3
Caerau, M. Glam 25 G5
Caerau, S. Glam 25 K7
Caerdeon 41 F2
Caergeiliog 50 C3
Caergwrle 52 E8
Caerleon 16 E1
Caer Llan 29 J7
Caernarfon 50 D4
Caerphilly 25 K6
Caersws 41 K4
Caerwent 17 F1
Caerwys 52 C6
Caethle 40 F4
Caio 28 A4
Cairnbaan 83 H2
Cairnborrow 98 D4
Cairnbrogie 99 H6
Cairncross, Tays 93 F4
Cairncross, Border 81 G3
Cairndow 84 D3
Cairneyhill 86 D6
Cairngaan 70 C8
Cairngarroch 70 B7
Cairnhill 98 F5
Cairnie 98 D4
Cairnorrie 99 H4
Cairnryan 70 B5
Caister-on-Sea 49 L6
Caistor 59 F6
Caistor St Edmund 49 H7
Caistron 75 F3
Calbost 108 D4
Calbourne 10 F6
Calcot 20 A6
Caldbeck 66 F3
Caldbergh 62 D1
Caldecote, Cambs 36 E2
Caldecote, Herts 37 F7
Caldecote, Cambs 37 G5
Caldecott, Leic 46 B8
Caldecott, Northnts 36 C4
Calderbank 78 E3
Calder Bridge 66 C6
Calderbrook 53 L2
Caldercruix 79 F3
Calder Mains 107 F4
Caldermill 78 D5
Calder Vale 61 J5
Caldicot 17 F2
Caldwell, N. Yks 68 D5
Caldwell, Derby 44 D6
Caldy 52 D5
Caledrhydiau 27 J2
Calfsound 112 C3
Calgary 88 E3
Califer 97 J3
California, Central 86 C7
California, Norf 49 L6
Calke 44 E5
Callaly 75 G3
Callander 85 J4
Callanish 108 C3
Callestick 2 F4
Calligarry 111 G4
Callington 4 D5
Callow 29 J4
Callow End 33 H6
Callow Hill, Wilts 18 B2
Callow Hill, H. & W. 33 G3
Callows Grave 33 E4
Calmore 10 E3
Calmsden 30 E7
Calne 18 B3
Caln St Dennis 30 E6
Calow 55 H6
Calshot 11 F4
Calstock 4 E5
Calstone Wellington 18 B4
Calthorpe 49 G4
Calthwaite 67 G2
Caltinish 110 C2
Calton, Staffs 44 C2
Calton, N. Yks 62 C4
Calveley 53 G8
Calver 55 F6
Calverhall 43 H4
Calver Hill 29 H3
Calverleigh 7 K4
Calverley 62 F6
Calvert 20 A1
Calverton, Notts 45 H3
Calverton, Bucks 35 J7
Calvine 91 K5
Cam 30 B8
Camas-luinie 94 E3
Camastianavaig 111 F3
Camasunary 111 F4
Camault Muir 96 D4
Camb 113 D2
Camber 14 F7
Camberley 20 C7
Camberwell 21 H6
Camblesforth 63 K7
Cambo 75 G5
Cambois 75 K5
Camborne 2 E5
Cambridge, Glos 30 B7
Cambridge, Cambs 37 H5
Cambus 86 B5
Cambusbarron 85 K5
Cambuskenneth 85 L5
Cambuslang 78 D4
Camden Town 21 G5
Camelford 4 B3
Camelon 86 B6
Camelsdale 11 K1
Camerory 97 J5
Camerton, Cumbr 66 C3
Camerton, Avon 17 H5
Cammachmore 93 L2
Cammeringham 56 D5
Campbelton 77 G2
Campbeltown 76 C5
Campsall 55 J2
Campsey Ash 39 J5
Camp, The 30 D7
Campton 36 E7
Camrose 26 D5
Camster 107 H5
Camusnagaul, Highld 90 B4
Camusnagaul, Highld 101 F3
Camusrory 94 D6
Camusteel 100 C7
Camusterrach 100 C7
Canada 10 D3
Canal Foot 61 G2
Candlesby 57 J7
Candy Mill 79 H5
Cane End 20 A6
Canewdon 22 E4
Canisbay 107 J2
Cann 9 K2
Cann Common 9 K2
Cannich 96 B5
Cannington 16 D7
Cannock 44 A6
Cannock Wood 44 B6
Canonbie 73 J6
Canon Bridge 29 J3
Canon Frome 33 F6
Canon Pyon 29 J3
Canons Ashby 35 G5
Canonstown 2 D6
Canterbury 15 H3
Cantley, Norf 49 J7
Cantley, S. Yks 55 K3
Cantlop 43 G7
Canton 25 K7
Cantraydoune 97 F4
Cantraywood 97 F4
Cantsfield 61 K2
Canvey Island 22 D5
Canwick 56 D7
Canworthy Water 4 C2
Caol 90 C4
Caolas 88 B4
Caol Ila 82 D5
Caolis 110 B5
Capel 12 E2
Capel Bangor 41 F5
Capel Betws Lleucu 27 L2
Capel Carmel 40 A1
Capel Coch 50 D2
Capel Curig 51 G5
Capel Cynon 27 H3
Capel Dewi 27 J3
Capel Garmon 51 H5
Capel Gwyn, Gwyn 50 C3
Capel Gwyn, Dyfed 27 J5
Capel Gwynfe 24 F2
Capel Hendre 27 K6
Capel Isaac 27 K5
Capel Iwan 27 G4
Capel-le-Ferne 15 J5
Capel Llanilltern 25 J6
Capel Parc 50 D2
Capel St Mary 39 F7
Capel-y-ffin 29 G4
Capenhurst 52 E6
Capernwray 61 J2
Capheaton 75 G5
Cappercleuch 79 K7
Capstone 14 D2
Capton 5 J6
Caputh 92 B7
Carbis Bay 2 D6
Carbost, Highld 110 E3
Carbost, Highld 111 F2
Carbrooke 48 E7
Carburton 55 K6
Car Colston 45 J3
Carcroft 55 J2
Cardenden 87 F5
Cardeston 42 E6
Cardiff 25 K7
Cardigan 27 F3
Cardington, Beds 36 D6
Cardington, Shrops 43 G8
Cardinham 3 K3
Cardow 98 A4
Cardrona 80 B6
Cardross 84 F7
Cardurnock 73 G8
Careby 46 D6
Carew 26 E7
Carew Cheriton 26 E7
Carew Newton 26 E7
Carey 29 K4
Carfrae 87 J8
Cargen 72 E6
Cargenbridge 72 E6
Cargo 73 J8
Cargreen 4 E5
Carham 81 F6
Carhampton 7 L1
Carharrack 2 F5
Carisbrooke 11 F6
Carishader 108 B3
Cark 61 G2
Carlby 46 D6
Carlecotes 54 E3
Carleton, N. Yks 62 C5
Carleton, Cumbr 67 G1
Carleton, Lancs 61 G5
Carleton Forehoe 49 F7
Carleton Rode 39 G1
Carlingcott 17 H5
Carlisle 67 G1
Carlops 79 J4
Carloway 108 C2
Carlton, Cambs 38 B5
Carlton, Beds 36 C5
Carlton, N. Yks 62 D1
Carlton, Leic 44 E7
Carlton, Cleve 69 F4
Carlton, S. Yks 55 G3
Carlton, W. Yks 63 G7
Carlton, Notts 45 H3
Carlton, Suff 39 J4
Carlton, N. Yks 63 K1
Carlton, N. Yks 63 K7
Carlton Colville 39 L1
Carlton Curlieu 45 H8
Carlton Husthwaite 63 H2
Carlton in Cleveland 69 H6
Carlton in Lindrick 55 J5
Carlton-le-Moorland 56 D8
Carlton Miniott 63 H1
Carlton-on-Trent 56 B7
Carlton Scroop 46 C3
Carluke 79 F4
Carmacoup 78 E7
Carmarthen 27 J5
Carmel, Gwyn 50 C2
Carmel, Clwyd 52 C6
Carmel, Gwyn 50 D5
Carmel, Dyfed 27 K6
Carminish 108 B5
Carmunnock 78 C4
Carmyle 78 D3
Carmyllie 93 G7
Carnaby 65 J6
Carnach, W. Isles 108 C5
Carnach, Highld 100 F2
Carnach, Highld 94 F3
Carnbee 87 J4
Carnbo 86 D4
Carn Brea Village 2 E5
Carnell 77 K4
Carnforth 61 H2
Carn-gorm 94 E3
Carnhedryn 26 C5
Carnhell Green 2 E6
Carnie 99 H8
Carnish 110 C1
Carno 41 J4
Carnock 86 D6
Carnon Downs 3 F5
Carnousie 98 F3
Carnoustie 87 J1
Carn Towan 2 B7
Carnwath 79 G5
Carnyorth 2 B6
Carperby 68 C8
Carpley Green 62 C1
Carradale 76 D4
Carragreich 108 B5
Carrbridge 97 H6
Carreglefn 50 C2
Carrick Castle 84 D5
Carriden 86 E6
Carrington, Lothn 80 B3
Carrington, Lincs 47 G2
Carrington, G. Man 53 J4
Carrog 42 C3
Carron, Grampn 98 B4
Carron, Central 86 B6
Carronbridge 72 D4
Carr Shield 67 L2
Carrutherstown 73 G6
Carr Vale 55 H7
Carrville 68 F2
Carsaig 89 G6
Carsegowan 71 F6
Carsethorn 72 E8
Carshalton 21 G7
Carsington 44 D2
Carskiey 76 B7
Carsluith 71 F6
Carsphairn 72 A4
Carstairs 79 G5
Carstairs Junction 79 G5
Carswell Marsh 31 H8
Carter's Clay 10 E2
Carterton 31 G7
Carterway Heads 68 C1
Carthew 3 J4
Carthorpe 63 G1
Cartington 75 G3
Cartland 79 F5
Cartmel 61 G2
Cartmel Fell 61 H1
Carway 27 J7
Cas 76 C1
Cascob 32 B4
Cashmoor 10 A3
Cassington 31 J6
Cassiobury Park 20 E4
Casswell's Bridge 46 E5
Castallack 2 C7
Castellau 25 J6
Castell Howell 27 J3
Castell-y-bwch 16 D1
Casterton 61 K2
Castle Acre 48 D6
Castle Ashby 36 B5
Castlebay 110 B5
Castle Bolton 68 C7
Castle Bromwich 34 C1
Castle Bytham 46 C6
Castlebythe 26 E5
Castle Caereinion 42 C7
Castle Camps 38 B6
Castle Carrock 67 H1
Castle Cary, Somer 9 H1
Castlecary, Strath 85 K7
Castle Combe 17 K3
Castlecraig, Highld 97 G2
Castlecraig, Border 79 J5
Castle Donington 45 F5
Castle Douglas 72 C7
Castle Eaton 31 F8
Castle Eden 69 G3
Castleford 63 H7
Castle Frome 33 F6
Castle Gresley 44 D6
Castle Heaton 81 H5
Castle Hedingham 38 C7
Castlehill, Strath 79 F4
Castle Hill, Suff 39 G6
Castle Kennedy 70 C6
Castlemartin 26 D8
Castlemilk 73 G6
Castlemorris 26 D4
Castlemorton 33 G7
Castle O'er 73 H4
Castle Rising 48 B5
Castle Side 68 C2
Castlethorpe 35 J6
Castleton, Strath 84 A6
Castleton, Gwent 16 D2
Castleton, N. Yks 64 D2
Castleton, Derby 54 E5
Castletown, Highld 107 G3
Castletown, T. & W. 75 K8
Castletown, I. of M. 60 P5
Caston 48 E8
Castor 46 E8
Catacol 76 E3
Catbrain 17 G2
Catcliffe 55 H5
Catcott 16 E7
Caterham 13 G1
Catesby 35 G5
Catfield 49 J5
Catfirth 113 D4
Catford 21 H6
Catforth 61 H6
Cath 26 B5
Cathcart 78 C3
Cathedine 25 K2
Catherington 11 J3
Catherton 33 F3
Catlodge 91 H2
Catlowdy 73 K6
Catmore 19 F2
Caton 61 J3
Catrine 77 L5
Cat's Ash 16 E1
Catsfield 13 L5
Catshill 34 A3
Cattal 63 H4
Cattawade 39 G7
Catterall 61 H5
Catterick 68 E7
Catterick Bridge 68 E7
Catterick Garrison 68 D7
Catterlen 67 G3
Catterline 93 K4
Catterton 63 J5
Catthorpe 35 G3
Cattistock 9 G5
Catton, N. Yks 63 G2
Catton, Norf 49 H6
Catton, Northum 67 L1
Catwick 59 F2
Catworth 36 D3
Caulcott 31 K5
Cauldhame 85 J5
Cauldon 44 B3
Caulkerbush 66 B1
Caulside 73 K5
Caunsall 33 H2
Caunton 56 B7
Causewayhead 85 L5
Causeyend 99 J7
Causey Park Bridge 75 H4
Cautley 67 J7
Cavendish 38 D6
Cavenham 38 C4
Caversfield 31 K5
Caversham 20 B6
Caversta 108 C4
Caverswall 44 A3
Cawdor 97 G3
Cawood 63 J6
Cawsand 4 E6
Cawston 49 G5
Cawthorne 55 F3
Cawton 63 K2
Caxton 37 G5
Caynham 32 E3
Caythorpe, Lincs 46 C3
Caythorpe, Notts 45 H3
Cayton 65 H4
Ceann-na-Cleithe 108 B5
Cefn Berain 51 J4
Cefn-brith 51 J5
Cefn Coch 41 L1
Cefn-coed-y-cymmer 25 J4
Cefn Cribwr 25 G6
Cefn Cross 25 G6
Cefn Einion 32 B2
Cefneithin 27 K6
Cefn Hengoed 25 K5
Cefn-mawr 42 D3
Cefn-y-bedd 42 E2
Cefn-y-pant 27 F5
Ceint 50 D3
Cellan 27 L3
Cellarhead 44 A3
Cemaes 50 C1
Cemmaes 41 H3
Cemmaes Road 41 H3
Cenarth 27 G3
Cennin 50 D6
Ceres 87 H3
Cerne Abbas 9 H4
Cerney Wick 30 E8
Cerrigceinwen 50 D3
Cerrigydrudion 51 J6
Cessford 80 F7
Chaceley 30 C4
Chacewater 2 F5
Chackmore 35 H7
Chacombe 35 F6
Chadderton 53 L3
Chaddesden 44 E4
Chaddesley Corbett 33 H3
Chaddleworth 18 F3
Chadlington 31 H5
Chadshunt 34 E5
Chad Valley 34 B2
Chadwell St Mary 22 C6
Chadwick End 34 D3
Chaffcombe 8 E3
Chagford 5 H3
Chailey 13 G5
Chainhurst 13 L2
Chalbury Common 10 B4
Chaldon 13 G1
Chaldon Herring or
East Chaldon 9 J6
Chale 11 F7
Chale Green 11 F7
Chalfont Common 20 E4
Chalfont St Giles 20 D4
Chalfont St Peter 20 E4
Chalford 30 C7
Chalgrove 20 A4
Chalk 14 C1
Challacombe 7 G1
Challoch 70 E5
Challock 15 G3
Chalton, Beds 36 D8
Chalton, Hants 11 J3
Chalvington 13 J6
Champany 86 D7
Chandler's Cross 20 E4
Chandler's Ford 10 F2
Channerwick 113 D5
Chantry, Suff 39 G6
Chantry, Somer 17 J6
Chapel 87 F5
Chapel Allerton, Somer 16 F5
Chapel Allerton, W. Yks 63 G6
Chapel Amble 3 H2
Chapel Brampton 35 J4
Chapel Chorlton 43 K4
Chapel-en-le-Frith 54 D5
Chapelgate 47 H5
Chapel Haddlesey 63 J7
Chapelhall 78 E3
Chapelhill, Tays 86 D1
Chapel Hill, Lincs 46 F2
Chapel Hill, Grampn 99 K5
Chapel Hill, Gwent 29 K7
Chapelknowe 73 J6
Chapel Lawn 32 C3
Chapel-le-Dale 61 L2
Chapel of Garioch 99 G6
Chapel Row 19 G4
Chapel Stile 66 F6
Chapel St Leonards 57 K6
Chapelton, Strath 78 D5
Chapelton, Devon 7 F3
Chapelton, Tays 93 H7
Chapeltown, Grampn 98 B6
Chapeltown, S. Yks 55 G4
Chapeltown, Lancs 53 J2
Chapmanslade 17 K6
Chapmans Well 4 D2
Chappel 38 D8
Chard 8 E4
Chardstock 8 E4
Charfield 17 J1
Charing 14 F4
Charing Heath 14 F4
Charingworth 34 D7
Charlbury 31 H6
Charlcombe 17 J4
Charlecote 34 D5
Charles 7 G2
Charleston 92 E7
Charlestown, Highld 100 D4
Charlestown, Fife 86 D6
Charlestown, Highld 96 E4
Charlestown, Dorset 9 H7
Charlestown, Corn 3 J4
Charlestown, Grampn 93 L1
Charlestown of Aberlour 98 B4
Charles Tye 38 F5
Charlesworth 54 D4
Charlton, Wilts 18 A2
Charlton, H. & W. 34 B6
Charlton, Wilts 10 C2
Charlton, Wilts 18 C5
Charlton, G. Lon 21 J6
Charlton, W. Susx 11 K3
Charlton, Northnts 31 K4
Charlton, Wilts 9 L2
Charlton Abbots 30 E5
Charlton Adam 9 G2
Charlton Horethorne 9 H2
Charlton Kings 30 D5
Charlton Mackrell 9 G2
Charlton Marshall 9 L4
Charlton Musgrove 9 J1
Charlton-on-Otmoor 31 K6
Charlwood 12 F2
Charlynch 8 D1
Charminster 9 H5
Charmouth 8 E5
Charndon 20 A1
Charney Bassett 31 H8
Charnock Richard 53 G2
Charsfield 39 H5
Charter Alley 19 G5
Charterhouse 17 F5
Chartershall 85 L5
Charterville Allotments 31 H6
Chartham 15 H3
Chartham Hatch 15 H3
Chartridge 20 D3
Chart Sutton 14 D3
Charwelton 35 G5
Chase Terrace 44 B7
Chasetown 44 B7
Chastleton 31 G5
Chatburn 62 A5
Chatcull 43 J4
Chatham 14 D2
Chathill 81 K7
Chattenden 14 D1
Chatteris 37 G2
Chattisham 39 F6
Chatto 74 D2
Chatton 81 J7
Chawleigh 7 H4
Chawston 36 E5
Chawton 11 J1
Cheadle, Staffs 44 B3
Cheadle, G. Man 53 K5
Cheadle Hulme 53 K5
Cheam 21 G7
Chearsley 20 B2
Chebsey 43 K5
Checkendon 20 A5
Checkley, Staffs 44 B4
Checkley, Ches 43 J3
Chedburgh 38 C5
Cheddar 17 F5
Cheddington 20 D2
Cheddleton 44 A2
Cheddon Fitzpaine 8 D2
Chedgrave 49 J8
Chedington 9 F4
Chediston 39 J3
Chedworth 30 E6
Chedzoy 8 E1
Cheeseman's Green 15 G5
Cheetham Hill 53 K3
Cheldon 7 H4
Chelford 53 K6
Chellaston 44 E4
Chellington 36 C5
Chelmarsh 33 G2
Chelmondiston 39 H7
Chelmorton 54 E6
Chelmsford 22 D3
Chelsea 21 G6
Chelsfield 21 J7
Chelsworth 38 E6
Cheltenham 30 D5
Chelveston 36 C4
Chelvey 17 F4
Chelwood 17 H4
Chelwood Gate 13 H4
Cheney Longville 32 D2
Chenies 20 E4
Chepstow 29 K8
Cherhill 18 B3
Cherington, Warw 34 D7
Cherington, Glos 30 D8
Cheriton, Devon 8 C4
Cheriton, W. Glam 24 C5
Cheriton, Hants 11 G2
Cheriton Bishop 5 H2
Cheriton Fitzpaine 7 J5
Cherrington 43 H5
Cherry Burton 58 D2
Cherry Hinton 37 H5
Cherry Willingham 56 E6
Chertsey 20 E7
Cheselbourne 9 J5
Chesham 20 D3
Chesham Bois 20 D4
Cheshunt 21 H3
Cheslyn Hay 44 A7
Chessington 21 F7
Chester 52 F7
Chesterblade 17 H6
Chesterfield, Staffs 44 C7
Chesterfield, Derby 55 G6
Chester-le-Street 68 E1
Chesters, Border 74 C2
Chesters, Border 80 E7
Chesterton, Warw 34 E5
Chesterton, Cambs 46 E8
Chesterton, Cambs 37 H4
Chesterton, Staffs 43 K3
Chesterton, Oxon 31 K5
Chestfield 15 H2
Cheswardine 43 J5
Cheswick 81 J5
Cheswick Green 34 C3
Chetnole 9 H4
Chettiscombe 7 K4
Chettisham 37 J2
Chettle 10 A3
Chetton 33 F1
Chetwode 35 H8
Chetwynd Aston 43 J6
Cheveley 38 B4
Chevening 13 H1
Chevington 38 C5
Chevington Drift 75 J4
Chevithorne 7 K4
Chew Magna 17 G4
hew Stoke 17 G4
Chewton Mendip 17 G5
Chicheley 36 C6
Chichester 11 K4
Chickerell 9 H6
Chicklade 9 L1
Chicksands 36 E7
Chicksgrove 10 A2
Chidden 11 H3
Chiddingfold 12 C3
Chiddingly 13 J5
Chiddingstone 13 H2
Chideock 9 F5
Chidham 11 J4
Chieveley 19 F3
Chignall Smealy 22 C2
Chignall St James 22 C3
Chigwell 21 J4
Chigwell Row 21 J4
Chilbolton 18 E7
Chilcombe 9 G5
Chilcompton 17 H5
Chilcote 44 D6
Childer Thornton 52 E6
Child Okeford 9 K3
Childrey 18 E2
Child's Ercall 43 H5
Childswickham 34 B7
Childwall 52 F5
Chilfrome 9 G5
Chilgrove 11 K3
Chilham 15 G3
Chillaton 4 E3
Chillenden 15 J3
Chillerton 11 F6
Chillesford 39 J5
Chillingham 81 J7
Chillington, Somer 8 E3
Chillington, Devon 5 H7
Chilmark 10 A1
Chilson 31 H6
Chilsworthy, Devon 6 D5
Chilsworthy, Corn 4 E4
Chilthorne Domer 9 G3
Chilton, Bucks 20 A2
Chilton, Durham 68 E4
Chilton, Oxon 19 F2
Chilton Cantelo 9 G2
Chilton Foliat 18 E3
Chilton Lane 68 F3
Chilton Polden 16 E7
Chilton Street 38 C6
Chilton Trinity 16 D7
Chilworth, Surrey 12 D2
Chilworth, Hants 10 F3
Chimney 31 H7
Chineham 19 H5
Chingford 21 H4
Chinley 54 D5
Chinley Head 54 D5
Chinnor 20 B3
Chipnall 43 J4
Chippenham, Cambs 38 B4
Chippenham, Wilts 17 L3
Chipperfield 20 E3
Chipping, Herts 37 G7
Chipping, Lancs 61 K5
Chipping Campden 34 C7
Chipping Hill 22 E2
Chipping Norton 31 H5
Chipping Ongar 21 K3
Chipping Sodbury 17 J2
Chipping Warden 35 F6
Chipstable 8 B2
Chipstead 12 F1
Chirbury 42 D8
Chirk 42 D4
Chirnside 81 G4
Chirnsidebridge 81 G4
Chirton 18 B5
Chisbury 18 D4
Chiselborough 9 F3
Chiseldon 18 C3
Chiselhampton 31 K8
Chislehurst 21 J6
Chislet 15 J2
Chiswell Green 21 F3
Chiswick 21 G6
Chisworth 54 C4
Chitcomb 11 G2
Chithurst 11 K2
Chittering 37 H3
Chitterne 18 A6
Chittlehamholt 7 G3
Chittlehampton 7 G3
Chittoe 18 A4
Chivenor 6 F2
Chobham 20 D7
Cholderton 18 D6
Cholesbury 20 D3
Chollerton 74 F6
Cholsey 19 G2
Cholstrey 29 J2
Choppington 75 J5
Chopwell 75 H8
Chorley, Staffs 44 B6
Chorley, Shrops 33 F2
Chorley, Lancs 53 G2
Chorley, Ches 43 G2
Chorleywood, Herts 20 E4
Chorleywood, Bucks 20 E4
Chorlton 43 J2
Chorlton-cum-Hardy 53 K4
Chorlton Lane 42 F3
Chowley 43 F2
Chrishall 37 H7
Christchurch, Dorset 10 C5
Christchurch, Cambs 47 H8
Christchurch, Glos 29 K6
Christian Malford 18 A3
Christleton 52 F7
Christmas Common 20 B4
Christon 16 E5
Christon Bank 81 L7
Christow 5 J3
Chudleigh 5 J4
Chudleigh Knighton 5 J4
Chulmleigh 7 G4
Chunal 54 D4
Church, N. Yks 69 J7
Church, Lancs 61 L7
Churcham 30 B6
Church Aston 43 J6
Church Brampton 35 J4
Church Broughton 44 D4
Church Cove 2 F8
Church Crookham 19 K5
Churchdown 30 C5
Church Eaton 43 K6
Church End, Wilts 18 B3
Churchend, Essex 22 C1
Church End, Beds 20 D1
Church End, Warw 34 D1
Church End, Beds 36 E7
Churchend, Essex 23 G4
Church End, Cambs 47 G7
Church End, Cambs 37 H5
Church End, Hants 19 H5
Church End, Essex 37 J6
Church Fenton 63 J6
Church Gresley 44 D6
Church Hanborough 31 J6
Churchill, Devon 8 D4
Churchill, Avon 17 F5
Churchill, Oxon 31 G5
Churchill, H. & W. 33 H3
Churchinford 8 D3
Church Knowle 10 A6
Church Langton 45 J8
Church Lawford 35 F3
Church Lawton 43 K2
Church Leigh 44 B4
Church Lench 34 B5
Church Minshull 53 H7
Church Norton 11 K5
Churchover 35 G2
Church Preen 43 G8
Church Pulverbatch 42 F7
Churchstanton 8 C3
Church Stoke 42 D8
Churchstow 5 H7
Church Stowe 35 H5
Church Street 14 D1
Church Stretton 42 F8
Churchtown, Mers 52 E2
Churchtown, Lancs 61 H5
Churchtown, I. of M. 60 R2
Church Village 25 J6
Church Warsop 55 J7
Churt 19 K7
Churton 42 F2
Churwell 63 F7
Chwilog 50 D7
Chyandour 2 C6
Cilan Uchaf 40 B1
Cilcain 52 C7
Cilcennin 27 K1
Cilfor 51 F7
Cilfrew 25 F4
Cilfynydd 25 J5
Cilgerran 27 F3
Cilgwyn 25 F2
Ciliau-Aeron 27 K2
Cilmaengwyn 25 F4
Cilmery 28 E2
Cilrhedyn 27 G4
Ciltwrch 29 F3
Cilwendeg 27 G4
Cilybebyll 25 F4
Cilycwm 28 B3
Cinderford 30 A6
Cirencester 30 E7
City Dulas 50 D2
City of London 21 H5
City, The 20 B4
Clachaig 84 D6
Clachan, Strath 76 C2
Clachan, Highld 111 F3
Clachan, Strath 89 J7
Clachan, Strath 89 K4
Clachan-a-Luib 110 C1
Clachan of Campsie 85 J7
Clachan of Glendaruel 84 B6
Clachan-Seil 89 J7
Clachbreck 83 G4
Clachtoll 104 B6
Clackmannan 86 C5
Clacton-on-Sea 23 H2
Claddach Kirkibost 110 B1
Claggan 89 H4
Claigan 110 E2
Claines 33 H5
Clandown 17 H5
Clanfield, Oxon 31 G7
Clanfield, Hants 11 H3
Clanville 18 E6
Clanyard 70 C8
Claonaig 76 D2
Claonel 102 D3
Clapgate 10 B4
Clapham, Beds 36 D5
Clapham, W. Susx 12 D6
Clapham, G. Lon 21 G6
Clapham, N. Yks 61 L3
Clappers 81 H4
Clappersgate 66 F6
Clapton 8 F4
Clapton-in-Gordano 17 F3
Clapton-on-the-Hill 31 F6
Clapworthy 7 G3
Clarbeston 26 E5
Clarbeston Road 26 E5
Clarborough 56 B5
Clardon 107 G3
Clare 38 C6
Clarebrand 72 C7
Clarencefield 73 F7
Clarkston 78 C4
Clashmore, Highld 104 B5
Clashmore, Highld 103 F5
Clashnessie 104 B5
Clashnoir 98 B6
Clatt 98 E6
Clatter 41 J4
Clatworthy 8 B1
Claughton, Lancs 61 J3
Claughton, Lancs 61 J5
Claverdon 34 C4
Claverham 17 F4
Clavering 37 H7
Claverley 43 J8
Claverton 17 J4
Clawdd-newydd 51 K5
Clawton 4 D2
Claxby, Lincs 57 F4
Claxby, Lincs 57 J6
Claxton, N. Yks 64 D6
Claxton, Norf 49 J7
Claybokie 92 B2
Claybrooke Magna 35 F2
Clay Coton 35 G3
Clay Cross 55 G7
Claydon, Oxon 35 F5
Claydon, Suff 39 G5
Claygate 21 F7
Claygate Cross 13 K1
Clayhanger, Devon 8 B2
Clayhanger, W. Mids 44 B7
Clayhidon 8 C3
Clayock 107 G4
Claypole 46 B3
Clayton, W. Yks 62 E6
Clayton, W. Susx 13 G5
Clayton, S. Yks 55 H3
Clayton, Staffs 43 K3
Clayton-le-Moors 61 L6
Clayton-le-Woods 61 J7
Clayton West 55 F2
Clayworth 56 B5
Cleadale 111 F5
Cleadon 75 K7
Clearwell 29 K7
Cleasby 68 E5
Cleat, W. Isles 110 B4
Cleat, Orkney 112 C5
Cleatlam 68 D5
Cleator 66 C5
Cleator Moor 66 C5
Cleckheaton 62 E7
Cleedownton 32 E2
Cleehill 33 E3
Clee St Margaret 32 E2
Cleethorpes 59 H6
Cleeton St Mary 33 F3
Cleeve 17 F4
Cleeve Hill 30 D5
Cleeve Prior 34 B6
Clehonger 29 J4
Cleigh 84 A2
Cleish 86 D5
Cleland 78 E4
Clench Common 18 C4
Clenchwarton 48 A5
Clent 33 J3
Cleobury Mortimer 33 F3
Cleobury North 33 F2
Clephanton 97 G3
Clerklands 80 D7
Clestrain 112 B4
Cleughbrae 73 F6
Clevancy 18 B3
Clevedon 16 F3
Cleveleys 61 G5
Cleverton 18 A2
Clewer 17 F5
Cley next the Sea 48 F3
Cliasmol 108 B4
Cliburn 67 H4
Cliddesden 19 H6
Cliffe, Kent 14 D1
Cliffe, N. Yks 63 K6
Cliff End 14 E7
Cliffe Woods 14 D1
Clifford, H. & W. 29 G3
Clifford, W. Yks 63 H5
Clifford Chambers 34 C5
Clifford's Mesne 30 B5
Cliffs End 15 K2
Clifton, Derby 44 C3
Clifton, Beds 36 E7
Clifton, Central 84 F1
Clifton, Avon 17 G3
Clifton, Notts 45 G4
Clifton, Cumbr 67 H4
Clifton, Lancs 61 H6
Clifton, H. & W. 33 H6
Clifton, Oxon 31 J4
Clifton, Northum 75 J5
Clifton Campville 44 D6
Clifton Hampden 31 K8
Clifton Reynes 36 C5
Clifton upon Dunsmore 35 G3
Clifton upon Teme 33 G4
Climping 12 C6
Clint 63 F4
Clint Green 48 F6
Clintmains 80 E6
Clippesby 49 K6
Clipsham 46 C6
Clipston, Notts 45 H4
Clipston, Northnts 35 J2
Clitheroe 61 L5
Clive 43 G5
Clivocast 113 E1
Clocaenog 51 K5
Clochan 98 D2
Clock Face 53 G4
Cloddymoss 97 H2
Clodock 29 H5
Clola 99 K4
Clophill 36 D7
Clopton, Northnts 36 D2
Clopton, Suff 39 H5
Clopton Green 38 C5
Closeburn 72 D4
Close Clark 60 P4
Clothall 37 F7
Clotton 53 G7
Clough Foot 62 C7
Cloughton 65 H3
Cloughton Newlands 65 H3
Clousta 113 C4
Clouston 112 B4
Clova 92 E4
Clovelly 6 D3
Clove Lodge 68 B5
Clovenfords 80 C6
Clovenstone 99 G7
Clovullin 90 B5
Clowne 55 H6
Clows Top 33 G3
Clubworthy 4 C2
Cluer 108 B5
Clun 32 C2
Clunbury 32 C2
Clunderwen 26 F6
Clunes 90 D3
Clungunford 32 C3
Clunie 98 F3
Clunton 32 C2
Cluny 87 F5
Clutton, Ches 42 F2
Clutton, Avon 17 H5
Clwyt-y-bont 50 E4
Clydach, W. Glam 24 E4
Clydach, Gwent 25 L3
Clydach Vale 25 H5
Clydebank 85 H8
Clydey 27 G4
Clyffe Pypard 18 B3
Clynder 84 E6
Clynelish 103 G3
Clynnog-fawr 50 D6
Clyro 29 G3
Clyst Honiton 5 K2
Clyst Hydon 8 B4
Clyst St George 5 K3
Clyst St Lawrence 8 B4
Clyst St Mary 5 K2
Clyth 107 H6
Cnoc Amhlaigh 108 D3
Cnwch Coch 41 F6
Coad's Green 4 C4
Coal Aston 55 G6
Coalbrookdale 43 H7
Coalburn 79 F6
Coalcleugh 67 L2
Coaley 30 B7
Coalpit Heath 17 H2
Coalport 43 H7
Coalsnaughton 86 C5
Coaltown of Balgonie 87 G5
Coaltown of Wemyss 87 G5
Coalville 44 F6
Coalway 29 K6
Coast 100 E2
Coatbridge 78 E3
Coatdyke 78 E3
Coate 18 B4
Coates, Glos 30 D7
Coates, Cambs 47 G8
Coatham 69 H4
Coatham Mundeville 68 E4
Coatsgate 73 F3
Cobbaton 7 G3
Coberley 30 D6
Cobham, Kent 14 C2
Cobham, Surrey 21 F7
Cobnash 29 J1
Cockayne 69 J7
Cockayne Hatley 37 F6
Cock Bridge 98 B8
Cockburnspath 80 F2
Cock Clarks 22 E3
Cockenzie and Port Seton 87 H7
Cockerham 61 H4
Cockerington 57 H4
Cockermouth 6 D3
Cockernhoe 21 F1
Cockfield, Durham 68 D4
Cockfield, Suff 38 E5
Cockfosters 21 G4
Cocking 11 K3
Cockington 5 J5
Cocklake 17 F6
Cockley Beck 66 E6
Cockley Cley 48 C7
Cockpole Green 20 B5
Cockshutt 42 F5
Cockthorpe 48 E3
Cockwood 5 K3
Cockyard 29 J4
Coddenham 39 G5
Coddington, Notts 46 B2
Coddington, Ches 42 F2
Coddington, H. & W. 33 G6
Codford St Mary 18 A7
Codford St Peter 18 A6
Codicote 21 G2
Codnor 44 F3
Codrington 17 J3
Codsall 43 K7
Codsall Wood 43 K7
Coedely 25 J6
Coedkernew 16 D2
Coedpoeth 42 D2
Coed-y-paen 29 H8
Coed Ystumgwern 40 E1
Coelbren 25 G3
Coffinswell 5 J5
Cofton Hackett 34 B3
Cogenhoe 36 B4
Coggeshall 22 E1
Coilacriech 92 E2
Coille Mhorgil 90 C1
Coillore 110 E3
Coity 25 H6
Col 108 D2
Colaboll 105 G7
Colan 3 G3
Colaton Raleigh 8 B6
Colbost 110 E2
Colburn 68 D7
Colby, Norf 49 H4
Colby, Cumbr 67 J4
Colby, I. of M. 60 P4
Colchester 38 F8
Cold Ash 19 G3
Cold Ashby 35 H3
Cold Ashton 17 J3
Cold Aston 31 F6
Coldbackie 105 H2
Coldblow 21 K6
Cold Brayfield 36 C5
Coldean 13 G6
Coldeast 5 J4
Colden Common 11 F2
Coldfair Green 39 K4
Cold Hanworth 56 E5
Coldharbour 12 E2
Cold Hesledon 69 G2
Cold Higham 35 H5
Coldingham 81 H3
Cold Kirby 63 J1
Cold Newton 45 J7
Cold Norton 22 E3
Cold Overton 46 B6
Coldred 15 J4
Coldridge 7 G5
Coldstream 81 G6
Coldwaltham 12 D5
Coldwells 99 L5
Coldwells Croft 98 E6
Cole 9 H1
Colebatch 32 C2
Colebrook 7 L5
Colebrooke 5 H1
Coleburn 68 E7
Coleby, Humbs 58 C5
Coleby, Lincs 56 D7
Coleford, Devon 7 H5
Coleford, Somer 17 H6
Coleford, Glos 29 K6
Colehill 10 B4
Coleman's Hatch 13 H3
Colemere 42 F4
Colenden 86 E2
Coleorton 44 F6
Colerne 17 K3
Colesbourne 30 D6
Colesden 36 E5
Coleshill, Warw 34 D2
Coleshill, Bucks 20 D4
Coleshill, Oxon 31 G8
Colgate 13 G3
Colgrain 84 F6
Colinsburgh 87 H4
Colinton 87 F8
Colintraive 84 C7
Colkirk 48 E5
Collafirth 113 C2
Collaton St Mary 5 J5
Collessie 87 F3
Collier Row 21 J4
Colliers End 21 H1
Collier Street 13 L2
Collieston 99 K6
Collin 72 F6
Collingbourne Ducis 18 D5
Collingbourne Kingston 18 D5
Collingham, Notts 56 C7
Collingham, W. Yks 63 G5
Collington 33 F4
Collingtree 35 J5
Colliston 93 H7
Collyweston 46 C7
Colmonell 70 C3
Colmworth 36 E5
Colnabaichin 98 B8
Colnbrook 20 E6
Colne, Lancs 62 B5
Colne, Cambs 37 G3
Colne Engaine 38 D7
Colney 49 G7
Colney Heath 21 G3
Colney Street 21 F3
Coln Rogers 30 E7
Coln St Aldwyns 31 F7
Coln St Dennis 30 E6
Colpy 98 F5
Colsterdale 62 E1
Colsterworth 46 C5
Colston Bassett 45 H4
Coltfield 97 K2
Coltishall 49 H5
Colton, Staffs 44 B5
Colton, Cumbr 61 G1
Colton, Norf 49 G7
Colton, N. Yks 63 J5
Col Uarach 108 D3
Colvend 71 K6
Colvister 113 D2
Colwall Green 33 G6
Colwall Stone 33 G6
Colwell 74 F6
Colwich 44 B5
Colwinston 25 H7
Colworth 12 C6
Colwyn Bay 51 H3
Colyford 8 D5
Colyton 8 D5
Combe, H. & W. 32 C4
Combe, Berks 18 E4
Combe, Oxon 31 J6
Combe Florey 8 C1
Combe Hay 17 J5
Combeinteignhead 5 K4
Combe Martin 7 F1
Combe Moor 32 C4
Combe Raleigh 8 C4
Comberbach 53 H6
Comberton 37 G5
Combe St Nicholas 8 E3
Combpyne 8 D5
Combrook 34 E5
Combs, Derby 54 D6
Combs, Suff 38 F5
Combs Fords 38 F5
Combwich 16 D6
Comers 98 F8
Commins Coch 41 H3
Commondale 69 J5
Common Moor 3 L3
Common Side 55 G6
Common, The 10 D1
Compstall 54 C4
Compton, Surrey 12 C2
Compton, Wilts 18 C5
Compton, Hants 11 F2
Compton, Berks 19 G3
Compton, W. Susx 11 J3
Compton, Devon 5 J5
Compton Abbas 9 K3
Compton Abdale 30 E6
Compton Bassett 18 B3
Compton Beauchamp 18 D2
Compton Bishop 16 E5
Compton Chamberlayne 10 B2
Compton Dando 17 H4
Compton Dundon 9 F1
Compton Martin 17 G5
Compton Pauncefoot 9 H2
Compton Valence 9 G5
Comrie 85 K2
Conchra 84 C6
Conderton 34 A7
Condicote 31 F5
Condorrat 85 K7
Condover 43 F7
Coneyhurst 12 E4
Coneysthorpe 64 E5
Coney Weston 38 E3
Congerstone 44 E7
Congham 48 C5
Congleton 53 K7
Congresbury 17 F4
Conicavel 97 H3
Coningsby 57 G8
Conington, Cambs 36 E2
Conington, Cambs 37 G4
Conisbrough 55 J4
Conisby 82 B5
Conisholme 57 J4
Coniston, Humbs 59 F3
Coniston, Cumbr 66 F7
Coniston Cold 62 C4
Conistone 62 C3
Connah's Quay 52 D7
Connel 84 B1
Connel Park 72 B2
Connor Downs 2 D6
Conon Bridge 96 D3
Cononley 62 C5
Conordan 111 F3
Consall 44 A3
Consett 68 D1
Constable Burton 68 D7
Constantine 2 F7
Contin 96 C3
Conwy 51 G3
Conyer 14 F2
Cookbury 6 E5
Cookham 20 C5
Cookham Dean 20 C5
Cookham Rise 20 C5
Cookhill 34 B5
Cookley, H. & W. 33 H2
Cookley, Suff 39 J3
Cookley Green 20 A4
Cookney 93 K2
Cooksbridge 13 H5
Cooksmill Green 22 C3
Coolham 12 E4
Cooling 14 D1
Coombe, Corn 6 C4
Coombe, Corn 3 H4
Coombe Bissett 10 C2
Coombe Hill 30 C5
Coombe Keynes 9 K6
Coombes 12 E6
Copdock 39 G6
Copford Green 22 F1
Copister 113 D3
Cople 36 E6
Copley 68 C4
Coplow Dale 54 E6
Copmanthorpe 63 J5
Coppathorne 4 C1
Coppenhall 43 L6
Copperhouse 2 D6
Coppingford 36 E2
Copplestone 7 H5
Coppull 53 G2
Copsale 12 E4
Copster Green 61 K6
Copt Heath 34 C3
Copt Hewick 63 G2
Copthorne 13 G3
Copt Oak 45 F6
Copythorne 10 E3
Corbridge 75 F7
Corby 36 B2
Corby Glen 46 C5
Coreley 33 F3
Corfe 8 D3
Corfe Castle 10 A6
Corfe Mullen 10 A5
Corfton 32 D2
Corgarff 98 B8
Corhampton 11 H2
Corlae 72 B4
Corley 34 E2
Corley Ash 34 D2
Corley Moor 34 D2
Cornelly 25 G6
Corney 66 D7
Cornforth 68 F3
Cornhill 98 E3
Cornhill-on-Tweed 81 G6
Cornholme 62 C7
Cornish Hall End 38 B7
Cornriggs 67 L2
Cornsay 68 D2
Corntown 96 D3
Cornwell 31 G5
Cornwood 5 G6
Cornworthy 5 J6
Corpach 90 B4
Corpusty 49 G4
Corran, Highld 90 B5
Corran, Highld 94 D5
Corrany 60 R3
Corrie 76 F3
Corrie Common 73 H5
Corriemoillie 96 B2
Corrimony 96 B5
Corringham, Lincs 56 C4
Corringham, Essex 22 D5
Corris 41 G3
Corris Uchaf 41 G3
Corry 111 G3
Corry of Ardnagrask 96 D4
Corscombe 9 G4
Corse 98 F4
Corsham 17 K3
Corsindae 99 F8
Corsley 17 K6
Corsley Heath 17 K6
Corsock 72 C6
Corston, Avon 17 H4
Corston, Wilts 17 L2
Corstorphine 87 E7
Corton, Wilts 18 A6
Corton, Suff 49 L8
Corton Denham 9 H2
Corwen 51 K6
Coryton, Essex 22 D5
Coryton, Devon 4 E3
Cosby 45 G8
Coseley 43 L8
Cosgrove 35 J6
Cosham 11 H4
Cosheston 26 E7
Cossall 45 F3
Cossington, Somer 16 E6
Cossington, Leic 45 H6
Costa 112 B3
Costessey 49 G6
Costock 45 G5
Coston 46 B5
Cotebrook 53 G7
Cotegill 67 J6
Cotehill 67 G1
Cotes, Leic 45 G5
Cotes, Staffs 43 K4
Cotesbach 35 G2
Cotgrave 45 H4
Cothall 99 H7
Cotham 45 J3
Cothelstone 8 C1
Cotherstone 68 C5
Cothill 31 J8
Cotleigh 8 D4
Coton, Staffs 44 A4
Coton, Northnts 35 H3
Coton, Cambs 37 H5
Coton Clanford 43 K5
Coton in the Elms 44 D6
Cott 5 H5
Cottam, Notts 56 C5
Cottam, Lancs 61 J6
Cottartown 97 J5
Cottenham 37 H4
Cotterdale 67 L7
Cottered 37 G8
Cotterstock 36 D1
Cottesbrooke 35 J3
Cottesmore 46 C6
Cottingham, Northnts 36 B1
Cottingham, Humbs 58 E3
Cottisford 31 K4
Cotton, Staffs 44 B3
Cotton, Suff 38 F4
Cotton End 36 D6
Cottown, Grampn 98 E6
Cottown, Grampn 99 G7
Cottown, Grampn 99 H4
Cotwalton 43 L4
Coughton, Warw 34 B4
Coughton, H. & W. 29 K5
Coulags 100 E7
Coull 93 G1
Coulport 84 E6
Coulsdon 21 H8
Coulston 18 A5
Coulter 79 H6
Coulton 6 K2
Cound 43 G7
Coundon 68 E4
Coundon Grange 68 E4
Countersett 62 C1
Countess Wear 5 K3
Countesthorpe 45 G8
Countisbury 7 H1
Coupar Angus 92 D7
Coupland 81 H6
Cour 76 D3
Courteachan 94 B6
Courteenhall 35 J5
Court Henry 27 K5
Courtsend 23 G4
Courtway 8 D1
Cousland 80 B3
Cousley Wood 13 K3
Cove, Highld 100 D2
Cove, Strath 84 E6
Cove, Devon 7 K4
Cove, Hants 19 K5
Cove Bay 93 L1
Covehithe 39 L2
Coven 43 L7
Coveney 37 H2
Covenham St Bartholomew 57 H4
Covenham St Mary 57 H4
Coventry 34 E3
Coverack 3 F8
Coverham 62 E1
Covington 36 D3
Cowan Bridge 61 K2
Cowbeech 13 K5
Cowbit 47 F6
Cowbridge 25 H7
Cowden 13 H2
Cowdenbeath 86 E5
Cowes 11 F5
Cowesby 69 G8
Cowfold 12 F4
Cowgill 62 A1
Cowick 63 K7
Cowie 85 L6
Cowley, Glos 30 D6
Cowley, G. Lon 20 E5
Cowley, Devon 5 K2
Cowley, Oxon 31 K7
Cowling, N. Yks 62 C5
Cowling, N. Yks 63 F1
Cowlinge 38 C5
Cowpen Bewley 69 G4
Cowplain 11 H3
Cowshill 67 L2
Cowstrandburn 86 D5
Coxbank 43 H3
Coxbench 44 E3
Cox Common 39 K2
Coxheath 13 L1
Coxhoe 68 F3
Coxley 17 G6
Coxwold 63 J2
Coychurch 25 H7
Coylumbridge 97 H7
Coynach 98 D8
Crabbs Cross 34 B4
Crabtree 12 F4
Crabtree Green 42 E3
Crackenthorpe 67 J4
Crackington 4 B2
Crackleybank 43 J6
Crackpot 68 B7
Cracoe 62 C3
Cradley 33 G6
Crafthole 4 D6
Craggan 97 J6
Cragg Vale 62 D7
Craghead 68 E1
Crai 25 G2
Craibstone 99 H7
Craichie 93 G7
Craig, D. & G. 72 B6
Craig, Highld 100 F7
Craig Castle 98 D6
Craigcefnparc 24 E4
Craigdam 99 H5
Craigearn 99 G7
Craigellachie 98 B4
Craigendoran 84 F6
Craighouse 82 E5
Craigie, Grampn 99 J7
Craigie, Strath 77 K4
Craiglockhart 87 F7
Craigmaud 99 H3
Craigmillar 87 F7
Craignant 42 D4
Craigneuk, Strath 78 E3
Craigneuk, Strath 78 E4
Craignure 89 J5
Craigo 93 H5
Craigrothie 87 G3
Craigton, Tays 92 E6
Craigton, Tays 93 G8
Craigton, Grampn 93 K1
Craigtown 106 D4
Craik 73 J3
Crail 87 K4
Crailing 80 E7
Crailinghall 80 E7
Crakehall 68 E7
Cramalt 79 J7
Crambe 64 E6
Cramlington 75 J6
Cramond 86 E7
Cramond Bridge 86 E7
Cranage 53 J7
Cranberry 43 K4
Cranborne 10 B3
Cranbourne 20 D6
Cranbrook 14 D5
Cranbrook Common 14 D5
Cranfield 36 C6
Cranford 21 F6
Cranford St Andrew 36 C3
Cranford St John 36 C3
Cranham, Glos 30 C6
Cranham, G. Lon 21 K5
Crank 53 G4
Cranleigh 12 D3
Cranmore, I. of W. 10 E5
Cranmore, Somer 17 H6
Cranna 98 F3
Crannoch 98 D3
Cranoe 45 J8
Cransford 39 J4
Cranshaws 80 E3
Cranstal 60 R1
Crantock 3 F3
Cranwell 46 D3
Cranwich 48 C8
Cranworth 48 E7
Crapstone 4 F5
Craskins 98 E8
Crask of Aigas 96 C4
Craster 75 J2
Craswall 29 G4
Cratfield 39 J3
Crathes 93 J2
Crathie, Grampn 92 D2
Crathie, Highld 91 G2
Crathorne 69 G6
Craven Arms 32 D2
Crawcrook 75 H7
Crawford 72 E1
Crawfordjohn 79 F7
Crawick 72 C2
Crawleside 68 B2
Crawley, Hants 10 F1
Crawley, W. Susx 12 F3
Crawley, Oxon 31 H6
Crawley Down 13 G3
Crawshawbooth 62 B7
Crawton 93 K4
Cray 62 B2
Crayford 21 K6
Crayke 63 J2
Crays Hill 22 D4
Cray's Pond 20 A5
Creacombe 7 J4
Creagorry 110 B2
Creaton 35 J3
Creca 73 H6
Credenhill 29 J3
Crediton 5 J1
Creech St Michael 8 D2
Creed 3 H5
Creekmouth 21 J5
Creeting St Mary 38 F5
Creeton 46 D5
Creetown 71 F6
Creggans 84 C4
Cregneish 60 N5
Cregrina 28 F2
Creich 87 G2
Creigiau 25 J6
Cressage 43 G7
Cresselly 26 E7
Cressing 22 D1
Cresswell, Staffs 44 A4
Cresswell, Northum 75 J4
Cresswell Quay 26 E7
Creswell 55 J6
Cretingham 39 H4
Cretshengan 76 C1
Crewe, Ches 42 F2
Crewe, Ches 43 J2
Crew Green 42 E6
Crewkerne 9 F4
Crews Hill 21 H4
Crianlarich 85 F2
Cribyn 27 K2
Criccieth 50 E7
Crich 44 E2
Crichie 99 J4
Crichton 80 B3
Crick, Gwent 17 F1
Crick, Northnts 35 G3
Crickadarn 28 E3
Cricket St Thomas 8 E4
Crickheath 42 D5
Crickhowell 25 L3
Cricklade 30 F8
Cridling Stubbs 63 J7
Crieff 86 B2
Criggion 42 D6
Crigglestone 55 G2
Crimond 99 K3
Crimplesham 48 B7
Crinan 83 G2
Cringleford 49 G7
Crinow 26 F6
Cripplesease 2 D6
Cripp's Corner 14 D6
Crockenhill 21 K7
Crockernwell 5 H2
Crockerton 17 K6
Crocketford or Ninemile Bar 72 D6
Crockey Hill 63 K5
Crockham Hill 13 H1
Crockleford Heath 38 F8
Croeserw 25 G5
Croes-goch 26 C4
Croesor 51 F6
Croesyceiliog 29 H8
Croesyceilog 27 J6
Croes-y-mwyalch 16 E1
Croft, Leic 45 G8
Croft, Ches 53 H4
Croft, Lincs 57 K7
Croftamie 85 G6
Crofton 55 G2
Croft-on-Tees 68 E6
Crofts of Benachielt 107 G6
Crofts of Haddo 99 H5
Crofts of Inverthernie 99 G4
Crofty 24 D5
Croggan 89 J6
Croglin 67 H2
Croir 108 B3
Cromarty 97 F2
Cromdale 97 J6
Cromer, Herts 37 F8
Cromer, Norf 49 H3
Cromford 44 D2
Cromhall 17 H1
Cromhall Common 17 H2

Cromore 108 D3
Cromra 91 G2
Cromwell 56 B7
Cronberry 78 D7
Crondall 19 J6
Cronk, The 60 Q2
Cronk-y-Voddy 60 Q3
Cronton 53 F5
Crook, Durham 68 D3
Crook, Cumbr 67 G7
Crookham, Berks 19 G4
Crookham, Northum 81 H6
Crookham Village 19 J5
Crookhouse 80 F7
Crooklands 61 J1
Crook of Devon 86 D4
Cropredy 35 F6
Cropston 45 G6
Cropthorne 34 A6
Cropton 64 E4
Cropwell Bishop 45 H4
Cropwell Butler 45 H4
Crosby, Cumbr 66 C3
rosby, Humbs 58 C5
Crosby, Mers 52 E4
Crosby, I. of M 60 Q4
Crosby Garrett 67 K6
Crosby-on-Eden 73 K8
Crosby Ravensworth 67 J5
Croscombe 17 G6
Cross, W. Isles 108 D1
Cross, Somer 16 F5
Crossapol 88 A4
Cross Ash 29 J6
Crossbost 108 C3
Crosscanonby 66 C3
Crossdale Street 49 H4
Crossdougal 110 B3
Crossens 52 E2
Crossford, Fife 86 D6
Crossford, Strath 79 F5
Crossgates, Fife 86 E6
Crossgates, Powys 41 K7
Crossgill 61 J3
Cross Green, Devon 4 D3
Cross Green, Suff 38 E5
Crosshands, Dyfed 27 F5
Cross Hands, Dyfed 27 K6
Crosshill, Fife 86 E5
Crosshill, Strath 77 J7
Crosshouse 77 J4
Cross Houses 43 G7
Crossings 74 B6
Cross in Hand 13 J4
Cross Inn, Dyfed 40 E7
Cross Inn, Dyfed 27 H2
Cross Inn, M. Glam 25 J6
Crosskeys 25 L5
Crosskirk 107 F2
Cross Lanes, Clwyd 42 E3
Crosslanes, Shrops 42 E6
Cross Lanes, N. Yks 63 J3
Crosslee 73 J2
Crossmichael 72 C7
Crossmoor 61 H6
Cross of Jackston 99 G5
Crossroads 93 J2
Crossway 29 J6
Crossway Green 33 H4
Crossways 9 J6
Crosswell 26 F4
Crosthwaite 67 G7
Croston 53 F2
Crostwick 49 H6
Crostwight 49 J4
Croughton 31 K4
Crovie 99 H2
Crowan 2 E6
Crowborough 13 J3
Crowcombe 8 C1
Crowdecote 54 E7
Crowfield, Suff 39 G5
Crowfield, Northnts 35 H6
Crow Hill 30 A5
Crowhurst, Surrey 13 G2
Crowhurst, E. Susx 13 L5
Crowland 47 F6
Crowlas 2 D6
Crowle, Humbs 58 B5
Crowle, H. & W 33 J5
Crowlista 108 B3
Crowmarsh Gifford 20 A5
Crownhill 4 E6
Crownthorpe 49 F7
Crowthorne 20 C7
Crowthorn School 53 J2
Crowton 53 G6
Croxall 44 C6
Croxdale 68 E3
Croxden 44 B4
Croxley Green 20 E4
Croxton, Norf 38 D2
Croxton, Humbs 59 E5
Croxton, Cambs 37 F5
Croxton, Staffs 43 J4
Croxton Kerrial 46 B5
Croxton Park 46 B5
Croy, Highld 97 F4
Croy, Strath 85 K7
Croyde 6 E2
Croydon, Cambs 37 G6
Croydon, G. Lon 21 H7
Cruckmeole 42 F7
Cruckton 42 F6
Cruden Bay 99 K5
Crudgington 43 H6
Crudwell 18 A1
Crug 32 A3
Crugmeer 3 H2
Crugybar 27 L4
Crulivig 108 B3
Crumlin 25 L5
Crundale, Dyfed 26 D6
Crundale, Kent 15 G4
Cruwys Morchard 7 J4
Crux Easton 18 F5
Crwbin 27 J6
Crymych 27 F4
Crynant 25 F4
Cuaig 100 C6
Cubbington 34 E4
Cubert 3 F4
Cublington 20 C1
Cuckfield 13 G4
Cucklington 9 J2
Cuckney 55 J6
Cuddesdon 31 K7
Cuddington, Bucks 20 B2
Cuddington, Ches 53 G6
Cuddington Heath 43 F3
Cuddy Hill 61 H6
Cudham 21 J8
Cudliptown 5 F4
Cudworth, Somer 8 E3
Cudworth, S. Yks 55 G3
Cuffley 21 H3
Culbo 96 E2
Culbokie 96 E3
Culburnie 96 C4
Culcabock 96 E4
Culcharry 97 G3
Culcheth 53 H4
Cul Doirlinn 89 H1
Culdrain 98 E5
Culduie 94 C1
Culford 38 D3
Culgaith 67 J4
Culham 31 K8
Culkein 104 B5
Culkerton 30 D8
Cullachie 97 H6
Cullen 98 E2
Cullercoats 75 K6
Cullicudden 96 E2
Cullingworth 62 D6
Cullipool 89 J7
Cullivoe 113 D1
Cullompton 8 B4
Culmington 32 D2
Culmstock 8 C3
Culnacraig 101 F1
Culnaknock 111 F1
Culrain 102 D4
Culross 86 C6
Culroy 77 J6
Culsh 99 H4
Culswick 113 C4
Cultercullen 99 J6
Cults, Grampn 98 E5
Cults, Grampn 99 H8
Culverstone Green 14 C2
Culverthorpe 46 D3
Culworth 35 G6
Cumbernauld 85 K7
Cumberworth 57 K6
Cuminestown 99 H3
Cummersdale 67 F1
Cummertrees 73 G7
Cummingstown 97 K2
Cumnock 72 A1
Cumnor 31 J7
Cumrew 67 H1
Cumwhinton 67 G1
Cumwhitton 67 H1
Cundall 63 H2
Cunninghamhead 77 J3
Cunnister 113 D2
Cupar 87 G3
Cupar Muir 87 G3
Curbar 55 F6
Curbridge, Hants 11 G3
Curbridge, Oxon 31 H7
Curdridge 11 G3
Curdworth 44 C8
Curland 8 D3
Currie 79 J3
Curry Mallet 8 E2
Curry Rivel 8 E2
Curtisden Green 13 L2
Curtisknowle 5 H6
Cury 2 E7
Cushnie 99 G2
Cushuish 8 C1
Cusop 29 G3
Cutiau 41 F2
Cutnall Green 33 H4
Cutsdean 30 E4
Cutthorpe 55 G6
Cuxham 20 A4
Cuxton 14 D2
Cuxwold 59 F6
Cwm, Clwyd 51 K3
Cwm, Gwent 25 K4
Cwmafan 25 F5
Cwmaman 25 J5
Cwmann 27 K3
Cwmavon 29 G7
Cwmbach, Dyfed 27 G5
Cwmbach, M. Glam 25 J4
Cwmbelan 41 J5
Cwmbran 29 G8
Cwmcarn 25 L5
Cwmcarvan 29 J7
Cwm-Cewydd 41 H2
Cwm-Cou 27 G3
Cwmdare 25 H4
Cwmdu, Powys 25 K2
Cwmdu, Dyfed 27 L4
Cwmduad 27 H4
Cwmfelin Boeth 27 F6
Cwmfelinfach 25 K5
Cwmfelin Mynach 27 G5
Cwmffrwd 27 J6
Cwmgwrach 25 G4
Cwm Irfon 28 C3
Cwmisfael 27 J6
Cwm-Llinau 41 H3
Cwmllynfell 25 F3
Cwmparc 25 H5
Cwmpengraig 27 H4
Cwmsychbant 27 J3
Cwmtillery 25 L4
Cwm-twrch Isaf 25 F3
Cwm-twrch Uchaf 25 F3
Cwm-y-glo 50 E4
Cwmyoy 29 H5
Cwmystwyth 41 G6
Cwrtnewydd 27 J3
Cwrt-y-cadno 28 A3
Cwrt-y-gollen 29 G6
Cyffylliog 51 K5
Cyfronydd 42 C7
Cymmer, W. Glam 25 G5
Cymmer, M. Glam 25 J5
Cynghordy 28 C4
Cynwyd 51 K6
Cynwyl Elfed 27 H5

D

Dacre, N. Yks 62 E3
Dacre, Cumbr 67 G4
Dacre Banks 62 E3
Daddry Shield 68 A3
Dadford 35 H7
Dadlington 44 F8
Dafen 24 D4
Dagenham 21 K5
Daglingworth 30 D7
Dagnall 20 D2
Dailly 77 H7
Dairsie or Osnaburgh 87 H3
Dalavich 84 B3
Dalbeattie 72 D7
Dalbeg 108 C2
Dalblair 72 B2
Dalby 60 P4
Dalchalloch 91 J5
Dalchenna 84 C4
Dalchreichart 96 A7
Dalderby 57 G7
Dale, Shetld 113 B4
Dale, Dyfed 26 C7
Dale, Derby 44 F4
Dale Head 67 G5
Dalgarven 77 H3
Dalgety Bay 86 E6
Dalginross 85 K2
Dalhalvaig 106 D4
Dalham 38 C4
Daliburgh 110 B3
Dalkeith 80 B3
Dallas 97 K3
Dalleagles 77 L6
Dallinghoo 39 H5
Dallington 13 K5
Dalmally 84 D2
Dalmary 85 H5
Dalmellington 77 K7
Dalmeny 86 E7
Dalmore, W. Isles 108 C2
Dalmore, Highld 96 E2
Dalnabreck 89 J2
Dalnacreich 96 B3
Dalnavie 102 E6
Dalreavoch 103 F3
Dalry 77 H3
Dalrymple 77 J3
Dalserf 78 E4
Dalston 66 F1
Dalswinton 72 E5
Dalton, N. Yks 68 D6
Dalton, Lancs 53 F3
Dalton, Northum 74 F8
Dalton, D. & G 73 G6
Dalton, N. Yks 63 H2
Dalton, S. Yks 55 H4
Dalton, Northum 75 H6
Dalton-in-Furness 60 F2
Dalton-le-Dale 69 G2
Dalton-on-Tees 68 E6
Dalton Piercy 69 G3
Dalwhinnie 91 H3
Dalwood 8 D4
Damerham 10 C3
Damgate 49 K7
Damnaglaur 70 C8
Danbury 22 D3
Danby 64 E2
Danby Wiske 68 F7
Danderhall 87 G8
Danebridge, Staffs 54 C7
Danebridge, Ches 54 C7
Dane End 21 H1
Danehill 13 H4
Darenth 21 K6
Daresbury 53 G5
Darfield 55 H3
Dargate 15 G2
Darite 3 L3
Darlaston 44 A8
Darlingscote 34 D6
Darlington 68 E5
Darliston 43 G4
Darlton 56 B6
Darowen 41 H3
Darra 99 G4
Darras Hall 75 H6
Darrington 55 H1
Darsham 39 K4
Dartford 21 K6
Dartington 5 H5
Dartmeet 5 G4
Dartmouth 5 J6
Darton 55 G2
Darvel 78 C6
Darwen 61 K7
Datchet 20 D6
Datchworth 21 G2
Daugh of Kinermony 98 B4
Dauntsey 18 A2
Davenham 53 H6
Daventry 35 G4
Davidstow 4 B3
Davington 73 H3
Daviot, Highld 97 F5
Daviot, Grampn 99 G6
Davoch of Grange 98 D3
Dawley 43 H7
Dawlish 5 K4
Dawlish Warren 5 K4
Dawn 51 H3
Daws Heath 22 E5
Dawsmere 47 H4
Daylesford 31 G5
Ddol Cownwy 41 K2
Deal 15 K3
Dean, Cumbr 66 C4
Dean, Hants 11 G3
Dean, Devon 5 H5
Dean, Somer 17 H6
Deanburnhaugh 73 J2
Deane 19 G5
Deanland 10 A3
Dean Prior 5 H5
Dean Row 53 K5
Deans 79 H3
Deanscales 66 C4
Deanshanger 35 J7
Deanston 85 K4
Dearham 66 C3
Dearne 55 H3
Debach 39 H5
Debden 37 J7
Debden Green 37 J7
Debenham 39 G4
Dechmont 86 D7
Deddington 31 J4
Dedham 38 F7
Dedridge 79 H3
Deene 36 C1
Deenethorpe 36 C1
Deepcut 19 L5
Deepdale 61 L1
Deeping Gate 46 E7
Deeping St James 46 E7
Deeping St Nicholas 46 F6
Deerhill 98 D3
Deerhurst 30 C5
Deerness 112 C4
Defford 33 J6
Defynnog 25 H2
Deganwy 51 G3
Deighton, N. Yks 69 F6
Deighton, N. Yks 63 K5
Deiniolen 50 E4
Delabole 4 A3
Delamere 53 G7
Delfrigs 99 J6
Dell 108 D1
Delliefure 97 J5
Delph 54 C3
Dembleby 46 D4
Denbigh 51 K4
Denbury 5 J5
Denby 44 E3
Denby Dale 55 F3
Denchworth 18 E1
Denford 36 C3
Dengie, Essex 23 F3
Dengie, Essex 23 F3
Denham, Suff 38 C4
Denham, Bucks 20 E5
Denham, Suff 39 G3
Denham Green 20 E5
Denham Street 39 G3
Denhead, Fife 87 H3
Denhead, Grampn 99 K3
Denholm 74 B2
Denholme 62 D6
Denmead 11 H3
Denmore 99 J7
Denne Park 12 E4
Dennington 39 H4
Denny 85 L6
Dennyloanhead 85 L6
Denshaw 54 C2
Denside 93 K2
Densole 15 J4
Denston 38 C5
Denstone 44 B3
Dent 61 L1
Den, The 77 J2
Denton, Lincs 46 B4
Denton, Northnts 36 B5
Denton, Cambs 36 E2
Denton, Durham 68 E5
Denton, N. Yks 62 E5
Denton, Norf 39 H2
Denton, E. Susx 13 H6
Denton, Kent 15 J4
Denton, Oxon 31 K7
Denton, G. Man 53 L4
Denver 48 B7
Denwick 75 J2
Deopham 48 F7
Deopham Green 48 F8
Depden Green 38 C5
Deptford, Wilts 10 B1
Deptford, G. Lon 21 H6
Derby 44 E4
Derbyhaven 60 P5
Deri 25 K4
Derrington 43 K5
Derryguaig 89 F5
Derry Hill 18 A3
Derrythorpe 58 C6
Dersingham 48 B4
Dervaig 89 F3
Derwen 51 K5
Desborough 35 K2
Desford 45 F7
Detchant 81 J6
Detling 14 D3
Deuddwr 42 D6
Devauden 29 J8
Devil's Bridge 41 G6
Devizes 18 B4
Devonport 4 E6
Devonside 86 C5
Devoran 3 F6
Dewlish 9 J5
Dewsbury 63 F7
Dhoon 60 R3
Dhoor 60 R2
Dhowin 60 R1
Dial Post 12 E5
Dibden 10 F4
Dibden Purlieu 10 F4
Dickleburgh 39 G2
Didbrook 30 E4
Didcot 19 G2
Diddington 36 E4
Diddlebury 32 E2
Didley 29 J4
Didmarton 17 K2
Didsbury 53 K4
Didworthy 5 G5
Digby 46 D2
Diggle 54 D3
Dihewyd 27 J2
Dilham 49 J5
Dilhorne 44 A3
Dilston 75 F7
Dilton Marsh 17 K6
Dilwyn 29 J2
Dinas, Gwyn 50 B7
Dinas, Dyfed 27 G4
Dinas Cross 26 E4
Dinas Mawddwy 41 H2
Dinas Powys 25 K7
Dinchope 32 D2
Dinder 17 G6
Dinedor 29 K4
Dines Green 33 H5
Dingestow 29 J6
Dingley 35 J2
Dingwall 96 D3
Dinnet 93 F2
Dinnington, Somer 8 F3
Dinnington, S. Yks 55 J5
Dinnington, T. & W 75 J6
Dinorwig 50 E4
Dinton, Wilts 10 B1
Dinton, Bucks 20 B2
Dinwoodie Mains 73 G4
Dinworthy 6 D4
Dippen 76 C4
Dippenhall 19 K6
Dippin 76 F5
Dipple 98 C3
Diptford 5 H6
Dipton 68 D1
Dirleton 87 J6
Discoed 32 B4
Diseworth 45 F5
Dishes 112 D3
Dishforth 63 G2
Disley 54 C5
Diss 39 G3
Disserth 28 E2
Distington 66 C4
Ditcheat 9 H1
Ditchingham 39 J1
Ditchling 13 G5
Dittisham 5 J6
Ditton, Ches 53 F5
Ditton, Kent 13 L1
Ditton Green 38 B5
Ditton Priors 33 F2
Dixton, Glos 30 D4
Dixton, Gwent 29 K6
Dlengrasco 111 F2
Dobwalls 3 L3
Doccombe 5 H3
Dochgarroch 96 E4
Docking 48 C4
Docklow 29 K2
Dockray 67 F4
Doddinghurst 22 B4
Doddington, Lincs 56 D6
Doddington, Shrops 33 F3
Doddington, Kent 14 F3
Doddington, Cambs 37 H1
Doddington, Northum 81 H6
Doddiscombsleigh 5 J3
Dodford, Northnts 35 H4
Dodford, H. & W 33 J3
Dodington 17 J2
Dodleston 52 E7
Dodworth 55 G3
Doe Lea 55 H7
Dogmersfield 19 J5
Dog Village 5 K2
Dogyke 46 F2
Dolanog 41 K2
Dolau 32 A4
Dolbenmaen 50 E6
Dolfach 41 J6
Dolfor 41 L5
Dolgarrog 51 G4
Dolgellau 41 G2
Dolgran 27 J4
Doll 103 G3
Dollar 86 C5
Dolley Green 32 B4
Dolphinholme 61 J4
Dolphinton 79 J5
Dolton 7 F4
Dolwen 51 H3
Dolwyddelan 51 G5
Dol-y-cannau 29 G3
Dolyhir 29 G2
Domgay 42 D6
Doncaster 55 J3
Donhead St Andrew 9 L2
Donhead St Mary 9 L2
Donibristle 86 E6
Donington 46 F4
Donington on Bain 57 G5
Donisthorpe 44 E6
Donkey Town 20 D7
Donnington, H. & W 30 B4
Donnington, Berks 19 F4
Donnington, Glos 31 F5
Donnington, Shrops 43 G7
Donnington, Shrops 43 J6
Donnington, W. Susx 11 K4
Donyatt 8 E3
Doonfoot 77 J6
Doonholm 77 J6
Dorchester, Dorset 9 H5
Dorchester, Oxon 31 K8
Dordon 44 D7
Dore 55 G5
Dores 96 D5
Dorking 12 E2
Dormansland 13 H2
Dormanstown 69 H4
Dormington 29 K3
Dorney 20 D6
Dornie 94 D3
Dornoch 103 F5
Dornock 73 H7
Dorridge 34 C3
Dorrington, Lincs 46 D2
Dorrington, Shrops 43 F7
Dorsington 34 C6
Dorstone 29 H3
Dorton 20 A2
Dosthill 44 D7
Doublebois 3 K3
Doughton 17 K1
Douglas, Strath 79 F6
Douglas, I. of M 60 Q4
Douglas and Angus 87 H1
Douglastown 92 F7
Douglas Water 79 F6
Doulting 17 H6
Dounby 112 B3
Doune 85 K4
Douneside 98 D8
Dounie 102 D4
Dousland 5 F5
Dove Holes 54 D6
Dovenby 66 C3
Dover 15 K4
Doverdale 33 H4
Doveridge 44 C4
Dowdeswell 30 E6
Dowland 7 F4
Dowlish Wake 8 E3
Down Ampney 30 F8
Downderry 4 D6
Downe 21 J7
Downend, Berks 19 F3
Downend, I. of W 11 G6
Downfield 87 G1
Downgate 4 D4
Downham, Lancs 62 A5
Downham, Essex 22 D4
Downham, Northum 81 G6
Downham Market 48 B7
Down Hatherley 30 C5
Downhead 17 H6
Downholme 68 D7
Downies 93 L2
Downley 20 C4
Downs 25 K7
Downside Abbey 17 H5
Down St Mary 7 H5
Downton, Wilts 10 C2
Downton, Hants 10 D5
Downton on the Rock 32 D3
Dowsby 46 E5
Dowsdale 47 F6
Dowthwaitehead 67 F4
Doxey 43 L5
Doynton 17 J3
Draethen 25 L6
Draffan 78 E5
Drakeland Corner 5 F6
Drakemyre 77 H2
Drakes Broughton 33 J6
Draughton, N. Yks 62 D4
Draughton, Northnts 35 J3
Drax 63 K7
Draycote 35 F3
Draycott, Glos 31 F4
Draycott, Derby 45 F4
Draycott, Somer 17 F5
Draycott in the Clay 44 C5
Draycott in the Moors 44 A3
Drayton, Leic 36 B1
Drayton, Somer 8 F2
Drayton, Oxon 34 F6
Drayton, Norf 49 G6
Drayton, Hants 11 H4
Drayton, H. & W 33 J3
Drayton, Oxon 31 J8
Drayton Bassett 44 C7
Drayton Camp 18 F6
Drayton Parslow 36 B8
Drayton St Leonard 31 K8
Drefach, Dyfed 27 H4
Dre-fach, Dyfed 27 K3
Drefach, Dyfed 27 K6
Drefelin 27 H4
Dreghorn 77 J4
Drem 87 J7
Drewsteignton 5 H2
Driby 57 H6
Driffield 30 E8
Drift 2 C7
Drigg 66 C7
Drighlington 63 F7
Drimnin 89 G3
Drimpton 8 F4
Drinishader 108 B5
Drinkstone 38 E4
Drinkstone Green 38 E4
Drointon 44 B5
Droitwich 33 H4
Droman 104 C3
Dron 86 E3
Dronfield 55 G6
Dronfield Woodhouse 55 G6
Drongan 77 K6
Droxford 11 H3
Droylsden 53 L4
Druid 51 K6
Druidston 26 C6
Druimarbin 90 B4
Druimdrishaig 83 G4
Druimindarroch 94 B7
Druimkinnerras 96 C5
Drum 86 D4
Drumbeg 104 C5
Drumblade 98 E4
Drumbuie 94 C2
Drumchapel 85 H7
Drumchardine 96 D4
Drumchork 100 D3
Drumclog 78 D6
Drumelzier 79 J6
Drumfearn 94 B4
Drumgley 92 F6
Drumguish 91 J2
Drumin 98 A5
Drumlassie 98 F8
Drumlemble 76 B6
Drumligair 99 J7
Drumlithie 93 J3
Drummore 70 C8
Drumnadrochit 96 D6
Drumnagorrach 98 E3
Drumoak 93 J2
Drums 99 J6
Drumsallie 90 A4
Drumsturdy 87 H1
Drumuie 111 F2
Drumuillie 97 H6
Drumwhindle 99 J5
Drury 52 D7
Drybeck 67 J5
Drybridge, Grampn 98 D2
Drybridge, Strath 77 J4
Drybrook 30 A6
Dry Doddington 46 B3
Dry Drayton 37 G4
Dryhope 79 K7
Drymen 85 G6
Drymuir 99 J4
Drynoch 111 F3
Dubford 99 G2
Dubwath 66 D3
Duckington 43 F2
Ducklington 31 H7
Duck's Cross 36 E5
Duddenhoe End 37 H7
Duddingston 87 F7
Duddington 46 C7
Duddo 81 H5
Duddon 53 G7
Duddon Bridge 60 E1
Dudleston Heath 42 E4
Dudley, W. Mids 33 J1
Dudley, T. & W 75 J6
Duffield 44 E3
Duffryn, Gwent 16 D2
Duffryn, W. Glam 25 G5
Dufftown 98 C4
Duffus 98 A2
Dufton 67 J4
Duggleby 65 F6
Duirinish 94 C2
Duisky 90 B4
Dukestown 25 K3
Dukinfield 53 L4
Dulas 50 D2
Dulcote 17 G6
Dulford 8 B4
Dull 91 K7
Dullingham 38 B5
Dulnain Bridge 97 H6
Duloe, Beds 36 E4
Duloe, Corn 3 L4
Dulsie 97 H4
Dulverton 7 K3
Dulwich 21 H6
Dumbarton 85 F7
Dumbleton 34 B7
Dumcrieff 73 G3
Dumfries 72 E6
Dumgoyne 85 H6
Dummer 19 G6
Dunan 111 F3
Dunball 16 E6
Dunbar 87 K7
Dunbeath 107 G7
Dunbeg 84 A1
Dunblane 85 K4
Dunbog 87 F3
Duncanston 96 D3
Duncanstone 98 E6
Dunchideock 5 J3
Dunchurch 35 F3
Duncote 35 H5
Duncow 72 E5
Duncrievie 86 E4
Duncton 12 C5
Dundee 87 H1
Dundon 9 F1
Dundonald 77 J4
Dundonnell 101 F3
Dundraw 66 E2
Dundreggan 96 B7
Dundrennan 71 J7
Dundry 17 G4
Dunecht 99 G8
Dunfermline 86 D6
Dunford Bridge 54 E3
Dunham-on-the-Hill 53 F6
Dunham on Trent 56 C6
Dunhampton 33 H4
Dunham Town 53 J5
Dunholme 56 E6
Dunino 87 J3
Dunipace 85 L6
Dunkeld 92 B7
Dunkeswell 8 C4
Dunkirk 15 G3
Dunk's Green 13 K1
Dunley 33 G4
Dunlop 77 K3
Dunmore, Central 86 B6
Dunmore, Strath 76 C1
Dunnet 107 H2
Dunnichen 93 G7
Dunning 86 D3
Dunnington, Warw 34 B5
Dunnington, Humbs 65 J7
Dunnington, N. Yks 63 K4
Dunnockshaw 62 B7
Dunollie 84 A1
Dunoon 84 D7
Dunragit 70 C6
Duns 81 F4
Dunsby 46 E5
Dun Scaich 111 F4
Dunscore 72 D5
Dunscroft 55 K3
Dunsden Green 20 B6
Dunsfold 12 D3
Dunsford 5 J3
Dunshelt 87 F3
Dunsley 69 L5
Dunsmore 20 C3
Dunsop Bridge 61 K4
Dunstable 20 E1
Dunstall 44 C5
Dunstall Green 38 C4
Dunstan 75 J2
Dunster 7 K1
Duns Tew 31 J5
Dunston, Lincs 56 E7
Dunston, Norf 49 H7
Dunston, T. & W 75 J7
Dunston, Staffs 43 L6
Dunsville 55 K3
Dunswell 58 E3
Dunterton 4 D4
Duntisbourne Abbots 30 D7
Duntisbourne Leer 30 D7
Duntisbourne Rouse 30 D7
Duntish 9 H4
Duntocher 85 G7
Dunton, Bucks 20 C1
Dunton, Norf 48 D4
Dunton, Beds 37 F6
Dunton Bassett 35 G1
Dunton Green 13 J1
Dunure 77 H6
Dunvant 24 D5
Dunvegan 110 E2
Dunwich 39 K3
Durdar 67 G1
Durham 68 E2
Durisdeer 72 D3
Durleigh 8 D1
Durley, Wilts 18 D4
Durley, Hants 11 G3
Durnamuck 100 F2
Durness 104 F2
Durno 99 G6
Durran 107 G3
Durrington, Wilts 18 C6
Durrington, W. Susx 12 E6
Dursley 30 B8
Durston 8 D2
Durweston 9 K4
Dury 113 D3
Duston 35 J4
Duthil 97 H6
Dutlas 32 B3
Duton Hill 38 B8
Dutton 53 G6
Duxford 37 H6
Dwygyfylchi 51 G3
Dwyran 50 D4
Dyce 99 H7
Dye House 74 F8
Dyffryn, Gwyn 50 B3
Dyffryn, M. Glam 25 G5
Dyffryn Ardudwy 40 E1
Dyffryn Ceidrych 24 F2
Dyffryn Cellwen 25 G3
Dyke, Devon 6 D3
Dyke, Lincs 46 E5
Dyke, Grampn 97 H3
Dykehead, Tays 92 E5
Dykehead, Strath 79 F4
Dylife 41 H4
Dymchurch 15 H6
Dymock 30 B4
Dyrham 17 J3
Dysart 87 G5
Dyserth 51 K3

E

Eagland Hill 61 H5
Eagle 56 C7
Eaglescliffe 69 G5
Eaglesfield, Cumbr 66 C4
Eaglesfield, D. & G 73 H6
Eaglesham 78 C4
Eairy 60 P4
Eakring 55 K7
Ealand 58 B5
Ealing 21 F5
Eamont Bridge 67 H4
Earby 62 C5
Earcroft 61 K7
Eardington 33 G1
Eardisland 29 J2
Eardisley 29 H3
Eardiston, Shrops 42 E5
Eardiston, H. & W 33 F4
Earith 37 G3
Earle 81 H7
Earlestown 53 G4
Earlham 49 G7
Earlish 110 E1
Earls Barton 36 B4
Earls Colne 38 D8
Earl's Croome 33 H6
Earlsdon 34 E3
Earlsferry 87 H5
Earlsford 99 H5
Earl's Green 38 F4
Earl Shilton 45 F8
Earl Soham 39 H4
Earl Sterndale 54 D7
Earlston, Border 80 D6
Earlston, Strath 77 K4
Earl Stonham 39 G5
Earlswood, Warw 34 C3
Earlswood, Gwent 29 J8
Earnley 11 K5
Earsary 110 B5
Earsdon 75 K6
Earsham 39 J2
Earswick 63 K4
Eartham 12 C6
Easby 69 H6
Easdale 89 J7
Easebourne 11 K2
Easenhall 35 F3
Easington, Bucks 20 A2
Easington, Humbs 59 H5
Easington, Cleve 69 K5
Easington, Northum 81 K6
Easington Lane 69 F2
Easingwold 63 J3
Easole Street 15 J3
Eassie 92 E7
East Aberthaw 25 J8
East Allington 5 H7
East Anstey 7 J3
East Ashling 11 K4
East Auchronie 99 H8
East Barkwith 57 F5
East Barming 13 L1
East Barnet 21 G4
East Barsham 48 E4
East Beckham 49 G4
East Bedfont 20 E6
East Bergholt 38 F7
East Bilney 48 E6
East Blatchington 13 H6
East Boldre 10 E4
Eastbourne 13 K7
East Brent 16 E5
Eastbridge 39 K4
East Bridgford 45 H3
East Buckland 7 G2
East Budleigh 8 B6
East Burrafirth 113 C4
East Burton 9 K6
Eastbury, Berks 18 E3
Eastbury, Herts 20 E4
East Calder 79 H3
East Carleton 49 G7
East Carlton 36 B2
East Chaldon or
Chaldon Herring 9 J6
East Challow 18 E2
East Chiltington 13 G5
East Chinnock 9 F3
East Chisenbury 18 C5
Eastchurch 15 F1
East Clandon 12 D1
East Claydon 35 J8
East Coker 9 G3
East Combe, Somer 8 C1
Eastcombe, Glos 30 C7
East Compton 17 H6
Eastcote, W. Mids 34 C3
Eastcote, G. Lon 21 F5
Eastcott, Wilts 18 B5
Eastcott, Corn 6 C4
East Cottingwith 58 B2
Eastcourt 18 A1
East Cowes 11 G5
East Cowton 68 F6
East Cramlington 75 J6
East Creech 10 A6
East Croachy 96 E6
East Davoch 98 D8
East Dean, Hants 10 D2
East Dean, E. Susx 13 J7
East Dean, W. Susx 11 L3
East Dereham 48 E6
East Down 7 G1
East Drayton 56 B6
East End, Dorset 10 A5
East End, Hants 10 E5
East End, Kent 14 E5
East End, Avon 17 F3
East End, Hants 18 F4
East End, Oxon 31 H6
East End, Herts 37 H8
Easter Ardross 102 E6
Easter Balmoral 92 D2
Easter Boleskine 96 D6
Easter Compton 17 G2
Easter Fearn 102 E5
Eastergate 12 C6
Easter Kinkell 96 D3
Easter Lednathie 92 E5
Easter Muckovie 97 F4
Eastern Green 34 D3
Easter Ord 99 H8
Easter Quaff 113 D5
Easter Skeld 113 C4
Easterton 18 B5
Eastertown 16 E5
East Farleigh 13 L1
East Farndon 35 J2
East Ferry 58 C7
Eastfield, Strath 79 F3
Eastfield, N. Yks 65 H4
Eastfield, Strath 85 K7
East Garston 18 E3
Eastgate, Durham 68 B3
Eastgate, Norf 49 G5
East Ginge 19 F2
East Goscote 45 H6
East Grafton 18 D4
East Grimstead 10 D2
East Grinstead 13 G3
East Guldeford 14 F6
East Haddon 35 H4
East Hagbourne 19 G2
East Halton 59 F5
Eastham, Mers 52 E5
East Ham, G. Lon 21 J5
Easthampstead 20 C7
East Hanney 18 F1
East Hanningfield 22 D3
East Hardwick 55 H2
East Harling 38 E2
East Harlsey 69 G7
East Harptree 17 G5
East Hartford 75 J6
East Harting 11 J3
East Hatley 37 F5
East Hauxwell 68 D7
East Haven 93 G8
East Heckington 46 E3
East Hedleyhope 68 D2
East Hendred 19 F2
East Heslerton 65 G5
East Hoathly 13 J5
Easthope 43 G8
Easthorpe 22 F1
East Horrington 17 G6
East Horsley 12 D1
East Huntspill 16 E6
East Hyde 21 F2
East Ilsley 19 F2
Eastington, Glos 30 B7
Eastington, Glos 31 F6
Eastington, Devon 7 H5
East Kennett 18 C4
East Keswick 63 G5
East Kilbride 78 D4
East Kirkby 57 H7
East Knighton 9 K6
East Knoyle 9 K1
East Lambrook 9 F3
East Lamington 103 F6
East Langdon 15 K4
East Langton 35 J1
East Langwell 103 F3
East Lavington 12 C5
East Layton 68 D6
Eastleach Martin 31 G7
Eastleach Turville 31 F7
East Leake 45 G5
Eastleigh, Hants 11 F3
East Leigh, Devon 7 G5
East Lexham 48 D6
East Lilburn 81 J7
Eastling 14 F3
East Linton 87 J7
East Liss 11 J2
East Lound 58 B7
East Lulworth 9 K6
East Mains 93 H2
East Malling 13 K1
East Marden 11 K3
East Markham 56 B6
East Marton 62 C4
East Meon 11 H2
East Mersea 23 G2
East Molesey 21 F7
East Morden 9 L5
East Morton 62 E5
Eastney 11 H5
Eastnor 33 G7
East Norton 45 J7
East Oakley 19 G5
Eastoft 58 C5
East Ogwell 5 J4
Easton, Lincs 46 C5
Easton, Cambs 36 E3
Easton, Hants 11 G1
Easton, Norf 49 G6
Easton, Somer 17 G6
Easton, Devon 5 H3
Easton, Suff 39 H5
Easton, Dorset 9 H7
Easton, Cumbr 73 K6
Easton Grey 17 K2
Easton-in-Gordano 17 G3
Easton Maudit 36 B5
Easton on the Hill 46 D7
Easton Royal 18 D4
East Ord 81 H4
East Panson 4 D2
East Peckham 13 K2
East Pennard 9 G1
East Perry 36 E4
East Portlemouth 5 H8
East Prawle 5 H8
East Preston 12 D6
East Putford 6 D4
East Quantoxhead 16 C6
East Rainton 68 F2
East Ravendale 59 G7
East Raynham 48 D5
Eastrea 47 F8
East Retford 56 B5
Eastriggs 73 H7
Eastrington 58 B3
East Rudham 48 D5
East Runton 49 G3
East Ruston 49 J5
Eastry 15 K3
East Saltoun 80 C3
Eastside, Orkney 112 C5
Eastside, Orkney 112 C5
East Sleekburn 75 J5
East Stockwith 56 B4
East Stoke, Notts 45 J3
East Stoke, Dorset 9 K6
East Stour 9 J2
East Stourmouth 15 J2
East Stratton 19 G7
East Studdal 15 K4
East Taphouse 3 K3
East-the-Water 6 E3
East Tilbury 22 C6
East Tisted 11 J1
East Torrington 57 F5
East Tuddenham 48 F6
East Tytherley 10 D2
East Tytherton 18 A3
East Village 7 J5
Eastville 57 J8
East Wall 43 G8
East Walton 48 C6
Eastwell 45 J5
East Wellow 10 E2
Eastwell Park 15 G4
East Wemyss 87 G5
East Whitburn 79 G3
Eastwick 21 J2
East Wickham 21 J6
East Williamston 26 E7
East Winch 48 B6
East Wittering 11 J5
East Witton 62 E1
Eastwood, W. Yks 62 C7
Eastwood, Essex 22 E5
Eastwood, Notts 45 F3
East Woodhay 18 F4
East Worldham 19 J7
Eathorpe 34 E4
Eaton, Notts 56 B6
Eaton, Shrops 32 C2
Eaton, Shrops 32 E1
Eaton, Ches 53 G7
Eaton, Norf 49 H7
Eaton, Leic 45 J5
Eaton, Oxon 31 J7
Eaton, Ches 53 K7
Eaton Bishop 29 J4
Eaton Bray 20 D1
Eaton Constantine 43 G7
Eaton Hastings 31 G8
Eaton Socon 36 E5
Eaton upon Tern 43 H5
Ebberston 65 F4
Ebbesbourne Wake 10 A2
Ebbw Vale 25 K4
Ebchester 68 D1
Ebford 5 K3
Ebrington 34 C6
Ecchinswell 19 F5
Ecclaw 80 F3
Ecclefechan 73 G6
Eccles, Kent 14 D2
Eccles, Border 80 F5
Eccles, G. Man 53 J4
Ecclesfield 55 G4
Eccleshall 43 K5
Ecclesmachan 86 D7
Eccles Road 38 F1
Eccleston, Mers 53 F4
Eccleston, Ches 52 F7
Eccleston, Lancs 53 G2
Eccup 63 F5
Echt 99 G8
Eckford 80 F7
Eckington, Derby 55 H6
Eckington, H. & W 33 J6
Ecton 36 B4
Edale 54 E5
Edburton 12 F5
Edderton 103 F5
Eddleston 79 K5
Edenbridge 13 H2
Edenfield 53 K2
Edenhall 67 H3
Edenham 46 D5
Eden Park 21 H7
Edensor 55 F6
Edenthorpe 55 K3
Edern 50 B7
Edgbaston 34 B2
Edgcott 20 A1
Edge 42 E7
Edgebolton 43 G5
Edge End 29 K6
Edgefield 49 F4
Edgefield Street 49 F4
Edgeworth 30 D7
Edgmond 43 J6
Edgmond Marsh 43 J5
Edgton 32 C2
Edgware 21 G4
Edgworth 53 J2
Edinbane 110 E2
Edinburgh 87 F7
Edingale 44 D6
Edingley 45 H2
Edingthorpe 49 J4
Edington, Somer 16 E7
Edington, Wilts 17 L5
Edithmead 16 E6
Edith Weston 46 C7
Edlesborough 20 D2
Edlingham 75 H3
Edlington 57 G6
Edmondsham 10 B3
Edmondsley 68 E2
Edmondthorpe 46 B6
Edmonstone 112 C3
Edmonton 21 H4
Edmundbyers 68 C1
Ednam 80 F6
Edrom 81 G4
Edstaston 43 G4
Edstone 34 C4
Edvin Loach 33 F5
Edwalton 45 G4
Edwardstone 38 E6
Edwinsford 27 L4
Edwinstowe 55 K7
Edworth 37 F6
Edwyn Ralph 33 F5
Edzell 93 H5
Efail Isaf 25 J6
Efailnewydd 50 C7
Efailwen 26 F5
Efenechtyd 42 C2
Effingham 12 E1
Effirth 113 C4
Efford 7 J5
Egerton, Kent 14 F4
Egerton, G. Man 53 J2
Egerton Forstal 14 E4
Eggesford Station 7 G4
Eggington 36 C8
Egginton 44 D5
Egglescliffe 69 G5
Eggleston 68 B4
Egham 20 E6
Egleton 46 B7
Eglingham 75 H2
Egloshayle 3 J2
Egloskerry 4 C3
Eglwysbach 51 H3
Eglwys-Brewis 25 J8
Eglwyswen 26 F4
Eglwyswrw 26 F4
Egmanton 56 B7
Egremont 66 C5
Egton 64 F2
Egton Bridge 64 F2
Eight Ash Green 38 E8
Eilanreach 94 D4
Eilean Darach 101 G3
Eilean Glas 108 C5
Eilean Iarmain or
Isleornsay 94 B4
Eishken 108 C4
Eisingrug 50 F7
Elan Village 41 J7
Elberton 17 H2
Elburton 5 F6
Elcombe 18 C2
Eldersfield 30 C4
Elderslie 77 K1
Eldroth 62 A3
Eldwick 62 E5
Elerch or Bont-goch 41 F5
Elford, Staffs 44 C6
Elford, Northum 81 K6
Elgin 98 B2
Elgol 111 F4
Elham 15 H4
Elie 87 H4
Elim 50 C2
Eling 10 E3
Elishader 111 F1
Elishaw 74 E4
Elkesley 56 A6
Elkstone 30 D6
Elland 62 E7
Ellary 83 G4
Ellastone 44 C3
Ellemford 80 F3
Ellenhall 43 K5
Ellen's Green 12 D3
Ellerbeck 69 G7
Ellerby 69 K5
Ellerdine Heath 43 H5
Ellerker 58 D4
Ellerton, Humbs 58 B3
Ellerton, Shrops 43 J5
Ellesborough 20 C3
Ellesmere 42 F4
Ellesmere Port 52 E6
Ellingham, Norf 39 J1
Ellingham, Northum 81 K7
Ellingstring 62 E1
Ellington, Cambs 36 E3
Ellington, Northum 75 J4
Ellisfield 19 H6
Ellistown 44 F6
Ellon 99 J5
Ellough 39 J2
Elloughton 58 D4
Ellwood 29 K7
Elm 47 H7
Elmbridge 33 J4
Elmdon, W. Mids 34 C2
Elmdon, Essex 37 H7
Elmdon Heath 34 C2
Elmesthorpe 45 F8
Elmhurst 44 C6
Elmley Castle 34 A6
Elmley Lovett 33 H4
Elmore 30 B6
Elmore Back 30 B6
Elm Park 21 K5
Elmscott 6 C3
Elmsett 38 F6
Elmstead Market 23 G1
Elmsted 15 H4
Elmstone 15 J2
Elmstone Hardwicke 30 D5
Elmswell 38 E4
Elmton 55 J6
Elphin 104 D7
Elphinstone 87 G7
Elrick 99 H8
Elrig 70 E7
Elsdon 74 F4
Elsecar 55 G3
Elsenham 37 J8
Elsfield 31 K6
Elsham 58 E5
Elsing 48 F6
Elslack 62 C5
Elsrickle 79 H5
Elstead 12 C2
Elsted 11 K3
Elston 45 J3
Elstone 7 G4
Elstow 36 D6
Elstree 21 F4
Elstronwick 59 G3
Elswick 61 H6
Elsworth 37 G4
Elterwater 66 F6
Eltham 21 J6
Eltisley 37 F5
Elton, Glos 30 B6
Elton, H. & W 32 D3
Elton, Cambs 46 D8
Elton, Ches 52 F6
Elton, Derby 55 F7
Elton, Cleve 69 G5
Elton, Notts 45 J4
Elvanfoot 72 E2
Elvaston 44 F4
Elveden 38 D3
Elvingston 87 H7
Elvington, N. Yks 58 B2
Elvington, Kent 15 J3
Elwick, Cleve 69 G3
Elwick, Northum 81 K6
Elworth 53 J7
Elworthy 8 B1
Ely, Cambs 37 J2
Ely, S. Glam 25 K7
Emberton 36 B6
Embleton, Cumbr 66 D3
Embleton, Northum 81 L7
Embo 103 G4
Emborough 17 H5
Embsay 62 D4
Emery Down 10 D4
Emley 55 F2
Emmer Green 20 B6
Emmington 20 B3
Emneth 47 H7
Emneth Hungate 47 J7
Empingham 46 C7
Empshott 11 J1
Emsworth 11 J4
Enaclete 108 B3
Enborne 18 F4
Enchmarsh 43 G8
Enderby 45 G8
Endmoor 61 J1
Endon 43 L2
Enfield 21 H4
Enford 18 C5
Engine Common 17 H2
Englefield 20 A6
Englefield Green 20 D6
English Bicknor 29 K6
Englishcombe 17 J4
English Frankton 42 F5
Enham-Alamein 18 E6
Enmore 8 D1
Ennerdale Bridge 66 C5
Enochdhu 92 B5
Ensbury 10 B5
Ensdon 42 F6
Ensis 7 F3
Enstone 31 H5
Enterkinfoot 72 D3
Enville 33 H2
Eochar 110 B2
Eoligarry 110 B4
Eorabus 88 E6
Eoropie 108 D1
Epperstone 45 H3
Epping 21 J3
Epping Green, Herts 21 G3
Epping Green, Essex 21 J3
Eppleby 68 D5
Epsom 21 G7
Epwell 34 E6
Epworth 58 B6
Erbistock 42 E3
Erbusaig 94 C3
Erdington 34 C1
Eredine 84 B4
Ericstane 73 F2
Eridge Green 13 J3
Eriswell 38 C3
Erith 21 K6
Erlestoke 18 A5
Ermington 5 G6
Erpingham 49 G4
Errogie 96 D6
Errol 87 F2
Erskine 85 G7
Erskine Bridge 85 G7
Erwarton 39 H7
Erwood 28 E3
Eryholme 68 F6
Eryrys 52 D8
Escomb 68 D3
Escrick 63 K5
Esgairgeiliog 41 G3
Esh 68 D2
Esher 21 F7
Eshott 75 J4
Eshton 62 C4
Esh Winning 68 D2
Eskadale 96 C5
Eskbank 80 B3
Eskdale Green 66 D6
Eskdalemuir 73 H4
Esknish 82 C5
Esprick 61 H6
Essendine 46 D6
Essendon 21 G3
Essich 96 E5
Essington 44 A7
Esslemont 99 J6
Eston 69 H5
Etal 81 H6
Etchilhampton 18 B4
Etchingham 13 L4
Etchinghill, Staffs 44 B6
Etchinghill, Kent 15 H5
Etherley 68 D4
Eton 20 D6
Etteridge 91 H2
Ettersgill 68 A4
Ettington 34 D6
Etton, Humbs 58 D2
Etton, Cambs 46 E7
Ettrick 73 H2
Ettrickbridge 80 B7
Etwall 44 D4
Euston 38 D3
Euxton 53 G2
Evanton 96 E2
Evedon 46 D3
Evelix 103 F4
Evenjobb 32 B4
Evenley 31 K4
Evenlode 31 G5
Evenwood 68 D4
Everbay 112 D3
Evercreech 17 H7
Everdon 35 G5
Everingham 58 C2
Everleigh 18 D5
Everley 65 F3
Evershot 9 G4
Evershott 36 C7
Eversley 20 B7
Eversley Cross 20 B7
Everton, Notts 56 A4
Everton, Hants 10 D5
Everton, Beds 36 F5
Evertown 73 J6
Evesbatch 33 F6
Evesham 34 B6
Evington 45 H7
Ewden Village 55 F4
Ewell 21 G7
Ewell Minnis 15 J4
Ewelme 20 A4
Ewen 30 E8
Ewenny 25 H7
Ewerby 46 E3
Ewes 73 J4
Ewhurst 12 D2
Ewhurst Green 14 D6
Ewloe 52 E7
Eworthy 4 E2
Ewshot 19 K6
Ewyas Harold 29 H5
Exbourne 7 G5
Exbury 10 F4
Exebridge 7 K3
Exelby 63 F1
Exeter 5 K2
Exford 7 J2
Exhall 34 C5
Exminster 5 K3
Exmouth 5 L3
Exnaboe 113 C6
Exning 38 B4
Exton, Leic 46 C6
Exton, Hants 11 H2
Exton, Somer 7 K2
Exton, Devon 5 K3
Eyam 54 F6
Eydon 35 G5
Eye, H. & W 32 D4
Eye, Cambs 47 F7
Eye, Suff 39 G3
Eyemouth 81 H3
Eyeworth 37 F6
Eyhorne Street 14 E3
Eyke 39 J5
Eynesbury 36 E5
Eynort 110 E3
Eynsford 21 K7
Eynsham 31 J7
Eype 9 F5
Eyre 111 F2
Eythorne 15 J4
Eyton, Shrops 32 C2
Eyton, H. & W 32 D4
Eyton upon the Weald Moors 43 H6

F

Faccombe 18 E5
Faceby 69 G6
Faddiley 43 G2
Fadmoor 69 J8
Faerdre 24 E4
Faifley 85 H7
Failand 17 G3
Failford 77 K5
Failsworth 53 K3
Fairbourne 40 F2
Fairburn 63 H7
Fairfield 34 A3
Fairford 31 F7
Fairlie 77 H2
Fairlight 14 E7
Fairmile 8 B5
Fairmilehead 79 K3
Fair Oak, Hants 11 F3
Fairoak, Staffs 43 J4
Fairseat 14 C2
Fairstead 22 D2
Fairwarp 13 H4
Fairy Cross 6 E3
Fakenham 48 E4
Fala 80 C3
Fala Dam 80 C3
Falahill 80 B4
Faldingworth 56 E5
Falfield 30 A8
Falkenham 39 H7
Falkirk 86 B6
Falkland 87 F4
Falla 74 D2
Fallin 85 L5
Falmer 13 G6
Falmouth 3 G6
Fanagmore 104 C4
Fangdale Beck 69 H7
Fangfoss 64 E7
Fanmore 88 F4
Fans 80 E5
Farcet 46 F8
Far Cotton 35 J5
Farden 32 E3
Fareham 11 G4
Farewell 44 B6
Far Forest 33 G3
Faringdon 31 G8
Farington 61 J7
Farleigh, Avon 17 F4
Farleigh, Surrey 21 H7
Farleigh Hungerford 17 K5
Farleigh Wallop 19 H6
Farlesthorpe 57 J6
Farleton 61 J1
Farley, Staffs 44 B3
Farley, Wilts 10 D2
Farley, Shrops 42 E7
Farley Green 12 D2
Farley Hill 20 B7
Farleys End 30 B6
Farlington 63 K3
Farlow 33 F2
Farmborough 17 H4
Farmcote 30 E5
Farmington 31 F6
Farmoor 31 J7
Farmtown 98 E3
Farnborough, Warw 34 F6
Farnborough, G. Lon 21 J7
Farnborough, Hants 19 K5
Farncombe 12 C2
Farndish 36 C4
Farndon, Ches 42 F2
Farndon, Notts 45 J2
Farnell 93 H6
Farnham, Dorset 10 A3
Farnham, N. Yks 63 G3
Farnham, Essex 21 J1
Farnham, Suff 39 J4
Farnham, Surrey 19 K6
Farnham Common 20 D5
Farnham Green 37 H8
Farnham Royal 20 D5
Farningham 21 K7
Farnley 62 F5
Farnley Tyas 54 E2
Farnsfield 45 H2
Farnworth, Ches 53 G5
Farnworth, G. Man 53 J3
Farr, Highld 106 C3
Farr, Highld 96 E5
Farringdon 8 B5
Farrington Gurney 17 H5
Farsley 62 F6
Farthinghoe 35 G7
Farthingstone 35 H5
Farway 8 C5
Fascadale 89 G1
Fasnacloich 90 B7
Fassfern 90 B4
Fattfield 68 F1
Fattahead 98 F3
Faugh 67 H1
Fauldhouse 79 G3
Faulkbourne 22 D2
Faulkland 17 J5
Fauls 43 G4
Faversham 15 G2
Favillar 98 B5
Fawfieldhead 54 D7
Fawkham Green 21 K7
Fawler 31 H6
Fawley, Bucks 20 B5
Fawley, Berks 18 E2
Fawley, Hants 11 F4
Fawley Chapel 29 K5
Faxfleet 58 C4
Faygate 12 F3
Fazeley 44 D7
Fearby 62 E1
Fearnan 91 J7
Fearnbeg 100 C6
Fearnhead 53 H4
Fearnmore 100 C5
Featherstone, W. Yks 63 H7
Featherstone, Staffs 43 L7
Feckenham 34 B4
Fedderate 99 H4
Feering 22 E1
Feetham 68 B7
Feizor 62 A3
Felbridge 13 G3
Felbrigg 49 H4
Felcourt 13 G2
Felden 20 E3
Felindre, Powys 32 A2
Felindre, W. Glam 24 E4
Felindre, Dyfed 24 F2
Felindre, Dyfed 27 H4
Felindre Farchog 26 F4
Felinfach 25 J1
Felinfoel 24 D4
Felington 81 H5
Felingwmuchaf 27 K5
Felixkirk 63 H1
Felixstowe 39 H7
Fell End 67 K7
Felling 75 J7
Fell Side 66 F3
Felmersham 36 C5
Felmingham 49 H5
Felpham 12 C7
Felsham 38 E5
Felstead 22 C1
Feltham 21 F6
Felthorpe 49 G6
Felton, Avon 17 G4
Felton, Northum 75 H3
Felton, H. & W 29 K3
Felton Butler 42 E6
Feltwell 38 C1
Fence 62 B6
Fence Houses 68 F1
Fencote 68 E7
Fen Ditton 37 H4
Fen Drayton 37 G4
Fen End 34 D3
Feniscowles 61 K7
Feniton 8 C5
Fenny Bentley 44 C2
Fenny Bridges 8 C5
Fenny Compton 34 F5
Fenny Drayton 44 E8
Fenny Stratford 36 B7
Fen Pitton 37 H4
Fenrother 75 H4
Fenstanton 37 G4
Fenton, Lincs 46 B2
Fenton, Lincs 56 C6
Fenton, Cambs 37 G3
Fenton, Northum 81 H6
Fenton, Staffs 43 K3
Fenwick, Northum 75 G6
Fenwick, S. Yks 55 J2
Fenwick, Northum 81 J5
Fenwick, Strath 77 K3
Feochaig 76 C6
Feock 3 G6
Feolin Ferry 82 D5
Feriniquarrie 110 D2
Fern 93 F5
Ferndale 25 J5
Ferndown 10 B4
Ferness 97 H4
Fernham 18 D1
Fernhill Heath 33 H5
Fernhurst 11 K2
Fernilea 110 E3
Fernilee 54 D6
Ferrensby 63 G3
Ferrindonald 111 G4
Ferring 12 D6
Ferrybridge 63 H7
Ferryden 93 J6
Ferryhill 68 E3
Ferryside 27 H6
Fersfield 38 F2
Fersit 90 E4
Feshiebridge 97 G8
Fetcham 12 E1
Fetterangus 99 J3
Fettercairn 93 H4
Fewston 62 E4
Ffairfach 27 L5
Ffaldybrenin 27 L3
Ffarmers 27 L3
Ffawyddog 25 L3
Ffestiniog 51 G6
Fforest 27 K7
Fforest-fach 24 E5
Ffostrasol 27 H3
Ffrith 42 D2
Ffynnon-ddrain 27 J5
Ffynnongroyw 52 C5
Fidden 88 E6
Fiddington, Glos 30 D4
Fiddington, Somer 16 D6
Fiddleford 9 K3
Fiddlers Hamlet 21 J3
Field 44 B4
Field Broughton 61 G1
Field Dalling 48 F4
Field Head 45 F7
Fifehead Magdalen 9 J2
Fifehead Neville 9 J3
Fifield, Berks 20 D6
Fifield, Oxon 31 G6
Figheldean 18 C6
Filby 49 K6
Filey 65 J4
Filgrave 36 B6
Filkins 31 G7
Filleigh, Devon 7 G3
Filleigh, Devon 7 H4
Fillingham 56 D5
Fillongley 34 D2
Filton 17 H3
Fimber 65 F6
Fincham 48 B7
Finchdean 11 J3
Finchhampstead 20 B7
Finchingfield 38 B7
Finchley 21 G4
Findern 44 E4
Findhorn 97 J2
Findochty 98 D2
Findon, W. Susx 12 E6
Findon, Grampn 93 L2
Finedon 36 C3
Fingal Street 39 H4
Fingask 99 G6
Fingest 20 B4
Finghall 68 D8
Fingland 72 C2
Fingringhoe 23 G1
Finmere 35 H7
Finningham 38 F4
Finningley 55 K4
Finnygaud 98 F3
Finsbay 108 B5
Finsbury 21 H5
Finsthwaite 61 G1
Finstock 31 H6
Finstown 112 B4
Fintry, Grampn 99 G3
Fintry, Strath 85 J6
Finzean 93 H2
Fionnphort 88 E6
Firbank 67 J7
Firbeck 55 J5
Firgrove 53 L2
Firsby 57 J7
Firth 113 D3
Fir Tree 68 D3
Fishbourne, I. of W 11 G5
Fishbourne, W. Susx 11 K4
Fishburn 69 F3
Fishcross 86 C5
Fisherford 98 F5
Fisher's Pond 11 F2
Fisherstreet 12 C3
Fisherton, Highld 97 F3
Fisherton, Strath 77 H6
Fishguard 26 D4
Fishlake 55 K2
Fishpool 53 K3
Fishtoft 47 G3
Fishtoft Drove 47 G3
Fishtown of Usan 93 J6
Fishwick 81 H4

Fiskavaig 110 E3
Fiskerton, Lincs 56 E6
Fiskerton, Notts 45 J2
Fittleton 18 C6
Fittleworth 12 D5
Fitton End 47 H6
Fitz 42 F6
Fitzhead 8 C2
Fitzwilliam 55 H2
Fiunary 89 H4
Five Ashes 13 J4
Fivehead 8 E2
Five Oak Green 13 K2
Five Oaks 12 D4
Five Penny Borve 108 D2
Five Penny Ness 108 D1
Five Roads 27 J7
Flackwell Heath 20 C5
Fladbury 34 A6
Fladdabister 113 D5
Flagg 54 E7
Flamborough 65 K5
Flamstead 20 E2
Flansham 12 C6
Flasby 62 C4
Flash 54 D7
Flashader 110 E2
Flaunden 20 E3
Flawborough 45 J3
Flawith 63 H3
Flax Bourton 17 G4
Flaxby 63 G4
Flaxfleet 58 C4
Flaxley 30 A6
Flaxpool 8 C1
Flaxton 64 D6
Fleckney 45 H8
Flecknoe 35 G4
Fleet, Lincs 47 G5
Fleet, Hants 19 K5
Fleet Hargate 47 G5
Fleetwood 61 G5
Fleggburgh or
Burgh St Margaret 49 K6
Flemingston 25 J7
Flemington 78 D4
Flempton 38 D3
Fletching 13 H4
Flexbury 6 C5
Flexford 12 C1
Flicham 48 C5
Flimby 66 C3
Flimwell 13 L3
Flint 52 D6
Flintham 45 J3
Flint Mountain 52 D6
Flinton 59 G3
Flitcham 48 C5
Flitton 36 D7
Flitwick 36 D7
Flixborough 58 C5
Flixton, N. Yks 65 H5
Flixton, Suf 39 J2
Flixton, G. Man 53 J4
Flockton 55 F2
Flodabay 108 B5
Flodden 81 H6
Flookburgh 61 G2
Flordon 49 G8
Flore 35 H4
Flotterton 75 F3
Flowton 39 F6
Flushing, Corn 3 G6
Flushing, Grampn 99 K4
Flyford Flavell 34 A5
Fobbing 22 D5
Fochabers 98 C3
Fochriw 25 K4
Fockerby 58 C5
Fodderty 96 D3
Foel 41 J2
Foggathorpe 58 B3
Fogo 80 F5
Foindle 104 C4
Folda 92 C5
Fole 44 B4
Foleshill 34 E2
Folke 9 H3
Folkestone 15 J5
Folkingham 46 D4
Folkington 13 J6
Folksworth 36 E2
Folkton 65 H5
Folla Rule 99 G5
Follifoot 63 G4
Folly Gate 5 F2
Fonthill Bishop 10 A1
Fonthill Gifford 9 L1
Fontmell Magna 9 K3
Fontwell 12 C6
Foolow 54 E6
Foots Cray 21 J6
Forcett 68 D5
Ford, Strath 84 A4
Ford, Staffs 44 B2
Ford, Bucks 20 B3
Ford, W. Susx 12 D6
Ford, Mers 52 E4
Ford, Glos 30 E5
Ford, Shrops 42 F6
Ford, Northum 81 H6
Ford, Devon 5 H7
Ford, Wilts 17 K3
Fordcombe 13 J2
Fordell 86 E6
Forden 42 D7
Ford End 22 C2
Forder Green 5 H5
Fordham, Cambs 38 B3
Fordham, Norf 48 B8
Fordham, Essex 38 E8
Fordingbridge 10 C3
Fordon 65 H5
Fordoun 93 J4
Ford Street, Somer 8 C3
Fordstreet, Essex 38 E8
Fordwells 31 H6
Fordwich 15 H3
Fordyce 98 E2
Foremark 44 E5
Forestburn Gate 75 G4
Forestfield 79 F3
Forest Gate 21 J5
Forest Green 12 E2
Forest Head 67 H1
Forest Hill 31 K7
Forest-in-Teesdale 68 A4
Forest Mill 86 C5
Forest Row 13 H3
Forestside 11 J3
Forest Town 55 J7
Forfar 92 F6
Forgandenny 86 D3
Forge Side 29 G7
Forgie 98 C3
Formby 52 D3
Forncett End 49 G8
Forncett St Mary 49 G8
Forncett St Peter 39 G1
Forneth 92 B7
Fornham All Saints 38 D4
Fornham St Martin 38 D4
Forres 97 J3
Forsbrook 44 A3
Forse 107 H6
Forsinard 106 D5
Forston 9 H5
Fort Augustus 96 B8
Forteviot 86 D3
Fort George 97 F3
Forth 79 G4
Forthampton 30 C4
Forth Road Bridge, Fife 86 E7
Forth Road Bridge, Lothn 86 E7
Fortingall 91 J7
Forton, Somer 8 E4
Forton, Shrops 42 F6
Forton, Lancs 61 H4
Forton, Staffs 43 J5
Fortrie, Grampn 98 F4
Fortrie, Grampn 99 J4
Fortrose 97 F3
Fortuneswell 9 H7
Fort William 90 C4
Forty Hill 21 H4
Forward Green 39 F5
Fosbury 18 E5
Fosdyke 47 G4
Foss 91 J6
Fossebridge 30 E6
Foss-y-ffin 27 J1
Foster Street 21 J3
Foston, Lincs 46 B3
Foston, Derby 44 C4
Foston, N. Yks 64 D6
Foston on the Wolds 65 J7
Fotherby 57 H4
Fotheringhay 46 D8
Foubister 112 C4
Foulden, Norf 48 C8
Foulden, Borde 81 H4
Foul Mile 13 K5
Foulridge 62 B5
Foulsham 48 F5
Fountainhall 80 C5
Four Ashes 38 F3
Four Crosses, Staffs 44 A7
Four Crosses, Powys 42 D6
Four Crosses, Powys 41 K3
Four Elms 13 H2
Four Forks 8 D1
Four Gotes 47 H6
Four Lanes 2 E6
Fourlanes End 53 K8
Four Marks 11 H1
Four Mile Bridge 50 B3
Four Oaks, W. Mids 44 C8
Four Oaks, W. Mids 34 D2
Four Oaks, E. Susx 14 E6
Fourpenny 103 G4
Fourstones 74 E7
Four Throws 14 D6
Fovant 10 B2
Foveran 99 J6
Fowey 3 K4
Fowlis 87 G1
Fowlis Wester 86 C2
Fowlmere 37 H6
Fownhope 29 K4
Foxdale 60 P4
Foxearth 38 D6
Foxfield 60 F1
Foxham 18 A3
Foxhole 3 H4
Foxholes 65 H5
Fox Lane 19 K5
Foxley, Norf 48 F5
Foxley, Wilts 17 K2
Foxt 44 B3
Foxton, Cambs 37 H6
Foxton, Leic 35 J2
Foxup 62 B2
Foxwist Green 53 H7
Foy 29 K5
Foyers 96 C6
Fraddon 3 H4
Fradley 44 C6
Fradswell 44 A4
Fraisthorpe 65 J6
Framfield 13 H4
Framingham Earl 49 H7
Framingham Pigot 49 H7
Framlingham 39 H4
Frampton, Lincs 47 G4
Frampton, Dorset 9 H5
Frampton Cotterell 17 H2
Frampton Mansell 30 D7
Frampton on Severn 30 B7
Frampton West End 47 G3
Framsden 39 G5
Framwellgate Moor 68 E2
Franche 33 H3
Frankby 52 D5
Frankley 34 A2
Frankton 34 F3
Frant 13 J3
Fraserburgh 99 J2
Frating Green 23 G1
Fratton 11 H4
Freathy 4 D6
Freckenham 38 B3
Freckleton 61 H7
Freeby 45 K5
Freeland 31 J6
Freester 113 D4
Freethorpe 49 K7
Freiston 47 G3
Fremington, N. Yks 68 C7
Fremington, Strath 78 D4
Fremington, Devon 6 F2
Frenchbeer 5 G3
Frensham 19 K6
Fresgoe 106 E3
Freshfield 52 D3
Freshford 17 J4
Freshwater 10 E6
Freshwater East 26 E8
Fressingfield 39 H3
Freston 39 G7
Freswick 107 J3
Fretherne 30 B7
Frettenham 49 H6
Freuchie 87 F4
Friar's Gate 13 H3
Friday Bridge 47 H7
Fridaythorpe 65 F7
Friern Barnet 21 G4
Friesthorpe 56 E5
Frieth 20 B4
Frilford 31 J8
Frilsham 19 G3
Frimley 20 C8
Frindsbury 14 D2
Fring 48 C4
Fringford 31 L5
Frinsted 14 E3
Frinton-on-Sea 23 J1
Friockheim 93 G7
Frisby on the Wreake 45 H6
Friskney 47 H2
Friston, E. Susx 13 J7
Friston, Suff 39 K4
Fritchley 44 E2
Fritham 10 D3
Frith Bank 47 G3
Frith Common 33 F4
Frithelstock 6 E4
Frithville 47 G2
Frittenden 14 E4
Fritton, Norf 39 H1
Fritton, Norf 49 K7
Fritwell 31 K5
Frizington 66 C5
Frocester 30 B7
Frodesley 43 G7
Frodsham 53 G6
Froggatt 55 F6
Froghall 44 B3
Frogmore 20 C7
Frolesworth 35 G1
Frome 17 J6
Fromes Hill 33 F6
Frome St Quintin 9 G4
Fron, Gwyn 50 C7
Fron, Powys 42 D7
Fron, Powys 41 K7
Froncysyllte 42 D3
Frongoch 51 J7
Frostenden 39 K2
Frosterley 68 C3
Froxfield 18 D4
Froxfield Green 11 J2
Fryerning 22 C3
Fryton 64 D5
Fulbeck 46 C2
Fulbourn 37 J5
Fulbrook 31 G6
Fulford, Staffs 44 A4
Fulford, Somer 8 D2
Fulford, N. Yks 63 K5
Fulham 21 G6
Fulking 12 F5
Fuller's Moor 43 F2
Fuller Street 22 D2
Fullerton 18 E7
Fulletby 57 G6
Full Sutton 64 E7
Fullwood 77 K2
Fulmer 20 D5
Fulmodeston 48 E4
Fulnetby 57 E6
Fulstow 57 H4
Fulwell 75 K8
Fulwood, S. Yks 55 G5
Fulwood, Lancs 61 J6
Funtington 11 K4
Funtley 11 G4
Funzie 113 E2
Furnace 84 C4
Furneux Pelham 37 H8
Furzebrook 10 A6
Fyfett 8 D3
Fyfield, Wilts 18 C4
Fyfield, Hants 18 D6
Fyfield, Glos 31 G7
Fyfield, Oxon 31 J8
Fyfield, Essex 21 K3
Fylingthorpe 65 G2
Fyvie 99 G5

G

Gabroc Hill 77 K2
Gaddesby 45 H6
Gadfa 50 D2
Gaer 25 K2
Gaer-fawr 29 J8
Gaerllwyd 29 J8
Gaerwen 50 D3
Gagingwell 31 J5
Gailey 43 L6
Gainford 68 D5
Gainsborough 56 C5
Gainsford End 38 C7
Gairloch 100 D4
Gairlochy 90 C3
Gaisgill 67 J6
Gaitsgill 67 F2
Galashiels 80 C6
Galby 45 H7
Galgate 61 H4
Galhampton 9 H2
Gallatown 87 F5
Galley Common 34 E1
Galleywood 22 D3
Gallowfauld 92 F7
Galltair 94 D3
Galmisdale 111 F5
Galmpton, Devon 5 G7
Galmpton, Devon 5 J6
Galphay 63 F2
Galson 108 D2
Galston 77 L4
Galtrigill 110 D2
Gamblesby 67 J3
Gamlingay 37 F5
Gamston, Notts 56 B6
Gamston, Notts 45 H4
Ganarew 29 K6
Ganllwyd 41 G1
Ganstead 59 F3
Ganthorpe 64 D5
Ganton 65 G5
Garbat 96 C2
Garbhallt 84 C5
Garboldisham 38 F2
Gardenstown 99 H2
Garderhouse 113 C4
Gare Hill 17 J6
Garelochhead 84 E5
Garenin 108 B2
Garford 31 J8
Garforth 63 H6
Gargrave 62 C4
Gargunnock 85 K5
Garlieston 71 F7
Garlogie 99 G8
Garmond 99 H3
Garmouth 98 C2
Garn 50 B7
Garnant 24 E3
Garn Dolbenmaen 50 D6
Garnett Bridge 67 H7
Garnkirk 78 D3
Garn-yr-erw 29 G6
Garrabost 108 D3
Garras 2 F7
Garreg 50 F6
Garreg Bank 42 D6
Garrigill 67 K2
Garros 111 F1
Garrygualach 90 C1
Garrynamonie 110 B4
Garsdale 67 K8
Garsdale Head 67 K7
Garsdon 18 A2
Garshall Green 44 A4
Garsington 31 K7
Garstang 61 H5
Garston 52 F5
Garswood 53 G4
Gartcosh 78 D3
Garth, Clwyd 42 D3
Garth, Powys 28 D3
Garth, M. Glam 25 G5
Garth, I. of M 60 Q4
Garthbrengy 25 J1
Gartheli 27 K2
Garthmyl 42 C8
Garthorpe, Leic 46 B5
Garthorpe, Humbs 58 C5
Gartly 98 E5
Gartmore 85 H5
Gartocharn 85 G6
Garton 59 G3
Garton-on-the-Wolds 65 G7
Gartymore 103 J2
Garvald 87 J7
Garvard 82 C2
Garve 96 B2
Garvestone 48 F7
Garvock 93 J4
Garway 29 J5
Garynahine 108 C3
Gastard 17 K4
Gasthorpe 38 E2
Gatcombe 11 F6
Gatebeck 61 J1
Gate Burton 56 C5
Gateforth 63 J7
Gatehead 77 J4
Gate Helmsley 64 D7
Gatehouse of Fleet 71 H6
Gatelawbridge 72 E4
Gateley 48 E5
Gatenby 63 G1
Gateshead 75 J7
Gatesheath 53 F7
Gateside, Fife 86 E4
Gateside, Tays 92 F7
Gateside, Strath 77 J2
Gathurst 53 G3
Gatley 53 K5
Gattonside 80 D6
Gauldry 87 G2
Gaunt's Common 10 B4
Gautby 57 F6
Gavinton 80 F4
Gawber 55 G3
Gawcott 35 H7
Gawsworth 53 K7
Gawthrop 61 K2
Gawthwaite 61 F1
Gaydon 34 E5
Gayhurst 36 B6
Gayle 68 A8
Gayles 68 D6
Gay Street 12 D4
Gayton, Staffs 44 A5
Gayton, Norf 48 C6
Gayton, Mers 52 D5
Gayton, Northnts 35 J5
Gayton le Marsh 57 J5
Gayton Thorpe 48 C6
Gaywood 48 B5
Gazeley 38 C4
Geary 110 E1
Gedding 38 E5
Geddington 36 B2
Gedintailor 111 F3
Gedney 47 H5
Gedney Broadgate 47 H5
Gedney Drove End 47 H5
Gedney Dyke 47 H5
Gedney Hill 47 G6
Gedney Marsh 45 F5
Gee Cross 54 C4
Geise 107 G3
Geldeston 39 J1
Gelli 26 E6
Gellifor 52 C7
Gelligaer 25 K5
Gellilydan 51 F7
Gellioedd 51 J6
Gellywen 27 G5
Gelston 72 C8
Gentleshaw 44 B6
Geocrab 108 B5
Georgeham 6 E2
George Nympton 7 H3
Georgetown 77 K1
Georgia 2 C6
Germansweek 4 E2
Germoe 2 D7
Gerrans 3 G6
Gerrards Cross 20 E5
Geshader 108 B3
Gestingthorpe 38 D7
Geuffordd 42 D6
Geufron 41 H5
Gibraltar 57 K8
Gidea Park 21 K4
Gidleigh 5 G3
Gifford 80 D3
Giggleswick 62 B3
Gilberdyke 58 C4
Gilchriston 80 C3
Gilcrux 66 D3
Gildersome 63 F7
Gildingwells 55 J5
Gileston 25 J8
Gilfach Goch 25 H6
Gilfachreda 27 J2
Gilgarran 66 C4
Gillamoor 64 D3
Gilling East 63 K2
Gillingham, Kent 14 D2
Gillingham, Norf 39 K1
Gillingham, Dorset 9 K2
Gilling West 68 D6
Gillow Heath 53 K8
Gills 107 J2
Gilmanscleuch 80 B7
Gilmerton, Lothn 80 A3
Gilmerton, Tays 86 B2
Gilmorton 35 G2
Gilsland 74 C7
Gilsland Spa 74 C7
Gilston 80 C4
Gilwern 29 G6
Gimingham 49 H4
Gipping 38 F4
Gipsey Bridge 47 F2
Girlsta 113 D4
Girsby 69 F6
Girthon 71 H6
Girton, Notts 56 C7
Girton, Cambs 37 H4
Girvan 70 C2
Gisburn 62 B5
Gisla 108 B3
Gisleham 39 L2
Gislingham 38 F3
Gissing 39 G2
Gittisham 8 C5
Gladestry 29 G2
Gladsmuir 87 H7
Glais 24 F4
Glaisdale 64 E2
Glame 111 F2
Glamis 92 E7
Glanaber Terrace 51 G6
Glanaman 24 E3
Glandford 48 F3
Glandwr, Dyfed 27 F5
Glandwr, Gwent 25 L4
Glangrwyney 29 G6
Glanmule 32 A1
Glanrhyd 26 F3
Glanton 75 G2
Glanton Pike 75 G2
Glanvilles Wootton 9 H4
Glan-y-don 52 C6
Glan-yr-afon 51 K6
Glapthorn 36 D1
Glapwell 55 H7
Glasbury 29 F4
Glascoed, Powys 42 C7
Glascoed, Gwent 29 H7
Glascoed, Clwyd 51 J3
Glascote 44 D7
Glascwm 28 F2
Glasdrum 90 B7
Glasfryn 51 J5
Glasgow 78 C3
Glasinfryn 50 E4
Glasnakille 111 F4
Glaspwll 41 G4
Glassburn 96 B5
Glasserton 71 F8
Glassford 78 E5
Glasshouse Hill 30 B5
Glasshouses 62 E3
Glasslaw 99 H3
Glasson, Lancs 61 H4
Glasson, Cumbr 73 H7
Glassonby 67 H3
Glaston 46 B7
Glastonbury 17 F7
Glatton 36 E2
Glazebury 53 H4
Glazeley 33 G2
Gleadless 55 G5
Gleadsmoss 53 K7
Gleaston 60 F2
Glemsford 38 D6
Glenancross 94 B6
Glenastle 82 C7
Glen Auldyn 60 R2
Glenbarr 76 B4
Glenbeg 89 G2
Glen Bernisdale 111 F2
Glenbervie 93 J3
Glenboig 78 E3
Glenborrodale 89 H2
Glenbreck 79 H7
Glenbuck 78 E7
Glenburn 77 K1
Glencaple 72 E7
Glencarse 87 E2
Glencoe 90 C6
Glencraig 86 E5
Glendevon 86 C4
Glendoebeg 96 C8
Glendon Hall 36 B2
Glenegedale 82 C6
Glenelg 94 D4
Glenfarg 86 E3
Glenfield 45 G7
Glenfinnan 94 E7
Glenfoot 86 E3
Glengarnock 77 J2
Glengrasco 111 F2
Glenkerry 73 H2
Glenkindie 98 D7
Glenlee 72 B5
Glenlivet 98 A6
Glenluce 70 C6
Glenmavis 78 E3
Glenmaye 60 P4
Glenmore 111 F2
Glenmoy 92 F5
Glen Parva 45 G8
Glenprosen Village 92 E5
Glenreasdell Mains 76 D2
Glenridding 67 F5
Glenrothes 87 F4
Glenside 108 C4
Glentham 56 E4
Glenton 98 F6
Glentress 79 K6
Glen Trool Lodge 70 F3
Glentrool Village 70 E4
Glentworth 56 C5
Glenuig 94 B8
Glen Village 86 B7
Glen Vine 60 Q4
Glespin 79 F7
Gletness 113 D4
Glewstone 29 K5
Glinton 46 E7
Glooston 45 J8
Glossop 54 D4
Gloster Hill 75 J3
Gloucester 30 C6
Gloup 113 D1
Glusburn 62 D5
Gluss 113 C3
Glympton 31 J5
Glynarthen 27 H3
Glyn Ceiriog 42 D4
Glyncoch 25 J5
Glyncorrwg 25 G5
Glynde 13 H6
Glyndebourne 13 H5
Glyndyfrdwy 42 C3
Glyn Neath 25 G4
Glynogwr 25 H6
Glyntaff 25 J6
Gnosall 43 K5
Gnosall Heath 43 K5
Goadby 45 J8
Goadby Marwood 45 J5
Goatacre 18 B3
Goathill 9 H3
Goathland 64 F2
Goathurst 8 D1
Gobowen 42 E4
Godalming 12 C2
Godington 35 H8
Godmanchester 37 F3
Godmanstone 9 H5
Godmersham 15 G3
Godney 17 F6
Godolphin Cross 2 E6
Godre'r-graig 25 F4
Godshill, Hants 10 C3
Godshill, I. of W 11 G6
Godstone 13 G1
Goff's Oak 21 H3
Gogar 86 E7
Goginan 41 F5
Golan 50 E6
Golant 3 K4
Golberdon 4 D4
Golborne 53 H4
Golcar 54 E2
Goldcliff 16 E2
Golden Cross 13 J5
Golden Green 13 K2
Golden Grove 27 K6
Goldenhill 43 K2
Golden Pot 19 J6
Golden Valley 30 D5
Golders Green 21 G5
Goldhanger 22 F3
Golding 43 G7
Goldsborough, N. Yks 63 G4
Goldsborough, N. Yks 69 L5
Goldsithney 2 D6
Goldthorpe 55 H3
Gollanfield 97 G3
Golspie 103 G3
Gomeldon 10 C1
Gomersal 62 F7
Gomshall 12 D2
Gonalston 45 H3
Gonfirth 113 C3
Good Easter 22 C2
Gooderstone 48 C7
Goodleigh 7 F2
Goodmanham 58 C2
Goodnestone, Kent 15 G2
Goodnestone, Kent 15 J3
Goodrich 29 K6
Goodrington 5 J6
Goodwick 26 D4
Goodworth Clatford 18 E6
Goodyers End 34 E2
Goole 58 B4
Goonbell 2 F5
Goonhavern 3 F4
Gooseham 6 C4
Goosetrey 53 J7
Goosey 18 E1
Goosnargh 61 J6
Goostrey 53 J6
Gordon 80 E5
Gordonbush 103 G3
Gordonstown, Grampn 98 E3
Gordonstown, Grampn 99 G5
Gorebridge 80 B3
Gorefield 47 H6
Goring 19 H2
Goring-by-Sea 12 E6
Gorleston-on-Sea 49 L7
Gorran Haven 3 J5
Gors 41 F6
Gorsedd 52 C6
Gorseinon 24 D5
Gorseness 112 C4
Gorsgoch 27 J2
Gorslas 27 K6
Gorsley 30 A5
Gorstan 96 B2
Gorton 53 K4
Gosbeck 39 G5
Gosberton 47 F4
Gosfield 38 C8
Gosforth, Cumbr 66 C6
Gosforth, T. & W 75 J7
Gosmor 36 E8
Gosport 11 H5
Gossabrough 113 D2
Goswick 81 J5
Gotham 45 G4
Gotherington 30 D5
Goudhurst 13 L3
Goulceby 57 G6
Gourdas 99 G4
Gourdon 93 K4
Gourock 84 E7
Govan 77 L1
Goveton 5 H7
Govig 108 B4
Govilon 29 G6
Gowanhill 99 K2
Gowdall 63 K7
Gowerton 24 D5
Gowkhall 86 D6
Goxhill, Humbs 59 F2
Goxhill, Humbs 59 F4
Graffham 12 C5
Grafham 36 E4
Grafton, H. & W 32 E4
Grafton, Oxon 31 G7
Grafton, N. Yks 63 H3
Grafton, H. & W 29 J4
Grafton Flyford 34 A5
Grafton Regis 35 J6
Grafton Underwood 36 C2
Grafty Green 14 E4
Graianrhyd 42 D2
Graig, Gwyn 51 H3
Graig, Clwyd 51 K3
Graig-fechan 42 C2
Graig Penllyn 25 H7
Grain 14 E1
Grainsby 59 G7
Grainthorpe 57 H4
Graizelound 58 B7
Grampound 3 H5
Grampound Road 3 H4
Gramsdale 110 C2
Granborough 35 J8
Granby 45 J4
Grandborough 35 F4
Grandtully 91 L6
Grange, Mers 52 D5
Grange, Cumbr 66 E5
Grange, N. Yks 69 H7
Grange Crossroads 98 D3
Grange Hill 21 J4
Grange Moor 55 F2
Grangemouth 86 C6
Grange-over-Sands 61 H2
Grangepans 86 D6
Grangetown 69 H4
Grange Villa 68 E1
Granish 97 G7
Gransmoor 65 J7
Granston 26 C4
Grantchester 37 H5
Grantham 46 C4
Grantley 63 F3
Grantlodge 99 G7
Granton 87 F7
Grantown-on-Spey 97 J6
Grantshouse 81 G3
Grappenhall 53 H5
Grasby 58 E6
Grasmere 66 F6
Grasscroft 54 C3
Grassendale 52 E5
Grassholme 68 B4
Grassington 62 D3
Grassmoor 55 H7
Grassthorpe 56 B7
Grateley 18 D6
Gratwich 44 B4
Graveley, Cambs 37 F4
Graveley, Herts 37 F8
Gravelly Hill 34 C1
Gravels 42 E7
Graveney 15 G2
Gravesend 14 C1
Gravir 108 C4
Grayingham 56 D4
Grayrigg 67 H7
Grays 14 C1
Grayshott 11 K1
Grayswood 12 C3
Graythorp 69 H4
Grazeley 20 A7
Greasbrough 55 H4
Greasby 52 D5
Great Abington 37 J6
Great Addington 36 C3
Great Alne 34 C5
Great Altcar 52 E3
Great Amwell 21 H2
Great Asby 67 J5
Great Ashfield 38 F4
Great Ayton 69 H5
Great Baddow 22 D3
Great Bardfield 38 B7
Great Barford 36 E5
Great Barr 44 B8
Great Barrington 31 G6
Great Barrow 53 F7
Great Barton 38 D4
Great Barugh 64 E5
Great Bavington 75 F5
Great Bealings 39 H6
Great Bedwyn 18 D4
Great Bentley 23 H1
Great Bircham 48 C4
Great Blakenham 39 G5
Great Bolas 43 H5
Great Bookham 12 E1
Great Bosullow 2 C6
Great Bourton 35 F6
Great Bowden 35 J2
Great Bradley 38 B5
Great Braxted 22 E2
Great Bricett 38 F5
Great Brickhill 36 C7
Great Bridgeford 43 K5
Great Brington 35 H4
Great Bromley 39 F8
Great Broughton 69 H6
Great Budworth 53 H6
Great Burdon 68 F5
Great Burstead 22 C4
Great Busby 69 H6
Great Canfield 22 B2
Great Carlton 57 J5
Great Casterton 46 D7
Great Chart 15 F4
Great Chatwell 43 J6
Great Chesterford 37 J6
Great Cheverell 18 A5
Great Chishill 37 H7
Great Clacton 23 H2
Great Clifton 66 C4
Great Coates 59 G5
Great Comberton 34 A6
Great Corby 67 G1
Great Cornard 38 D6
Great Coxwell 31 G8
Great Cransley 36 B3
Great Cressingham 48 D7
Great Crosby 52 E4
Great Cubley 44 C4
Great Dalby 45 J6
Great Doddington 36 B4
Great Driffield 65 H7
Great Dunham 48 D6
Great Dunmow 22 C1
Great Durnford 18 C7
Great Easton, Leic 46 B8
Great Easton, Essex 38 B8
Great Eccleston 61 H5
Great Edstone 64 E4
Great Ellingham 48 F8
Great Elm 17 J6
Great Eversden 37 G5
Great Finborough 38 F5
Greatford 46 D6
Great Fransham 48 D6
Great Gaddesden 20 E2
Great Gidding 36 E2
Great Givendale 64 F7
Great Glemham 39 J4
Great Glen 45 H8
Great Gonerby 46 B4
Great Gransden 37 F5
Great Green 38 E5
Great Habton 64 E5
Great Hallingbury 21 K2
Greatham, Cleve 69 G4
Greatham, Hants 11 J1
Great Hampden 20 C3
Great Harrowden 36 B3
Great Harwood 61 L6
Great Haseley 20 A3
Great Hatfield 59 F2
Great Haywood 44 B5
Great Heck 63 J7
Great Henny 38 D7
Great Hinton 17 L5
Great Hockham 38 E1
Great Holland 23 J2
Great Horkesley 38 E8
Great Hormead 37 H7
Great Horwood 35 J7
Great Houghton, S. Yks 55 H3
Great Houghton, Northnts 35 J5
Great Hucklow 54 E6
Great Kelk 65 J7
Great Kingshill 20 C4
Great Langton 68 E7
Great Leighs 22 D2
Great Limber 59 F6
Great Linford 36 B6
Great Livermere 38 D3
Great Longstone 54 F6
Great Lumley 68 E2
Great Lyth 42 F7
Great Malvern 33 G6
Great Maplestead 38 D7
Great Marton 61 G6
Great Massingham 48 C5
Great Milton 20 A3
Great Missenden 20 C3
Great Mitton 61 L6
Great Mongeham 15 K3
Great Moulton 39 G1
Great Musgrave 67 K5
Great Ness 42 E6
Great Oakley, Northnts 36 B2
Great Oakley, Essex 39 G8
Great Offley 36 E8
Great Ormside 67 K5
Great Orton 66 F1
Great Oxendon 35 J2
Great Palgrave 48 D6
Great Parndon 21 J3
Great Paxton 36 F4
Great Plumstead 49 J6
Great Ponton 46 C4
Great Preston 63 H7
Great Raveley 37 F2
Great Rissington 31 F6
Great Rollright 31 H4
Great Ryburgh 48 E5
Great Ryle 75 G2
Great Saling 38 C8
Great Salkeld 67 H3
Great Sampford 38 B7
Great Sankey 53 G5
Great Saxham 38 C4
Great Shefford 18 E3
Great Shelford 37 H5
Great Smeaton 69 F6
Great Snoring 48 E4
Great Somerford 18 A2
Great Stainton 68 F4
Great Stambridge 22 F4
Great Staughton 36 E4
Great Steeping 57 J7
Great Stonar 15 K3
Greatstone-on-Sea 15 G6
Great Strickland 67 H4
Great Stukeley 37 F3
Great Sturton 57 G6
Great Swinburne 74 F6
Great Tew 31 H5
Great Tey 38 D8
Great Thurlow 38 B5
Great Torrington 6 E4
Great Tosson 75 G3
Great Totham, Essex 22 E2
Great Totham, Essex 22 E2
Great Wakering 22 F5
Great Waldingfield 38 E6
Great Walsingham 48 E4
Great Waltham 22 C2
Great Warley 21 K4
Great Washbourne 30 D4
Great Welnetham 38 D5
Great Wenham 38 F7
Great Whittington 75 G6
Great Wigborough 23 F2
Great Wilbraham 37 J5
Great Wishford 10 B1
Great Witcombe 30 D6
Great Witley 33 G4
Great Wolford 31 G4
Greatworth 35 G6
Great Wratting 38 B6
Great Wymondley 36 F8
Great Wyrley 44 A7
Great Wytheford 43 G6
Great Yarmouth 49 L7
Great Yeldham 38 C7
Greenburn 79 G3
Greendykes 81 J7
Greenfield, Oxon 20 B4
Greenfield, G. Man 54 C3
Greenfield, Clwyd 52 C6
Greenfield, Highld 90 D1
Greenfield, Beds 36 D7
Greenford 21 F5
Greengairs 85 K7
Greenham 19 F4
Green Hammerton 63 H4
Greenhaugh 74 D5
Greenhead 74 C7
Green Hill, Wilts 18 B2
Greenhill, G. Lon 21 F5
Greenhill, S. Yks 55 G5
Greenhill, Central 85 L7
Greenhithe 14 B1
Greenholm 77 L4
Greenholme 67 H6
Greenhouse 80 D7
Greenhow Hill 62 E3
Greenigoe 112 C4
Greenland 107 H3
Greenlaw 80 F5
Greenloaning 85 L4
Greenmount 53 J2
Greenmow 113 D5
Greenock 84 E7
Greenodd 61 G1
Green Ore 17 G5
Greenside 75 H7
Greensidehill 75 F2
Greenskares 99 G2
Greens Norton 35 H6
Greenstead Green 38 D8
Greensted 21 K3
Green Street 21 F4
Green Street Green 21 J7
Green, The, Cumbr 60 E1
Green, The, Wilts 9 K1
Greenwich 21 J6
Greet 30 E4
Greete 32 E3
Greetham, Leic 46 C6
Greetham, Lincs 57 H6
Greetland 62 D7
Gregson Lane 61 J7
Greinton 8 F1
Grendon, Northnts 36 B4
Grendon, Warw 44 D8
Grendon Common 44 D8
Grendon Green 29 K2
Grendon Underwood 20 A1
Grenitote 108 A6
Grenoside 55 G4
Gresford 42 E2
Gresham 49 G4
Greshornish 110 E2
Gressenhall 48 E6
Gressingham 61 J3
Greta Bridge 68 C5
Gretna 73 J7
Gretna Green 73 J7
Gretton, Northnts 46 B8
Gretton, Glos 30 E4
Gretton, Shrops 43 G8
Grewelthorpe 63 F2
Greygarth 62 E2
Greysouthen 66 C4
Greystoke 67 G3
Greystone 93 G7
Greywell 19 J5
Griais 108 D2
Grianan 108 D3
Griff 34 E2
Griffithstown 29 G8
Grike 66 C5
Grimeford Village 53 H2
Grimethorpe 55 H3
Griminish 110 B2
Grimister 113 D2
Grimley 33 H4
Grimness 112 C5
Grimoldby 57 H5
Grimsargh 61 J6
Grimsby 59 G6
Grimscote 35 H5
Grimscott 6 C5
Grimshader 108 D3
Grimsthorpe 46 D5
Grimston, Norf 48 C5
Grimston, Leic 45 H5
Grimstone 9 H5
Grindale 65 J5
Grindiscol 113 D5
Grindleford 55 F6
Grindleton 62 A5
Grindley Brook 43 G3
Grindlow 54 E6
Grindon, Staffs 44 B2
Grindon, Northum 81 H5
Gringle 43 J7
Gringley on the Hill 56 B4
Grinsdale 73 J8
Grinshill 43 G5
Grinton 68 C7
Grisedale 67 K7
Gristhorpe 65 H4
Griston 48 E8
Grittenham 18 B2
Grittleton 17 K2
Grizebeck 60 F1
Grizedale 66 F7
Grobister 112 D3
Groby 45 G7
Groes, W. Glam 25 F6
Groes, Clwyd 51 K4
Groes-faen 25 J6
Groeslon 50 D5
Grogport 76 D3
Gromford 39 J5
Gronant 51 K2
Groombridge 13 J3
Grosebay 108 B5
Grosmont, N. Yks 64 F2
Grosmont, Gwent 29 J5
Groton 38 E6
Grove, Notts 56 B6
Grove, Oxon 18 F1
Grove, Dorset 9 H7
Grove, Kent 15 J2
Grove Park 21 J6
Grovesend 24 D4
Grudie 96 B2
Gruids 102 D3
Gruline 89 G4
Grumbla 2 C7
Grunasound 113 C5
Grundisburgh 39 H5
Gruting 113 C4
Guardbridge 87 H3
Guarlford 33 H6
Guestling Green 14 E7
Guestwick 48 F5
Guide Post 75 J5
Guilden Morden 37 F6
Guilden Sutton 52 F7
Guildford 12 C2
Guildtown 86 E1
Guilsborough 35 H3
Guilsfield 42 D6
Guisborough 69 J5
Guiseley 62 E5
Guist 48 E5
Guiting Power 30 E5
Gullane 87 H6
Gulval 2 C6
Gumfreston 26 F7
Gumley 35 H1
Gunby, Humbs 58 B3
Gunby, Lincs 46 C5
Gundleton 11 H1
Gunn 7 G2
Gunnerside 68 B7
Gunnerton 74 F6
Gunness 58 C5
Gunnislake 4 E4
Gunnista 113 D4
Gunthorpe, Norf 48 F4
Gunthorpe, Notts 45 H3
Gunwalloe Fishing Cove 2 E7
Gurnard 11 F5
Gurney Slade 17 H6
Gurnos 25 F4
Gussage All Saints 10 B3
Gussage St Michael 10 A3
Guston 15 K4
Gutcher 113 D2
Guthrie 93 G6
Guyhirn 47 G7
Guy's Head 47 H5
Guy's Marsh 9 K2
Guyzance 75 J3
Gwaelod-y-Garth 25 K6
Gwaenysgor 51 K2
Gwalchmai 50 C3
Gwaun-Cae-Gurwen 24 F3
Gwbert 26 F2
Gweek 2 F7
Gwehelog 29 H7
Gwenddwr 28 E3
Gwennap 2 F5
Gwenter 2 F8
Gwernaffield 52 D7
Gwernesney 29 J7
Gwernogle 27 K4
Gwernymynydd 52 D7
Gwersyllt 42 E2
Gwespyr 52 C5
Gwinear 2 D6
Gwithian 2 D5
Gwyddelwern 51 K6
Gwyddgrug 27 J4
Gwytherin 51 H4

H

Habberley, Shrops 42 E7
Habberley, H. & W 33 H3
Habost, W. Isles 108 C4
Habost, W. Isles 108 D1
Habrough 59 F5
Haccombe 5 J4
Haceby 46 D4
Hacheston 39 J5
Hackford 48 F7
Hackforth 68 E7
Hacklete 108 B3
Hackleton 35 K5
Hackness, Orkney 112 B5
Hackness, N. Yks 65 G3
Hackney 21 H5
Hackthorn 56 D5
Hackthorpe 67 H4
Haconby 46 E5
Hadden 81 F6
Haddenham, Bucks 20 B3
Haddenham, Cambs 37 H3
Haddington 87 J7
Haddiscoe 49 K8
Haddon 36 E1
Hademore 44 C7
Hadfield 54 D4
Hadham Cross 21 J2
Hadham Ford 21 J1
Hadleigh, Essex 22 E5
Hadleigh, Suff 38 F6
Hadley 43 H6
Hadley End 44 C5
Hadlow 13 K1
Hadlow Down 13 J4
Hadnall 43 G5
Hadstock 37 J6
Hadzor 33 J4
Haffenden Quarter 14 E4
Hafod-Dinbych 51 H5
Haggbeck 73 K6
Haggerston 81 J5
Hagley, H. & W 33 J2
Hagley, H. & W 29 K3
Hagworthingham 57 H7
Haigh 53 H3
Haighton Green 61 J6
Haile 66 C6
Hailes 30 E4
Hailey, Herts 21 H2
Hailey, Oxon 31 H6
Hailsham 13 J6
Hail Weston 36 E4
Hainault 21 J4
Hainford 49 H6
Hainton 57 F5
Haisthorpe 65 J6
Halam 45 H2
Halberton 7 L4
Halcro 107 H3
Hale, Hants 10 C3
Hale, Lincs 46 E3
Hale, Ches 52 F5
Hale, G. Man 53 J5
Hale Bank 53 F5
Halebarns 53 J5
Hales, Staffs 43 J4
Hales, Norf 49 J8
Halesowen 34 A2
Hales Place 15 H3
Hale Street 13 K2
Halesworth 39 J3
Halewood 52 F5
Halford, Shrops 32 D2
Halford, Warw 34 D6
Halfpenny Green 33 H1
Halfway, Berks 18 F4
Halfway, Powys 25 G1
Halfway, Dyfed 27 L4
Halfway House 42 E6
Halfway Houses 14 F1
Halifax 62 D7
Halistra 110 E2
Halket 77 K2
Halkirk 107 G4
Halkyn 52 D6
Halland 13 J5
Hallaton 45 J8
Hallatrow 17 H5
Hallbankgate 74 B8
Hall Dunnerdale 66 E7
Hallen 17 G2
Hall Green 34 C2
Halliburton 80 E5
Hallin 110 E2
Halling 14 D2
Hallington 75 F6
Hall of the Forest 32 B2
Halloughton 45 H2
Hallow 33 H5
Hallrule 74 B2
Halls 87 K7
Hallsands 5 J8
Hallthwaites 60 E1
Hallworthy 4 B3
Hallyne 79 J5
Halmer End 43 K3
Halmore 30 A7
Halmyre Mains 79 J5
Halnaker 12 C6
Halsall 52 E2
Halse, Somer 8 C2
Halse, Northnts 35 G6
Halsetown 2 D6
Halsham 59 G4
Halsinger 6 F2
Halstead, Essex 38 D7
Halstead, Leic 45 J7
Halstead, Kent 21 J7
Halstock 9 G4
Haltham 57 G7
Haltoft End 47 G3
Halton, Bucks 20 C2
Halton, Clwyd 42 E4
Halton, Ches 53 G5
Halton, Lancs 61 J3
Halton East 62 D4
Halton Gill 62 B2
Halton Holegate 57 J7
Halton Lea Gate 74 C8
Halton West 62 B4
Haltwhistle 74 D7
Halvergate 49 K7
Halwell 5 H6
Halwill 4 E2
Halwill Junction 4 E1
Ham, Shetld 113 A5
Ham, Glos 30 A8
Ham, Wilts 18 E4
Ham, G. Lon 21 F6
Ham, Highld 107 H2
Hamble 11 F4
Hambleden 20 B5
Hambledon, Surrey 12 C3
Hambledon, Hants 11 H3
Hambleton, Lancs 61 G5
Hambleton, N. Yks 63 J6
Hambridge 8 E2
Hambrook, Avon 17 H3
Hambrook, W. Susx 11 J4
Hameringham 57 H7
Hamerton 36 E3
Hametoun 113 A5
Ham Green, H. & W 34 B4
Ham Green, Avon 17 G3
Hamilton 78 E4
Hammersmith 21 G6
Hammerwich 44 B7
Hammond Street 21 H3
Hammoon 9 K3
Hamnavoe, Shetld 113 C5
Hamnavoe, Shetld 113 D2
Hampden Park 13 K6
Hampnett 30 F6
Hampole 55 J2
Hampreston 10 B5
Hampstead 21 G5
Hampstead Norreys 19 G3
Hampsthwaite 63 F4
Hampton, H. & W 34 B6
Hampton, G. Lon 21 F7
Hampton, Shrops 33 G2
Hampton Bishop 29 K4
Hampton Heath 43 F3
Hampton in Arden 34 D2
Hampton Lovell 33 H4
Hampton Lucy 34 D5
Hampton on the Hill 34 D4
Hampton Poyle 31 K6
Hamsey 13 H5
Hamstall Ridware 44 C6
Hamstead, W. Mids 44 B8
Hamstead, I. of W 10 E5
Hamstead Marshall 18 F4
Hamsterley, Durham 68 D1
Hamsterley, Durham 68 D3
Ham Street, Somer 9 G1
Hamstreet, Kent 15 G5
Hamworthy 10 A5
Hanbury, H. & W 34 A4
Hanbury, Staffs 44 C5
Hanchurch 43 K3
Handbridge 52 F7
Handcross 12 F4
Handforth 53 K5
Handley 52 F8
Handsacre 44 B6
Handsworth, W. Mids 34 B1
Handsworth, S. Yks 55 H5
Hanford 43 K3
Hanging Langford 10 B1
Hanham 17 H3
Hankelow 43 H3
Hankerton 18 A1
Hankham 13 K6
Hanley 43 K3
Hanley Castle 33 H6
Hanley Child 33 F4
Hanley Swan 33 H6
Hanley William 33 F4
Hanlith 62 C3
Hanmer 42 F4
Hannington, Northnts 36 B3
Hannington, Wilts 31 F8
Hannington, Hants 19 G5
Hannington Wick 31 F8
Hanslope 35 K6
Hanthorpe 46 D5
Hanwell 34 F6
Hanwood 42 F7
Hanworth, G. Lon 21 F6
Hanworth, Norf 49 G4
Happendon 79 F6
Happisburgh 49 J4
Happisburgh Common 49 J5
Hapsford 53 F6
Hapton, Lancs 62 A6
Hapton, Norf 49 G8
Harberton 5 H6
Harbertonford 5 H6
Harbledown 15 H3
Harborne 34 B2
Harborough Magna 35 F3
Harbottle 74 F3
Harbury 34 E5
Harby, Notts 56 C6
Harby, Leic 45 J4
Harcombe 8 C5
Harden 62 D6
Hardendale 67 H5
Hardgate 93 J1
Hardham 12 D5
Hardhorn 61 G6
Hardingham 48 F7
Hardingstone 35 J5
Hardings Wood 43 K2
Hardington 17 J5
Hardington Mandeville 9 G3
Hardington Marsh 9 G4
Hardley 10 F4
Hardley Street 49 J7
Hardmead 36 C6
Hardraw 68 A7
Hardstoft 55 H7
Hardway, Hants 11 H4
Hardway, Somer 9 J1
Hardwick, Northnts 36 B4
Hardwick, Bucks 20 C2
Hardwick, Cambs 37 G5
Hardwick, Norf 39 H2
Hardwick, Oxon 31 H7
Hardwick, Oxon 31 K5
Hardwicke, Glos 30 C6
Hardwicke, Glos 30 D5
Hardwicke, H. & W 29 G3
Hareby 57 H7
Hareden 61 K4
Harefield 20 E4
Hare Hatch 20 C6
Harehope 75 G1
Harescombe 30 C6
Haresfield 30 C6
Hare Street 37 G8
Harewood 63 G5
Harford 5 G6
Hargrave, Suff 38 C5
Hargrave, Northnts 36 D3
Hargrave, Ches 53 F7
Harker 73 J7
Harkstead 39 G7
Harlaston 44 D6
Harlaxton 46 B4
Harlech 50 E7
Harlesden 21 G5
Harleston, Suff 38 F4
Harleston, Norf 39 H2
Harleston, Devon 5 H7
Harlestone 35 J4
Harle Syke 62 B6
Harley 43 G7
Harlington 36 D7
Harlosh 110 E2
Harlow 21 J2
Harlow Hill 75 G7
Harlsey 69 G7
Harlthorpe 58 B3
Harlton 37 G5
Harman's Cross 10 A6
Harmby 68 D8
Harmer Green 21 G2
Harmer Hill 43 F5
Harmston 56 D7
Harnham 10 C2
Harnhill 30 E7
Harold Hill 21 K4
Haroldston West 26 C6
Haroldswick 113 E1
Harold Wood 21 K4
Harome 63 K1
Harpenden 21 F2
Harpford 8 B5
Harpham 65 H6
Harpley, Norf 48 C5
Harpley, H. & W 33 F4
Harpole 35 H4
Harpsdale 107 G4
Harpsden 20 B5
Harpswell 56 D5
Harpurhey 53 K3
Harpur Hill 54 D6
Harrapool 111 G3
Harrietfield 86 C2
Harrietsham 14 E3
Harrington, Cumbr 66 B4
Harrington, Lincs 57 H6
Harrington, Northnts 35 J2
Harringworth 46 C8
Harriseahead 43 K2
Harrogate 63 G4
Harrold 36 C5
Harrow 21 F5
Harrowbarrow 4 E4
Harrowden 36 D6
Harrow on the Hill 21 F5
Harston, Leic 46 B4
Harston, Cambs 37 H5
Hart 69 G3
Hartburn 75 G5
Hartest 38 D5
Hartfield 13 H3
Hartford, Cambs 37 F3
Hartford, Ches 53 H6
Hartfordbridge 19 J5
Hartford End 22 C2
Hartforth 68 D6
Harthill, Ches 43 G2
Harthill, Strath 79 G3
Harthill, S. Yks 55 H5
Hartington 54 E7
Hartland 6 C3
Hartland Quay 6 C3
Hartlebury 33 H3
Hartlepool 69 H3
Hartley, Kent 14 C2
Hartley, Northum 75 K6
Hartley, Cumbr 67 K6
Hartley, Kent 13 L3
Hartley Wespall 20 A8
Hartley Wintney 19 J5
Hartlip 14 E2
Harton, Shrops 32 D2
Harton, N. Yks 64 E6
Harton, T. & W 75 K7
Hartpury 30 C5
Hartshill 44 E8
Hartshorne 44 E5
Hartsop 67 G5
Hartwell 35 J5
Hartwood 79 F4
Harvel 14 C2
Harvington 34 B6
Harwell 19 F2
Harwich 39 H7
Harwood, G. Man 53 J2
Harwood, Durham 67 L3
Harwood Dale 65 G3
Harworth 55 K4
Hascombe 12 C2
Haselbech 35 J3
Haselbury Plucknett 9 F3
Haseley 34 D4
Haselor 34 C5
Hasfield 30 C5
Hasguard 26 C7
Haskayne 52 E3
Hasketon 39 H5
Hasland 55 G7
Haslemere 11 L1
Haslingden 62 A7
Haslingden Grane 62 A7
Haslingfield 37 H5
Haslington 43 J2
Hassall 53 J8
Hassall Green 53 J8
Hassell Street 15 G4
Hassendean 74 B1
Hassingham 49 J7
Hassocks 13 G5
Hassop 55 F6
Hastigrow 107 H3
Hastingleigh 15 G4
Hastings 14 E8
Hastingwood 21 J3
Hastoe 20 D3
Haswell 69 F2
Hatch, Beds 36 E6
Hatch, Hants 19 H5
Hatch, Wilts 9 L2
Hatch Beauchamp 8 E2
Hatch End 21 F4
Hatching Green 21 F2
Hatchmere 53 G6
Hatcliffe 59 G6
Hatfield, Herts 21 G3
Hatfield, H. & W 29 K2
Hatfield, S. Yks 55 K3
Hatfield Broad Oak 21 K2
Hatfield Heath 21 K2
Hatfield Peverel 22 D2
Hatfield Woodhouse 58 A6
Hatford 31 H8
Hatherden 18 E5
Hatherleigh 7 F5
Hathern 45 G5
Hatherop 31 F7
Hathersage 55 F5
Hatherton, Staffs 44 A6
Hatherton, Ches 43 H3
Hatley St George 37 F5
Hatt 4 D5
Hattingley 11 H1
Hatton, Shrops 32 D1
Hatton, Warw 34 D4
Hatton, Derby 44 D4
Hatton, G. Lon 20 E6
Hatton, Lincs 57 F6
Hatton, Ches 53 G5
Hatton, Grampn 99 K5
Hattoncrook 99 H6
Hatton Heath 52 F7
Hatton of Fintray 99 H7
Haugham 57 H5
Haugh Head 81 J7
Haughley 38 F4
Haughley Green 38 F4
Haugh of Glass 98 D5
Haugh of Urr 72 D7
Haughton, Shrops 42 E5
Haughton, Shrops 43 G6
Haughton, Shrops 43 H8
Haughton, Staffs 43 K5
Haughton, Notts 55 K6
Haughton Green 53 L4
Haughton Moss 43 G2
Haunton 44 D6
Hauxley 75 J3
Hauxton 37 H5
Havant 11 J4
Haven 29 J2
Havenstreet 11 G5
Haverfordwest 26 D6
Haverhill 38 B6
Haverigg 60 E2
Havering-atte-Bower 21 K4
Haversham 36 B6
Haverthwaite 61 G1
Hawarden 52 E7
Hawes 68 A8
Hawford 33 H4
Hawick 74 B2
Hawkchurch 8 E4
Hawkedon 38 C5
Hawkeridge 17 K5
Hawkerland 8 B6
Hawkesbury 17 J2
Hawkesbury Upton 17 J2
Hawkes End 34 D2
Hawkhill 75 J2
Hawkhurst 13 L3
Hawkinge 15 J4
Hawkley 11 J2
Hawkridge 7 J2
Hawkshead 66 F7
Hawksland 79 F6
Hawkswick 62 C2
Hawksworth, W. Yks 62 E5
Hawksworth, Notts 45 J3
Hawkwell 22 E4
Hawley, Hants 20 C8
Hawley, Kent 21 K6
Hawling 30 E5
Hawnby 69 H8
Haworth 62 D6
Hawsker 65 G2
Hawstead 38 D5
Hawthorn 69 G2
Hawthorn Hill 20 C6
Hawton 45 J2
Haxby 63 K4
Haxey 58 B7
Haydock 53 G4
Haydon 9 H3
Haydon Bridge 74 E7
Haydon Wick 18 C2
Haye 4 D5
Hayes, G. Lon 20 E5
Hayes, G. Lon 21 J7
Hayfield 54 D5
Hayle 2 D6
Haynes 36 D6
Haynes Church End 36 D6
Hay-on-Wye 29 G3
Hayscastle 26 C5
Hayscastle Cross 26 D5
Hayton, Notts 56 B5
Hayton, Humbs 58 C2
Hayton, Cumbr 66 D2
Hayton, Cumbr 67 H1
Hayton's Bent 32 E2
Haytor Vale 5 H4
Haywards Heath 13 G4
Haywood Oaks 45 H2
Hazelbank 79 F5
Hazelbury Bryan 9 J4
Hazeley 20 B8
Hazel Grove 53 L5
Hazelrigg 81 J6
Hazelslade 44 B6
Hazelwood 44 E3
Hazlemere 20 C4
Hazlerigg 75 J6
Hazleton 30 E6
Heacham 48 B4
Headbourne Worhty 11 F1
Headcorn 14 E4
Headingley 63 F6
Headington 31 K7
Headlam 68 D5
Headless Cross 34 B4
Headley, Surrey 12 F1
Headley, Hants 19 G4
Headley, Hants 11 K1
Head of Muir 85 L6
Headon 56 B6
Heads Nook 67 G1
Heage 44 E2
Healaugh, N. Yks 68 C7
Healaugh, N. Yks 63 J5
Heald Green 53 K5
Healey, N. Yks 62 E1
Healey, Northum 75 G8
Healey, Lancs 53 K2
Healeyfield 68 C2
Healing 59 G5
Heamoor 2 C6
Heanish 88 B4
Heanor 44 F3
Heanton Punchardon 6 F2
Heapham 56 C5
Hearthstane 79 J7
Heasley Mill 7 H2
Heast 111 G4
Heath, Derby 55 H7
Heath, S. Glam 25 K7
Heath and Reach 36 C8
Heathcote 54 E7
Heath End, Hants 19 G4
Heath End, Hants 19 K6
Heather 44 E6
Heathfield, Somer 8 C2
Heathfield, Strath 77 J1
Heathfield, Devon 5 J4
Heathfield, E. Susx 13 J4
Heath Hayes 44 B6
Heath Hill 43 J6
Heath House 16 F6
Heath, The 49 G5
Heathton 33 H1
Heatley 53 J5
Heaton, Staffs 54 C7
Heaton, Lancs 61 H3
Heaton, T. & W 75 J7
Heaton Moor 53 K4
Heaverham 21 K8
Heaviley 53 L5
Hebburn 75 K7
Hebden 62 D3
Hebden Bridge 62 C7
Hebden Green 53 H7
Hebron 75 H5
Heckfield 20 B7
Heckington 46 E3
Heckmondwike 62 F7
Heddington 18 A4
Heddle 112 B4
Heddon-on-the-Wall 75 H7
Hedenham 49 J8
Hedge End 11 F3
Hedgerley 20 D5
Hedging 8 E2
Hedley on the Hill 75 G8
Hednesford 44 B6
Hedon 59 F4
Hedsor 20 D5
Hegdon Hill 29 K2
Heggerscales 67 L5
Heglibister 113 C4
Heighington, Durham 68 E4
Heighington, Lincs 56 E7
Heights of Brae 96 D2
Heights of Kinlochewe 101 F5
Heiton 80 F6
Helbeck 67 K5
Hele, Devon 7 F1
Hele, Devon 7 K5
Helensburgh 84 E6
Helford 2 F7
Helhoughton 48 D5
Helions Bumpstead 38 B6
Helland 3 J2
Hellesdon 49 H6
Hellidon 35 G5
Hellifield 62 B4
Hellingly 13 J5
Hellington 49 J7
Hellister 113 C4
Helmdon 35 G6
Helmingham 39 G5
Helmsdale 103 J2
Helmshore 62 A7
Helmsley 63 K1
Helperby 63 H3
Helperthorpe 65 G5
Helpringham 46 E3
Helpston 46 E7
Helsby 53 F6
Helston 2 E7
Helstone 3 J1
Helton 67 H4
Helwith Bridge 62 B3
Hemblington 49 J6
Hemel Hempstead 20 E3
Hemingbrough 63 K6
Hemingby 57 G6
Hemingford Abbots 37 F3
Hemingford Grey 37 F3
Hemingstone 39 G5
Hemington, Northnts 36 D2
Hemington, Somer 17 J5
Hemley 39 H6
Hempholme 65 H7
Hempnall 49 H8
Hempnall Green 49 H8
Hempriggs 97 K2
Hempstead, Essex 38 B7
Hempstead, Norf 49 G4
Hempstead, Norf 49 K5
Hempsted 30 C6
Hempton, Norf 48 E5
Hempton, Oxon 31 J4
Hemsby 49 K6
Hemswell 56 D4
Hemsworth 55 H2
Hemyock 8 C3
Henbury, Avon 17 G3
Henbury, Ches 53 K6
Hendon, T. & W 69 G1
Hendon, G. Lon 21 G5
Hendre 52 C7
Hendy 27 K7
Heneglwys 50 D3
Henfield 12 F5
Hengoed, Shrops 42 D4
Hengoed, Powys 29 G2
Hengoed, M. Glam 25 K5
Hengrave 38 D4
Henham 37 J8
Henley, Shrops 32 E3
Henley, Somer 9 F1
Henley, Suff 39 G5
Henley, W. Susx 11 K2
Henley-in-Arden 34 C4
Henley-on-Thames 20 B5
Henley Park 12 C1
Henllan, Dyfed 27 H3
Henllan, Clwyd 51 K4
Henllan Amgoed 27 F5
Henllys 29 G8
Henlow 36 E7
Hennock 5 J3
Henryd 51 G3
Henry's Moat 26 E5
Hensall 63 J7

Henshaw 74 D7
Henstead 39 K2
Henstridge 9 J3
Henstridge Marsh 9 J2
Henton, Oxon 20 B3
Henton, Somer 16 F5
Henwood 4 C4
Heogan 113 D4
Heol Senni 25 H2
Heol-y-Cyw 25 H6
Hepburn 81 J7
Hepple 75 F3
Hepscott 75 J5
Heptonstall 62 C7
Hepworth, Suff 38 E3
Hepworth, W. Yks 54 E3
Herbrandston 26 C7
Hereford 29 K3
Hergest 29 G2
Heriot 80 B4
Hermitage, Border 74 B4
Hermitage, Berks 19 G3
Hermitage, Dorset 9 H4
Hermitage, W. Susx 11 J4
Hermitage, The 12 F1
Hermon, Gwyn 50 C4
Hermon, Dyfed 27 G4
Hermon, Dyfed 27 H4
Herne 15 H2
Herne Bay 15 H2
Herner 7 F3
Hernhill 15 G2
Herodsfoot 3 L3
Herongate 22 C4
Heronsgate 20 E4
Herriard 19 H6
Herringfleet 49 K8
Herringswell 38 C4
Herrington 69 F1
Hersden 15 J2
Hersham 21 F7
Herstmonceux 13 K5
Herston 112 C5
Hertford 21 H2
Hertford Heath 21 H2
Hertingfordbury 21 H2
Hesketh Bank 61 H7
Hesketh Lane 61 K5
Hesket Newmarket 66 F3
Heskin Green 53 G2
Hesleden 69 G3
Heslington 63 K4
Hessay 63 J4
Hessenford 4 D6
Hessett 38 E4
Hessle 58 E4
Hest Bank 61 H3
Heston 21 F6
Heswall 52 D5
Hethe 31 K5
Hethersett 49 G7
Hethersgill 73 K7
Hethpool 81 G7
Hett 68 E3
Hetton 62 C4
Hetton-le-Hole 69 F2
Heugh 75 G6
Heugh-head 98 C7
Heveningham 39 J3
Hever 13 H2
Heversham 61 H1
Hevingham 49 G5
Hewelsfield 29 K7
Hewish, Somer 9 F4
Hewish, Avon 16 F4
Hexham 74 F7
Hextable 21 K6
Hexton 36 E7
Hexworthy 5 G4
Heybridge, Essex 22 C4
Heybridge, Essex 22 E3
Heybridge Basin 22 E3
Heybrook Bay 4 E7
Heydon, Norf 49 G5
Heydon, Cambs 37 H6
Heydour 46 D4
Heylipol 88 A4
Heylor 113 C2
Heysham 61 H3
Heyshott 11 K3
Heytesbury 17 L6
Heythrop 31 H5
Heywood, G. Man 53 K2
Heywood, Wilts 17 K5
Hibaldstow 58 D6
Hickleton 55 H3
Hickling, Notts 45 H5
Hickling, Norf 49 K5
Hickling Green 49 K5
Hickling Heath 49 K5
Hidcote Boyce 34 C6
High Ackworth 55 H2
Higham, Lancs 62 B6
Higham, Suff 38 C4
Higham, Kent 14 D1
Higham, Suff 38 F7
Higham, Derby 55 G8
Higham Dykes 75 H6
Higham Ferrers 36 C4
Higham Gobion 36 E7
Higham on the Hill 44 E8
Highampton 6 E5
Higham Wood 13 K2
High Beach 21 J4
High Bentham 61 K3
High Bickington 7 G3
High Birkwith 62 B2
High Blantyre 78 D4
High Bonnybridge 85 L7
High Borve 108 D2
High Bradfield 55 F4
Highbridge, Highld 90 C3
Highbridge, Somer 16 E6
Highbrook 13 G3
Highburton 54 E2
Highbury 17 H6
High Buston 75 J3
High Callerton 75 H6
High Catton 64 E7
Highclere 18 F4
Highcliffe 10 D5
High Cogges 31 H7
High Coniscliffe 68 E5
High Cross, Herts 21 H2
High Cross, Hants 11 J2
High Cross Bank 44 D6
High Dougarie 76 D4
High Easter 22 C2
High Ellington 62 E1
Higher Ashton 5 J3
Higher Ballam 61 G6
High Ercall 43 G6
Higher End 53 G3
Higher Penwortham 61 J7
Higher Poynton 53 L5
Higher Tale 8 B4
Higher Town 2 P2
Higher Walreddon 4 E4
Higher Walton, Ches 53 G5
Higher Walton, Lancs 61 J7
Higher Wych 43 F3
Highfield, T. & W 75 H8
Highfield, Strath 77 J2
Highfields 37 G5
High Garrett 38 C8
High Grange 68 D3
High Green, S. Yks 55 G4
High Green, Norf 49 G7
High Green, H. & W 33 H6
High Halden 14 E5
High Halstow 14 D1
High Ham 9 F1
High Hatton 43 H5
High Hesket 67 G2
High Hoyland 55 F2
High Hunsley 58 D3
High Hurstwood 13 H4
High Lane, G. Man 54 C5
High Lane, H. & W 33 F4
High Laver 21 K3
Highleadon 30 B5
High Legh 53 J5
Highleigh 11 K5
Highley 33 G2
High Littleton 17 H5
High Melton 55 J3
Highmoor Cross 20 B5
Highmoor Hill 17 F2
Highnam 30 B6
High Newton 61 H1
High Newton-by-the-Sea 81 L7
High Offley 43 J5
High Ongar 21 K3
High Onn 43 K6
High Roding 22 C2
High Salvington 12 E6
High Shaw 68 A7
High Spen 75 H8
Highsted 14 F2
High Street, Corn 3 H4
High Street, Suff 39 K5
High Street Green 38 F5
Hightae 73 F6
Hightown, Mers 52 E3
Hightown, Ches 53 K7
High Toynton 57 G7
High Trewhill 75 G3
Highway, Wilts 18 B3
Highway, Corn 3 K4
Highworth 18 D1
High Wray 67 F7
High Wych 21 J2
High Wycombe 20 C4
Hilborough 48 D7
Hildenborough 13 J2
Hildersham 37 J6
Hilderstone 43 L4
Hilderthorpe 65 J6
Hilgay 48 B8
Hill 30 A8
Hillam 63 J7
Hillberry 60 Q4
Hillbrae, Grampn 98 F4
Hillbrae, Grampn 99 G6
Hill Brow 11 J2
Hilldyke 47 G3
Hill End, Durham 68 C3
Hill End, Fife 86 D5
Hillend, Fife 86 E6
Hillerton 5 H2
Hillesden 35 H8
Hillesley 17 J2
Hillfarrance 8 C2
Hill Head, Hants 11 G4
Hillhead, Strath 77 K6
Hillhead, Devon 5 K6
Hillhead of Auchentumb 99 J3
Hillhead of Cocklaw 99 K4
Hilliard's Cross 44 C6
Hilliclay 107 G3
Hillingdon 20 E5
Hillington 48 C5
Hillmorton 35 G3
Hill Mountain 26 D7
Hillockhead 98 C8
Hill of Beath 86 E5
Hill of Fearn 103 G6
Hill Ridware 44 B6
Hillside, Shetld 113 D3
Hillside, Tays 93 J5
Hillside, Grampn 93 L2
Hillswick 113 C3
Hill, The 60 E1
Hilmarton 18 B3
Hilperton 17 K5
Hilsea 11 H4
Hilton, Durham 68 D4
Hilton, Derby 44 D4
Hilton, Cambs 37 F4
Hilton, Cleve 69 G5
Hilton, Dorset 9 J4
Hilton, Grampn 99 J5
Hilton, Shrops 43 J8
Hilton, Cumbr 67 K4
Hilton of Cadboll 103 G6
Himbleton 33 J5
Himley 33 H1
Hincaster 61 J1
Hinckley 44 F8
Hinderclay 38 F3
Hinderwell 69 K5
Hindford 42 E4
Hindhead 11 K1
Hindley 53 H3
Hindley Green 53 H3
Hindlip 33 H5
Hindolveston 48 F5
Hindon 9 L1
Hindringham 48 E4
Hingham 48 F7
Hinkley Point Power Station 16 D6
Hinstock 43 H5
Hintlesham 39 F6
Hinton, Hants 10 D5
Hinton, Shrops 42 F7
Hinton, Northnts 35 G5
Hinton, Avon 17 J3
Hinton Ampner 11 G2
Hinton Blewett 17 G5
Hinton Charterhouse 17 J5
Hinton-in-the-Hedges 35 G7
Hinton Martell 10 B4
Hinton on the Green 34 B6
Hinton Parva 18 D2
Hinton St George 9 F3
Hinton St Mary 9 J3
Hinton Waldrist 31 H8
Hints, Staffs 44 C7
Hints, Shrops 33 F3
Hinwick 36 C4
Hinxhill 15 G4
Hinxton 37 H6
Hinxworth 37 F6
Hipperholme 62 E7
Hirn 93 J1
Hirnant 41 K1
Hirst 75 J5
Hirst Courtney 63 K7
Hirwaun 25 H4
Hiscott 7 F3
Histon 37 H4
Hitcham 38 E5
Hitchin 36 E8
Hither Green 21 H6
Hittisleigh 5 H2
Hixon 44 B5
Hoaden 15 J3
Hoaldalbert 29 H5
Hoar Cross 44 C5
Hoarwithy 29 K5
Hoath 15 J2
Hobarris 32 C3
Hobbister 112 B4
Hobkirk 74 B2
Hobson 68 D1
Hoby 45 H6
Hockering 48 F6
Hockerton 45 J2
Hockley 22 E4
Hockley Heath 34 C3
Hockliffe 36 C8
Hockwold cum Wilton 38 C2
Hockworthy 8 B3
Hoddesdon 21 H3
Hoddlesden 61 L7
Hodgeston 26 E8
Hodnet 43 H5
Hodthorpe 55 J6
Hoe 48 E6
Hoe Gate 11 H3
Hoff 67 J5
Hoggeston 35 K8
Hoghton 61 K7
Hognaston 44 D2
Hogsthorpe 57 K6
Holbeach 47 G5
Holbeach Bank 47 G5
Holbeach Drove 47 G6
Holbeach Hurn 47 G5
Holbeach St Johns 47 G6
Holbeach St Marks 47 G4
Holbeach St Matthew 47 H4
Holbeck 55 J6
Holberrow Green 34 B5
Holbeton 5 G6
Holborn 21 H5
Holbrook, Derby 44 E3
Holbrook, Suff 39 G7
Holburn 81 J6
Holbury 10 F4
Holcombe, Somer 17 H6
Holcobe, Devon 5 K4
Holcombe Rogus 8 B3
Holcot 35 J4
Holden 62 A5
Holdenby 35 H4
Holdgate 32 E2
Holdingham 46 D3
Hole in the Wall 29 L5
Holemoor 6 E5
Holford 16 C6
Holker 61 G2
Holkham 48 D3
Hollacombe, Devon 6 D5
Hollacombe, Devon 7 G4
Holland-on-Sea 23 J2
Hollandstoun 112 D2
Hollesley 39 J6
Hollinfare 53 H4
Hollingbourne 14 E3
Hollington, Staffs 44 B4
Hollington, Derby 44 D4
Hollington, E. Susx 14 D7
Hollingworth 54 D4
Hollins 53 K3
Hollinsclough 54 D7
Hollinwood 43 G4
Holloway 44 E2
Hollowell 35 H3
Holl's Green 37 F8
Hollybush, H. & W 33 G7
Hollybush, Strath 77 J6
Hollybush, Gwent 25 K4
Holly End 47 H7
Hollym 59 H4
Hollywood 34 B3
Holm 108 D3
Holmbury St Mary 12 E2
Holme, Notts 56 C8
Holme, Cambs 36 E2
Holme, W. Yks 54 E3
Holme, Cumbr 61 J2
Holme Chapel 62 B7
Holme Hale 48 D7
Holme Lacy 29 K4
Holme Marsh 29 H2
Holme next the Sea 48 C3
Holme-on-Spalding-Moor 58 C3
Holme on the Wolds 58 D2
Holmer 29 K3
Holmer Green 20 D4
Holmes Chapel 53 J7
Holmesfield 55 G6
Holmeswood 52 F2
Holmeood 55 H7
Holmfirth 54 E3
Holmhead 72 A1
Holmpton 59 H4
Holmrook 66 C7
Holmsgarth 113 D4
Holne 5 H5
Holnest 9 H4
Holsworthy 6 D5
Holsworthy Beacon 6 D5
Holt, Dorset 10 B4
Holt, Clwyd 42 F2
Holt, Norf 48 F4
Holt, H. & W 33 H4
Holt, Wilts 17 K4
Holtby 63 K4
Holt End 34 B4
Holt Heath 33 H4
Holton, Somer 9 H2
Holton, Suff 39 K3
Holton, Oxon 31 L7
Holton cum Beckering 57 F5
Holton le Clay 59 G6
Holton le Moor 58 E7
Holton St Mary 38 F7
Holwell, Herts 36 E7
Holwell, Oxon 31 G7
Holwell, Dorset 9 J3
Holwell, Leic 45 J5
Holwick 68 B4
Holworth 9 J6
Holybourne 19 J6
Holy Cross 33 J3
Holyhead 50 B2
Holy Island 81 K5
Holymoorside 55 G7
Holyport 20 C6
Holystone 74 F3
Holytown 78 E3
Holywell, Clwyd 52 C6
Holywell, Corn 2 F4
Holywell, Cambs 37 G3
Holywell, Dorset 9 G4
Holywell Green 54 D2
Holywell Lake 8 C2
Holywell Row 38 C3
Holywood 72 E5
Homer 43 H7
Homersfield 39 H2
Hom Green 29 K5
Homington 10 C2
Honeybourne 34 C6
Honeychurch 7 G5
Honey Hill 15 H2
Honiley 34 D3
Honing 49 J5
Honingham 49 G6
Honington, Lincs 46 C3
Honington, Warw 34 D6
Honington, Suff 38 E3
Honiton 8 C4
Honley 54 E2
Hoo 39 H5
Hooe, Devon 4 F6
Hooe, E. Susx 13 K6
Hook, Wilts 18 B2
Hook, Humbs 58 B4
Hook, Dyfed 26 D6
Hook, G. Lon 21 F7
Hook, Hants 19 J5
Hooke 9 G4
Hookgate 43 J4
Hook Norton 31 H4
Hookway 5 J2
Hookwood 12 F2
Hoole 52 F7
Hoo St Werburgh 14 D1
Hooton Levitt 55 J4
Hooton Pagnell 55 H3
Hooton Roberts 55 H4
Hope, Powys 42 D7
Hope, Derby 54 E5
Hope, Shrops 42 E7
Hope, Clwyd 52 E8
Hope, Devon 5 G7
Hope Bagot 33 E3
Hope Bowdler 32 D1
Hopeman 97 K2
Hope Mansell 29 L6
Hopesay 32 C2
Hope under Dinmore 29 K2
Hopton, Suff 38 E3
Hopton, Staffs 43 L5
Hopton Cangeford 32 E2
Hopton Castle 32 C3
Hopton on Sea 49 L7
Hopton Wafers 33 F3
Hopwas 44 C7
Hopwood 34 B3
Horam 13 J5
Horbling 46 E4
Horbury 55 F2
Horden 69 G2
Horderley 32 D2
Hordle 10 D5
Hordley 42 E4
Horeb, Dyfed 27 H3
Horeb, Dyfed 27 K5
Horham 39 H3
Horkstowe 58 D5
Horley, Surrey 13 F2
Horley, Oxon 34 F6
Hornblotton Green 9 G1
Hornby, N. Yks 69 F6
Hornby, Lancs 61 J3
Horncastle 57 G7
Hornchurch 21 K5
Horncliffe 81 H5
Horndean 11 J3
Horndon on the Hill 22 C5
Horne 13 G2
Horn Hill 20 E4
Horning 49 J6
Horninghold 45 K8
Horninglow 44 D5
Horningsea 37 H4
Horningsham 17 K6
Horningtoft 48 E5
Hornish Point 110 B2
Hornsby 67 H1
Hornsea 59 G2
Hornsey 21 H5
Hornton 34 E6
Horrabridge 4 F5
Horringer 38 D4
Horse Bridge, Staffs 44 A2
Horsebridge, Hants 10 E1
Horsebridge, E. Susx 13 J5
Horsebrook 43 K6
Horsehay 43 H7
Horseheath 38 B6
Horsehouse 62 D1
Horsell 20 D8
Horseman's Green 42 F3
Horseway 37 H2
Horsey 49 K5
Horsford 49 G6
Horsforth 63 F6
Horsham, W. Susx 12 E3
Horsham, H. & W 33 G5
Horsham St Faith 49 H6
Horsington, Lincs 57 F7
Horsington, Somer 9 J2
Horsley, Glos 30 C8
Horsley, Derby 44 E3
Horsley, Northum 74 E4
Horsley, Northum 75 G7
Horsley Cross 39 G8
Horsleycross Street 39 G8
Horsleyhill 74 B2
Horsley Woodhouse 44 E3
Horsmonden 13 L2
Horspath 31 K7
Horstead 49 H6
Horsted Keynes 13 G4
Horton, Lancs 62 B4
Horton, Wilts 18 B4
Horton, Dorset 10 B4
Horton, Northnts 36 B5
Horton, W. Glam 24 C6
Horton, Bucks 20 D2
Horton, Somer 8 E3
Horton, Berks 20 E6
Horton, Avon 17 J2
Horton, Northum 81 J6
Horton, Staffs 53 L8
Horton-cum-Studley 31 K6
Horton Green 42 F3
Horton Heath 11 F3
Horton in Ribblesdale 62 B2
Horton Kirby 21 K7
Horwich 53 H2
Horwood 6 F3
Hose 45 J5
Hosh 86 B2
Hostingfield 37 H5
Hoswick 113 D5
Hotham 58 C3
Hothfield 14 F4
Hoton 45 G5
Hough 43 J2
Hougham 46 B3
Hougharry 110 B1
Hough Green 52 F5
Hough-on-the-Hill 46 C3
Houghton, W. Susx 12 D5
Houghton, Dyfed 26 D7
Houghton, Hants 10 E1
Houghton, Cambs 37 F3
Houghton, Cumbr 73 K8
Houghton Conquest 36 D6
Houghton-le-Spring 69 F2
Houghton on the Hill 45 H7
Houghton Regis 20 E1
Houghton St Giles 48 E4
Houlsyke 64 E2
Hound Green 20 B8
Houndslow 80 E5
Houndwood 81 G3
Hounslow 21 F6
Housetter 113 C2
Houston 77 K1
Houstry 107 G6
Hove 13 F6
Hoveringham 45 H3
Hoveton 49 J6
Hovingham 63 K2
How 67 H1
How Caple 29 L4
Howden 58 B4
Howden-le-Wear 68 D3
Howe, Norf 49 H8
Howe, Highld 107 J3
Howe Green 22 D3
Howell 46 E3
Howe of Teuchar 99 G4
Howe Street, Essex 38 B7
Howe Street, Essex 22 C2
Howe, The, Cumbr 61 H1
Howe, The, I. of M 60 N5
Howey 28 E2
Howgate, Border 74 D1
Howgate, Lothn 79 K4
Howick 75 J2
Howle 43 H5
Howlett End 38 A7
Howmore 110 B3
Hownam 74 D2
Howsham, Humbs 58 E6
Howsham, N. Yks 64 E6
Howton 29 J5
Howtown 67 G5
Howwood 77 J1
Hoxa 112 C5
Hoxne 39 G3
Hoylake 52 D5
Hoyland Nether 55 G3
Hoyland Swaine 55 F3
Hubberholme 62 C2
Hubbert's Bridge 47 F3
Huby, N. Yks 63 F5
Huby, N. Yks 63 J3
Hucclecote 30 C6
Hucking 14 E3
Hucknall 45 G3
Huddersfield 54 E2
Huddington 33 J5
Hudswell 68 D6
Huggate 65 F7
Hughenden Valley 20 C4
Hughley 43 G8
Hugh Town 2 P2
Huish, Wilts 18 C4
Huish, Devon 7 F4
Huish Champflower 8 B2
Huish Episcopi 9 F2
Hulcott 20 C2
Hulland 44 D3
Hulland Ward 44 D3
Hullavington 17 K2
Hullbridge 22 E4
Hulme End 54 E8
Hulme Walfield 53 K7
Hulver Street 39 K2
Humber Bridge 58 E4
Humber Court 29 K2
Humberston 59 H6
Humbie 80 C3
Humbleton, Humbs 59 G3
Humbleton, Northum 81 H7
Hume 80 F5
Humshaugh 74 F6
Huna 107 J2
Huncoat 62 A6
Huncote 45 G8
Hundalee 74 C2
Hunderthwaite 68 B4
Hundleby 57 H7
Hundleton 26 D7
Hundon 38 C6
Hundred Acres 11 G3
Hundred End 61 H7
Hundred House 28 F2
Hundred, The 32 E4
Hungarton 45 H7
Hungerford 18 E4
Hungerford Newtown 18 E3
Hunmanby 65 H5
Hunningham 34 E4
Hunsdon 21 J2
Hunsingore 63 H4
Hunsonby 67 H3
Hunspow 107 H2
Hunstanton 48 B3
Hunstanworth 68 B2
Hunston, Suff 38 E4
Hunston, W. Susx 11 K4
Hunstrete 17 H4
Hunt End 34 B4
Hunter's Quay 84 D7
Huntingdon 37 F3
Huntingfield 39 J3
Huntington, Staffs 44 A6
Huntington, H. & W 29 G2
Huntington, Lothn 87 H7
Huntington, N. Yks 63 K4
Huntingtower 86 D2
Huntley 30 B6
Huntly 98 E4
Hunton, N. Yks 68 D7
Hunton, Kent 13 L2
Hunt's Cross 52 F5
Huntsham 7 L3
Huntspill 16 E6
Huntworth 8 E1
Hunwick 68 D3
Hunworth 48 F4
Hurdsfield 53 L6
Hurley, Berks 20 C5
Hurley, Warw 44 D8
Hurlford 77 K4
Hurliness 112 B5
Hurn 10 C5
Hursley 10 F2
Hurst, Berks 20 B6
Hurst, N. Yks 68 C6
Hurst, G. Man 53 L3
Hurstbourne Priors 18 F6
Hurstbourne Tarrant 18 E5
Hurst Green, Surrey 13 G1
Hurst Green, Lancs 61 K6
Hurst Green, E. Susx 13 L4
Hurstpierpoint 13 F5
Hurworth-on-Tees 68 F5
Hury 68 B5
Husbands Bosworth 35 H2
Husborne Crawley 36 C7
Hushinish 108 A4
Husthwaite 63 J2
Hutcherleigh 5 H6
Huthwaite 55 H8
Huttoft 57 K6
Hutton, Essex 22 C4
Hutton, Avon 16 E5
Hutton, N. Yks 64 E6
Hutton, Cumbr 67 G4
Hutton, Border 81 H4
Hutton, Lancs 61 H7
Hutton Bonville 68 F6
Hutton Buscel 65 G4
Hutton Conyers 63 G2
Hutton Cranswick 65 H7
Hutton End 67 G3
Hutton Henry 69 G3
Hutton-le-Hole 64 E3
Hutton Magna 68 D5
Hutton Roof, Cumbr 67 F3
Hutton Roof, Cumbr 61 J2
Hutton Rudby 69 G6
Hutton Sessay 63 H2
Hutton Wandesley 63 J4
Huxley 53 G7
Huyton-with-Roby 52 F4
Hycemoor 66 C8
Hyde, Hants 10 C3
Hyde, Glos 30 C7
Hyde, G. Man 53 L4
Hyde Heath 20 D3
Hydestile 12 C2
Hynish 88 A5
Hyssington 42 E8
Hythe, Hants 10 F4
Hythe, Kent 15 H5
Hythe End 20 E6
Hythie 99 K3

I

Ibberton 9 J4
Ible 55 F8
Ibsley 10 C4
Ibstock 44 F6
Ibstone 20 B4
Ibthorpe 18 E5
Ibworth 19 G5
Ickburgh 48 D8
Ickenham 20 E5
Ickford 20 A3
Ickham 15 J3
Ickleford 36 E7
Icklesham 14 E7
Ickleton 37 H6
Icklingham 38 C3
Ickwell Green 36 E6
Icomb 31 G5
Idbury 31 G5
Iddesleigh 7 F5
Ide 5 J2
Ideford 5 J4
Ide Hill 13 H1
Iden 14 F6
Iden Green 14 E5
Idlicote 34 D6
Idmiston 10 D1
Idridgehay 44 D3
Idrigill 110 E1
Idstone 18 D2
Ifield 12 F3
Ifold 12 D3
Iford 13 H6
Ifton Heath 42 E4
Ightfield 43 G4
Ightham 13 J1
Iken 39 K5
Ilam 44 C2
Ilchester 9 G2
Ilderton 81 J7
Ilford 21 J5
Ilfracombe 6 F1
Ilkeston 45 F3
Ilketshall St Andrew 39 J2
Ilketshall St Lawrence 39 J2
Ilketshall St Margaret 39 J2
Ilkley 62 E5
Illey 34 A2
Illingworth 62 D7
Illogan 2 E5
Illston on the Hill 45 J8
Ilmer 20 B3
Ilmington 34 D6
Ilminster 8 E3
Ilsington 5 H4
Ilston 24 D5
Ilton, N. Yks 62 E2
Ilton, Somer 8 E3
Immingham 59 F5
Impington 37 H4
Ince 52 F6
Ince Blundell 52 E3
Ince-in-Makerfield 53 G3
Inchbare 93 H5
Inchberry 98 C3
Incheril 100 F5
Inchina 100 E2
Inchinnan 85 G8
Inchlaggan 90 C1
Inchmore 96 B4
Inchnadamph 104 D6
Inchture 87 F2
Indian Queens 3 H4
Ingatestone 22 C4
Ingbirchworth 55 F3
Ingestre 44 A5
Ingham, Suff 38 D3
Ingham, Lincs 56 D5
Ingham, Norf 49 J5
Ingleby Arncliffe 69 G6
Ingleby Greenhow 69 H6
Inglesbatch 17 J4
Inglesham 31 G8
Ingleton, Durham 68 D4
Ingleton, N. Yks 61 K2
Inglewhite 61 J5
Ingliston 86 E7
Ingoe 75 G6
Ingoldisthorpe 48 B4
Ingoldmells 57 K7
Ingoldsby 46 D4
Ingram 75 G2
Ingrave 22 C4
Ings 67 G7
Ingst 17 G2
Ingworth 49 G5
Inkberrow 34 B5
Inkhorn 99 J5
Inkpen 18 E4
Inkstack 107 H2
Innellan 84 D7
Innerleithen 80 B6
Innerleven 87 G4
Innermessan 70 B5
Innerwick, Lothn 80 F2
Innerwick, Tays 91 G7
Innsworth 30 C5
Insch 98 F6
Insh 91 K1
Inskip 61 H6
Instow 6 E2
Inver, Grampn 92 D2
Inver, Highld 103 G5
Inverailort 94 C7
Inveralligin 100 D6
Inverallochy 99 K2
Inveramsay 99 G6
Inveran 102 D4
Inveraray 84 C4
Inverarish 111 F3
Inverarity 92 F7
Inverarnan 84 F3
Inverasdale 100 D3
Inverbervie 93 K4
Inverbrough 97 G5
Inverchoran 96 A3
Invercreran 90 B7
Inverdruie 97 H7
Inverebrie 99 J5
Inveresk 87 G7
Inverey 92 B3
Inverfarigaig 96 D6
Invergarry 90 E1
Invergordon 97 F2
Invergowrie 87 G1
Inverguhomery 99 K4
Inverguseran 94 C5
Inverharroch 98 C5
Inverie 94 C6
Inverinate 94 E3
Inverkeilor 93 H7
Inverkeithing 86 E6
Inverkeithny 98 F4
Inverkip 84 E7
Inverkirkaig 104 B7
Invermoidart 89 H1
Invermoriston 96 C7
Invernaver 105 J2
Inverness 96 E4
Inverroy 90 D3
Invershin 102 D4
Inverugie 99 L4
Inveruglas 84 F4
Inverurie 99 G6
Invervar 91 H7
Inwardleigh 5 F2
Inworth 22 E2
Iping 11 K2
Ipplepen 5 J5
Ipsden 20 A5
Ipstones 44 B2
Ipswich 39 G6
Irby 52 D5
Irby in the Marsh 57 J7
Irby upon Humber 59 F6
Irchester 36 C4
Ireby, Cumbr 66 E3
Ireby, Lancs 61 K2
Ireland, Orkney 112 B4
Ireland, Shetld 113 C5
Ireleth 60 F2
Ireshopeburn 68 A3
Irlam 53 J4
Irnham 46 D5
Iron Acton 17 H2
Iron-Bridge 43 H7
Iron Cross 34 B5
Ironmacannie 72 B6
Ironside 99 H3
Ironville 44 F2
Irstead 49 J5
Irthington 73 K7
Irthlingborough 36 C3
Irton 65 H4
Irvine 77 J4
Isauld 106 E3
Isbister, Orkney 112 B3
Isbister, Orkney 112 B4
Isbister, Shetld 113 D3
Isfield 13 H5
Isham 6 B3
Isle Abbotts 8 E2
Isle Brewers 8 E2
Isleham 38 B3
Isle of Whithrn 71 F8
Isleornsay or Eilean Iarmain 94 B4
Islesburgh 113 C3
Isleworth 21 F6
Isley Watton 44 F5
Islington 21 H5
Islip, Northnts 36 C3
Islip, Oxon 31 K6
Islivig 108 A3
Istead Rise 14 C2
Isycoed 42 F2
Itchen Abbas 11 G1
Itchen Stoke 11 G1
Itchingfield 12 E4
Itchington 17 H2
Itteringham 49 G4
Itton 5 G2
Itton Common 29 J8
Ivegill 67 G2
Ivelet 68 B7
Iver 20 E5
Iver Heath 20 E5
Iveston 68 D1
Ivinghoe 20 D2
Ivinghoe Aston 20 D2
Ivington 29 J2
Ivington Green 29 J2
Ivybridge 5 G6
Ivychurch 15 G6
Ivy Hatch 13 J1
Iwade 14 E2
Iwerne Courtney or Shroton 9 K3
Iwerne Minster 9 K3
Ixworth 38 E3
Ixworth Thorpe 38 E3

J

Jack Hill 62 F4
Jackstown 99 G5
Jackton 78 C4
Jacobstow 4 B2
Jacobstowe 7 F5
Jameston 26 E8
Jamestown, Highld 96 C3
Jamestown, Strath 85 F6
Jamestown, D. & G 73 H4
Jarrow 75 K7
Jawcraig 86 B7
Jayes Park 12 E2
Jaywick 23 H2
Jedburgh 74 C1
Jeffreyston 26 E7
Jemimaville 97 F2
Jevington 13 J6
Johnby 67 G3
John o' Groats 107 J2
Johnshaven 93 J5
Johnston 26 D6
Johnstone 77 K1
Johnstonebridge 73 G4
Jordans 20 D4
Jordanston 26 D4
Jump 55 G3
Juniper Green 79 J3
Jurby East 60 Q2
Jurby West 60 Q2

K

Kaber 67 K5
Kaimes 79 K3
Kalnakill 100 B6
Kames, Strath 84 B7
Kames, Strath 78 D7
Kea 3 G5
Keadby 58 C5
Keal 57 H7
Keal Cotes 57 H7
Kearsley 53 J3
Kearstwick 61 K1
Kearton 68 B7
Keasden 61 L3
Keddington 57 H5
Kedington 38 C6
Kedleston 44 E3
Keelby 59 F6
Keele 43 K3
Keeley Green 36 D6
Keeston 26 D6
Keevil 17 L5
Kegworth 45 F5
Kehelland 2 E5
Keig 98 F7
Keighley 62 D5
Keilarsbrae 86 B5
Keilhill 99 G3
Keillmore 83 F3
Keills 82 D5
Keils 82 E5
Keinton Mandeville 9 G1
Keir Mill 72 D4
Keisby 46 D5
Keiss 107 J3
Keith 98 D3
Keithock 93 H5
Kelbrook 62 C5
Kelby 46 D3
Keld, N. Yks 68 A6
Keld, Cumbr 67 H5
Keldholme 64 E4
Kelfield 63 J6
Kelham 45 J2
Kellan 89 G4
Kellas, Grampn 98 A3
Kellas, Tays 87 H1
Kellaton 5 J8
Kelleth 67 J6
Kelling 49 F3
Kellington 63 J7
Kelloe 69 F3
Kelly 4 D3
Kelly Bray 4 D4
Kelmarsh 35 J3
Kelmscot 31 G8
Kelsale 39 J4
Kelsall 53 G7
Kelshall 37 G7
Kelso 80 F6
Kelston 17 J4
Keltneyburn 91 J7
Kelton 68 B4
Kelton Hill or Rhonehouse 72 C8
Kelty 86 E5
Kelvedon 22 E2
Kelvedon Hatch 21 K4
Kelynack 2 B7
Kemacott 7 G1
Kemback 87 H3
Kemberton 43 J7
Kemble 30 D8
Kemerton 33 J7
Kemeys Commander 29 H7
Kemnay 99 G7
Kempley 30 A5
Kempsey 33 H6
Kempsford 31 F8
Kempston 36 D6
Kempston Hardwick 36 D6
Kempton 32 C2
Kemp Town 13 G6
Kemsing 21 K8
Kenardington 15 F5
Kenchester 29 J3
Kencot 31 G7
Kendal 67 H7
Kenfig 25 G6
Kenfig Hill 25 G6
Kenilworth 34 D3
Kenley, Shrops 43 G7
Kenley, G. Lon 21 H8
Kenmore, Highld 100 C6
Kenmore, Tays 91 J7
Kenn, Avon 16 F4
Kenn, Devon 5 K3
Kennacraig 76 D1
Kennerleigh 7 J5
Kennet 86 C5
Kennethmont 98 E6
Kennett 38 C4
Kennford 5 K3
Kenninghall 38 F2
Kennington, Kent 15 G4
Kennington, Oxon 31 K7
Kennoway 87 G4
Kenny Hill 38 B3
Kennythorpe 64 E6
Kenovdy 88 A4
Kensaleyre 111 F2
Kensington 21 G6
Kensworth 20 E2
Kentchurch 29 J5
Kentford 38 C4
Kentisbeare 8 B4
Kentisbury 7 G1
Kentmere 67 G6
Kenton, G. Lon 21 F5
Kenton, Suff 39 G4
Kenton, Devon 5 K3
Kentra 89 H2
Kents Bank 61 G2
Kent's Green 30 B5
Kent's Oak 10 E2
Kenwick 42 F4
Kenwyn 3 G5
Kenyon 53 H4
Keoldale 104 E2
Keose 108 C3
Keppoch 94 D3
Kepwick 69 G7
Keresley 34 E2
Kerne Bridge 29 K6
Kerridge 53 L6
Kerris 2 C7
Kerry 32 A1
Kerrycroy 77 G1
Kerrysdale 100 D4
Kerry's Gate 29 H4
Kersall 56 B7
Kersey 38 F6
Kershader 108 C3
Kershopefoot 73 K5
Kersoe 34 A6
Kerswell 8 B4
Kerswell Green 33 H6
Kesgrave 39 H6
Kessingland 39 L2
Kestle Mill 3 G4
Keston 21 J7
Keswick, Cumbr 66 E4
Keswick, Norf 49 H7
Keswick, Norf 49 J4
Kettering 36 B3
Ketteringham 49 G7
Kettins 92 D8
Kettlebaston 38 E5
Kettlebrook 44 D7
Kettleburgh 39 H4
Kettleness 69 L5
Kettleshulme 54 C6
Kettlesing Bottom 63 F4
Kettlestone 48 E4
Kettlethorpe 56 C6
Kettletoft 112 D3
Kettlewell 62 C2
Ketton 46 C7
Kew 21 F6
Kewstoke 16 E4
Kexbrough 55 G3
Kexby, Lincs 56 C5
Kexby, N. Yks 64 E7
Key Green 53 K7
Keyham 45 H7
Keyhaven 10 E5
Keyingham 59 G4
Keymer 13 G5
Keynsham 17 H4
Keysoe 36 D4
Keysoe Row 36 D4
Keyston 36 D3
Keyworth 45 H4
Kibblesworth 68 E1
Kibworth Beauchamp 45 H8
Kibworth Harcourt 45 H8
Kidbrooke 21 J6
Kiddemore Green 43 K7
Kidderminster 33 H3
Kiddington 31 J5
Kidlington 31 J6
Kidmore End 20 A6
Kidsgrove 43 K2
Kidstones 62 C1
Kidwelly 27 J7
Kielder 74 C4
Kiells 82 D5
Kilbarchan 77 K1
Kilbeg 111 G4
Kilberry 76 C1
Kilbirnie 77 J2
Kilbride, Strath 84 A2
Kilbride, W. Isles 110 B4
Kilbride, Highld 111 F3
Kilburn, Derby 44 E3
Kilburn, N. Yks 63 J2
Kilby 45 H8
Kilchattan 82 C2
Kilchattan Bay 77 G2
Kilchenzie 76 B5
Kilchiaran 82 B5
Kilchoan 89 F2
Kilchoman 82 B5
Kilchrenan 84 C2
Kilconquhar 87 H4
Kilcot 30 A5
Kilcoy 96 D3
Kilcreggon 84 E6
Kildale 69 J6
Kildalloig 76 C6
Kildary 103 F6
Kildonan 76 F5
Kildonnan 111 F5
Kildrummy 98 D7
Kildwick 62 D5
Kilfinan 84 B7
Kilfinnan 90 D2
Kilgetty 26 F7
Kilgwrrwg Common 29 J8
Kilham, Northum 81 G6
Kilham, Humbs 65 H6
Kilkenneth 88 A4
Kilkhampton 6 C4
Killamarsh 55 H5
Killay 24 E5
Killchianaig 83 F3
Killean 76 B3
Killearn 85 H6
Killen 96 E3
Killerby 68 D5
Killichonan 91 G6
Killiechonate 90 D3
Killiechronan 89 G4
Killiecrankie 91 L5
Killiemor 89 F5
Killilan 94 E2
Killimster 107 J4
Killin 85 H1
Killinghall 63 F4
Killingholme 59 F5
Killington 61 K1
Killochyett 80 C5
Killundine 89 G4
Kilmacolm 77 J1
Kilmahumaig 83 G2
Kilmaluag 108 D6
Kilmany 87 G2
ilmarie 111 F4
Kilmarnock 77 K4
Kilmartin 83 H2
Kilmaurs 77 K3
Kilmelford 89 K7
Kilmeny 82 C5
Kilmersdon 17 H5
Kilmeston 11 G2
Kilmichael Glassary 84 A5
Kilmichael of Inverlussa 83 G3
Kilmington, Devon 8 D5
Kilmington, Wilts 9 J1
Kilmonivaig 90 C3
Kilmorack 96 C4
Kilmore 111 G4
Kilmory, Strath 76 E5
Kilmory, Highld 89 G1
Kilmory, Strath 83 G4
Kilmuir, Highld 110 E1
Kilmuir, Highld 110 E2
Kilmuir, Highld 96 E4
Kilmuir, Highld 103 F6
Kilnave 82 B4
Kilncadzow 79 F5
Kilndown 13 L3
Kilnhurst 55 H4
Kilninian 88 F4
Kilninver 89 K6
Kiln Pit Hill 68 C1
Kilnsea 59 J5
Kilnsey 62 C3
Kilnwick 65 G8
Kiloran 82 C2
Kilpatrick 76 E5
Kilpeck 29 J4
Kilpheder 110 B4
Kilphedir 103 H2
Kilpin 58 B4
Kilrenny 87 J4
Kilsby 35 G3
Kilspindie 87 F2
Kilsyth 85 K7
Kiltarlilty 96 D4
Kilton 16 C6
Kilvaxter 110 E1
Kilve 16 C6
Kilvington 45 K3
Kilwinning 77 J3
Kimberley, Norf 48 F7
Kimberley, Notts 45 G3
Kimble 20 C3
Kimblesworth 68 E2
Kimble Wick 20 C3
Kimbolton, Cambs 36 D4
Kimbolton, H. & W 32 E4
Kimcote 35 G2
Kimmeridge 9 L7
Kimmerston 81 H6
Kimpton, Wilts 18 D6
Kimpton, Herts 21 F2
Kinbrace 106 D6
Kinbuck 85 K4
Kincaple 87 H3
Kincardine, Fife 86 C6
Kincardine, Highld 102 E5
Kincardine O'Neil 93 G2
Kincorth 99 J8
Kincraig 97 G8
Kineton, Warw 34 E5
Kineton, Glos 30 E5
Kinfauns 86 E2
Kingarth 77 F2
Kingcoed 29 J7
Kingforth 58 E5
Kingham 31 G5
Kingholm Quay 72 E6
Kinghorn 87 F6
Kinglassie 87 F5
Kingoodie 87 G2
King's Acre 29 J3
Kingsand 4 E6
Kingsbarns 87 J3
Kingsbridge, Devon 5 H7
Kingsbridge, Somer 7 K2
King's Bromley 44 C6
Kingsburgh 111 E2
Kingsbury, Warw 44 D8
Kingsbury, G. Lon 21 F5
Kingsbury Episcopi 9 F2
King's Caple 29 K5
Kingsclere 19 G5
King's Cliffe 46 D8
Kingscote 17 K1
Kingscott 7 F4
King's Coughton 34 B5
Kingscross 76 F5
Kingsdon 9 G2
Kingsdown 15 K4
Kingseat 86 E5
Kingsey 20 B3
Kingsfold 12 E3
Kingsford 33 H2
Kingshall Street 38 E4
King's Heath 34 B2
Kingskerswell 5 J5
Kingskettle 87 G4
Kingsland 32 D4
Kings Langley 20 E3
Kingsley, Staffs 44 B3
Kingsley, Ches 53 G6
Kingsley, Hants 19 J7
Kingsley Green 11 K1
King's Lynn 48 B6
King's Meaburn 67 J4
Kingsmuir, Tays 93 F7
Kings Muir, Border 79 K6
Kingsnorth, Kent 14 E1
Kingsnorth, Kent 15 G5
King's Norton, W. Mids 34 B3
King's Norton, Leic 45 H7
King's Nympton 7 G4
King's Pyon 29 J2
Kings Ripton 37 F3
King's Somborne 10 E1
King's Stag 9 J3
King's Stanley 30 C7
Kings Sutton 35 F7
Kingstanding 44 B8
Kingsteignton 5 J4
King Sterndale 54 D6
Kingsthorne 29 J4
Kingsthorpe 35 J4
Kingston, Grampn 98 C2
Kingston, Hants 10 C4
Kingston, I. of W 11 F6
Kingston, Cambs 37 G5
Kingston, Devon 5 G7
Kingston, Kent 15 H3
Kingston, Dorset 9 J4
Kingston, Lothn 87 J6
Kingston Bagpuize 31 J8
Kingston Blount 20 B4
Kingston by Sea 12 F6
Kingston Deverill 9 K1
Kingstone, Staffs 44 B5
Kingstone, Somer 8 E3
Kingstone, H. & W 29 J4
Kingston Lisle 18 E2
Kingston near Lewes 13 G6
Kingston on Soar 45 G5
Kingston Seymour 16 F4
Kingston St Mary 8 D2
Kingston upon Hull 59 E4
Kingston upon Thames 21 F7
Kingstown 73 J8
King's Walden 21 F1
Kingswear 5 J6
Kingswells 99 H8
Kingswinford 33 H2
Kingswood, Bucks 20 A2
Kingswood, Warw 34 C3
Kingswood, Powys 42 D7
Kingswood, Kent 14 E3
Kingswood, Surrey 12 F1
Kingswood, H. & W 29 G2
Kingswood, Avon 17 H3
Kingswood, Glos 17 J1
King's Worty 11 F1
Kington, H. & W 34 A5
Kington, Dorset 10 A7
Kington, Powys 29 H2
Kington Langley 17 L3
Kington Magna 9 J2
Kington St Michael 17 L3
Kingussie 91 J1
Kingweston 9 G1
Kinharrachie 99 J5
Kinknockie 99 K4
Kinlet 33 G2
Kinloch, Highld 94 B4
Kinloch, Tays 92 C7
Kinloch, Highld 104 E5
Kinloch, Highld 111 F5
Kinlochard 85 G4
Kinlochbervie 104 D3
Kinlocheil 90 A4
Kinlochewe 100 F5
Kinloch Hourn 94 E5
Kinloch Laggan 91 G3
Kinlochleven 90 C5
Kinlochmore 90 C5
Kinloch Rannoch 91 H6
Kinlochspelve 89 H6
inloss 97 J2
Kinmel Bay 51 J2
Kinmuck 99 H7
Kinmundy 99 H7
Kinnadie 99 J4
Kinnaird 87 F2
Kinnell 93 H6
Kinnerley 42 E5
Kinnersley, H. & W 29 H3
Kinnersley, H. & W 33 H6
Kinnerton, Powys 32 B4
Kinnerton, Clwyd 52 E7
Kinnesswood 86 E4
Kinninvie 68 C4
Kinoulton 45 H4
Kinross 86 E4
Kinrossie 86 E1
Kinsham 32 C4
Kinsley 55 H2
Kintarvie 108 C4
Kintbury 18 E4
Kintessack 97 J2
Kintillo 86 E3
Kintocher 98 E8
Kintore 99 G7
Kintour 82 D6
Kinuachdrachd 83 G2
Kinveachy 97 H7
Kinver 33 H2
Kippax 63 H6
Kippen 85 J5
Kippford or Scaur 71 K6
Kirbister 112 B4
Kirbuster 112 B3
Kirby Bedon 49 H7
Kirby Bellars 45 J6
Kirby Cane 49 J8
Kirby Cross 23 J1
Kirby Grindalythe 65 G6
Kirby Hill, N. Yks 68 D6
Kirby Hill, N. Yks 63 G3
Kirby Knowle 63 H1
Kirby-le-Soken 23 J1
Kirby Mills 64 E4
Kirby Misperton 64 E5
Kirby Muxloe 45 G7
Kirby Row 39 J1
Kirby Sigston 69 G7
Kirby Underdale 64 F7
Kirby Wiske 63 G1
Kirdford 12 D4
Kirivick 108 B2
Kirk 107 H4
Kirkandrews-on-Eden 73 J8
Kirkbampton 66 F1
Kirkbean 72 E8
Kirk Bramwith 55 K2
Kirkbride 66 E1
Kirkbuddo 93 G7
Kirkburn 65 G7
Kirkburton 54 E2
Kirkby, Lincs 56 E4
Kirkby, Mers 52 F4
Kirkby, N. Yks 69 H6
Kirkby Fleetham 68 E7
Kirkby Green 56 E8
Kirkby in Ashfield 45 G2
Kirkby-in-Furness 60 F1
Kirkby la Thorpe 46 D3
Kirkby Lonsdale 61 K2
Kirkby Malham 62 B3
Kirkby Mallory 45 F7
Kirkby Malzeard 63 F2
Kirkby Mills 64 E4
Kirkbymoorside 64 D4
Kirkby on Bain 57 G7
Kirkby Overblow 63 G5
Kirkby Stephen 67 K6
Kirkby Thore 67 J4
Kirkby Underwood 46 D5
Kirkcaldy 87 F5
Kirkcambeck 74 B7
Kirkcarswell 71 J7
Kirkcolm 70 B5
Kirkconnel 72 C2
Kirkcowan 70 E5
Kirkcudbright 71 H6
Kirk Deighton 63 G4
Kirk Ella 58 E4
Kirkfieldbank 79 F5
Kirkgunzeon 72 D7
Kirkham, N. Yks 64 E6
Kirkham, Lancs 61 H6
Kirkhamgate 63 F7
Kirk Hammerton 63 H4
Kirkharle 75 G5
Kirkheaton, W. Yks 54 E2
Kirkheaton, Northum 75 G6
Kirkhill, Highld 96 D4
Kirkhill, Tays 93 H5
Kirkhope 80 B7
Kirkhouse 80 B6
Kirkibost, W. Isles 108 B3
Kirkibost, Highld 111 F4
Kirkinner 71 F6
Kirkintilloch 85 J7
Kirk Ireton 44 D2
Kirkland, D. & G 72 C2
Kirkland, Cumbr 66 C5
Kirkland, D. & G 72 D4
Kirkland, Cumbr 67 J3
Kirk Langley 44 D4
Kirkleatham 69 H4
Kirklevington 69 G6
Kirkley 39 L1
Kirklington, N. Yks 63 G1
Kirklington, Notts 55 K8
Kirklinton 73 K7
Kirkliston 86 E7
Kirkmaiden 70 C8
Kirk Merrington 68 E3
Kirkmichael, Tays 92 B5
Kirkmichael, Strath 77 J7
Kirk Michael, I. of M 60 Q2
Kirkmond le Mire 57 F4
Kirkmuirhill 78 E5
Kirknewton, Northum 81 H6
Kirknewton, Lothn 79 J3
Kirk of Shotts 79 F3
Kirkoswald, Cumbr 67 H2
Kirkoswald, Strath 77 H7
Kirkpatrick Durham 72 C6
Kirkpatrick-Fleming 73 H6
Kirk Sandall 55 K3
Kirksanton 60 E1
Kirk Smeaton 55 J2
Kirkstile 98 E5
Kirkton, Border 74 B2
Kirkton, Highld 94 D3
Kirkton, D. & G 72 E5
Kirkton, Highld 100 E7
Kirkton, Grampn 99 F3
Kirkton, Highld 103 F4
Kirkton, Grampn 98 F6
Kirkton, Tays 92 F7
Kirkton, Fife 87 G2
Kirkton, Strath 79 G7
Kirkton, Grampn 99 L3
Kirkton Manor 79 K6
Kirkton of Auchterhouse 92 E8
Kirkton of Auchterless 99 G4
Kirkton of Barevan 97 G4
Kirkton of Bourtie 99 H6
Kirkton of Collace 87 E1
Kirkton of Craig 93 J6
Kirkton of ulsalmond 98 F5
Kirkton of Durris 93 J2
Kirkton of Glenbuchat 98 C7
Kirkton of Glenisla 92 D5
Kirkton of Kingoldrum 92 E6
Kirkton of Largo 87 H4
Kirkton of Lethendy 92 C7
Kirkton of Logie Buchan 99 J6
Kirkton of Maryculter 93 K2
Kirkton of Menmuir 93 G5
Kirkton of Monikie 93 G8
Kirkton of Rayne 99 F5
Kirkton of Skene 99 H8
Kirkton of Strathmartine 87 G1
Kirkton of Tough 98 F7
Kirktown 99 K3
Kirktown of Alvah 98 F2
Kirktown of Bourtie 99 H6
Kirktown of Deskford 98 E2
Kirktown of Fetteresso 93 K3
Kirkwall 112 C4
Kirkwhelpington 75 F5
Kirk Yetholm 81 G7
Kirmington 59 F5
Kirmond le Mire 57 F4
Kirn 84 D7
Kirriemuir 92 E6
Kirstead Green 49 H8
Kirtlebridge 73 H6
Kirtling 38 B5
Kirtling Green 38 B5
Kirtlington 31 K6
Kirtomy 106 C3
Kirton, Notts 56 A7
Kirton, Lincs 47 G4
Kirton, Suff 39 H7
Kirton End 47 F3
Kirton Holme 47 F3
Kirton in Lindsey 58 D7
Kislingbury 35 H5
Kites Hardwick 35 F4
Kittybrewster 99 J8
Kitwood 11 H1
Kiveton Park 55 H5
Knaith 56 C5
Knaphill 20 D8
Knapp 8 E2
Knapton, N. Yks 65 F5
Knapton, Norf 49 J4
Knapton, N. Yks 63 J4
Knapwell 37 G4
Knaresborough 63 G4
Knarsdale 67 J1
Knaven 99 H4
Knayton 63 H1
Knebworth 21 G1
Kneep 108 B3
Kneesall 55 K7
Kneesworth 37 G6
Kneeton 45 J3
Knelston 24 C6
Knightacott 7 G2
Knightcote 34 E5
Knighton, Powys 32 B3
Knighton, Devon 5 F7
Knighton, Leic 45 H7
Knighton, Staffs 43 J5
Knighton, Staffs 53 J7
Knightwick 33 G5
Knill 29 G1
Knipton 46 B4
Knitsley 68 D2
Kniveton 44 D2
Knochenkelly 76 F5
Knock, W. Isles 108 D3
Knock, Grampn 98 E3
Knock, Strath 89 G5
Knock, Cumbr 67 J4
Knockally 107 G7
Knockan 104 D7
Knockandhu 98 B6
Knockando 98 A4
Knockbain 96 E3
Knockbrex 71 G7
Knock Castle 77 G1
Knockdee 107 G3
Knockenkelly 76 F5
Knockentiber 77 J4
Knockholt 21 J8
Knockholt Pound 21 J8
Knockin 42 E5
Knocknaha 76 B6
Knockrome 82 E4
Knocksharry 60 P3
Knodishall 39 K4
Knolls Green 53 K6
Knolton 42 E4
Knook 18 A6
Knossington 45 K7
Knott End-on-Sea 61 G5
Knotting 36 D4
Knottingley 63 H7
Knotty Green 20 D4
Knowbury 32 E3
Knowehead 72 B4
Knowesgate 75 F5
Knoweside 77 H6
Knowes of Elrick 98 F3
Knowetownhead 74 B2
Knowle, W. Mids 34 C3
Knowle, Devon 6 E2
Knowle, Shrops 3 E3
Knowle, Avon 17 H3
Knowle, Devon 7 H5
Knowle Green 61 K6
Knowl Hill 20 C6
Knowlton 15 J3
Knowsley 52 F4
Knowsley Hall 52 F4
Knowstone 7 J3
Knucklas 32 B3
Knutsford 53 J6
Knypersley 43 K2
Kuggar 2 F8
Kyleakin 94 C3
Kyle of Lochalsh 94 C3
Kylerhea 94 C3
Kylesmorar 94 D6
Kyles Scalpay 108 C5
Kyles Stockinish 108 B5
Kylestrome 104 D5
Kyloe 81 J5
Kynnersley 43 H6
Kyre Park 33 F4

L

Labost 108 C2
Laceby 59 G6
Lacey Green 20 C3
Lach Dennis 53 J6
Lache 52 E7
Lackalee 108 B5
Lackford 38 C3
Lacock 17 L4
Ladbroke 34 F5
Laddingford 13 K2
Lade Bank 47 G2
Ladock 3 G4
Ladybank 87 G4
Ladykirk 81 G5
Ladysford 99 J2
Lagavulin 82 D7
Lagg, Strath 82 E4
Lagg, Strath 76 E5
Laggan, Highld 90 D2
Laggan, Highld 91 H2
Lagganulva 89 F4
Laide 100 E2
Laindon 22 C5
Lair 100 F7
Lairg 102 D3
Lairgmore 96 D5
Lake 18 C7
Lakenham 49 H7
Lakenheath 38 C2
Lakesend 47 J8
Lakeside 61 G1
Laleham 20 E7
Laleston 25 G7
Lamarsh 38 D7
Lamas 49 H5
Lambden 80 F5
Lamberhurst 13 K3
Lamberton 81 H4
Lambeth 21 H6
Lambfell Moar 60 P3
Lambley, Northum 74 C8
Lambley, Notts 45 H3
Lambourn 18 E3
Lambourne End 21 J4
Lambs Green 12 F3
Lambston 26 D6
Lamerton 4 E4
Lamesley 75 J8
Lamington, Highld 103 F6
Lamington, Strath 79 G6
Lamlash 76 F4
Lamonby 67 G3
Lamorna 2 C7
Lamorran 3 G5
Lampeter 27 K3
Lampeter Velfry 26 F6
Lamphey 26 E7
Lamplugh 66 C4
Lamport 35 J3
Lamyatt 9 H1
Lana 4 D2
Lanark 79 F5
Lancaster 61 H3
Lanchester 68 D2
Landbeach 37 H4
Landcross 6 E3
Landerberry 99 G8
Landford 10 D3
Landimore 24 C5
Landkey 7 F2
Landore 24 E5
Landrake 4 D5
Landscove 5 H5
Landshipping 26 E6
Landulph 4 E5
Landwade 38 B4
Landywood 44 A7
Laneast 4 C3
Lane End 20 C4
Lane Green 44 A7
Laneham 56 C6
Laneshaw Bridge 62 C5
Langar 45 J4
Langbank 85 F7
Langbar 62 D4
Langcliffe 62 B3
Langdale End 65 G3
Langdon Beck 67 L3
Langdon Hills 22 C5
Langenhoe 23 G2
Langford, Devon 8 B4
Langford, Notts 56 C8
Langford, Essex 22 E3
Langford, Beds 36 E6
Langford, Oxon 31 G7
Langford Budville 8 C2
Langham, Leic 46 B6
Langham, Suff 38 E4
Langham, Norf 48 F3
Langham, Essex 38 F7
Langho 61 L6
Langholm 73 J5
Langleeford 81 H7
Langley, Warw 34 C4
Langley, Kent 14 E3
Langley, Berks 20 E6
Langley, Hants 11 F4
Langley, Herts 21 G1
Langley, Essex 37 H7
Langley, W. Susx 11 K2
Langley, Ches 53 L6
Langley Burrell 18 A3
Langley Marsh 8 B2
Langley Park 68 E2
Langley Street 49 J7
Langney 13 K6
Langold 55 J5
Langore 4 D3
Langport 9 F2
Langrick 47 F3
Langridge 17 J4
Langrigg 66 D2
Langrish 11 J2
Langsett 54 F3
Langshaw 80 D6
Langstone 11 J4
Langthorne 68 E7
Langthorpe 63 G3
Langthwaite 68 C4
Langtoft, Lincs 46 E6
Langtoft, Humbs 65 H6
Langton, Durham 68 D5
Langton, N. Yks 64 E6
Langton, Lincs 57 G7
Langton, Lincs 57 H6
Langton by Wragby 57 F6
Langton Green 13 J3
Langton Herring 9 H6
Langton Matravers 10 B7
Langtree 6 E4
Langwathby 67 H3
Langwith 55 J7
Langwith Junction 55 J7
Langworth 56 E6
Lanivet 3 J3
Lanlivery 3 J4
Lanner 2 F6
Lanreath 3 K4
Lansallos 3 K4
Lanton, Border 80 E7
Lanton, Northum 81 H6
Lapford 7 H5
Laphroaig 82 C7
Lapley 43 K6
Lapworth 34 C3
Larbert 86 B6
Largie 98 F5
Largoward 87 H4
Largs 77 H2
Largybeg 76 F5
Largymore 76 F5
Larkfield 84 E7
Larkhall 78 E4
Larkhill 18 C6
Larling 38 E2
Larrston 74 B4
Lartington 68 C5
Lary 92 E1
Lasham 19 H6
Laskentyre 108 B5
Lassodie 86 E5
Lastingham 64 E3
Latchingdon 22 E3
Latchley 4 E4
Lately Common 53 H4
Lathbury 36 B6
Latheron 107 H6
Latheronwheel 107 G6
Lathones 87 H4
Latimer 20 E4
Latteridge 17 H2
Lattiford 9 H2
Latton 30 E8
Lauchintilly 99 G7
Lauder 80 D5
Laugharne 27 H6
Laughterton 56 C6
Laughton, Lincs 56 C4
Laughton, Leic 35 H2
Laughton, E. Susx 13 J5
Laughton-en-le-Morthen 55 J5
Launcells 6 C5
Launceston 4 D3
Launde Abbey 45 J7
Launton 31 L5
Laurencekirk 93 J4
Laurieston 72 B7

Lavant 11 K4
Lavendon 36 C5
Lavenham 38 E6
Laverhay 73 G4
Laverstock 10 C1
Laverstoke 19 F6
Laverton, Glos 30 E4
Laverton, N. Yks 63 F2
Laverton, Somer 17 J5
Law 79 F4
Lawers 91 H8
Lawford 39 F7
Lawhitton 4 D3
Lawkland 62 A3
Lawley 43 H7
Lawnhead 43 K5
Lawrenny 26 E7
Lawshall 38 D5
Lawton 29 J2
Laxay 108 C3
Laxdale 108 D3
Laxey 60 R3
Laxfield 39 H3
Laxfirth 113 D4
Laxo 113 D3
Laxobigging 113 D3
Laxton, Humbs 58 B4
Laxton, Notts 56 B7
Laxton, Northnts 46 C8
Laycock 62 D5
Layer Breton 22 F2
Layer de la Haye 23 F1
Layham 38 F6
Laytham 58 B3
Lazenby 69 H5
Lazonby 67 H3
Lea, Wilts 18 A2
Lea, H. & W 30 A5
Lea, Shrops 32 C2
Lea, Lincs 56 C5
Lea, Shrops 42 F7
Lea, Derby 55 G8
Leac Eskadale 108 C5
Leachkin 96 E4
Leadburn 79 K4
Leadenham 46 C2
Leaden Roding 22 B2
Leadgate, Durham 68 D1
Leadgate, Cumbr 67 K2
Leadhills 72 D2
Leafield 31 H6
Leagrave 20 E1
Leake Commonside 47 G2
Leake Hurn's End 47 H3
Lealholm 64 E2
Lealt, Highld 111 F1
Lealt, Strath 83 F2
Lea Marston 44 D8
Leamington Hastings 35 F4
Leamington Spa, Royal 34 E4
Leargybreck 82 E4
Learmouth 81 G6
Leasgill 61 H1
Leasingham 46 D3
Leask 99 K5
Leatherhead 12 E1
Leathley 63 F5
Leaton 42 F6
Lea Town 61 H6
Leaveland 15 G3
Leavenheath 38 E7
Leavening 64 E6
Leaves Green 21 J7
Lebberston 65 H4
Lechlade 31 G8
Leckford 10 E1
Leckfurin 105 J3
Leckhampstead, Berks 18 F3
Leckhampstead, Bucks 35 J7
Leckhampton 30 D6
Leckmelm 101 G2
Leckwith 25 K7
Leconfield 58 E2
Ledaig 89 L5
Ledburn 20 D1
Ledbury 33 G7
Ledgemoor 29 J2
Ledicot 32 D4
Ledmore 104 D7
Lednagullin 106 D3
Ledsham, Ches 52 E6
Ledsham, W. Yks 63 H7
Ledston 63 H7
Ledwell 31 J5
Lee, Devon 6 E1
Lee, Hants 10 E3
Lee, Shrops 42 F4
Lee, Strath 88 F6
Lee, Lancs 61 J4
Leebotten 113 D5
Leebotwood 43 F8
Lee Brockhurst 43 G5
Leece 60 F3
Lee Clump 20 D3
Leeds, Kent 14 E3
Leeds, N. Yks 63 G6
Leedstown 2 E6
Leek 44 A2
Leekhampstead 18 F3
Leek Wootton 34 D4
Lee Mill 5 F6
Leeming 68 E8
Leeming Bar 68 E7
Lee Moor 5 F5
Lee-on-the-Solent 11 G4
Lees, G. Man 54 C3
Lees, Derby 44 D4
Leeswood 52 D7
Lee, The 20 D3
Legbourne 57 H5
Legburthwaite 66 F5
Legerwood 80 D5
Legsby 57 F5
Leicester 45 G7
Leicester Forest East 45 G7
Leigh, Wilts 18 B1
Leigh, Shrops 42 E7
Leigh, Surrey 12 F2
Leigh, H. & W 33 G5
Leigh, G. Man 53 H3
Leigh, Dorset 9 H4
Leigh, Kent 13 J2
Leigh Beck 22 E5
Leigh Delamere 17 K3
Leigh Green 14 F5
Leigh-on-Sea 22 E5
Leigh Sinton 33 G5
Leighterton 17 K1
Leigh, The 30 C5
Leighton, Powys 42 D7
Leighton, Shrops 43 H7
Leighton, Somer 17 J6
Leighton Bromswold 36 E3
Leighton Buzzard 36 C8
Leigh upon Mendip 17 H6
Leigh Woods 17 G3
Leinthall Earls 32 D4
Leinthall Starkes 32 D4
Leintwardine 32 D3
Leire 35 G1
Leirinmore 105 F2
Leiston 39 K4
Leith 87 F7
Leitholm 81 F5
Lelant 2 D6
Lelley 59 G3
Lem Hill 33 G3
Lemmington Hall 75 H2
Lempitlaw 81 F6
Lemreway 108 C4
Lendalfoot 70 C3
Lenham 14 E3
Lenham Heath 14 F4
Lenie 96 D6
Lennel 81 G5
Lennoxtown 85 J7
Lenton 46 D4
Lenwade 49 F6
Lenzie 85 J7
Leochel Cushnie 98 E7
Leominster 29 J2
Leonard Stanley 30 C7
Lepe 11 F5
Lephin 110 D2
Lephinmore 84 B5
Leppington 64 E6
Lepton 54 F2
Lerryn 3 K4
Lerwick 113 D4
Lesbury 75 J2
Leslie, Grampn 98 E6
Leslie, Fife 87 F4
Lesmahagow 79 F6
Lesnewth 4 B2
Lessingham 49 J5
Lessonhall 66 E1
Leswalt 70 B5
Letchmore Heath 21 F4
Letchworth 36 F7
Letcombe Bassett 18 E2
Letcombe Regis 18 E2
Letham, Fife 87 G3
Letham, Tays 93 G7
Lethenty 99 H4
Letheringham 39 H5
Letheringsett 48 F4
Lettaford 5 H3
Letterewe 100 E4
Letterfearn 97 G6
Letterfinlay 90 D2
Letters 101 G3
Letterston 26 D5
Lettoch 97 K5
Letton, H. & W 32 C3
Letton, H. & W 29 H3
Letty Green 21 G2
Letwell 55 J5
Leuchars 87 H2
Leurbost 108 C3
Levedale 43 L6
Leven, Humbs 59 F2
Leven, Fife 87 G4
Levens 61 H1
Levenshulme 53 K4
Levenwick 113 D5
Leverburgh 108 B5
Leverington 47 H6
Leverton 47 G3
Levington 39 H7
Levisham 64 F3
Levishie 96 C7
Lew 31 H7
Lewannick 4 C3
Lewdown 4 E3
Lewes 13 H5
Leweston 26 D5
Lewisham 21 H6
Lewiston 96 D6
Lewknor 20 B4
Leworthy 7 G2
Lewtrenchard 4 E3
Ley 3 K3
Leybourne 14 C3
Leyburn 68 D7
Leycett 43 J3
Leyland 61 J7
Leylodge 99 G7
Leys 99 K3
Leysdown on Sea 15 G1
Leysmill 93 H7
Leys of Cossans 92 E7
Leysters 32 E4
Leyton 21 H5
Lezant 4 D4
Lhanbryde 98 B2
Lhen, The 60 Q1
Libanus 25 H2
Libberton 79 G5
Liberton 87 F7
Lichfield 44 C7
Lickey 34 A3
Lickey End 34 A3
Lickfold 12 C4
Liddington 18 D2
Lidgate 38 C5
Lidlington 36 C7
Lieurary 107 F3
Liff 87 G1
Lifton 4 D3
Lighthorne 34 E5
Lightwater 20 D7
Lightwood 43 L3
Lightwood Green 42 E3
Likisto 108 B5
Lilbourne 35 G3
Lilburn Tower 81 J7
Lilleshall 43 J6
Lilley 36 E8
Lilliesleaf 80 D7
Lillingstone Dayrell 35 J7
Lillingstone Lovell 35 J6
Lillington 9 H3
Lilstock 16 C6
Limbrick 53 H2
Limefield 53 K2
Limekilns 86 D6
Limerigg 86 B7
Limington 9 G2
Limmerhaugh 78 D7
Limpenhoe 49 J7
Limpley Stoke 17 J4
Limpsfield 13 G1
Linby 45 G2
Linchmere 11 K1
Lincoln 56 D6
Lincomb 33 H4
Lindale 61 H1
Lindal in Furness 60 F2
Linden 80 C6
Lindfield 13 G4
Lindford 11 K1
Lindores 87 F3
Lindridge 33 F4
Lindsell 38 B8
Lindsey 38 E6
Linford, Hants 10 C4
Linford, Essex 22 C6
Lingague 60 P4
Lingdale 69 J5
Lingen 32 C4
Lingfield 13 G2
Lingwood 49 J7
Liniclate 110 B2
Linicro 110 E1
Linkenholt 18 E5
Linkinhorne 4 D4
Linklater 112 C5
Linksness 112 C4
Linktown 87 F5
Linley 32 C1
Linley Green 33 F5
Linlithgow 86 D7
Linlithgow Bridge 86 C7
Linshader 108 C3
Linshiels 74 E3
Linsidemore 102 D4
Linslade 36 C8
Linstead Parva 39 J3
Linstock 73 K8
Linthwaite 54 E2
Lintlaw 81 G4
Lintmill 98 E2
Linton, H. & W 30 A5
Linton, N. Yks 62 C3
Linton, Derby 44 D6
Linton, Border 80 F7
Linton, Cambs 37 J6
Linton, Kent 13 L2
Linton-on-Ouse 63 H3
Linwood, Hants 10 C4
Linwood, Lincs 57 F5
Linwood, Strath 77 K1
Lionel 108 D1
Liphook 11 K1
Liskeard 3 L3
Liss 11 J2
Lissett 65 J7
Liss Forest 11 J2
Lissington 57 F5
Lisvane 25 K6
Litcham 48 D6
Litchborough 35 H5
Litchfield 19 F5
Litherland 52 E4
Litlington, Cambs 37 G6
Litlington, E. Susx 13 J6
Little Abington 37 J6
Little Addington 36 C3
Little Alne 34 C4
Little Asby 67 J6
Little Aston 44 B7
Little Atherfield 11 F6
Little Ayre 112 B5
Little Ayton 69 H5
Little Baddow 22 D3
Little Badminton 17 K2
Little Bardfield 38 B7
Little Barford 36 E5
Little Barningham 49 G4
Little Barrington 31 G6
Little Barugh 64 E5
Little Bealings 39 H6
Little Bedwyn 18 D4
Little Bentley 39 G8
Little Berkhamsted 21 G3
Little Birch 29 K4
Little Blakenham 39 G6
Littleborough, Notts 56 C5
Littleborough, G. Man 53 L2
Littlebourne 15 J3
Little Bowden 35 J2
Little Bradley 38 B5
Little Brampton 32 C3
Little Brechin 93 G5
Littlebredy 9 G6
Little Brickhill 36 C7
Little Brington 35 H4
Little Bromley 39 F8
Little Budworth 53 G7
Little Burstead 22 C4
Littlebury 37 J7
Littlebury Green 37 H7
Little Bytham 46 D6
Little Carlton 57 J5
Little Casterton 46 D7
Little Cawthorpe 57 H5
Little Chalfont 20 D4
Little Chart 14 F4
Little Chesterford 37 J6
Little Cheverell 18 A5
Little Chishill 37 H7
Little Clacton 23 H2
Little Comberton 34 A6
Little Common 13 L6
Little Compton 31 G4
Little Cowarne 29 L2
Little Coxwell 31 G8
Little Cressingham 48 D7
Little Dalby 45 J6
Littledean 30 A6
Little Dens 99 K4
Little Dewchurch 29 K4
Little Downham 37 J2
Little Dunham 48 D6
Little Dunkeld 92 B7
Little Dunmow 22 C1
Little Easton 22 C1
Little Eaton 44 E3
Little Ellingham 48 F8
Little End 21 K3
Little Eversden 37 G5
Little Fakenham 38 E3
Little Faringdon 31 G7
Little Fenton 63 J6
Littleferry 103 G4
Little Fransham 48 E6
Little Gaddesden 20 D2
Little Gidding 36 E2
Little Glemham 39 J5
Little Gransden 37 F5
Little Gruinard 100 E3
Little Habton 64 E5
Little Hadham 21 J1
Little Hallingbury 21 K2
Littleham, Devon 8 B6
Littleham, Devon 6 E3
Little Hampden 20 C3
Littlehampton 12 D6
Little Harrowden 36 B3
Little Haseley 20 A3
Little Hautbois 49 H5
Little Haven 26 C6
Little Hay 44 C7
Little Haywood 44 B5
Littlehempston 5 J5
Little Hereford 32 E4
Little Horkesley 38 E7
Little Horsted 13 H5
Little Horwood 35 J7
Littlehoughton, Northum 75 J2
Little Houghton, Northnts 35 K5
Little Hucklow 54 E6
Little Hulton 53 J3
Little Kingshill 20 C4
Little Langdale 66 F6
Little Langford 10 B1
Little Laver 21 K3
Little Leigh 53 H6
Little Leighs 22 D2
Little Lever 53 J3
Little London, Hants 20 A8
Little London, Hants 18 E6
Little London, Lincs 47 F5
Little London, E. Susx 13 J5
Little Longstone 54 E6
Little Malvern 33 G6
Little Maplestead 38 D7
Little Marcle 33 F7
Little Marlow 20 C5
Little Massingham 48 D5
Little Melton 49 G7
Littlemill, Highld 97 H3
Little Mill, Gwent 29 H7
Littlemill, Strath 77 K6
Little Milton 20 A3
Little Missenden 20 D4
Littlemore 31 K7
Little Ness 42 F6
Little Newcastle 26 D5
Little Newsham 68 D5
Little Oakley, Northnts 36 B2
Little Oakley, Essex 39 H8
Little Orton 66 F1
Littleover 44 E4
Little Paxton 36 E4
Little Petherick 3 H2
Little Plumstead 49 J6
Littleport 37 J2
Little Raveley 37 F2
Little Ribston 63 G4
Little Rissington 31 F6
Little Ryburgh 48 E5
Little Ryle 75 G2
Little Salkeld 67 H3
Little Sampford 38 B7
Little Saxham 38 C4
Little Scatwell 96 B3
Little Shelford 37 H5
Little Smeaton 55 J2
Little Snoring 48 E4
Little Somerford 18 A2
Little Stainton 69 F4
Little Stanney 52 F6
Little Staughton 36 E4
Little Steeping 57 J7
Littlestone-on-Sea 15 G6
Little Stonham 39 G4
Little Stretton, Shrops 42 F8
Little Stretton, Leic 45 H7
Little Strickland 67 H5
Little Stukeley 36 F3
Little Tew 31 H5
Little Thetford 37 J3
Littlethorpe 63 G3
Little Thurlow 38 B5
Little Thurrock 22 C6
Littleton, Surrey 20 E7
Littleton, Hants 11 F1
Littleton, Somer 9 F1
Littleton, Ches 52 F7
Littleton Drew 17 K2
Littleton-on-Severn 17 G1
Littleton Panell 18 B5
Little Torrington 6 E4
Little Totham 22 E2
Little Town, Cumbr 66 E5
Littletown, Durham 69 F2
Little Wakering 22 F5
Little Walden 37 J6
Little Waldingfield 38 E6
Little Walsingham 48 E4
Little Waltham 22 D2
Little Warley 22 C4
Little Weighton 58 D3
Little Welnetham 38 D4
Little Wenlock 43 H7
Little Whittingham Green 39 H3
Littlewick Green 20 C6
Little Wilbraham 37 J5
Little Witley 33 G4
Little Wittenham 31 K8
Little Wolford 31 G4
Littleworth, Staffs 44 B6
Littleworth, H. W 33 H5
Littleworth, Oxon 31 H8
Little Wymondley 36 F8
Little Wyrley 44 B7
Little Yeldham 38 C7
Litton, Derby 54 E6
Litton, Somer 17 G5
Litton Cheney 9 G5
Liverpool 52 E4
Liversedge 62 E7
Liverton 69 K5
Livingston 79 H3
Livingston Village 79 H3
Lixwm 52 C6
Lizard 2 F8
Llanaber 40 F6
Llanaelhaearn 50 C7
Llanafan 41 F6
Llanafan-fawr 28 D2
Llanallgo 50 D3
Llanarmon 50 D7
Llanarmon Dyffryn Ceiriog 42 C4
Llanarmon-yn-Ial 42 C2
Llanarth, Gwent 29 H6
Llanarth, Dyfed 27 J2
Llanarthney 27 K5
Llanasa 52 C5
Llanbabo 50 C3
Llanbadarn Fawr 40 F5
Llanbadarn Fynydd 41 L6
Llanbadarn-y-garreg 28 F3
Llanbadrig 50 C1
Llanbeder 29 H8
Llanbedr, Gwyn 40 E3
Llanbedr, Powys 28 F3
Llanbedr, Powys 29 G5
Llanbedr-Dyffryn-Clwyd 52 C8
Llanbedrgoch 50 D3
Llanbedrog 50 C8
Llanbedr-y-cennin 51 G4
Llanberis 50 E4
Llanbister 32 A3
Llanblethian 25 H7
Llanboidy 27 F5
Llanbradach 25 K5
Llanbrynmair 41 J3
Llancarfan 25 J7
Llancayo 29 H7
Llancynfelyn 41 F4
Llandaff 25 K7
Llandawke 27 F6
Llanddaniel Fab 50 D3
Llanddarog 27 K6
Llanddeiniol 40 E6
Llanddeiniolen 50 E4
Llandderfel 51 J7
Llanddeusant, Gwyn 50 C2
Llanddeusant, Dyfed 25 F2
Llanddew 25 J1
Llanddewi 24 C6
Llanddewi Brefi 41 G8
Llanddewi'r Cwm 28 E3
Llanddewi Rhydderch 29 H6
Llanddewi Velfrey 26 F6
Llanddewi Ystradenni 32 A4
Llanddoged 51 H4
Llanddona 50 E3
Llanddowror 27 G6
Llanddulas 51 J3
Llandefaelog, Powys 25 J1
Llandefaelog, Dyfed 27 J6
Llandefaelog tre'r-graig 25 K1
Llandefalle 25 K1
Llandegai 50 E3
Llandegfan 50 E3
Llandegla 42 C2
Llandegley 32 A4
Llandegveth 29 H8
Llandeilo 27 L5
Llandeilo Graban 28 E3
Llandeilo'r-Fan 25 G1
Llandeloy 26 C5
Llandenny 29 J7
Llandevenny 16 F2
Llandinabo 29 K5
Llandinam 41 K5
Llandissilio 26 F5
Llandogo 29 K7
Llandough, S. Glam 25 H7
Llandough, S. Glam 25 K7
Llandovery 25 F1
Llandow 25 H7
Llandre, Dyfed 28 A3
Llandre, Dyfed 41 F5
Llandrillo 51 K7
Llandrillo-yn-Rhos 51 H2
Llandrindod Wells 41 K7
Llandrinio 42 D6
Llandudno 51 G2
Llandudno Junction 51 H3
Llandudwen 50 B7
Llandwrog 50 D5
Llandybie 27 L6
Llandyfan 27 L6
Llandyfriog 27 H3
Llandyfrydog 50 D2
Llandygwydd 27 G3
Llandyrnog 51 L4
Llandyssil 42 C8
Llandysul 27 J3
Llanegryn 40 F3
Llanegwad 27 K5
Llaneilian-yn-Rhos 51 H3
Llanelidan 51 L5
Llanelieu 25 K1
Llanellen 29 H6
Llanelli 24 D4
Llanelltyd 41 G2
Llanelly 29 G6
Llanelly Hill 25 L3
Llanelwedd 28 E2
Llanelwy or St Asaph 51 K3
Llanenddwyn 40 E1
Llanengan 40 B1
Llanerchymedd 50 D2
Llanerfyl 41 K3
Llanfachraeth 50 C2
Llanfachreth 41 K2
Llanfaelog 50 C3
Llanfaes 50 F3
Llanfaethlu 50 C2
Llanfaglan 50 D4
Llanfair 50 E8
Llanfair Caereinion 41 L3
Llanfair Clydogau 27 L2
Llanfair Dyffryn Clwyd 42 C2
Llanfairfechan 51 F3
Llanfair Kilgeddin 29 H7
Llanfair-Nant-Gwyn 26 F4
Llanfairpwllgwyngyll 50 E3
Llanfair Talhaiarn 51 J3
Llanfair Waterdine 32 B3
Llanfairyneubwll 50 C3
Llanfairynghornwy 50 C1
Llanfallteg 27 F6
Llanfaredd 28 E2
Llanfarian 40 E6
Llanfechain 42 C5
Llanfechan 28 D2
Llanfechell 50 C1
Llanfendigaid 40 E3
Llanferres 52 C7
Llanfflewyn 50 C2
Llanfihangel-ar-arth 27 J4
Llanfihangel Crucorney 29 H5
Llanfihangel Glyn Myfyr 51 J6
Llanfihangel Nant Bran 25 H1
Llanfihangel-nant-Melan 29 F2
Llanfihangel Rhydithon 32 A4
Llanfihangel Rogiet 17 F2
Llanfihangel Tal-y-llyn 25 K2
Llanfihangel-uwch-Gwili 27 K5
Llanfihangel-y-Creuddyn 41 F6
Llanfihangel-yng-Ngwynfa 41 K2
Llanfihangel yn Nhowyn 50 C3
Llanfihangel-y-pennant Gwyn 50 E6
Llanfihangel-y-pennant Gwyn 41 F3
Llanfihangel Ystum Llewern 29 J6
Llanfihangel-y-traethau 50 E7
Llanfilo 25 K1
Llanfoist 29 G6
Llanfor 51 J7
Llanfrechfa 29 H8
Llanfrothen 51 F6
Llanfrynach 25 J2
Llanfwrog, Gwyn 50 C2
Llanfwrog, Clwyd 52 C8
Llanfyllin 42 C6
Llanfynydd, Clwyd 42 D2
Llanfynydd, Dyfed 27 K5
Llanfyrnach 27 G4
Llangadfan 41 K2
Llangadog 24 F2
Llangadwaladr, Gwyn 50 C4
Llangadwaladr, Clwyd 42 C4
Llangaffo 50 D4
Llangain 27 H6
Llangammarch Wells 28 D3
Llangan 25 H7
Llangarron 29 K5
Llangasty-Talyllin 25 K2
Llangathen 27 K5
Llangattock 25 L3
Llangattock Lingoed 29 H5
Llangattock-Vibon-Avel 29 J6
Llangedwyn 42 C5
Llangefni 50 D3
Llangeinor 25 H6
Llangeitho 27 L2
Llangeler 27 H4
Llangelynin 40 E3
Llangendeirne 27 J6
Llangennech 24 D4
Llangennith 24 C5
Llangenny 29 G6
Llangernyw 51 H4
Llangian 50 B8
Llangiwg 24 F4
Llanglydwen 27 F5
Llangoed 51 F3
Llangoedmor 27 G3
Llangollen 42 C3
Llangolman 26 F5
Llangorse 25 K2
Llangorwen 40 F5
Llangovan 29 J7
Llangower 51 J7
Llangranog 27 H2
Llangristiolus 50 D3
Llangrove 29 K6
Llangua 29 H5
Llangunllo 32 B3
Llangunnor 27 J5
Llangurig 41 J6
Llangwm, Dyfed 26 D7
Llangwm, Clwyd 51 K6
Llangwm, Gwent 29 J7
Llangwnnadl 50 B7
Llangwyfan 52 C7
Llangwyllog 50 D3
Llangwyryfon 40 F7
Llangybi, Gwyn 50 D6
Llangybi, Dyfed 27 L2
Llangybi, Gwent 29 H8
Llangyfelach 24 E5
Llangynhafal 52 C7
Llangynidr 25 K3
Llangynin 27 G6
Llangynog, Dyfed 27 H6
Llangynog, Powys 41 K1
Llangynwyd 25 G6
Llanhamlach 25 J2
Llanharan 25 J6
Llanharry 25 J6
Llanhennock 16 E1
Llanhilleth 25 L4
Llanidloes 41 J5
Llaniestyn 50 B7
Llanigon 29 G3
Llanilar 41 F6
Llanild 25 H6
Llanishen, Gwent 29 J7
Llanishen, S. Glam 25 K6
Llanllechid 51 F4
Llanlleonfel 28 D3
Llanllowell 29 J8
Llanllugan 41 K3
Llanllwch 27 H6
Llanllwchaiarn 32 A1
Llanllwni 27 J3
Llanllyfni 50 D5
Llanmadoc 24 C5
Llanmaes 25 H8
Llanmartin 16 E2
Llanmerewig 42 C8
Llanmihangel 25 H7
Llanmiloe 27 G7
Llanmorlais 24 D5
Llannefydd 51 J3
Llannon 27 K7
Llannor 50 C7
Llanon 40 E7
Llanover 29 H7
Llanpumsaint 27 J5
Llanrhaeadr 51 K4
Llanrhaeadr-ym-Mochnant 42 C5
Llanrhian 26 C4
Llanrhidian 24 C5
Llanrhos 51 G2
Llanrhyddlad 50 C2
Llanrhystud 40 E7
Llanrothal 29 J6
Llanrug 50 E4
Llanrwst 51 H4
Llansadurnen 27 G6
Llansadwrn, Dyfed 24 E1
Llansadwrn, Gwyn 50 E3
Llansaint 27 H7
Llansamlet 24 E5
Llansannan 51 J4
Llansannor 25 H7
Llansantffraed, Dyfed 40 E7
Llansantffraed, Powys 25 K2
Llansantffraed-Cwmdeuddwr 41 J7
Llansantffraed-in-Elwel 28 E2
Llansantffraid Glan Conwy 51 H3
Llansantffraid-ym-Mechain 42 C5
Llansawel 27 L4
Llansilin 42 D5
Llansoy 29 J7
Llanspyddid 25 J2
Llanstadwell 26 D7
Llanstephan, Powys 28 F3
Llanstephan, Dyfed 27 H6
Llanthony 29 G5
Llantilio Crossenny 29 H6
Llantilio Pertholey 29 H6
Llantrisant, Gwyn 50 C2
Llantrisant, Gwent 29 H8
Llantrisant, M. Glam 25 J6
Llantrithyd 25 J7
Llantwit Fardre 25 J6
Llantwit Major 25 H8
Llanuwchllyn 51 H7
Llanvaches 17 F1
Llanvair Discoed 17 F1
Llanvapley 29 H6
Llanvetherine 29 H6
Llanveynoe 29 H4
Llanvihangel Crucorney 29 H5
Llanvihangel Gobion 29 H7
Llanvihangel-Ystern-Llewern 29 J6
Llanwarne 29 K5
Llanwddyn 41 K2
Llanwenog 27 J3
Llanwern 16 E2
Llanwinio 27 G5
Llanwnda, Dyfed 26 D4
Llanwnda, Gwyn 50 D5
Llanwnnen 27 K3
Llanwnog 41 K4
Llanwrda 24 F1
Llanwrin 41 G3
Llanwrthwl 41 J7
Llanwrtyd 41 H8
Llanwrtyd-Wells 28 C3
Llanwyddelan 41 K3
Llanyblodwel 42 D5
Llanybri 27 H6
Llanybydder 27 K3
Llanycefn 26 E5
Llanychaer 26 D4
Llanycrwys 27 L3
Llanymawddwy 41 J2
Llanymynech, Shrops 42 D5
Llanymynech, Powys 42 D5
Llanynghenedl 50 C2
Llanynys 51 L4
Llan-y-pwll 42 E2
Llanyre 41 K7
Llanystumdwy 50 D7
Llanywern 25 K2
Llawhaden 26 E6
Llawnt 42 D4
Llawr Dref 40 B1
Llawryglyn 41 J4
Llay 42 E2
Llechcynfarwy 50 C2
Llechfaen 25 J2
Llechryd, Dyfed 27 G3
Llechryd, M. Glam 25 K4
Lledrod 41 F6
Llidiad-Nenog 27 K4
Llithfaen 50 C6
Llong 52 D7
Llowes 29 F3
Llwchmynydd 40 A1
Llwn-y-groes 27 K2
Llwydcoed 25 H4
Llwyn 32 B2
Llwyncelyn 27 J2
Llwyndafydd 27 H2
Llwynderw 42 C7
Llwyndyrys 50 C6
Llwyngwril 40 E3
Llwynhendy 24 D5
Llwynmawr 42 D4
Llwynypia 25 H5
Llynclys 42 D5
Llynfaes 50 D3
Llysfaen 51 H3
Llysworney 25 H7
Llys-y-fran 26 E5
Llywel 25 G1
Loanend 81 H4
Loanhead 79 K3
Loans 77 J4
Lochaber 90 C2
Lochailort 94 C7
Lochaline 89 H4
Lochans 70 B6
Locharbriggs 72 E5
Lochawe 84 D2
Lochboisdale 110 B4
Lochbuie 89 H6
Lochcarnan 110 C2
Lochcarron 94 E4
Lochdon 89 J5
Lochearnhead 85 H2
Lochee 87 G1
Lochend, Highld 96 D5
Lochend, Highld 107 H3
Lochepport 110 C1
Locherben 72 E4
Lochfoot 72 D6
Lochgair 84 B5
Lochganvich 108 C3
Lochgarthside 96 D7
Lochgelly 86 E5
Lochgilphead 84 A6
Lochgoilhead 84 D4
Lochhill 98 B2
Lochinver 104 B6
Lochlane 85 K2
Lochluichart 96 B2
Lochmaben 73 F5
Lochmaddy 110 C1
Lochore 86 E5
Lochranza 76 E2
Lochside 93 J5
Lochskipport 110 C3
Lochslin 103 G5
Lochton 93 J2
Lochwinnoch 77 J2
Lochwood 78 D3
Lockengate 3 J3
Lockerbie 73 G5
Lockeridge 18 C4
Lockerley 10 D2
Locking 16 E5
Lockington, Humbs 58 D2
Lockington, Leic 45 F5
Lockleywood 43 H5
Locks Heath 11 G4
Lockton 64 F4
Loddington, Northnts 36 B3
Loddington, Leic 45 J7
Loddiswell 5 H7
Loddon 49 J8
Lode 37 J4
Loders 9 F5
Lodsworth 12 C4
Lofthouse 62 E3
Lofthouse Gate 63 G7
Loftus 69 K5
Logan 72 A1
Loggerheads 43 J4
Loggie 101 G2
Logie, Fife 87 H2
Logie, Tays 93 H5
Logie Coldstone 98 D8
Logie Hill 103 F6
Logie Newton 98 F5
Logie Pert 93 H5
Logierait 92 A6
Login 26 F5
Lolworth 37 G4
Lonbain 100 B6
Londesborough 58 C2
London 21 H6
London (Gatwick) Airport 12 F2
London (Heathrow) Airport 20 E6
London Colney 21 F3
Londonderry 63 G1
Londonthorpe 46 C4
Londubh 100 D3
Lonemore, Highld 100 C4
Lonemore, Highld 103 F5
Long Ashton 17 G3
Long Bennington 46 B3
Longbenton 75 J7
Longborough 31 F5
Long Bredy 9 G5
Longbridge, W. Mids 34 B3
Longbridge, Warw 34 D4
Longbridge Deverill 17 K6
Long Buckby 35 H4
Longburton 9 H3
Long Clawson 45 J5
Longcliffe 44 D2
Long Common 11 G3
Long Compton, Warw 31 G4
Long Compton, Staffs 43 K5
Longcot 18 D1
Long Crendon 20 A3
Long Crichel 10 A3
Longcroft 85 K7
Longden 42 F7
Long Ditton 21 F7
Longdon, Staffs 44 B6
Longdon, H. & W 33 H7
Longdon on Tern 43 H6
Longdown 5 J2
Longdowns 2 F6
Long Drax 58 A4
Long Duckmanton 55 H6
Long Eaton 45 F4
Longfield 14 C2
Longford, Glos 30 C5
Longford, Derby 44 D4
Longford, W. Mids 34 E2
Longford, G. Lon 20 E6
Longford, Shrops 43 H4
Longford, Shrops 43 J6
Longforgan 87 G1
Longformacus 80 E4
Longframlington 75 H3
Long Gill 62 A4
Longham, Dorset 10 B5
Longham, Norf 48 E6
Long Hanborough 31 J6
Long Hermiston 86 E7
Longhirst 75 J5
Longhope, Glos 30 A6
Longhope, Orkney 112 B5
Longhorsley 75 H4
Longhoughton 75 J2
Long Itchington 34 F4
Long Lawford 35 F3
Longley Green 33 G5
Long Load 9 F2
Longmanhill 99 G2
Long Marston, Herts 20 C2
Long Marston, Warw 34 C6
Long Marston, N. Yks 63 J4
Long Marton 67 J4
Long Melford 38 D6
Longmoor Camp 11 J1
Longmorn 98 B3
Long Newnton 17 L1
Longnewton, Border 80 D7
Longnewton, Cleve 69 F5
Longney 30 B6
Longniddry 87 H7
Longnor, Staffs 54 D7
Longnor, Shrops 43 F7
Longparish 18 F6
Long Preston 62 B4
Longridge, Lothn 79 G3
Longridge, Lancs 61 K6
Longriggend 85 L7
Long Riston 59 F2
Longsdon 44 A2
Longside 99 K4
Longslow 43 H4
Longstanton 37 G4
Longstock 10 E1
Longstowe 37 G5
Long Stratton 39 G1
Long Street 35 J6
Long Sutton, Somer 9 F2
Long Sutton, Lincs 47 H5
Long Sutton, Hants 19 J6
Longthorpe 46 E8
Longton, Lancs 61 H7
Longton, Staffs 43 L3
Longtown, H. & W 29 H5
Longtown, Cumbr 73 J7
Longville in the Dale 43 G8
Long Whatton 45 F5
Longwick 20 B3
Long Wittenham 31 K8
Longwitton 75 G4
Longworth 31 H8
Longyester 80 D3
Lonmore 110 E2
Looe 3 L4
Loose 13 L1
Loosley Row 20 C3
Lopcombe Corner 10 D1
Lopen 9 F3
Loppington 43 F5
Lopwell 4 E5
Lorbottle 75 G3
Lorton 66 D4
Loscoe 44 E3
Lossiemouth 98 B1
Lostock Gralam 53 H6
Lostock Junction 53 K3
Lostwithiel 3 K4
Lothbeg 103 H2
Lothersdale 62 C5
Lothmore 103 H2
Loudwater 20 D4
Loughborough 45 G6
Loughor 24 D5
Loughton, Bucks 36 B7
Loughton, Shrops 32 F2
Loughton, Essex 21 J4
Lound, Notts 56 A5
Lound, Lincs 46 D6
Lound, Suff 49 L8
Lount 44 E6
Louth 57 H5
Love Clough 62 B7
Lover 10 D2
Loversall 55 J4
Loves Green 22 C3
Loveston 26 E7
Lovington 9 G1
Low Bentham 61 K3
Low Bradfield 55 F4
Low Bradley 62 D5
Low Braithwaite 67 G2
Low Brunton 74 F6
Low Burnham 58 B6
Lowca 66 C4
Low Catton 64 E7
Low Coniscliffe 68 E5
Low Crosby 73 K8
Lowdham 45 H3
Low Dinsdale 69 F5
Low Dovengill 67 K7
Low Eggborough 63 J7
Low Ellington 62 F1
Lower Aisholt 8 D1
Lower Ashton 5 J3
Lower Assendon 20 B5
Lower Basildon 20 A6
Lower Beeding 12 F4
Lower Benefield 36 C2
Lower Boddington 35 F5
Lower Breinton 29 J4
Lower Bullingham 29 K4
Lower Cam 30 B7
Lower Chapel 25 J1
Lower Chute 18 E5
Lower Darwen 61 K7
Lower Diabaig 100 C5
Lower Down 32 C2
Lower Dunsforth 63 H3
Lower Farringdon 11 J1
Lower Frankton 42 E4
Lower Froyle 19 J6
Lower Gledfield 102 D4
Lower Green 48 E4
Lower Halstow 14 E2
Lower Hardres 15 H3
Lower Heyford 31 J5
Lower Higham 14 D1
Lower Hordley 42 E5
Lower Killeyan 82 B7
Lower Langford 17 F4
Lower Largo 87 H4
Lower Lemington 31 G4
Lower Lye 32 D4
Lower Machen 25 L6
Lower Maes-coed 29 H4
Lower Moor 34 A6
Lower Nazeing 21 H3
Lower Penarth 25 K7
Lower Penn 43 K8
Lower Pennington 10 E5
Lower Peover 53 J6
Lower Quinton 34 C6
Lower Shelton 36 C6
Lower Shiplake 20 B6
Lower Shuckburgh 35 F4
Lower Slaughter 31 F5
Lower Stanton St Quintin 17 L2
Lower Stoke 14 E1
Lower Stondon 36 E7
Lower Sundon 36 D8
Lower Swanwick 11 G4
Lower Swell 31 F5
Lower Thurlton 49 K8
Lower Tysoe 34 E6
Lower Upham 11 G3
Lower Vexford 8 C1
Lower Weare 16 F5
Lower Welson 29 G3
Lower Wield 19 H6
Lower Winchendon or Nether Winchendon 20 B2
Lower Woodend 20 C5
Lower Woodford 10 C1
Lowesby 45 J7
Lowestoft 49 L8
Loweswater 66 D4
Lowford 39 F7
Low Gate 74 F7
Lowgill, Cumbr 67 J7
Lowgill, Lancs 61 K3
Low Ham 9 F2
Low Hesket 67 G2
Low Hesleyhurst 75 G4
Lowick, Northnts 36 C2
Lowick, Cumbr 61 F1
Lowick, Northum 81 J6
Low Leighton 54 D5
Low Mill 69 J7
Low Moor 61 L5
Lownie Moor 93 F7
Lowood 80 D6
Low Row, Cumbr 74 B7
Low Row, N. Yks 68 B7
Lowsonford 34 C4
Lowther 67 H4
Lowthorpe 65 H6
Lowton 53 H4
Lowton Common 53 H4
Low Torry 86 D6
Low Waters 78 E4
Low Worsall 69 F6
Loxbeare 7 K4
Loxhill 12 D3
Loxhore 7 F2
Loxley 34 D5
Loxton 16 E5
Loxwood 12 D3
Lubenham 35 J2
Luccombe 7 J1
Luccombe Village 11 G7
Lucker 81 K6
Luckett 4 D4
Luckington 17 K2
Lucklawhill 87 H2
Luckwell Bridge 7 K2
Lucton 32 D4
Ludag 110 B4
Ludborough 57 G4
Ludchurch 26 F6
Luddenden 62 D7
Luddesdown 14 C2
Luddington 58 C5
Ludford 57 F5
Ludgershall, Bucks 20 A2
Ludgershall, Wilts 18 D5
Ludgvan 2 D6
Ludham 49 J6
Ludlow 32 E3
Ludwell 9 L2
Ludworth 69 F2
Luffincott 4 D2
Lugar 78 C7
Luggate Burn 87 K7
Luggiebank 85 K7
Lugton 77 K2
Lugwardine 29 K3
Luib 111 F3
Lulham 29 J3
Lullington, Derby 44 D6
Lullington, Somer 17 J5
Lulsgate Bottom 17 G4
Lulsley 33 G5
Lulworth Camp 9 K6
Lumb 62 D7
Lumby 63 H6
Lumphanan 98 E8
Lumphinnans 86 E5
Lumsdaine 81 G3
Lumsden 98 D6
Lunan 93 H6
Lunanhead 93 F6
Luncarty 86 D2
Lund, Humbs 58 D2
Lund, N. Yks 63 K6
Lundie 92 D8
Lundin Links 87 H4
Lunna 113 D3
Lunning 113 D3
Lunsford's Cross 13 L5
Lunt 52 E3
Luntley 29 H2
Luppitt 8 C4
Lupton 61 J1
Lurgashall 12 C4
Lusby 57 H7
Luskentyre 108 B5
Luss 84 F5
Lusta 110 E2
Lustleigh 5 H3
Luston 32 D4
Luthermuir 93 H5
Luthrie 87 G3
Luton, Kent 14 D2
Luton, Beds 20 E1
Luton, Devon 5 K4
Lutterworth 35 G2
Lutton, Northnts 36 E2
Lutton, Devon 5 F6
Lutton, Lincs 47 H5
Lutworthy 7 H4
Luxborough 7 K2
Luxulyan 3 J4
Lybster 107 H6
Lydbury North 32 C2
Lydd 15 G6
Lydden 15 J4
Lyddington 46 B8
Lydd-on-Sea 15 G6
Lydeard St Lawrence 8 C1
Lydford, Devon 4 F3
Lydford, Somer 9 G1
Lydgate 62 C7
Lydham 32 C1
Lydiard Millicent 18 B2
Lydiate 52 E3
Lydlinch 9 J3
Lydney 29 L7
Lydstep 26 E8
Lye 33 J2
Lye Green 20 D3
Lyford 31 H8
Lymbridge Green 15 H4
Lyme Regis 8 E5
Lyminge 15 H4
Lymington 10 E5
Lyminster 12 D6
Lymm 53 H5
Lymore 10 D5
Lympne 15 H5
Lympsham 16 E5
Lympstone 5 K3
Lynchat 91 J1
Lyndhurst 10 E4
Lyndon 46 C7
Lyne 20 E7
Lyneal 42 F4
Lyneham, Wilts 18 B3
Lyneham, Oxon 31 G5
Lynemouth 75 J4
Lyne of Gorthleck 96 D6
Lyne of Skene 99 G7
Lyness 112 B5
Lyng, Somer 8 E2
Lyng, Norf 48 F6
Lynmouth 7 H1
Lynsted 14 F2
Lynton 7 H1
Lyon's Gate 9 H4
Lyonshall 29 H2
Lytchett Matravers 10 A5
Lytchett Minster 10 A5
Lyth 107 H3
Lytham 61 G7
Lytham St Anne's 61 G7
Lythe 69 L5
Lythes 112 C5

M

Maaruig 108 C4
Mabe Burnthouse 2 F6
Mabie 72 E6
Mablethorpe 57 K5
Macclesfield 53 L6
Macclesfield Forest 54 C6
Macduff 99 G2
Macharioch 76 C7
Machen 25 L6
Machrihanish 76 B5
Machynlleth 41 G3
Mackworth 44 E4
Macmerry 87 H7
Madderty 86 C2
Maddiston 86 C7
Madehurst 12 C5
Madeley 43 J3
Madeley Heath 43 J3
Madingley 37 G4
Madley 29 J4
Madresfield 33 H6
Madron 2 C6
Maenaddwyn 50 D2
Maenclochog 26 E5
Maendy 25 J7
Maentwrog 51 F6
Maer 43 J4
Maerdy, M. Glam 25 H5
Maerdy, Clwyd 51 K6
Maesbrook 42 E5
Maesbury Marsh 42 E5
Maes-glas 16 E2
Maeshafn 52 D7
Maesllyn 27 H3
Maesmynis 28 E3
Maesteg 25 G5
Maesybont 27 K6
Maesycrugiau 27 J3
Maesycwmmer 25 K5
Magdalen Laver 21 K3
Maggieknockater 98 C4
Magham Down 13 K5
Maghull 52 E3
Magor 17 F2
Maiden Bradley 17 K7
Maidencombe 5 K5
Maidenhead 20 C5
Maiden Law 68 D2
Maiden Newton 9 G5
Maidens 77 G7
Maidenwell 3 K2
Maidford 35 H5
Maids Moreton 35 J7
Maidstone 13 L1
Maidwell 35 J3
Mail 113 D5
Main 96 C5
Mains of Clunas 97 G4
Mains of Drum 93 K2
Mains of Loch 98 C2
Mainstone 32 B2
Maisemore 30 C5
Makerstoun 80 E6
Malash 15 G3
Malborough 5 H8
Maldon 22 E3
Malham 62 C3
Mallaig 94 B6
Malltraeth 50 D4
Mallwyd 41 H2
Malmesbury 18 A2
Malpas, Gwent 16 E1
Malpas, Ches 43 F3
Malpas, Corn 3 G5
Maltby, Cleve 69 G5
Maltby, S. Yks 55 J4
Maltby le Marsh 57 J5
Maltman's Hill 14 F4
Malton 64 E5
Malvern Link 33 G6
Malvern Wells 33 G6
Mamble 33 F3
Manaccan 2 F7
Manafon 42 C7
Manaton 5 H3
Manby 57 H5
Mancetter 44 E8
Manchester 53 K4
Mancot 52 E7
Mandally 90 D1
Manea 37 H2
Manfield 68 E5
Mangersta 108 B3
Mangotsfield 17 H3
Manish 108 B5
Mankinholes 62 C7
Manley 53 G6
Manmoel 25 K4
Mannal 88 A4
Manningford Bohune 18 C5
Manningford Bruce 18 C5
Mannings Heath 12 F4
Mannington 10 B4
Manningtree 39 G7
Mannofield 99 J8
Manorbier 26 E8
Manordeilo 24 E1
Manorowen 26 D4
Mansell Gamage 29 H3
Mansell Lacy 29 J3
Mansfield, Strath 72 B2
Mansfield, Notts 55 J7
Mansfield Woodhouse 55 J7
Manston 9 K3
Manswood 10 A4
Manthorpe 46 D6
Manton, Leic 46 B7
Manton, Wilts 18 C4
Manton, Humbs 58 D6
Manuden 37 H8
Maplebeck 56 B7
Maple Cross 20 E4
Mapledurham 20 A6
Mapledurwell 19 H5
Maplehurst 12 E4
Mapleton 44 C3
Mapperley 44 F3
Mapperton 9 G5
Mappleborough Green 34 B4
Mappleton 59 G2
Mappowder 9 J4
Marazion 2 D6
Marbury 43 G3
March 37 H1
Marcham 31 J8
Marchamley 43 G5
Marchington 44 C5
Marchington Woodlands 44 C5
Marchwiel 42 E3
Marchwood 10 E4
Marcross 25 H8
Marden, Wilts 18 B5
Marden, H. & W 29 K3
Marden, Kent 13 L2
Mardy 29 H6
Marefield 45 J7
Mareham le Fen 57 G7
Mareham on the Hill 57 G7
Maresfield 13 H4
Marfleet 59 F4
Margam 25 F6
Margaret Marsh 9 K3
Margaret Roding 22 B2
Margaretting 22 C3
Margate 15 K1
Margnaheglish 76 F4
Marham 48 C7
Marhamchurch 6 C5
Marholm 46 E7
Marian-glas 50 E2
Mariansleigh 7 H3
Marishader 111 F1
Mark 16 E6
Markbeech 13 H2
Markby 57 J6
Mark Causeway 16 E6
Mark Cross 13 J3
Market Bosworth 44 F7
Market Deeping 46 E7
Market Drayton 43 H4
Market Harborough 35 J2
Markethill 92 D8
Market Lavington 18 B5
Market Overton 46 B6
Market Rasen 57 F5
Market Stainton 57 G6
Market Weighton 58 C2
Market Weston 38 E3
Markfield 45 F6
Markham 25 K4
Markinch 87 F4
Markington 63 F3
Marksbury 17 H4
Marks Tey 22 F1
Markyate 20 E2
Marlborough 18 C4
Marlcliff 34 B5
Marldon 5 J5
Marlesford 39 J5
Marley Green 43 G3
Marlingford 49 G7
Marloes 26 B7
Marlow 20 C5
Marlpit Hill 13 H2
Marnhull 9 J3
Marnoch 98 E3
Marnock 85 K8
Marple 54 C5
Marr 55 J3
Marrick 68 C7
Marros 27 G7
Marsden 54 D2
Marsett 62 C1
Marsh 8 D3
Marshalls Heath 21 F2
Marsham 49 G5
Marshaw 61 J4
Marsh Baldon 31 K8
Marshborough 15 K3
Marshbrook 32 D2
Marshchapel 59 H7
Marshfield, Gwent 16 D2
Marshfield, Avon 17 J3
Marshgate 4 B2
Marsh Gibbon 20 A1
Marsh Green, Devon 8 B5
Marsh Green, Kent 13 H2
Marsh Green, Shrops 43 H6
Marshland St James 47 J7
Marshside, Mers 52 E2
Marshside, Kent 15 J2
Marsh, The 42 E8
Marshwood 8 E5
Marske 68 D6
Marske-by-the-Sea 69 J4
Marston, Wilts 18 A5
Marston, Lincs 46 B3
Marston, Warw 44 D8
Marston, H. & W 29 H2
Marston, Ches 53 H6
Marston, Staffs 43 K6
Marston, Oxon 31 K7
Marston, Staffs 43 L5
Marston Green 34 C2
Marston Magna 9 G2
Marston Meysey 31 F8
Marston Montgomery 44 C4
Marston Moretaine 36 C6
Marston on Dove 44 D5
Marston Stannett 29 K2
Marston St Lawrence 35 G6
Marston Trussell 35 H2
Marstow 29 K6
Marsworth 20 D2
Marten 18 D4
Marthall 53 K6
Martham 49 K6
Martin, Hants 10 B3
Martin, Lincs 57 F8
Martindale 67 G5
Martin Dales 57 F7
Martin Drove End 10 B2
Martinhoe 7 G1
Martin Hussingtree 33 H4
Martinscroft 53 H5
Martinstown 9 H6
Martlesham 39 H6
Martlesham Heath 39 H6
Martletwy 26 E6
Martley 33 G4
Martock 9 F3
Marton, Lincs 56 C5
Marton, Shrops 42 D7
Marton, N. Yks 64 E4
Marton, Warw 34 F4
Marton, N. Yks 63 H3
Marton, Cleve 69 H5
Marton, Ches 53 K7
Marvig 108 D4
Marwick 112 B3
Marwood 7 F2
Marybank 96 C3
Maryburgh 96 D3
Marygold 81 G3
Marykirk 93 H5
Marylebone, G. Man 53 G3
Marylebone, G. Lon 21 G5
Marypark 98 A5
Maryport, Cumbr 66 C3
Maryport, D. & G 70 C8
Marystow 4 E3
Mary Tavy 4 F4
Maryton 93 H6
Marywell, Grampn 93 G2
Marywell, Tays 93 H7
Masham 63 F1
Mashbury 22 C2
Masongill 61 K2
Mastrick 99 J8
Matching 21 K2
Matching Green 21 K2
Matching Tye 21 K2
Matfen 75 G6
Matfield 13 K2
Mathern 17 G1
Mathon 33 G6
Mathry 26 C4
Matlaske 49 G4
Matlock 55 F7
Matlock Bath 55 F8
Matson 30 C6
Matterdale End 67 F4
Mattersey 56 A5
Mattingley 19 J5
Mattishall 48 F6
Mattishall Burgh 48 F6
Mauchline 77 K5
Maud 99 J4
Maugersbury 31 G5
Maughold 60 R2
Maulden 36 D7
Maulds Meaburn 67 J5
Maunby 63 G1
Maund Bryan 29 K2
Mautby 49 K6
Mavesyn Ridware 44 B6
Mavis Enderby 57 H7
Mawbray 66 C2
Mawdesley 53 F2
Mawgan 2 F7
Mawla 2 F5
Mawnan 2 F7
Mawnan Smith 2 F7
Maxey 46 E7
Maxstoke 44 D8
Maxton 80 E6
Maxwellheugh 80 F6
Maybole 77 H7
Mayfield, Lothn 80 B3
Mayfield, Staffs 44 C3
Mayfield, Surrey 12 D1
Mayfield, E. Susx 13 J4
Mayford 12 C1
Mayland 22 F3
Maypole 29 J6
Maypole Green 49 K8
Maywick 113 C5
Meadle 20 C3
Meadowtown 42 E7
Meal Bank 67 H7
Mealista 108 A3
Mealsgate 66 E2
Mearbeck 62 B3
Mearbrook 62 B3
Meare 17 F6
Meare Green 8 E2
Mears Ashby 36 B4
Measham 44 E6
Meathop 61 H1
Meaux 59 E3
Meavag 108 B5
Meavaig 108 B3
Meavy 5 F5
Medbourne 45 K8
Medburn 75 H6
Meddon 6 C4
Meden Vale 55 J7
Medmenham 20 C5
Medstead 11 H1
Meerbrook 54 C7
Meer End 34 D3
Meesden 37 H7
Meeth 7 F5
Meggethead 79 J7
Meidrim 27 G5
Meifod 42 C6
Meigle 92 E7
Meikle Earnock 78 E4
Meikleour 92 C8
Meikle Tarty 99 J6
Meikle Wartle 99 G5
Meinciau 27 J6
Meir 43 L3
Meir Heath 43 L3
Melbost 108 D3
Melbost Borve 108 D2
Melbourn 37 G6
Melbourne, Humbs 58 B2
Melbourne, Derby 44 E5
Melbury Abbas 9 K2
Melbury Bubb 9 G4
Melbury Osmond 9 G4
Melby 113 B4
Melchbourne 36 D4
Melcombe Bingham 9 J4
Meldon, Devon 5 F2
Meldon, Northum 75 H5
Meldreth 37 G6
Melfort 89 K7
Melgarve 91 F2
Meliden 51 K2
Melincourt 25 G4
Melin-y-coed 51 H4
Melin-y-ddol 41 K3
Melin-y-grug 41 K3
Melin-y-Wig 51 K6
Melkinthorpe 67 H4
Melkridge 74 D7
Melksham 17 L4
Melldalloch 84 B7
Melling, Mers 52 E3
Melling, Lancs 61 K2
Mellis 39 G3
Mellon Charles 100 D2
Mellon Udrigle 100 D2
Mellor, G. Man 54 C5
Mellor, Lancs 61 K6
Mellor Brook 61 K6
Mells, Somer 17 J6
Mells, Suff 39 K3
Melmerby, N. Yks 63 G2
Melmerby, N. Yks 62 D1
Melmerby, Cumbr 67 J3
Melplash 9 F5
Melrose 80 D6
Melsonby 68 D6
Meltham 54 D2
Melton 39 H5
Meltonby 64 E7
Melton Constable 48 F4
Melton Mowbray 45 J6
Melton Ross 58 E5
Melvaig 100 C3
Melverley 42 E6
Melvich 106 D3
Membury 8 D4
Memsie 99 J2
Menai Bridge 50 E3
Mendham 39 H2
Mendlesham 39 G4
Mendlesham Green 39 F4
Menheniot 4 C5
Mennock 72 D3
Menston 62 E5
Menstrie 85 L5
Mentmore 20 D2
Meole Brace 43 F6
Meonstoke 11 H3
Meopham 14 C2
Meopham Station 14 C2
Mepal 37 H2
Meppershall 36 E7
Merbach 29 H3
Mere, Ches 53 J5
Mere, Wilts 9 K1
Mere Brow 52 F2
Mereclough 62 B6
Mere Green 44 C8
Mereworth 13 K1
Mergie 93 J3
Meriden 34 D2
Merkadale 110 E3
Merlin's Bridge 26 D6
Merrington 43 F5
Merriott 9 F3
Merrivale 5 F4
Merrymeet 4 C5
Mersham 15 G5
Merstham 13 F1
Merston 11 K4
Merstone 11 G6
Merther 3 G5
Merthyr 27 H5
Merthyr Cynog 28 D4
Merthyr Dyfan 25 K7
Merthyr Mawr 25 G7
Merthyr Tydfil 25 J4
Merthyr Vale 25 J5
Merton, Norf 48 E8
Merton, Devon 7 F4
Merton, G. Lon 21 G7
Merton, Oxon 31 K6
Mervinslaw 74 C2
Meshaw 7 H4
Messing 22 E2
Messingham 58 C6
Metfield 39 H2
Metheringham 56 E7
Methil 87 G5
Methley 63 G7
Methlick 99 H5
Methven 86 D2
Methwold 48 C8
Methwold Hythe 48 C8
Mettingham 39 J2
Mevagissey 3 J5
Mexborough 55 H3
Mey 107 H2
Meysey Hampton 31 F7
Miavaig 108 B3
Michael 60 Q2
Michaelchurch 29 K5
Michaelchurch Escley 29 H4
Michaelchurch-on-Arrow 29 G2
Michaelstone-y-Fedw 16 D2
Michaelston-le-Pit 25 K7
Michaelstow 3 J2
Micheldever 19 G7
Michelmersh 10 E2
Mickfield 39 G4
Mickleby 69 L5
Micklefield 63 H6
Mickleham 12 E1
Mickleover 44 E4
Mickleton, Durham 68 B4
Mickleton, Glos 34 C6
Mickle Trafford 52 F7
Mickley 63 F2
Mickley Square 75 G7
Mid Ardlaw 99 J2
Midbea 112 C2
Mid Beltie 93 H1
Mid Culbeuchly 98 F2
Middle Assendon 20 B5
Middle Aston 31 J5
Middle Barton 31 J5
Middlebie 73 H6
Middle Claydon 35 J8
Middleham 62 E1
Middlehope 32 D2
Middle Littleton 34 B6
Middle Maes-coed 29 H4
Middlemarsh 9 H4
Middle Mill 26 C5
Middle Rasen 56 E5
Middlesbrough 69 G4
Middlesmoor 62 D2
Middleston Moor 68 E3
Middlestown 55 F2
Middleton, Strath 88 A4
Middleton, Northnts 36 B1
Middleton, Lothn 80 B4
Middleton, Norf 48 B6
Middleton, Warw 44 C8
Middleton, Derby 44 D2
Middleton, Essex 38 D7
Middleton, Shrops 42 D8
Middleton, Shrops 32 E3
Middleton, H. & W 32 E4
Middleton, N. Yks 64 E4
Middleton, Shrops 42 E5
Middleton, N. Yks 62 E5
Middleton, Derby 54 E7
Middleton, Hants 18 F6
Middleton, Northum 75 G5
Middleton, W. Yks 63 G7
Middleton, Lancs 61 H4
Middleton, Northum 81 J7
Middleton, Cumbr 61 K1
Middleton, G. Man 53 K3
Middleton, Suff 39 K4
Middleton, Northum 81 K6
Middleton Cheney 35 F6
Middleton Green 44 A4
Middleton Hall 81 H7
Middleton-in-Teesdale 68 B4
Middleton-on-Sea 12 C7
Middleton on the Hill 32 E4
Middleton-on-the-Wolds 65 G8
Middleton Priors 33 F1
Middleton Scriven 33 F2
Middleton St George 69 F5
Middleton Stoney 31 K5
Middleton Tyas 68 E6
Middletown 42 E6
Middle Tysoe 34 E6
Middle Wallop 10 D1
Middlewich 53 J7
Middle Winterslow 10 D1
Middle Woodford 10 C1
Middlewood Green 39 F4
Middlezoy 8 E1
Middridge 68 E4
Midfield 105 G2
Midge Hall 61 J7
Midgeholme 74 C8
Midgham 19 G4
Midgley 62 D7
Midhopestones 55 F4
Midhurst 12 C4
Midlem 80 D7
Midsomer Norton 17 H5
Midtown 100 D3
Midtown of Buchromb 98 C4
Midville 57 H8
Mid Walls 113 C4
Mid Yell 113 D2
Migvie 98 D8
Milborne Port 9 H3
Milborne St Andrew 9 K5
Milborne Wick 9 H2
Milbourne 75 H6
Milburn 67 J4
Milbury Heath 17 H1
Milcombe 31 J4
Milden 38 E6
Mildenhall, Suff 38 C3
Mildenhall, Wilts 18 D4
Milebrooke 32 C3
Milebush 13 L2
Mile Elm 18 A4
Mile End, Essex 38 E8
Mile End, Glos 29 K6
Mileham 48 E6
Milesmark 86 D6
Milfield 81 H6
Milford, Staffs 44 A5
Milford, Surrey 12 C2
Milford, Devon 6 C3
Milford, Derby 44 E3
Milford Haven 26 D7
Milford on Sea 10 D5
Milkwall 29 K7
Milland 11 K2
Mill Bank 62 D7
Millbounds 112 C3
Millbreck 99 K4

Millbrex 99 H4
Millbridge 19 K6
Millbrook, Beds. 36 D7
Millbrook, Hants. 10 E3
Millbrook, Corn. 4 E6
Millburn 77 K5
Mill Corner 14 E6
Mill End, Bucks. 20 B5
Mill End, Herts. 37 G7
Millerhill 87 G8
Miller's Dale 54 E6
Mill Green, Essex 22 C3
Mill Green, Shrops. 43 H5
Millheugh 78 E4
Mill Hill 21 G4
Millholme 67 H7
Millhouse, Strath. 84 B7
Millhouse, Cumbr. 66 F3
Millikenpark 77 K1
Millington 64 F7
Mill Lane 19 J5
Millmeece 43 K4
Millom 60 E1
Millport 77 G2
Mill Side 61 H1
Mill Street 48 F6
Millthrop 67 J7
Milltimber 93 K1
Millton of Corsindale 99 F8
Millton of Murtle 93 K1
Milltown, Devon 7 F2
Milltown, Derby 55 G7
Milltown, D. & G. 73 J6
Milltown of Aberdalgie 86 D2
Milltown of Auchindoun 98 C4
Milltown of Craigston 99 G3
Milltown of Edinvillie 98 B4
Milltown of Rothiemay 98 E4
Milltown of Towie 98 D7
Milnathort 86 E4
Milngavie 85 H7
Milnrow 53 L2
Milnthorpe 61 H1
Milovaig 110 D2
Milson 33 F3
Milstead 14 F3
Milston 18 C6
Milton, Highld 96 B3
Milton, Cumbr. 74 B7
Milton, Highld 96 C5
Milton, Highld 96 D4
Milton, D. & G. 70 D6
Milton, D. & G. 72 D6
Milton, Grampn. 98 E2
Milton, Dyfed 26 E7
Milton, Oxon 19 F1
Milton, Highld 103 F6
Milton, Strath. 85 G7
Milton, Highld 97 H3
Milton, Cambs 37 H4
Milton, Central 85 H4
Milton, Highld 107 J4
Milton, Oxon 31 J4
Milton, Staffs 43 L2
Milton Abbas 9 K4
Milton Abbot 4 E4
Milton Bridge 79 K3
Milton Bryan 36 C7
Milton Clevedon 9 H1
Milton Coldwells 99 J5
Milton Combe 4 E5
Milton Damerel 6 D4
Miltonduff 98 A2
Milton Ernest 36 D5
Milton Green 52 F8
Milton Hill 19 F1
Milton Keynes 36 B7
Milton Keynes Village 36 B7
Milton Libourne 18 C4
Milton Malsor 35 J5
Milton of Auchinhove 98 E8
Milton of Balgonie 87 G4
Milton of Buchanan 85 G5
Milton of Campfield 93 H1
Milton of Campsie 85 J7
Milton of Corsindae 99 F8
Milton of Cushnie 98 E7
Milton of Lesmore 98 D6
Milton of Murtle 93 K1
Milton of Tullich 92 E2
Milton on Stour 9 K2
Milton Regis 14 F2
Milton-under-Wychwood 31 G6
Milverton 8 C2
Milwich 44 A4
Milwr 52 C6
Minard 84 B5
Minchinhampton 30 C7
Mindrum 81 G6
Minehead 7 K1
Minera 42 D2
Minety 18 B1
Minffordd 50 E7
Mingary 110 B3
Miningsby 57 H7
Minions 4 C4
Minishant 77 J6
Minllyn 41 H2
Minnes 99 J6
Minnigaff 70 F5
Minskip 63 G3
Minstead 10 D3
Minster, Kent 14 F1
Minster, Kent 15 K2
Minsteracres 68 C1
Minsterley 42 E7
Minster Lovell 31 H6
Minsterworth 30 B6
Minterne Magna 9 H4
Minting 57 F6
Mintlaw 99 K4
Minto 74 B1
Minton 32 D1
Minwear 26 E6
Minworth 34 C1
Mirbister 112 B4
Mireland 107 J3
Mirfield 54 F2
Miserden 30 D7
Miskin 25 J6
Misson 56 A4
Misterton, Notts. 56 B4
Misterton, Somer. 9 F4
Misterton, Leic. 35 G2
Mistley 39 G7
Mitcham 21 G7
Mitcheldean 30 A6
Mitchell 3 G4
Mitchel Troy 29 J6
Mitford 75 H5
Mithian 2 F4
Mitton 43 K6
Mixbury 31 L4
Mixon 54 D8
Mobberley 53 J6
Moccas 29 H3
Mochdre, Clwyd 51 H3
Mochdre, Powys 41 K5
Mochrum 70 E7
Mockerkin 66 C4
Modbury 5 G6
Moddershall 43 L4
Moelfre, Clwyd 42 C5
Moelfre, Gwyn 50 E2
Moffat 73 F3
Mogerhanger 36 E6
Moira 44 E6
Mol-chlach 111 F4
Mold 52 D7
Molehill Green 21 K1
Molescroft 58 E2
Molesworth 36 D3
Molland 7 J3
Mollington, Ches. 52 E6
Mollington, Oxon. 35 F6
Mollinsburn 85 K7
Monachty 40 E7
Mondynes 93 J4
Monewden 39 H5
Moniaive 72 C4
Monifieth 87 H1
Monikie 93 F8
Monimail 87 F3
Monington 26 F3
Monken Hadley 21 G4
Monk Fryston 63 J7
Monkhopton 43 H8
Monkland 29 J2
Monkleigh 6 E3
Monknash 25 H7
Monkokehampton 7 F5
Monks Eleigh 38 E6
Monk's Heath 53 K6
Monk Sherborne 19 H5
Monkshill 99 G4
Monksilver 8 B1
Monks Kirby 35 F2
Monk Soham 39 H4
Monks Risbrough 20 C3
Monkswood 29 H7
Monkton, Devon 8 C4
Monkton, Kent 15 J2
Monkton, Strath. 77 J5
Monkton, T. & W. 75 K7
Monkton Combe 17 J4
Monkton Deverill 9 K1
Monkton Farleigh 17 K4
Monkton Heathfield 8 D2
Monkton Up Wimborne 10 B3
Monkwood 11 H1
Monmouth 29 K6
Monnington on Wye 29 H3
Monreith 70 E7
Monreith Mains 70 E7
Montacute 9 F3
Montford 42 F6
Montgarrie 98 E7
Montgomery 42 D8
Montgreenan 77 J3
Montrose 93 J6
Montsale 23 G4
Monxton 18 E6
Monyash 54 E7
Monymusk 99 F7
Monzie 86 B2
Moorby 57 G7
Moorcot 29 H2
Moor Crichel 10 A4
Moordown 10 B5
Moore 53 G5
Moorends 58 A5
Moorhall 55 G6
Moorhampton 29 H3
Moorhouse, Notts 56 B7
Moorhouse, Cumbr 66 F1
Moorland or Northmoor Green 8 E1
Moorlinch 8 E1
Moor Monkton 63 J4
Moor of Balvack 98 F7
Moorsholm 69 J5
Moorside 54 C3
Moor, The 13 L4
Moortown, Lincs 58 E7
Moortown, I. of W. 10 F6
Morangie 103 F5
Morar 94 B6
Morborne 36 E1
Morchard Bishop 7 H5
Morcombelake 8 F5
Morcott 46 C7
Morda 42 D5
Morden, G. Lon 21 G7
Morden, Dorset 9 L5
Mordiford 29 K4
Mordon 68 F4
More 32 C1
Morebath 7 K3
Morebattle 80 F7
Morecambe 61 H3
Morefield 101 G2
Moreleigh 5 H6
Moresby 66 B4
Moresby Parks 66 B5
Morestead 11 G2
Moreton, Oxon 20 A3
Moreton, Mers. 52 D4
Moreton, Staffs. 43 J6
Moreton, Essex 21 K3
Moreton, Dorset 9 K6
Moreton Corbet 43 G5
Moretonhampstead 5 H3
Moreton-in-Marsh 31 G4
Moreton Jeffries 29 L3
Moreton Morrell 34 E5
Moreton on Lugg 29 K3
Moreton Pinkney 35 G6
Moreton Say 43 H4
Moreton Valence 30 B7
Morfa Bychan 50 E7
Morfa Glas 25 G4
Morfa Nefyn 50 B6
Morgan's Vale 10 C2
Morland 67 H4
Morley, Durham 68 D4
Morley, Derby 44 E3
Morley, W. Yks 63 F7
Morley Green 53 K5
Morley St Botolph 48 F8
Morningside 87 F7
Morningthorpe 39 H1
Morpeth 75 H5
Morrey 44 C6
Morriston 24 E5
Morston 48 F3
Morthoe 6 E1
Mortimer 20 A7
Mortimer Cross 32 D4
Mortimer West End 20 A7
Mortlake 21 G6
Morton, Lincs 56 C4
Morton, Lincs 46 D5
Morton, Shrops 42 D5
Morton, Norf 49 G6
Morton, Avon 17 H1
Morton, Derby 55 H7
Morton Bagot 34 C4
Morton-on-Swale 68 F7
Morvah 2 C6
Morval 4 C6
Morvich, Highld 94 E3
Morvich, Highld 103 F3
Morville 43 H8
Morwenstow 6 C4
Mosborough 55 H5
Moscow 77 K3
Mosedale 66 F3
Moseley, W. Mids 34 B2
Moseley, H. & W 33 H5
Moss, Strath 88 A4
Moss, Clwyd 42 E2
Moss, S. Yks 55 J2
Mossat 98 D7
Mossbank, Shetld 113 D3
Moss Bank, Mers 53 G4
Mossblown 77 K5
Mossburnford 74 C2
Mossdale 72 B6
Mossend 78 E3
Mossgiel 77 K5
Mossley 54 C3
Moss of Barmuckity 98 B2
Moss Side 61 G6
Mosstodloch 98 C2
Mosterton 9 F4
Mostyn 52 C5
Motcombe 9 K2
Motherwell 78 E4
Mottingham 21 J6
Mottisfont 10 E2
Mottistone 10 F6
Mottram in Longdendale 54 C4
Mottram St Andrew 53 K6
Mouldsworth 53 G6
Moulin 91 L6
Moulsecoomb 13 G6
Moulsford 19 G2
Moulsoe 36 C6
Moulton, Suff 38 B4
Moulton, N. Yks 68 E6
Moulton, Lincs 47 G5
Moulton, Ches 53 H7
Moulton, Northnts 35 J4
Moulton Chapel 47 F6
Moulton Seas End 47 G5
Mount, Corn 3 F4
Mount, Corn 3 K3
Mountain Ash 25 J5
Mountain Cross 79 J5
Mountbenger 80 B7
Mount Bures 38 E7
Mountfield 13 L4
Mountgerald 96 D2
Mount Hawke 2 F5
Mountjoy 3 G3
Mountnessing 22 C4
Mounton 29 K8
Mountsorrel 45 G6
Mousehole 2 C7
Mousen 81 K6
Mouswald 73 F6
Mow Cop, Staffs 53 K8
Mow Cop, Ches 53 K8
Mowhaugh 74 E1
Mowsley 35 H2
Moy 90 F3
Moyles Court 10 C4
Moylgrove 26 F3
Muasdale 76 B3
Muchalls 93 L2
Much Birch 29 K4
Much Cowarne 29 L3
Much Dewchurch 29 J4
Muchelney 9 F2
Much Hadham 21 J1
Much Hoole 61 H7
Muchlarnick 3 L4
Much Marcle 30 A4
Much Wenlock 43 H8
Mucking 22 C5
Mucklestone 43 J4
Muckleton 43 G5
Muckletown 98 E6
Muckton 57 H5
Mudale 105 G5
Muddiford 7 F2
Mudeford 10 C5
Mudford 9 G3
Mudgley 17 F6
Mugdock 85 H7
Mugeary 111 F3
Mugginton 44 D3
Muggleswick 68 C1
Muie 102 E3
Muir 92 B3
Muirden 99 G3
Muirdrum 93 G8
Muirhead, Tays 87 G1
Muirhead, Strath 85 J8
Muirhouses 86 D6
Muirkirk 78 D7
Muir of Fairburn 96 C3
Muir of Fowlis 98 E7
Muir of Lochs 98 C2
Muir of Ord 96 D3
Muirshearlich 90 C3
Muirtack, Grampn 99 H4
Muirtack, Grampn 99 J5
Muirton, Tays 86 C3
Muirton, Highld 97 F2
Muirton of Ardblair 92 C7
Muirton of Ballochy 93 H5
Muker 68 B7
Mulbarton 49 G7
Mulben 98 C3
Mullion 2 E8
Mullion Cove 2 E8
Mumbles, The 24 E6
Mumby 57 K6
Munderfield Row 33 F5
Munderfield Stocks 33 F5
Mundesley 49 J4
Mundford 48 D8
Mundham, Norf 49 J8
Mundham, W. Susx 11 K4
Mundon 22 E3
Munerigie 90 D1
Mungasdale 100 E2
Mungrisdale 66 F3
Munlochy 96 E3
Munsley 33 F6
Munslow 32 E2
Murch 25 K7
Murcott 31 K6
Murkle 107 G3
Murlaggan, Highld 90 B2
Murlaggan, Highld 90 E3
Murrow 47 G7
Mursley 36 B8
Murthly 92 B8
Murton, Durham 69 F2
Murton, Northum 81 H5
Murton, Cumbr 67 K4
Murton, N. Yks 63 K4
Musbury 8 D5
Muscoates 64 D4
Musdale 84 B2
Musselburgh 87 G7
Muston, Leic 46 B4
Muston, N. Yks 65 H5
Mustow Green 33 H3
Mutford 39 K2
Muthill 86 B3
Mutterton 8 B4
Mybster 107 G4
Myddfai 25 F1
Myddle 42 F5
Mydroilyn 27 J2
Mylor Bridge 3 G6
Mynachlog-ddu 26 F4
Myndtown 32 C2
Mynydd-bach 29 J8
Mynydd Isa 52 D7
Mynydd Llandegai 50 F4
Mynydd Mechell 50 C1
Mynytho 50 C7
Myrebird 93 J2
Mytchett 19 K5
Mytholm 62 C7
Mytholmroyd 62 D7
Myton-on-Swale 63 H3

N

Naast 100 D3
Naburn 63 K5
Nackington 15 H3
Nacton 39 H6
Nafferton 65 H7
Nailsea 17 F3
Nailstone 44 F7
Nailsworth 30 C8
Nairn 97 G3
Nancegollan 2 E6
Nancledra 2 C6
Nanhoron 50 B7
Nannerch 52 C7
Nanpantan 45 G6
Nanpean 3 H4
Nanstallon 3 J3
Nant-ddu 25 J3
Nanternis 27 H2
Nantgaredig 27 J5
Nantgarw 25 K6
Nant Glas 41 J7
Nantglyn 51 K4
Nantlle 50 E5
Nantmawr 42 D5
Nantmel 41 K7
Nantmor 50 F6
Nant Peris 50 F5
Nantwich 43 H2
Nant-y-derry 29 H7
Nantyffyllon 25 G5
Nantyglo 25 K3
Nant-y-moel 25 H5
Naphill 20 C4
Nappa 62 B4
Napton on the Hill 35 F4
Narberth 26 F6
Narborough, Norf 48 C6
Narborough, Leic 45 G8
Nasareth 50 D5
Naseby 35 H3
Nash, H. & W 32 C4
Nash, Gwent 16 E2
Nash, Shrops 33 F3
Nash, Bucks 35 J7
Nash Lee 20 C3
Nassington 46 D8
Nasty 21 H1
Nateby, Lancs 61 H5
Nateby, Cumbr 67 K6
Natland 67 H8
Naughton 38 F6
Naunton, Glos 30 F5
Naunton, H. & W 33 H7
Naunton Beauchamp 34 A5
Navenby 56 D8
Navestock Heath 21 K4
Navestock Side 21 K4
Nawton 63 K1
Nayland 38 E7
Nazeing 21 J3
Neacroft 10 C5
Neal's Green 34 E2
Neasham 68 F5
Neath 25 F5
Neatishead 49 J5
Nebo, Gwyn 50 D5
Nebo, Dyfed 40 E7
Nebo, Gwyn 51 H5
Necton 48 D7
Nedd 104 C5
Nedging Tye 38 F6
Needham 39 H2
Needham Market 39 F5
Needingworth 37 G3
Neen Savage 33 F3
Neen Sollars 33 F3
Neenton 33 F2
Nefyn 50 C6
Neilston 77 K2
Nelson, Lancs 62 B6
Nelson, M. Glam 25 K5
Nelson Village 75 J6
Nemphlar 79 F5
Nempnett Thrubwell 17 G4
Nenthead 67 K2
Nenthorn 80 E6
Nercwys 52 D7
Nereabolls 82 B6
Nerston 78 D4
Nesbit 81 H6
Ness, N. Yks 64 D5
Ness, Ches 52 E6
Nesscliffe 42 E6
Neston, Ches 52 D6
Neston, Wilts 17 K4
Nether Alderley 53 K6
Netheravon 18 C6
Nether Blainslie 80 D5
Netherbrae 99 G3
Nether Broughton 45 H5
Netherburn 79 F5
Nether Burrow 61 K2
Netherbury 9 F5
Nether Cerne 9 H5
Nether Compton 9 G3
Nether Dallachy 98 C2
Netherend 29 K7
Nether Exe 5 K1
Netherfield 13 L5
Netherhampton 10 C2
Nether Haugh 55 H4
Nether Heyford 35 H5
Nether Howecleuch 73 F2
Nether Kellet 61 J3
Nether Kinmundy 99 K4
Nether Kirkton 77 K2
Netherlaw 71 J7
Netherley 93 K2
Nethermill 73 F5
Nether Padley 55 F6
Netherplace 77 L2
Nether Poppleton 63 J4
Nether Row 66 F3
Netherseal 44 D6
Nether Silton 69 G7
Nether Stowey 16 C7
Netherthird 72 A2
Netherthong 54 E3
Netherton, H. & W 34 A6
Netherton, Mers 52 E3
Netherton, W. Yks 55 F2
Netherton, Northum 75 F3
Netherton, Tays 93 G6
Netherton, Central 85 H7
Netherton, Devon 5 J4
Nethertown, Cumbr 66 B6
Nethertown, Highld 107 J2
Nether Wallop 10 E1
Nether Wasdale 66 D6
Nether Whitacre 44 D8
Nether Winchendon or Lower Winchendon 20 B2
Netherwitton 75 G4
Nether Worton 31 J4
Nethy Bridge 97 J6
Netley 11 F4
Netley Marsh 10 E3
Nettlebed 20 B5
Nettlebridge 17 H6
Nettlecombe 9 G5
Nettleden 20 E2
Nettleham 56 E6
Nettlestead 13 K1
Nettlestead Green 13 K1
Nettlestone 11 H5
Nettleton, Lincs 59 F6
Nettleton, Wilts 17 K3
Neuk, The 93 J2
Nevendon 22 D4
Nevern 26 E3
New Abbey 72 E7
New Aberdour 99 H2
New Addington 21 H7
New Alresford 11 G1
New Ayth 92 D7
Newark, Orkney 112 D2
Newark, Cambs 46 F7
Newark-on-Trent 45 J2
Newarthill 78 E4
New Ash Green 14 C2
New Bewick 75 G1
Newbiggin, Durham 68 B4
Newbiggin, N. Yks 68 B7
Newbiggin, N. Yks 62 C1
Newbiggin, Cumbr 60 F3
Newbiggin, Cumbr 67 G4
Newbiggin, Cumbr 67 H2
Newbiggin, Cumbr 67 J4
Newbiggin-by-the-Sea 75 K5
Newbigging, Tays 92 D7
Newbigging, Tays 93 F8
Newbigging, Tays 92 F8
Newbigging, Strath 79 H5
Newbiggin-on-Lune 67 K6
Newbold, Leic 44 F6
Newbold, Derby 55 G6
Newbold on Avon 35 F3
Newbold-on-Stour 34 D6
Newbold Pacey 34 D5
Newbold Verdon 45 F7
New Bolingbroke 57 H8
Newborough, Staffs 44 C5
Newborough, Cambs 46 F7
Newbottle 35 G7
Newbourne 39 H6
New Brancepeth 68 E2
Newbridge, Corn 2 C6
Newbridge, Clwyd 42 D3
Newbridge, Hants 10 D3
Newbridge, Lothn 86 E7
Newbridge, I. of W 10 F6
Newbridge, Gwent 25 L5
Newbridge-on-Usk 29 H8
Newbridge-on-Wye 28 E2
New Brighton 52 E4
New Brinsley 45 F2
Newbrough 74 E7
New Buckenham 39 F1
Newbuildings 7 H5
Newburgh, Lancs 53 F2
Newburgh, Fife 87 F3
Newburgh, Grampn 99 J6
Newburn 75 H7
Newbury 19 F4
Newby, Cumbr 67 H4
Newby, N. Yks 69 H5
Newby, N. Yks 61 L2
Newby Bridge 61 G1
Newby East 73 K8
New Byth 99 H3
Newby West 66 F1
Newby Wiske 63 G1
Newcastle, Shrops 32 B2
Newcastle, Gwent 29 J6
Newcastle Emlyn 27 H3
Newcastleton 73 K5
Newcastle-under-Lyme 43 K3
Newcastle upon Tyne 75 J7
Newchapel, Surrey 13 G2
Newchapel, Dyfed 27 G4
Newchapel, Staffs 43 K2
New Cheriton 11 G2
Newchurch, Powys 29 G2
Newchurch, Kent 15 G5
Newchurch, I. of W 11 G6
Newchurch, Dyfed 27 H5
Newchurch, Gwent 29 J8
Newchurch in Pendle 62 B6
New Clipstone 55 J7
New Costessey 49 G7
Newcott 8 D4
New Cross 41 F6
New Cumnock 72 B2
New Deer 99 H4
Newdigate 12 E2
New Duston 35 J4
New Earswick 63 K4
New Edlington 55 J4
New Ellerby 59 F3
Newell Green 20 C6
New Eltham 21 J6
New End 34 C4
Newenden 14 E6
Newent 30 B5
New Farnley 63 F6
New Ferry 52 E5
Newfield, Durham 68 E3
Newfield, Highld 103 F6
New Fryston 63 H7
Newgale 26 C5
New Galloway 72 B6
Newgate 48 F3
Newgate Street 21 H3
New Grimsby 2 N2
Newhall, Derby 44 D5
Newhall, Ches 43 H3
Newham 81 K7
Newham Hall 81 K7
New Hartley 75 K6
Newhaven 13 H6
New Hedges 26 F7
Newhey 53 L2
New Holland 58 E4
Newholm 69 L5
New Horndean 81 G5
New Houghton, Norf 48 C5
New Houghton, Derby 55 H7
Newhouse 78 E3
New Houses 62 B2
New Hutton 67 H7
New Hythe 14 D3
Newick 13 H4
Newington, Kent 14 E2
Newington, Kent 15 H5
Newington, Oxon 31 L8
New Inn, Gwent 29 H8
New In, Dyfed 27 J4
New Inn, Gwent 29 J7
New Invention 32 B3
New Kelso 100 E7
New Lanark 79 F5
Newland, N. Yks 58 A4
Newland, H. & W 33 G6
Newland, Glos 29 K7
Newlandrig 80 B3
Newlands, Northum 68 C1
Newlands, Grampn 98 C3
Newlands, Highld 97 F4
Newlands of Geise 107 F3
New Lane 52 F2
New Leake 57 J8
New Leeds 99 J3
New Longton 61 J7
New Luce 70 C5
Newlyn 2 C7
Newmachar 99 H7
Newmains 79 F4
New Malden 21 G7
Newmarket, Suff 38 B4
Newmarket, W. Isles 108 D3
New Marske 69 J4
New Marton 42 E4
Newmill, Border 74 B2
New Mill, Corn 2 C6
New Mill, Herts 20 D2
Newmill, Grampn 98 D3
New Mill, W. Yks 54 E3
Newmill of Inshewan 92 F5
New Mills, Derby 54 C5
New Mills, Corn 3 H4
New Mills, Ches 53 J5
New Mills, Powys 41 K3
New Mills, Gwent 29 K7
Newmiln 86 E1
Newmilns 77 L4
New Milton 10 D5
New Moat 26 E5
Newnham, Glos 30 A6
Newnham, Kent 14 F3
Newnham, H. & W 33 F4
Newnham, Herts 37 F7
Newnham, Northnts 35 G5
Newnham, Hants 19 J5
New Ollerton 55 K7
New Pitsligo 99 H3
New Polzeath 3 H2
Newport, Glos 30 B8
Newport, Humbs 58 C3
Newport, Gwent 16 E2
Newport, Dyfed 26 E4
Newport, Devon 7 F2
Newport, I. of W 11 G6
Newport, Highld 107 G7
Newport, Shrops 43 J6
Newport, Essex 37 J7
Newport, Norf 49 L6
Newport-on-Tay 87 H2
Newport Pagnell 36 B6
Newpound Common 12 D4
New Prestwick 77 J5
Newquay, Corn 3 G3
New Quay, Dyfed 27 H2
New Rackheath 49 H6
New Radnor 32 B4
New Romney 15 G6
New Rossington 55 K4
New Row 61 K6
New Sauchie 86 B5
New Scone 86 E2
Newseat 99 G5
Newsham, N. Yks 68 D5
Newsham, Northum 75 K6
Newsholme, Lancs 62 B4
Newsholme, Humbs 58 B4
New Silksworth 69 F1
Newstead, Border 80 D6
Newstead, Notts 45 G2
Newstead, Northum 81 K7
New Stevenston 78 E4
Newthorpe 63 H6
New Tolsta 108 D2
Newton, Grampn 98 A2
Newton, W. Isles 108 A6
Newton, Border 74 B1
Newton, Northnts 36 B2
Newton, Staffs 44 B5
Newton, Strath 84 C5
Newton, Wilts 10 D2
Newton, Strath 78 D3
Newton, S. Glam 16 D3
Newton, Lincs 46 D4
Newton, Norf 48 D6
Newton, Lothn 86 D7
Newton, W. Glam 24 E6
Newton, Suff 38 E6
Newton, Highld 97 F2
Newton, Cumbr 60 F2
Newton, Highld 97 F4
Newton, Warw 35 G3
Newton, D. & G 73 G4
Newton, Strath 79 G6
Newton, Ches 53 G6
Newton, Northum 75 G7
Newton, M. Glam 25 G7
Newton, Ches 53 G8
Newton, H. & W 29 H4
Newton, Lancs 61 H6
Newton, Cambs 47 H6
Newton, Cambs 37 H6
Newton, Lancs 61 J2
Newton, Highld 107 J5
Newton, H. & W 29 K2
Newton, Lancs 61 K4
Newton Abbot 5 J4
Newton Arlosh 66 D1
Newton Aycliffe 68 E4
Newton Bewley 69 G4
Newton Blossomville 36 C5
Newton Bromswold 36 C4
Newton Burgoland 44 E7
Newton by Toft 56 E5
Newton Ferrers 5 F7
Newtonferry 108 A6
Newton Flotman 49 H8
Newtongrange 80 B3
Newton Harcourt 45 H8
Newtonhill 93 L2
Newton Kyme 63 H5
Newton-le-Willows, N., Yks. 68 E8
Newton-le-Willows, Mers 53 G4
Newton Longville 36 B7
Newton Mearns 77 L2
Newtonmore 91 J2
Newton of Ardtoe 89 H1
Newton of Balcanquhal 86 E3
Newton-on-Ouse 63 J4
Newton-on-Rawcliffe 64 F3
Newton on the Moor 75 H3
Newton on Trent 56 C6
Newton Poppleford 8 B6
Newton Purcell 35 H7
Newton Regis 44 D7
Newton Reigny 67 G3
Newton Solney 44 D5
Newton Stacey 18 F6
Newton St Cyres 5 J2
Newton Stewart 70 F5
Newton St Faith 49 H6
Newton St Loe 17 J4
Newton St Petrock 6 E4
Newton Tony 18 D6
Newton Tracey 7 F3
Newton under Roseberry 69 H5
Newton upon Derwent 64 E8
Newton Valence 11 J1
Newtown, Powys 32 A1
Newtown, Dorset 10 B5
Newtown, Cumbr 74 B7
Newtown, Highld 96 B8
Newtown, Ches 54 C5
Newtown, Derby 54 C5
Newtown, Hants 10 D3
Newtown, Hants 10 E2
Newtown, Hants 19 F4
Newtown, Shrops 43 F4
Newtown, I. of W 10 F5
Newtown, Northum 75 G3
Newtown, Ches 43 H3
Newtown, Hants 11 H3
Newtown, Northum 81 H6
New Town, Lothn 87 H7
Newtown, Northum 81 J7
Newtown, Grampn 97 K2
Newtown, Wilts 9 L2
Newtown, H. & W 29 L3
Newtown, Staffs 53 L7
Newtown, I. of M 60 Q4
Newtown-in-St-Martin 2 F7
Newtown Linford 45 G6
Newtown St Boswells 80 D6
New Tredegar 25 K4
Newtyle 92 D7
New Waltham 59 G6
New Wimpole 37 G6
New Winton 87 H7
New Yatt 31 H6
New York, Lincs 47 F2
New York, T. & W 75 K6
Neyland 26 D7
Nibley 17 H2
Nicholashayne 8 C3
Nicholaston 24 D6
Nidd 63 G3
Nigg, Highld 103 G6
Nigg, Grampn 99 J8
Nine Ashes 22 B3
Ninebanks 67 K1
Ninemile Bar or Crocketford 72 D6
Ninfield 13 L5
Ningwood 10 F6
Nisbet 80 E7
Niton 11 G7
Nitshill 77 L1
Niwbwrch 50 D4
Noak Hill 21 K4
Nobottle 35 H4
Nocton 56 E7
Noke 31 K6
Nolton 26 C6
Nolton Haven 26 C6
No Man's Heath, Warw 44 D7
No Man's Heath, Ches 43 G3
Nomansland, Wilts 10 D3
Nomansland, Devon 7 J4
Noneley 43 F5
Nonington 15 J3
Noonsbrough 113 C4
Nor 81 H5
Noranside 93 F5
Norbury, Derby 44 C3
Norbury, Shrops 42 E8
Norbury, Ches 43 G3
Norbury, Staffs 43 J5
Nordelph 47 J7
Norden, Dorset 10 A6
Norden, G. Man 53 K2
Nordley 43 H8
Norham 81 H5
Norley 53 G6
Norleywood 10 E5
Normanby, Humbs 58 C5
Normanby, N. Yks 64 E4
Normanby-by-Spital 56 E5
Normanby le Wold 57 F4
Norman Cross 36 E1
Normandy 12 C1
Norman's Green 8 B4
Normanton, Lincs 6 C3
Normanton, Derby 44 E4
Normanton, W. Yks 63 G7
Normanton, Notts 45 J2
Normanton le Heath 44 E6
Normanton on Soar 45 G5
Normanton-on-the-Wolds 45 H4
Normanton on Trent 56 B7
Normoss 61 G6
Norrington Common 17 K4
Norris Hill 44 E6
Northallerton 69 F7
Northam, Devon 6 E3
Northam, Hants 10 F3
Northampton 35 J4
North Ascot 20 D7
North Aston 31 J5
Northaw 21 G3
North Baddesley 10 E3
North Ballachulish 90 B5
North Barrow 9 H2
North Barsham 48 E4
North Benfleet 22 D5
North Berwick 87 J6
North Boarhunt 11 H3
Northborough 46 E7
Northbourne 15 K3
North Bovey 5 H3
North Bradley 17 K5
North Brentor 4 E3
North Buckland 6 E1
North Burlingham 49 J6
North Cadbury 9 H2
North Cairn 70 A4
North Carlton 56 D6
North Cave 58 C3
North Cerney 30 E7
Northchapel 12 C4
North Charford 10 C3
North Charlton 81 K7
North Cheriton 9 H2
North Chideock 9 F5
Northchurch 20 D3
North Cliffe 58 C3
North Clifton 56 C6
North Cotes 59 H6
Northcott 4 D2
North Cove 39 K2
North Cowton 68 E6
North Crawley 36 C6
North Cray 21 J6
North Creake 48 D4
North Curry 8 E2
North Dalton 65 G7
North Deighton 63 G4
North Duffield 58 A3
Northdyke 112 B3
North Elkington 57 G4
North Elmham 48 E5
Northend, Bucks 20 B4
Northend, Warw 34 E5
North End, W. Susx 12 E6
North End, Hants 18 F4
North End, Avon 16 F4
North End, Hants 11 H4
Northend, Avon 17 J4
North Erradale 100 C3
North Fearns 111 F3
North Ferriby 58 D4
Northfield, W. Mids 34 B3
Northfield, Border 81 H3
Northfield, Grampn 99 J8
Northfleet 14 C1
North Frodingham 65 H7
North Gorley 10 C3
North Green 39 H2
North Greetwell 56 E6
North Grimston 64 F6
North Hayling 11 J4
North Heasley 7 H2
North Heath 12 D4
North Hill, Corn 4 C4
North Hill, Cambs 37 H3
North Hinksey Village 31 J7
North Holmwood 12 E2
North Huish 5 H6
North Hykeham 56 D7
Northiam 14 E6
Northill 36 E6
Northington 11 G1
North Kelsey 58 E6
North Kessock 96 E4
North Kilvington 63 H1
North Kilworth 35 H2
North Kyme 46 E2
North Lancing 12 E6
Northlands 47 G2
Northleach 30 F6
North Lee 20 C3
Northleigh, Devon 8 C5
North Leigh, Oxon 31 H6
North Leverton with Habblesthorpe 56 B5
Northlew 4 F2
North Littleton 34 B6
North Lopham 38 F2
North Luffenham 46 C7
North Marden 11 K3
North Marston 20 B1
North Middleton 80 B4
North Molton 7 H3
Northmoor 31 J7
Northmoor Green or Moorland 8 E1
North Moreton 19 G2
Northmuir 92 E6
North Muskham 56 B8
North Newbold 58 D3
North Newington 34 F7
North Newnton 18 C5
North Newton 8 D1
North Nibley 30 B8
North Oakley 19 G5
North Ockendon 22 B5
Northolt 21 F5
Northop 52 D7
Northop Hall 52 D7
North Ormsby 57 G4
Northorpe, Lincs 56 C4
Northorpe, Lincs 46 D6
North Otterington 69 F8
North Owersby 56 E4
Northowram 62 E7
North Perrott 9 F4
North Petherton 8 D1
North Petherwin 4 C3
North Pickenham 48 D7
North Piddle 34 A5
Northpunds 113 D5
North Queensferry 86 E6
Northrepps 49 H4
North Rigton 63 F5
North Rode 53 K7
North Roe 113 C2
North Runcton 48 B6
North Sandwick 113 D2
North Scale 60 E2
North Scarle 56 C7
North Seaton 75 J5
North Shields 75 K7
North Shoebury 22 F5
North Shore 61 G6
North Side 47 F8
North Somercotes 57 J4
North Stainley 63 F2
North Stainmore 67 L5
North Stifford 22 C5
North Stoke, Oxon 20 A5
North Stoke, W. Susx 12 D5
North Stoke, Avon 17 J4
North Street 11 H1
North Sunderland 81 L6
North Tamerton 4 D2
North Tawton 7 G5
North Thoresby 59 G7
North Tidworth 18 D6
North Tolsta 108 D2
Northton 108 A5
Northtown 112 C5
North Tuddenham 48 F6
Northwall 112 D2
North Walsham 49 H4
North Waltham 19 G6
North Warnborough 19 J5
North Water Bridge 93 H5
North Watten 107 H4
Northway 30 D4
North Weald Bassett 21 J3
North Wheatley 56 B5
North Whilborough 5 J5
Northwich 53 H6
Northwick, Avon 17 G2
North Wick, Avon 17 G4
North Widcombe 17 G5
North Willingham 57 F5
North Wingfield 55 H7
North Witham 46 C5
Northwold 48 C8
Northwood, G. Lon 20 E4
Northwood, Shrops 42 F4
Northwood, I. of W 11 F5
Northwood, Derby 55 F7
Northwood Green 30 B6
North Wootton, Norf 48 B5
North Wootton, Somer 17 G6
North Wootton, Dorset 9 H3
North Wraxall 17 K3
Norton, H. & W 34 B6
Norton, Powys 32 C4
Norton, Glos 30 C5
Norton, W. Susx 12 C6
Norton, Shrops 32 D2
Norton, Suff 38 E4
Norton, N. Yks 64 E5
Norton, I. of W 10 E6
Norton, Herts 37 F7
Norton, Cleve 69 G4
Norton, Ches 53 G5
Norton, S. Yks 55 G5
Norton, Shrops 43 G7
Norton, Northnts 35 H4
Norton, H. & W 33 H5
Norton, S. Yks 55 J2
Norton, Notts 55 J6
Norton, Shrops 43 J7
Norton, Wilts 17 K2
Norton Bavant 17 L6
Norton Bridge 43 K4
Norton Canes 44 B7
Norton Canon 29 H3
Norton Disney 56 C8
Norton East 44 B7
Norton Ferris 9 J1
Norton Fitzwarren 8 C2
Norton Green 10 E6
Norton Hawkfield 17 G4
Norton Heath 22 C3
Norton in Hales 43 J4
Norton-in-the-Moors 43 K2
Norton-Juxta-Twycross 44 E7
Norton-le-Clay 63 H2
Norton Lindsey 34 D4
Norton Malreward 17 H4
Norton Mandeville 21 K3
Norton St Philip 17 J5
Norton Subcourse 49 K8
Norton sub Hamdon 9 F3
Norwell 56 B7
Norwell Woodhouse 56 B7
Norwich 49 H7
Norwick 113 E1
Norwood Green 21 F6
Norwood Hill 12 F2
Noseley 45 J8
Noss Mayo 5 F7
Nosterfield 63 F1
Nostie 94 D3
Notgrove 30 F5
Nottage 25 G7
Nottingham 45 G3
Nottington 9 H6
Notton, W. Yks 55 G2
Notton, Wilts 17 L4
Nounsley 22 D2
Noutard's Green 33 G4
Nox 42 F6
Nuffield 20 A5
Nunburnholme 64 F8
Nuneaton 34 E1
Nuneham Courtenay 31 K8
Nun Monkton 63 J4
Nunney 17 J6
Nunnington 63 K2
Nunnykirk 75 G4
Nunthorpe 69 H5
Nunton, W. Isles 110 B2
Nunton, Wilts 10 C2
Nursling 10 E3
Nursted 11 J2
Nutbourne 12 D5
Nutfield 13 G2
Nuthall 45 G3
Nuthampstead 37 H7
Nuthurst 12 E4
Nutley 13 H4
Nutwell 55 K3
Nybster 107 J3
Nyetimber 11 K5
Nyewood 11 K2
Nymet Rowland 7 H5
Nymet Tracey 5 H1
Nympsfield 30 C7
Nynehead 8 C2
Nyton 12 C6

O

Oadby 45 H7
Oad Street 14 E2
Oakamoor 44 B3
Oakbank 79 H3
Oakdale 25 K5
Oake 8 C2
Oaken 43 K7
Oakenclough 61 J5
Oakengates 43 J6
Oakenshaw, Durham 68 E3
Oakenshaw, W. Yks 62 E7
Oakford, Dyfed 27 J2
Oakford, Devon 7 K3
Oakgrove 53 L7
Oakham 46 B7
Oakhanger 11 J1
Oakhill 17 H6
Oakington 37 H4
Oaklands 21 G2
Oakle Street 30 B6
Oakley, Bucks 20 A2
Oakley, Beds 36 D5
Oakley, Fife 86 D6
Oakley, Suff 39 G3
Oakley, Hants 19 G5
Oakley Green 20 D6
Oakley Park 41 J5
Oakridge 30 D7
Oaks 42 F7
Oaksey 30 D8
Oakthorpe 44 E6
Oakwoodhill 12 E3
Oakworth 62 D6
Oare, Wilts 18 C4
Oare, Kent 15 G2
Oasby 46 D4
Oathlaw 93 F6
Oban 84 A1
Oborne 9 H3
Occlestone Green 53 H7
Occold 39 G3
Ochiltree 77 L5
Ockbrook 44 F4
Ockham 12 D1
Ockle 89 G1
Ockley 12 E2
Ocle Pychard 29 K3
Odcombe 9 G3
Odd Down 17 J4
Oddendale 67 H5
Oddingley 33 J5
Oddington, Glos 31 G5
Oddington, Oxon 31 K6
Odell 36 C5
Odiham 19 J5
Odstock 10 C2
Odstone 44 E7
Offchurch 34 E4
Offenham 34 B6
Offham, E. Susx 13 H5
Offham, Kent 13 K1
Offord Cluny 37 F4
Offord D'Arcy 37 F4
Offton 38 F6
Offwell 8 C5
Ogbourne Maizey 18 C3
Ogbourne St Andrew 18 C3
Ogbourne St George 18 D3
Ogle 75 H6
Ogmore 25 G7
Ogmore-by-Sea 25 G7
Ogmore Vale 25 H5
Okeford Fitzpaine 9 K3
Okehampton 5 F2
Okehampton Camp 5 F2
Old 35 J3
Old Aberdeen 99 J8
Old Alresford 11 G1
Old Basing 19 H5
Oldberrow 34 C4
Old Bewick 81 J7
Old Bolngbroke 57 H7
Oldborough 7 H5
Old Brampton 55 G6
Old Bridge of Urr 72 C7
Old Buckenham 38 F1
Old Burghclere 19 F5
Oldbury, W. Mids 34 A2
Oldbury, Warw 44 E8
Oldbury, Shrops 33 G1
Oldbury-on-Severn 17 H1
Oldbury on the Hill 17 K2
Old Byland 63 J1
Oldcastle 29 H5
Old Cleeve 16 B6
Old Clipstone 55 K7
Old Colwyn 51 H3
Oldcotes 55 J5
Old Dailly 70 D2
Old Dalby 45 H5
Old Deer 99 J4
Old Denaby 55 H4
Old Ellerby 59 F3
Old Felixstowe 39 J7
Oldfield 33 H4
Oldford 17 J5
Old Hall, The 59 G5
Oldham 53 L3
Oldhamstocks 80 F2
Old Heath 23 G1
Oldhurst 37 G3
Old Hutton 61 J1
Old Kea 3 G5
Old Kilpatrick 85 G7
Old Kinnernie 99 G8
Old Knebworth 21 G1
Oldland 17 H3
Old Leake 47 H2
Old Malton 64 E5
Oldmeldrum 99 H6
Old Milverton 34 E4
Old Monkland 78 E3
Old Newton 38 F4
Old Park 43 H7
Old Philpstoun 86 D7
Old Radnor 29 G2
Old Rayne 98 F6
Old Romney 15 G6
Oldshore Beg 104 C3
Oldshoremore 104 D3
Old Sodbury 17 J2
Old Somerby 46 C4
Oldstead 63 J1
Old Swarland 75 H3
Old Town, Northum 74 E4
Old Town, Cumbr 61 J1
Oldtown of Ord 98 F3
Old Warden 36 E6
Oldways End 7 J3
Old Weston 36 E3
Oldwhat 99 H3
Old Windsor 20 D6
Old Wives Lees 15 G3
Olgrinmore 107 F4
Oliver's Battery 11 F2
Ollaberry 113 C2
Ollach 111 F3
Ollerton, Shrops 43 H5
Ollerton, Ches 53 J6
Ollerton, Notts 55 K7
Olney 36 B5
Olton 34 C2
Olveston 17 H2
Ombersley 33 H4
Ompton 56 A7
Onchan 60 R4
Onecote 44 B2
Ongar Hill 47 J5
Ongar Street 32 C4
Onibury 32 D3
Onich 90 B5
Onllwyn 25 G3
Onneley 43 J3
Onslow Village 12 C2
Opinan, Highld 100 C4
Opinan, Highld 100 D2
Orby 57 J7
Orchard 9 K3
Orchard Portman 8 D2
Orcheston 18 B6
Orcop 29 J5
Orcop Hill 29 J5
Ord 111 G4
Ordhead 98 F7
Ordie 92 F1
Ordiquish 98 C3
Ore 14 E7
Oreton 33 F2
Orford, Ches 53 H4
Orford, Suff 39 K5
Orgreave 44 C6
Orinsay 108 C4
Orleton, H. & W 32 D4
Orleton, H. & W 33 F4
Orlingbury 36 B3
Ormesby 69 H5
Ormesby St Margaret 49 K6
Ormesby St Michael 49 K6
Ormiscaig 100 D2
Ormiston 87 H8
Ormsaigmore 89 F2
Ormskirk 52 F3
Orpington 21 J7
Orrell 53 G3
Orroland 71 J7
Orsett 22 C5
Orslow 43 K6
Orston 45 J3
Orton, Cumbr 67 J6
Orton, Northnts 35 K3
Orton Longueville 46 E8
Orton-on-the-Hill 44 E7
Orwell 37 G5
Osbaldeston 61 K6
Osbaston 44 F7
Osbournby 46 D4
Oscroft 53 G7
Ose 110 E2
Osgathorpe 44 F6
Osgodby, Lincs 56 E4
Osgodby, N. Yks 65 H4
Osgodby, N. Yks 63 K6
Oskaig 111 F3
Oskamull 89 F4
Osmaston 44 D3
Osmington 9 J6
Osmington Mills 9 J6
Osmotherley 69 G7
Osnaburgh or Dairsie 87 H3
Ospisdale 103 F5
Ospringe 15 G2
Ossett 55 F1
Ossington 56 B7
Ostend 22 F4
Oswaldkirk 63 K2
Oswaldtwistle 61 L7
Oswestry 42 D5
Otford 21 K8
Otham 14 D3
Othery 8 E1
Otley, W. Yks 62 F5
Otley, Suff 39 H5
Otterbourne 11 F2
Otterburn, N. Yks 62 B4
Otterburn, Northum 74 E4
Otterburn Camp 74 E4
Otterden Place 14 F3
Otter Ferry 84 B6
Otterham 4 B2
Ottershaw 20 E7
Otterswick 113 D2
Otterton 8 B6
Ottery St Mary 8 C5
Ottringham 59 G4
Oughtershaw 62 B1
Oughtibridge 55 G4
Oulston 63 J2
Oulton, Cumbr 66 E1
Oulton, Norf 49 G5
Oulton, W. Yks 63 G7
Oulton, Staffs 43 L4
Oulton, Suff 49 L8
Oulton Broad 39 L1
Oulton Street 49 G5
Oundle 36 D2
Ousby 67 J3
Ousden 38 C5
Ouseburn 63 H3
Ousefleet 58 C4
Ouston 68 E1
Outertown 112 B4
Outgate 66 F7
Outhgill 67 K6
Outlane 54 D2
Out Newton 59 H4
Out Rawcliffe 61 H5
Outwell, Cambs 47 J7
Outwell, Norf 47 J7
Outwood, Surrey 13 G2
Outwood, W. Yks 63 G7
Ovenden 62 D7
Over, Avon 17 G2
Over, Cambs 37 G3
Overbister 112 D2
Overbury 34 A7
Overcombe 9 H6
Over Haddon 54 F7
Over Kellet 61 J3
Over Kiddington 31 J5
Over Norton 31 H5
Overseal 44 D6
Over Silton 69 G7
Overstone 35 K4
Overstrand 49 H3
Overton, Clwyd 42 E3
Overton, Shrops 32 E3
Overton, D. & G 72 E7
Overton, Hants 19 G6
Overton, Lncs 61 H4
Overton, Grampn 99 H7
Overtown 79 F4
Over Wallop 18 D7
Over Whitacre 34 D1
Oving, Bucks 20 A1
Oving, W. Susx 11 L4
Ovingdean 13 G6
Ovingham 75 G7
Ovington, Essex 38 C6
Ovington, Durham 68 D5
Ovington, Norf 48 E7
Ovington, Hants 11 G1
Ovington, Northum 75 G7
Ower 10 E3
Owermoigne 9 J6
Owlswick 20 B3
Owmby-by-Spital 56 E5
Owslebury 11 G2
Owston 45 J7
Owston Ferry 58 C6
Owstwick 59 G3
Owthorpe 45 H4
Oxborough 48 C7
Oxenholme 67 H8
Oxenhope 62 D6
Oxen Park 61 G1
Oxenton 30 D4
Oxenwood 18 E5
Oxford 31 K7
Oxhill 34 E6
Oxley 43 L7
Oxley's Green 13 K4
Oxnam 74 C2
Oxshott 21 F7
Oxspring 55 F3
Oxted 13 G1
Oxton, Border 80 C4
Oxton, Notts 45 H2
Oxwich 24 C6
Oxwick 48 E5
Oykel Bridge 101 J1
Oyne 98 F6

P

Pabail Iarach 108 D3
Pabail Varach 108 D3
Packington 44 E6
Padanaram 92 F6
Padbury 35 J7
Paddington 21 G5
Paddlesworth 15 H5
Paddockhaugh 98 B3
Paddock Wood 13 K2
Paddolgreen 43 G4
Padeswood 52 D7
Padiham 62 B6
Padstow 3 H2
Padworth 20 A7
Pagham 11 K5
Paglesham Churchend 22 F4
Paglesham Eastend 22 F4
Paible, W. Isles 110 B1
Paible, W. Isles 108 B5
Paignton 5 J5
Pailton 35 F2
Painscastle 29 F3
Painshawfield 75 G7
Painswick 30 C7
Paisley 77 K1
Pakefield 39 L1
Pakenham 38 E4
Pale 51 J7
Palestine 18 D6
Paley Street 20 C6
Palgrave 39 G3
Palmerstown 25 K8
Palnackie 71 K6
Palnure 71 F5
Palterton 55 H7
Pamber End 20 A8
Pamber Green 19 H5
Pamber Heath 20 A7
Pamphill 10 A4
Pampisford 37 H6
Panbride 87 J1
Pancrasweek 6 C5
Pandy, Clwyd 42 C4
Pandy, Gwent 29 H5
Pandy, Powys 41 J3
Pandy Tudur 51 H4
Panfield 38 C8
Pangbourne 20 A6
Pannal 63 G4
Pant 42 D5
Panteg 26 D4
Pant Glas 50 D6
Pantgwyn 27 G3
Pant Mawr 41 H5
Panton 57 F6
Pant-pastynog 51 K4
Pantperthog 41 G3
Pant-y-dwr 41 J6
Pant-y-ffridd 42 C7
Pantyffynnon 27 L6
Panxworth 49 J6
Papcastle 66 D3
Papple 87 J7
Papplewick 45 G2
Papworth Everard 37 F4
Papworth St Agnes 37 F4
Par 3 J4
Parbold 53 F2
Parbrook 9 G1
Parcllyn 27 G2
Parc Seymour 16 F1
Pardshaw 66 C4
Parham 39 J4
Park Corner 20 A5
Park End, Northum 74 E6
Parkend, Glos 29 L7
Parkeston 39 H7
Parkgate, Ches 52 D6
Parkgate, Surrey 12 F2
Parkgate, D. & G 72 F5
Park Gate, Hants 11 G4
Parkgate, Ches 53 J6
Parkham 6 D3
Parkham Ash 6 D3
Parkhouse 29 K7
Parkhurst 11 F5
Parkmill 24 D6
Parkstone 10 B5
Parley Cross 10 B5
Parracombe 7 G1
Parrog 26 E4
Parson Cross 55 G4
Parson Drove 47 G7
Partick 77 L1
Partington 53 J4
Partney 57 J7
Parton, Cumbr 66 B4
Parton, D. & G 72 B6
Partridge Green 12 E5
Parwich 44 C2
Passenham 35 J7
Paston 49 J4
Patcham 13 G6
Patching 12 D6
Patchole 7 G1
Patchway 17 H2
Pateley Bridge 62 E3
Pathfinder Village 5 J2
Pathhead, Strath 72 B2
Pathhead, Lothn 80 B3
Pathhead, Fife 87 F5
Patmore Heath 37 H8
Patna 77 K6
Patney 18 B5
Patrick 60 P3
Patrick Brompton 68 E7
Patrington 59 H4
Patrixbourne 15 J3
Patterdale 67 F5
Pattingham 43 K8
Pattisball 35 H5
Patton Bridge 67 H7
Paul 2 C7
Paulerspury 35 J6
Paull 59 F4
Paulton 17 H5
Pavenham 36 B5
Pawston 81 G6
Paxford 34 C7
Paxton 81 H4
Payhembury 8 B4
Paythorne 62 B4
Pcaston 80 C3
Peacehaven 13 H6
Peak Dale 54 D6
Peak Forest 54 E6
Peakirk 46 E7
Peanmeanach 94 C7
Peasedown St John 17 J5
Peasemore 19 F3
Peasenhall 39 J4
Peaslake 12 D2
Peasmarsh 14 E6
Peaston Bank 80 C3
Peathill 99 J2
Peatling Magna 35 G1
Peatling Parva 35 G2
Peaton 32 E2
Pebmarsh 38 D7
Pebworth 34 C6
Pecket Well 62 C7
Peckforton 43 G2
Peckleton 45 F7
Pedmore 33 J2
Pedwell 9 F1
Peebles 79 K5
Peel 60 P3
Pegswood 75 J5
Peinchorran 111 F3
Peinlich 111 F2
Pelaw 75 J7
Pelcomb Cross 26 D6
Peldon 23 F2
Pelsall 44 B7
Pelton 68 E1
Peluth 66 D2
Pelynt 3 L4
Pembrey 24 C4
Pembridge 29 H2
Pembroke 26 D7
Pembroke Dock 26 D7
Pembury 13 K2
Penallt 29 K7
Penally 26 F8
Penare 3 H5
Penarth 25 K7
Pen-bont Rhydybeddau 41 F5
Penbryn 27 G2
Pencader 27 J4
Pencaitland 80 C3
Pencarreg 27 K3
Pencelli 25 J2
Pen-Clawdd 24 D5
Pencoed 25 H6
Pencombe 29 L2
Pencoyd 29 K5
Pencraig, Powys 41 K1
Pencraig, H. & W 29 K5
Pendeen 2 B6
Penderyn 25 H4
Pendine 27 G7
Pendlebury 53 J3
Pendleton 62 A6
Pendock 30 B4
Pendoggett 3 J2
Pendoylan 25 J7
Penegoes 41 G3
Penffordd 26 E5
Penffridd 50 E5
Penge 21 H6
Pengorffwysfa 50 D1
Penhalurick 2 F6
Penhow 17 F1
Penhurst 13 K5
Penicuik 79 K3
Peniel 51 K4
Penifiler 111 F2
Peninver 76 C5
Penisa'r Waun 50 E4
Penistone 55 F3
Penketh 53 G5
Penkridge 43 L6
Penley 42 F3
Penllergaer 24 E5
Pen-llyn, Gwyn 50 C2
Penllyn, S. Glam 25 H7
Penmachno 51 G5
Penmaen, W. Glam 24 D6
Penmaen, Gwent 25 K5
Penmaenmawr 51 G3
Penmaenpool 41 F2
Penmark 25 J8
Penmon 51 F2
Penmorfa 50 E6
Penmynydd 50 E3
Penn 20 D4
Pennal 41 G3
Pennan 99 H2
Pennant, Dyfed 40 E7
Pennant, Powys 41 H4
Pennant-Melangell 41 K1
Pennard 24 D6
Pennerley 42 E8
Pennington 60 F2
Pennjerick 2 F6
Penn Street 20 D4
Penny Bridge 61 G1
Pennycross 89 G6
Pennymoor 7 J4
Penparc 27 G3
Penparcau 40 E5
Penperlleni 29 H7
Penpillick 3 J4
Penpol 3 G6
Penpoll 3 K4
Penpont 72 D4
Penrherber 27 G4
Penrhiwceiber 25 J5
Penrhiw-llan 27 H3
Penrhiwpal 27 H3
Penrhos, Gwyn 50 B2
Penrhos, Gwyn 50 C7
Penrhos, Powys 25 G3
Penrhos, Gwent 29 J6
Penrhyn Bay 51 H2
Penrhyn-coch 41 F5
Penrhyndeudraeth 50 F7
Penrhyn-side 51 H2
Penrice 24 C6
Penrith 67 H3
Penrose 3 G2
Penruddock 67 G4
Penryn 3 F6
Pen-sarn, Gwyn 50 D6
Pen-sarn, Gwyn 40 E1
Pensarn, Clwyd 51 J3
Pensax 33 G4
Pensby 52 D5
Penselwood 9 J1
Pensford 17 H4
Penshaw 68 F1
Penshurst 13 J2
Pensilva 4 C5
Pentewan 3 J5
Pentir 50 E4
Pentire 3 F3
Pentney 48 C6
Penton Mewsey 18 E6
Pentraeth 50 E3
Pentre, Clwyd 42 C4
Pentre, Clwyd 42 D3
Pentre, Shrops 42 E6
Pentre, Clwyd 51 K4
Pentre, Powys 41 K5
Pentre-bach, Powys 25 H1
Pentrebach, M. Glam 25 J4
Pentre-bach, Dyfed 27 K3
Pentre Berw 50 D3
Pentre-bont 51 G5
Pentre-celyn 42 C2
Pentre-cwrt 27 H4
Pentre Dolau Honddu 28 D3
Pentredwr, Clwyd 42 C3
Pentre-dwr, W. Glam 24 E5
Pentrefelin, Gwyn 50 E7
Pentrefelin, Gwyn 51 H3
Pentrefoelas 51 H5
Pentregat 27 H2
Pentre-Gwenlais 27 L6
Pentre Halkyn 52 D6
Pentre-llyn-cymmer 51 J5
Pentre'r beirdd 42 C6
Pentre'r-felin 25 H1
Pentre-tafarn-y-fedw 51 H4
Pentre-ty-gwyn 25 G1
Pentrich 44 E2
Pentridge 10 B3
Pentwyn 25 L6
Pentyrch 25 K6
Penuwch 40 E7
Penwithick 3 J4
Penybanc 27 L5
Penybont, Powys 32 A4
Pen-y-bont, Clwyd 42 D5
Pen-y-bont, Dyfed 27 H5
Penybontfawr 41 K1
Pen-y-bryn, Gwyn 41 F2
Penybryn, M. Glam 25 K5
Penycae, Clwyd 42 D3
Pen-y-cae, Powys 25 G3
Pen-y-cae-mawr 29 J8
Pen-y-cefn 52 C6
Pen-y-clawdd 29 J7
Pen-y-coedcae 25 J6
Penycwm 26 C5
Penyffordd 52 E7
Pen-y-garn 27 K4
Pen-y-garnedd, Gwyn 50 E3
Penygarnedd, Powys 41 L1
Penygraig 25 H5
Penygroes, Gwyn 50 D5
Pen-y-groes, Dyfed 27 K6
Penysarn 50 D1
Pen-y-stryt 42 C2
Penywaun 25 H4
Penzance 2 C6
Peopleton 33 J5
Peover Heath 53 J6
Peper Harow 12 C2
Peplow 43 H5
Percie 93 G2
Percyhorner 99 J2
Perivale 21 F5
Perlethorpe 55 K6
Perranarworthal 2 F6
Perranporth 2 F4
Perranuthnoe 2 D7
Perranzabuloe 2 F4
Perry Barr 34 B1
Perry Green 21 J2
Pershore 33 J6
Pert 93 H5
Pertenhall 36 D4
Perth 86 E2
Perthy 42 E4
Perton 43 K8
Peterborough 46 E8
Peterburn 100 C3
Peterchurch 29 H4
Peterculter 93 K1
Peterhead 99 L4
Peterlee 69 G2
Petersfield 11 J2
Peter's Green 21 F2
Peters Marland 6 E4
Peterstone Wentlooge 16 D2
Peterston-super-Ely 25 J7
Peterstow 29 K5
Peter Tavy 4 F4
Petertown 112 B4
Petham 15 H3

Petrockstowe 6 F5
Pett 14 E7
Pettaugh 39 G5
Pettinain 79 G5
Pettistree 39 H5
Petton, Shrops 42 F5
Petton, Devon 7 L3
Petty 99 G5
Pettycur 87 F6
Pettymuic 99 J6
Petworth 12 C4
Pevensey Bay 13 K6
Pewsey 18 C4
Philham 6 C3
Philiphaugh 80 C7
Phillack 2 D6
Philleigh 3 G6
Philpstoun 86 D7
Phoenix Green 19 J5
Pica 66 C4
Piccotts End 20 E3
Pickering 64 E4
Picket Piece 18 E6
Pickhill 63 G1
Picklescott 42 F8
Pickmere 53 H6
Pickwell, Devon 6 E1
Pickwell, Leic 45 J6
Pickworth, Leic 46 C6
Pickworth, Lincs 46 D4
Picton, Ches 52 F6
Picton, N. Yks 69 G6
Piddinghoe 13 H6
Piddington, Oxon 20 A2
Piddington, Northnts 35 K5
Piddlehinton 9 J5
Piddletrenthide 9 J5
Pidley 37 G3
Piercebridge 68 E5
Pierowall 112 C2
Pigdon 75 H5
Pikehall 54 E8
Pilgrims Hatch 21 K4
Pilham 56 C4
Pill 17 G3
Pillaton 4 D5
Pillerton Hersey 34 E6
Pillerton Priors 34 D6
Pilleth 32 B4
Pilley 55 G3
Pilling 61 H5
Pilling Lane 61 G5
Pilning 17 G2
Pilsbury 54 E7
Pilsdon 8 F5
Pilsley, Derby 55 F6
Pilsley, Derby 55 H7
Pilton, Leic 46 C7
Pilton, Northnts 36 D2
Pilton, Somer 17 G6
Pimperne 9 L4
Pinchbeck 47 F5
Pinchbeck West 46 F5
Pinfold 52 E2
Pinhoe 5 K2
Pinmore 70 D2
Pinner 21 F5
Pinvin 34 A6
Pinwherry 70 C3
Pinxton 45 F2
Pipe and Lyde 29 K3
Pipe Gate 53 J7
Piperhill 97 G3
Pipewell 36 B2
Pippacot 7 F2
Pirbright 12 C1
Pirnmill 76 D3
Pirton, Herts 36 E7
Pirton, H. & W. 33 H6
Pishill 20 B5
Pistyll 50 C6
Pitblae 99 J2
Pitcairngreen 86 D2
Pitcalnie 103 G6
Pitcaple 99 G6
Pitchcombe 30 C7
Pitchcott 20 B1
Pitchford 43 G7
Pitch Green 20 B3
Pitch Place 12 C1
Pitcombe 9 H1
Pitcox 87 K7
Pitfichie 98 F7
Pitfour Castle 87 E2
Pitgrudy 103 F4
Pitlessie 87 G4
Pitlochry 91 L6
Pitmedden 99 H6
Pitminster 8 D3
Pitmunie 98 F7
Pitney 9 F2
Pitscottie 87 H3
Pitsea 22 D5
Pitsford 35 J4
Pitstone Green 20 D2
Pittendreich 98 A2
Pittentrail 103 F3
Pittenweem 87 J4
Pittington 68 F2
Pitton 10 D1
Pixey Green 39 H3
Place Newton 65 F5
Plains 78 E3
Plaish 43 G8
Plaistow 12 D3
Plaitford 10 D3
Plas 27 J5
Plas Gogerddan 41 F5
Plas Isaf 51 K6
Plas Llwyngwern 41 G3
Plas Llysyn 41 J4
Plastow Green 19 G4
Platt 13 K1
Plawsworth 68 E2
Plaxtol 13 K1
Playden 14 F6
Playford 39 H6
Play Hatch 20 B6
Playing Place 3 G5
Plealey 42 F7
Plean 85 L6
Pleasington 61 K7
Pleasley 55 J7
Plenmeller 74 D7
Pleshey 22 C2
Plockton 94 D2
Plocrapool 108 B5
Plocrapool Point 108 B5
Ploughfield 29 H3
Plowden 32 C2
Ploxgreen 42 E7
Pluckley 14 F4
Plumbland 66 D3
Plumley 53 J6
Plumpton, Cumbr 67 G3
Plumpton, E. Susx 13 G5
Plumpton, Lancs 61 G6
Plumpton Green 13 G5
Plumpton Head 67 H3
Plumstead 49 G4
Plumtree 45 H4
Plungar 45 J4
Plush 9 J4
Plwmp 27 H2
Plymouth 4 E6
Plymstock 4 F6
Plymtree 8 B4
Pockley 63 K1
Pocklington 64 F8
Pode Hole 46 F5
Podimore 9 G2
Podington 36 C4
Podmore 43 J4
Point Clear 23 G2
Pointon 46 E4
Pokesdown 10 C5
Polapit Tamar 4 D3
Polbain 104 A7
Polbathic 4 D6
Polbeth 79 H3
Polebrook 36 D2
Polegate 13 J6
Pole of Itlaw, The 98 F3
Polesworth 44 D7
Polglass 100 F1
Polgooth 3 H4
Poling 12 D6
Polkerris 3 J4
Pollington 55 K2
Polloch 89 J2
Pollokshaws 77 L1
Pollokshields 78 C3
Polmassick 3 H5
Polnessan 77 K6
Polperro 3 L4
Polruan 3 K4
Polsham 17 G6
Polstead 38 E7
Poltalloch 83 H2
Poltimore 5 K2
Polwarth 80 F4
Polyphant 4 C3
Polzeath 3 H2
Ponde 28 F4
Pondersbridge 37 F1
Ponders End 21 H4
Ponsanooth 2 F6
Ponsworthy 5 H4
Pontamman 27 L6
Pontantwn 27 J6
Pontardawe 24 F4
Pontardulais 27 K7
Pontarsais 27 J5
Pont Creuddyn 27 K2
Pont Cyfyng 51 G5
Pontefract 63 H7
Ponteland 75 H6
Ponterwyd 41 G5
Pontesbury 42 F7
Pontfadog 42 D4
Pontfaen, Dyfed 26 E4
Pont-faen, Powys 25 H1
Pont Henri 27 J7
Ponthir 16 E1
Ponthirwaun 27 G3
Pontllanfraith 25 K5
Pontlliw 24 E4
Pont Llogel 41 K2
Pontllyfni 50 D5
Pontlottyn 25 K4
Pontneddfechan 25 H4
Pont Pen-y-benglog 51 F4
Pontrhydfendigaid 41 G7
Pont Rhyd-y-cyff 25 G6
Pontrhydyfen 25 F5
Pont-rhyd-y-groes 41 G6
Pontrilas 29 H5
Pontrobert 42 C6
Pont-rug 50 E4
Ponts Green 13 K5
Pontshill 30 A5
Pont-Sian 27 J3
Pontsticill 25 J3
Pontyates 27 J7
Pontyberem 27 K6
Pontybodkin 52 D8
Pontyclun 25 J6
Pontycymer 25 H5
Pontymister 16 D1
Pont-y-pant 51 G5
Pontypool 29 G7
Pontypridd 25 J5
Pontywaun 25 L5
Pooksgreen 10 E3
Pool 2 E5
Poole 10 B5
Poole Keynes 30 E8
Poolewe 100 D3
Pooley Bridge 67 G4
Poolhill 30 B5
Pool of Muckhart 86 D4
Pool Quay 42 D6
Poorton 9 G5
Popham 19 G6
Poplar 21 H5
Porchfield 11 F5
Poringland 49 H7
Porkellis 2 E6
Porlock 7 J1
Port-an-eorna 94 C2
Port Ann 84 B6
Port Appin 89 L4
Port Askaig 82 D5
Port Bannatyne 77 F1
Portbury 17 G3
Port Carlisle 73 H7
Port Charlotte 82 B6
Portchester 11 H4
Portclair 96 C7
Port Driseach 84 B7
Port Ellen 82 C7
Port Elphinstone 99 G6
Portencross 77 G3
Port Erin 60 N5
Portesham 9 H6
Port e Vullen 60 R2
Port-Eynon 24 C6
Portfield Gate 26 D6
Portgate 4 E3
Portgaverne 3 J1
Port Glasgow 84 F7
Portgordon 98 C2
Portgower 103 J2
Porth 25 J5
Porthallow 3 F7
Porthcawl 25 G7
Porthcurno 2 B7
Port Henderson 100 C4
Porthgain 26 C4
Porthkerry 25 J8
Porthleven 2 E7
Porthllechog 50 D1
Porthmadog 50 E7
Porthmeor 2 C6
Porth Navas 2 F7
Portholland 3 H5
Porthoustock 3 G7
Porthpean 3 J4
Porthtowan 2 E5
Porthyrhyd, Dyfed 28 B4
Porthyrhyd, Dyfed 27 K6
Portincaple 84 E5
Portington 58 B3
Port Isaac 3 H1
Portishead 17 F3
Portknockie 98 D2
Portlethen 93 L2
Portlethen Village 93 L2
Portloe 3 H6
Port Logan 70 B7
Portmahomack 103 H5
Portmeirion 50 E7
Portmellon 3 J5
Port Mholair 108 D3
Port Mor 111 F6
Portmore 10 E5
Port Mulgrave 69 K5
Portnacroish 89 L4
Portnahaven 82 A6
Portnalong 110 E3
Portnaluchaig 94 B7
Portnancon 105 F2
Portobello 87 G7
Port of Menteith 85 H4
Port of Ness 108 D1
Porton 10 C1
Portpatrick 70 B6
Portquin 3 H1
Portreath 2 E5
Portree 111 F2
Portrye 77 G2
Portscatho 3 G6
Portsea 11 H4
Portskerra 106 D3
Portskewett 17 G2
Portslade 12 F6
Portslade-by-Sea 12 F6
Portsmouth 11 H4
Portsoy 98 E2
Port St Mary 60 P5
Port Sunlight 52 E5
Portswood 10 F3
Port Talbot 25 F5
Portway 34 B3
Port Wemyss 82 A6
Port William 70 E7
Portwrinkle 4 D6
Poslingford 38 C6
Postbridge 5 G4
Postcombe 20 B4
Postling 15 H5
Postwick 49 H7
Potarch 93 H2
Potsgrove 36 C8
Potten End 20 E3
Potterhanworth 56 E7
Potter Heigham 49 K6
Potteries, The 43 K3
Potterne 18 A5
Potterne Wick 18 B5
Potters Bar 21 G3
Potter's Cross 33 H2
Potterspury 35 J6
Potter Street 21 J3
Potterton 99 J7
Potto 69 G6
Potton 37 F6
Pott Row 48 C5
Pott Shrigley 53 L6
Poughill, Corn 6 C5
Poughill, Devon 7 J5
Poulshot 18 A5
Poulton 30 F7
Poulton-le-Fylde 61 G6
Pound Bank 33 G3
Pound Hill 13 F3
Poundon 35 H8
Poundsgate 5 H4
Poundstock 4 C2
Powburn 75 G2
Powderham 5 K3
Powerstock 9 G5
Powfoot 73 G7
Powick 33 H5
Powmill 86 D5
Poxwell 9 J6
Poyle 20 E6
Poynings 12 F5
Poyntington 9 H2
Poynton 53 L5
Poynton Green 43 G6
Poystreet Green 38 E5
Praa Sands 2 D7
Pratt's Bottom 21 J7
Praze-an-Beeble 2 E6
Predannack Wollas 2 E8
Prees 43 G4
Preesall 61 G5
Prees Green 43 G4
Preesgweene 42 D4
Prees Higher Heath 43 G4
Prendwick 75 G2
Pren-gwyn 27 J3
Prenteg 50 E6
Prenton 52 E5
Prescot 53 F4
Prescott 42 F5
Pressen 81 G6
Prestatyn 51 K2
Prestbury, Glos 30 D5
Prestbury, Ches 53 L6
Presteigne 32 C4
Presthope 43 G8
Prestleigh 17 H6
Preston, Glos 30 A4
Preston, Wilts 18 B3
Preston, Leic 46 B7
Preston, Glos 30 E7
Preston, Herts 21 F1
Preston, Humbs 59 F3
Preston, Border 81 F4
Preston, E. Susx 13 G6
Preston, Kent 15 J2
Preston, Devon 5 J4
Preston, Dorset 9 J6
Preston, Lancs 61 J7
Preston, Lothn 87 J7
Preston, Northum 81 K7
Preston Bagot 34 C4
Preston Bissett 35 H8
Preston Brockhurst 43 G5
Preston Brook 53 G5
Preston Candover 19 H6
Preston Capes 35 G5
Preston Gubbals 43 F6
Preston on Stour 34 D5
Preston on Wye 29 H3
Prestonpans 87 G7
Preston St Mary 38 E5
Preston-under-Scar 68 C7
Preston upon the Weald Moors 43 H6
Preston Wynne 29 K3
Prestwich 53 K3
Prestwick, Northum 75 H6
Prestwick, Strath 77 J5
Prestwood 20 C3
Price Town 25 H5
Prickwillow 38 A2
Priddy 17 G5
Priest Hutton 61 J2
Priest Weston 42 D8
Primethorpe 45 G8
Primrose Green 48 F6
Princes Risborough 20 C3
Princethorpe 34 F3
Princetown 5 F4
Priors Hardwick 35 F5
Priors Marston 35 F5
Priory, The 11 H5
Priory Wood 29 G3
Priston 17 H4
Prittlewell 22 E5
Privett 11 H2
Prixford 7 F2
Probus 3 G5
Proncy 103 F4
Prospect 66 D2
Prudhoe 75 G7
Puckeridge 21 H1
Puckington 8 E3
Pucklechurch 17 J3
Puddington, Ches 52 E6
Puddington, Devon 7 J4
Puddledock 38 F1
Puddletown 9 J5
Pudleston 29 K2
Pudsey 63 F6
Pulborough 12 D5
Puleston 43 J5
Pulford 52 E8
Pulham 9 J4
Pulham Market 39 G2
Pulham St Mary 9 H2
Pulloxhill 36 D7
Pulverbatch 42 F7
Pumsaint 27 L3
Puncheston 26 E5
Puncknowle 9 G6
Punnett's Town 13 K4
Purbrook 11 H4
Purfleet 21 K6
Puriton 16 E6
Purleigh 22 E3
Purley, Berks 20 A6
Purley, G. Lon 21 H7
Purlogue 32 B3
Purls Bridge 37 H2
Purse Caundle 9 H3
Purslow 32 C2
Purston Jaglin 55 H2
Purton, Glos 30 A7
Purton, Glos 30 A7
Purton, Wilts 18 B2
Purton Stoke 18 B1
Pury End 35 J6
Pusey 31 H8
Putley 33 F7
Putney 21 G6
Puttenham, Herts 20 C2
Puttenham, Surrey 12 C2
Puxton 16 F4
Pwll 24 C4
Pwllcrochan 26 D7
Pwllheli 50 C7
Pwllmeyric 17 G1
Pwll-y-glaw 25 F5
Pyecombe 13 F5
Pye Corner 16 E2
Pyle, I. of W. 11 F7
Pyle, M. Glam 25 G6
Pylle 17 H7
Pymore 37 H2
Pyrford 20 E8
Pyrton 20 A4
Pytchley 36 B3
Pyworthy 6 D5

Q

Quabbs 32 B2
Quadring 47 F4
Quaintn 20 B1
Quarley 18 D6
Quarndon 44 E3
Quarrier's Homes 77 J1
Quarrington 46 D3
Quarrington Hill 68 F3
Quarry Bank 33 J2
Quarry, The 30 B8
Quarrywood 98 A2
Quarter 78 E4
Quatford 33 G1
Quatt 33 G2
Quebec 68 D2
Quedgeley 30 C6
Queen Adelaide 37 J2
Queenborough 14 F1
Queen Camel 9 G2
Queen Charlton 17 H4
Queensbury 62 E6
Queensferry, Clwyd 52 E7
Queensferry, Lothn 86 E7
Queenzieburn 85 J7
Quendale 113 C6
Quendon 37 J7
Queniborough 45 H6
Quenington 31 F7
Quernmore 61 J3
Quethiock 4 D5
Quholm 112 B4
Quidenham 38 F2
Quidhampton, Wilts 10 C1
Quidhampton, Hants 19 G5
Quidnish 108 B5
Quilquox 99 J5
Quindry 112 C5
Quine's Hill 60 Q4
Quinton 35 J5
Quoditch 4 E2
Quorndon or Quorn 45 G6
Quorn or Quorndon 45 G6
Quothquan 79 G6
Quoyloo 112 B3
Quoys 113 E1

R

Raby 52 E5
Rachub 51 F4
Rackenford 7 J4
Rackham 12 D5
Rackheath 49 H6
Racks 73 F6
Rackwick, Orkney 112 B5
Rackwick, Orkney 112 C2
Radcliffe, Northum 75 J3
Radcliffe, G. Man 53 J3
Radcliffe on Trent 45 H4
Radclive 35 H7
Radcot 31 G8
Raddery 97 F3
Radernie 87 H4
Radford Semele 34 E4
Radlett 21 F4
Radley 31 K8
Radnage 20 B4
Radstock 17 H5
Radstone 35 G6
Radway 34 E6
Radway Green 43 J2
Radwell 37 F7
Radwinter 38 B7
Radyr 25 K6
Rafford 97 J3
Ragdale 45 H6
Raglan 29 J7
Ragnall 56 C6
Rainford 53 F3
Rainham, Kent 14 E2
Rainham, G. Lon 21 K5
Rainhill 53 F4
Rainow 54 C6
Rainton 63 G2
Rainworth 55 J8
Raisbeck 67 J6
Rait 87 F2
Raithby, Lincs 57 H5
Raithby, Lincs 57 H7
Rake 11 K2
Rame, Corn 4 E7
Rame, Corn 2 F6
Ram Lane 14 F4
Rampisham 9 G4
Rampside 60 F3
Rampton, Notts 56 B6
Rampton, Cambs 37 H4
Ramsbottom 53 J2
Ramsbury 18 D3
Ramscraigs 107 G7
Ramsdean 11 J2
Ramsdell 19 G5
Ramsden 31 H6
Ramsden Bellhouse 22 D4
Ramsden Heath 22 D4
Ramsey, Cambs 37 F2
Ramsey, Essex 39 H7
Ramsey, I. of M. 60 R2
Ramseycleuch 73 H2
Ramsey Forty Foot 37 G2
Ramsey Island 22 F3
Ramsey Mereside 37 F2
Ramsey St Mary's 37 F2
Ramsgate 15 K2
Ramsgill 62 E2
Ramshorn 44 B3
Ranby 55 K5
Rand 57 F6
Randwick 30 C7
Ranfurly 84 F8
Rangemore 44 C5
Rangeworthy 17 H2
Ranish 108 D3
Rankinston 77 K6
Ranskill 55 K5
Ranton 43 K5
Ranworth 49 J6
Rapness 112 C2
Rascarrel 71 J7
Raskelf 63 H2
Rassau 25 K3
Rastrick 62 E7
Ratagan 94 E4
Ratby 45 G7
Ratcliffe Culey 44 E8
Ratcliffe on the Wreake 45 H6
Rathen 99 K2
Rathillet 87 G2
Rathmell 62 B4
Ratho 86 E7
Ratho Station 86 E7
Rathven 98 D2
Ratley 34 E6
Ratlinghope 42 F8
Rattar 107 H2
Ratten Row 61 H5
Rattery 5 H5
Rattlesden 38 E5
Rattray 92 C7
Rauceby 46 D3
Raughton Head 67 F2
Raunds 36 C3
Ravenfield 55 H4
Ravenglass 66 C7
Raveningham 49 J8
Ravenscar 65 G2
Ravensdale 60 Q2
Ravensden 36 D5
Ravenseat 68 A6
Ravenshead 45 G2
Ravensmoor 43 H2
Ravensthorpe, W. Yks 55 F1
Ravensthorpe, Northnts 35 H3
Ravenstone, Bucks 36 B5
Ravenstone, Leic 44 F6
Ravenstonedale 67 K6
Ravenstown 61 G2
Ravenstruther 79 G5
Ravensworth 68 D6
Raw 65 G2
Rawcliffe, Humbs 58 A4
Rawcliffe, N. Yks 63 J4
Rawcliffe Bridge 58 B4
Rawdon 62 F6
Rawmarsh 55 H4
Rawreth 22 D4
Rawridge 8 D4
Rawtenstall 62 B7
Raydon 38 F7
Raylees 74 F4
Rayleigh 22 E4
Rayne 22 D1
Reach 37 J4
Read 62 A6
Reading 20 B6
Reading Street 14 F5
Reagill 67 J5
Rearquhar 103 F4
Rearsby 45 H6
Rease Heath 43 H2
Reaster 107 H3
Reawick 113 C4
Reay 106 E3
Rechullin 100 D6
Reculver 15 J2
Redberth 26 E7
Redbourn 21 F2
Redbourne 58 D6
Redbrook 29 K6
Redburn 97 H4
Redcar 69 J4
Redcastle 96 D4
Redcliff Bay 17 F3
Red Dial 66 E2
Redding 86 C7
Reddingmuirhead 86 C7
Reddish 53 K4
Redditch 34 B4
Rede 38 D5
Redenhall 39 H2
Redesdale Camp 74 E4
Redesmouth 74 E5
Redford, Durham 68 C3
Redford, Tays 93 G7
Redfordgreen 73 J2
Redgrave 38 F3
Redhill, Surrey 13 F1
Redhill, Avon 17 F4
Redhill, Grampn 99 F5
Redhill, Grampn 99 G8
Redisham 39 K2
Redland, Orkney 112 B3
Redland, Avon 17 G3
Redlingfield 39 G3
Redlynch, Wilts 10 D2
Redlynch, Somer 9 J1
Redmarley D'Abitot 30 B4
Redmarshall 69 F4
Redmile 45 J4
Redmire 68 C7
Redmoor 3 J3
Rednal 42 E5
Redpath 80 D6
Redpoint 100 C5
Red Rock 53 G3
Red Roses 27 G6
Red Row 75 J4
Redruth 2 F5
Red Street 43 K2
Red Wharf Bay 50 E2
Redwick, Gwent 16 F2
Redwick, Avon 17 G2
Redworth 68 E4
Reed 37 G7
Reedham 49 K7
Reedness 58 B4
Reef 108 B3
Reepham, Lincs 56 E6
Reepham, Norf 49 G5
Reeth 68 C7
Regaby 60 R2
Regoul 97 G3
Reiff 104 A7
Reigate 12 F1
Reighton 65 J5
Reiss 107 J4
Rejerrah 3 G4
Relubbus 2 D6
Relugas 97 H4
Remenham 20 B5
Remenham Hill 20 B5
Rempstone 45 G5
Rendcomb 30 E7
Rendham 39 J4
Rendlesham 39 J5
Renfrew 77 L1
Renhold 36 D5
Renishaw 55 H6
Rennington 75 J2
Renton 85 F7
Renwick 67 H2
Repps 49 K6
Repton 44 E5
Resolis 96 E2
Resolven 25 G4
Reston 81 G3
Retew 3 H4
Rettendon 22 D4
Rettendon Place 22 D4
Revesby 57 G7
Rewe 5 K2
Reydon 39 K3
Remerston 48 F7
Reynalton 26 E7
Reynoldston 24 C5
Rhadmad 40 E6
Rhandirmwyn 28 B3
Rhayader 41 J7
Rhedyn 50 B7
Rheindown 96 D4
Rhes-y-cae 52 C6
Rhewl, Clwyd 42 C3
Rhewl, Clwyd 51 L4
Rhiconich 104 D3
Rhicullen 102 E6
Rhigos 25 H4
Rhilochan 103 F3
Rhiroy 101 G3
Rhiw 40 B1
Rhiwbryfdir 51 F6
Rhiwderin 16 D2
Rhiwlas, Clwyd 42 C4
Rhiwlas, Gwyn 50 E4
Rhiwlas, Gwyn 51 J7
Rhodesia 55 J5
Rhodes Minnis 15 H4
Rhondda 25 H5
Rhonehouse or Kelton Hill 72 C8
Rhoose 25 J8
Rhos, W. Glam 25 F4
Rhos, Dyfed 27 H4
Rhoscefnhir 50 E3
Rhoscolyn 50 B3
Rhoscrowther 26 D7
Rhosesmor 52 D7
Rhos-fawr 50 C7
Rhosgadfan 50 E5
Rhosgoch, Gwyn 50 D2
Rhos-goch, Powys 29 F3
Rhoshirwaun 50 A8
Rhoslan 50 D6
Rhoslefain 40 E3
Rhosllanerchrugog 42 D3
Rhosmeirch 50 D3
Rhosneigr 50 C3
Rhosnesni 42 E2
Rhos-on-Sea 51 H2
Rhossili 24 C6
Rhosson 26 B5
Rhostrehwfa 50 D3
Rhostryfan 50 D5
Rhostyllen 42 E3
Rhosybol 50 D2
Rhos-y-brithdir 42 C5
Rhos-y-gwaliau 51 J7
Rhos-y-llan 50 B7
Rhu 84 E6
Rhuallt 51 K3
Rhuddlan 51 K3
Rhue 101 F2
Rhulen 28 F3
Rhunahaorine 76 C3
Rhyd, Gwyn 51 F6
Rhyd, Powys 41 J3
Rhydargaeau 27 J5
Rhydcymerau 27 K4
Rhydd 33 H6
Rhyd-Ddu 50 E5
Rhydding 25 F5
Rhyd-foel 51 J3
Rhydlewis 27 H3
Rhydlios 50 A7
Rhydlydan 51 H5
Rhydowen 27 J3
Rhyd-Rosser 40 E7
Rhydtalog 42 D2
Rhydwyn 50 C2
Rhyd-y-clafdy 50 C7
Rhydycroesau 42 D4
Rhydyfelin, Dyfed 40 E6
Rhydyfelin, M. Glam 25 J6
Rhyd-y-fro 24 F4
Rhydymain 41 H1
Rhydymwyn 52 D7
Rhyd-yr-onen 40 F3
Rhyl 51 K2
Rhymney 25 K4
Rhynd 86 E2
Rhynie 98 D6
Ribbesford 33 G3
Ribble Head 62 A2
Ribbleton 61 J6
Ribchester 61 K6
Riby 59 F6
Riccall 63 K6
Riccarton 77 K4
Richards Castle 32 D4
Richmond, N. Yks 68 D6
Richmond, G. Lon 21 F6
Rickarton 93 K3
Rickinghall 38 F3
Rickling 37 H7
Rickmansworth 20 E4
Riddell 80 D7
Riddlecombe 7 G4
Riddlesden 62 D5
Ridge, Wilts 10 A1
Ridge, Dorset 10 A6
Ridge, Herts 21 G3
Ridgehill 17 G4
Ridge Lane 44 D8
Ridgeway Cross 33 G6
Ridgewell 38 C6
Ridgewood 13 H5
Ridgmont 36 C7
Riding Mill 75 G7
Ridlington, Leic 46 B7
Ridlington, Norf 49 J4
Ridsdale 74 F5
Rienachait 104 B5
Rievaulx 63 J1
Rigg 73 H7
Riggend 85 K7
Rigside 79 F6
Rileyhill 44 C6
Rilla Mill 4 C4
Rillington 64 F5
Rimington 62 B5
Rimpton 9 H2
Rimswell 59 H4
Rinaston 26 D5
Ringasta 113 C6
Ringford 71 H6
Ringland 49 G6
Ringmer 13 H5
Ringmore 5 G7
Ringorm 98 B4
Ring's End 47 G7
Ringsfield 39 K2
Ringshall, Herts 20 D2
Ringshall, Suff 38 F5
Ringshall Stocks 38 F5
Ringstead, Norf 48 C3
Ringstead, Northnts 36 C3
Ringwood 10 C4
Ringwould 15 K4
Rinmore 98 D7
Rinnigill 112 B5
Rinsey 2 D7
Ripe 13 J5
Ripley, Hants 10 C5
Ripley, Surrey 12 D1
Ripley, Derby 44 F2
Ripley, N. Yks 63 F3
Riplingham 58 D3
Ripon 63 G2
Rippingale 46 D5
Ripple, H. & W. 33 H7
Ripple, Kent 15 K3
Ripponden 54 D2
Rireavach 100 F2
Risabus 82 C7
Risbury 29 K2
Risby, Suff 38 D4
Risby, Humbs 58 D5
Risca 16 D1
Rise 59 F2
Risegate 46 F5
Riseley, Berks 20 B7
Riseley, Beds 36 D4
Rishangles 39 G4
Rishton 61 L6
Rishworth 54 D2
Rising Bridge 62 A7
Risley, Derby 45 F4
Risley, Ches 53 H4
Risplith 63 F3
Rivar 18 E4
Rivenhall End 22 E2
River Bank 37 J4
Riverhead 13 J1
Rivington 53 H2
Roachill 7 J3
Roade 35 J5
Roadmeetings 79 F5
Roadside 107 G3
Roadside of Catterline 93 K4
Roadside of Kinneff 93 K4
Roadwater 16 B7
Roag 110 E2
Roath 25 K7
Robbingworth 21 K3
Roberton, Strath 79 G7
Roberton, Border 73 K2
Robertsbridge 13 L4
Roberttown 62 E7
Robeston Cross 26 C7
Robeston Wathen 26 E6
Robin Hood's Bay 65 G2
Roborough 7 F4
Roby Mill 53 G3
Rocester 44 C4
Roch 26 C5
Rochdale 53 K2
Roche 3 H3
Rochester, Kent 14 D2
Rochester, Northum 74 E4
Rochford, Essex 22 E4
Rochford, H. & W. 33 F4
Rock, H. & W. 33 G3
Rock, Corn 3 H2
Rock, Northum 75 J1
Rockbeare 8 B5
Rockbourne 10 C3
Rockcliffe, Cumbr 73 J7
Rockcliffe, D. & G. 71 K6
Rock Ferry 52 E5
Rockfield, Highld 103 H5
Rockfield, Gwent 29 J6
Rockhampton 30 A8
Rockingham 36 B1
Rockland All Saints 48 E8
Rockland St Mary 49 J7
Rockland St Peter 48 E8
Rockley 18 C3
Rockwell End 20 B5
Rodbourne 18 A2
Rodd 32 C4
Roddam 75 G1
Rodden 9 H6
Rode 17 K5
Rodeheath, Ches 53 K7
Rode Heath, Ches 53 K8
Rodel 108 B5
Roden 43 G6
Rodhuish 16 B7
Rodington 43 G6
Rodley 30 B6
Rodmarton 30 D8
Rodmell 13 H6
Rodmersham 14 F2
Rodney Stoke 17 F5
Rodsley 44 D3
Roecliffe 63 G3
Roehampton 21 G6
Roewen 51 G3
Roffey 12 E3
Rogart 103 F3
Rogate 11 K2
Rogerstone 16 D2
Rogerton 78 D4
Rogiet 17 F2
Roker 75 L8
Rollesby 49 K6
Rolleston, Staffs 44 D5
Rolleston, Notts 45 J2
Rolleston, Leic 45 J7
Rolston 59 G2
Rolvenden 14 E5
Rolvenden Layne 14 E5
Romaldkirk 68 B4
Romanby 69 F7
Romannobridge 79 J5
Romansleigh 7 H3
Romford 21 K5
Romiley 53 L4
Romsey 10 E2
Romsley, H. & W. 34 A2
Romsley, Shrops 33 G2
Ronague 60 P4
Rookby 67 L5
Rookhope 68 B2
Rookley 11 G6
Rooks Bridge 16 E5
Roos 59 G3
Roosebeck 60 F3
Rootpark 79 G4
Ropley 11 H1
Ropley Deane 11 H1
Ropsley 46 C4
Rora 99 K3
Rorrington 42 E7
Rose 2 F4
Roseacre 61 H6
Rose Ash 7 H3
Rosebank 79 F5
Rosebush 26 E5
Rosedale Abbey 64 E3
Roseden 81 J7
Rosehall 102 C3
Rosehearty 99 J2
Rosehill 43 H4
Roseisle 97 K2
Rosemarket 26 D7
Rosemarkie 97 F3
Rosemary Lane 8 C3
Rosemount, Tays 92 C7
Rosemount, Strath 77 J5
Rosenannon 3 H3
Rosewell 79 K3
Roseworthy 2 E6
Rosgill 67 H5
Roshven 94 C8
Roskhill 110 E2
Rosley 66 F2
Roslin 79 K3
Rosliston 44 D6
Rosneath 84 E6
Ross, D. & G. 71 H7
Ross, Northum 81 K6
Rossett 52 E8
Rossington 55 K4
Rossland 85 G7
Ross-on-Wye 29 L5
Roster 107 H6
Rostherne 53 J5
Rosthwaite 66 E5
Roston 44 C3
Rosyth 86 E6
Rothbury 75 G3
Rotherby 45 H6
Rotherfield 13 J4
Rotherfield Greys 20 B5
Rotherfield Peppard 20 B5
Rotherham 55 H4
Rothersthorpe 35 J5
Rotherwick 19 J5
Rothes 98 B4
Rothesay 77 F1
Rothiebrisbane 99 G5
Rothienorman 99 G5
Rothiesholm 112 D3
Rothley 45 G6
Rothmaise 99 F5
Rothwell, Northnts 36 B2
Rothwell, Lincs 59 F7
Rothwell, W. Yks 63 G7
Rotsea 65 H7
Rottal 92 E5
Rottingdean 13 G6
Rottington 66 B5
Roud 11 G6
Rougham 48 D5
Rougham Green 38 E4
Roughburn 90 E3
Rough Close 43 L4
Rough Common 15 H3
Roughlee 62 B5
Roughley 44 C8
Roughsike 74 B6
Roughton, Lincs 57 G7
Roughton, Norf 49 H4
Roughton, Shrops 43 J8
Roundhay 63 G6
Roundstreet Common 12 D4
Roundway 18 B4
Rounton 69 G6
Rousdon 8 D5
Rous Lench 34 B5
Routh 59 E2
Row, Cumbr 67 G8
Row, Corn 3 J2
Rowanburn 73 K6
Rowde 18 A4
Rowfoot 74 C7
Rowhedge 23 G1
Rowhook 12 E3
Rowington 34 D4
Rowland 54 F6
Rowland's Castle 11 J3
Rowlands Gill 75 H8
Rowledge 19 K6
Rowlestone 29 H5
Rowley 42 E7
Rowley Regis 34 A2
Rowly 12 D2
Rowney Green 34 B3
Rownhams 10 E3
Rowsham 20 C2
Rowsley 55 F7
Rowston 46 D2
Rowton, Ches 52 F7
Rowton, Shrops 43 H6
Roxburgh 80 E6
Roxby, Humbs 58 D5
Roxby, N. Yks 69 K5
Roxton 36 E5
Roxwell 22 C3
Royal British Legion Village 13 L1
Royal Leamington Spa 34 E4
Royal Tunbridge Wells 13 J3
Roybridge 90 D3
Roydon, Norf 48 C5
Roydon, Norf 39 F2
Roydon, Essex 21 J2
Royston, S. Yks 55 G2
Royston, Herts 37 G6
Royton 53 L3
Ruabon 42 E3
Ruaig 88 B4
Ruan Lanihorne 3 G5
Ruan Minor 2 F8
Ruardean 29 L6
Ruardean Woodside 29 L6
Rubery 34 A3
Rubha Bhocaig 108 B5
Rubha Cam nan Gall 110 C2
Rubha Crago 108 C5
Rubh' Aird-mhicheil 110 B3
Ruckcroft 67 H2
Ruckinge 15 G5
Ruckland 57 H6
Ruckley 43 G7
Ruddington 45 G4
Rudge 17 K5
Rudgeway 17 H2
Rudgwick 12 D3
Rudhall 29 L5
Rudloe 17 K4
Rudry 25 K6
Rudston 65 H6
Rudyard 54 C8
Rufford 52 F2
Rufforth 63 J4
Rugby 35 G3
Rugeley 44 B6
Ruilick 96 D4
Ruisgarry 108 A5
Ruishton 8 D2
Ruislip 20 E5
Ruislip Common 20 E5
Rumbling Bridge 86 D5
Rumburgh 39 J2
Rumford 3 G2
Rumney 25 L7
Runcorn 53 G5
Runcton 11 K4
Runcton Holme 48 B7
Runfold 19 K6
Runhall 48 F7
Runham 49 K6
Runnington 8 C2
Runswick Bay 69 L5
Runwell 22 D4
Ruscombe 20 B6
Rushall, H. & W. 30 A4
Rushall, W. Mids 44 B7
Rushall, Wilts 18 C5
Rushall, Norf 39 G2
Rushbrooke 38 D4
Rushbury 32 E1
Rushden, Northnts 36 C4
Rushden, Herts 37 G7
Rushford 38 E2
Rush Green 21 K5
Rushlake Green 13 K5
Rushmere 39 K2
Rushmere St Andrew 39 H6
Rushmoor 19 K6
Rushock 33 H3
Rusholme 53 K4
Rushton, Northnts 36 B2
Rushton, Ches 53 G7
Rushton, Shrops 43 H7
Rushton Spencer 53 L7
Rushwick 33 H5
Rushyford 68 E4
Ruskie 85 J4
Ruskington 46 D2
Rusland 61 G1
Rusper 12 F3
Ruspidge 30 A6
Russell's Water 20 B5
Rustington 12 D6
Ruston Parva 65 H6
Ruswarp 65 F2
Rutherford 80 E6
Rutherglen 78 C3
Ruthernbridge 3 J3
Ruthin 52 C8
Ruthrieston 99 J8
Ruthven, Tays 92 D7
Ruthven, Grampn 98 E4
Ruthven, Highld 97 G5
Ruthvoes 3 H3
Ruthwell 73 G7
Ruyton-XI-Towns 42 E5
Ryal 75 G6
Ryal Fold 61 K7
Ryall 8 F5
Ryarsh 14 C3
Rydal 66 F6
Ryde 11 G5
Rye 14 F6
Rye Foreign 14 F6
Rye Harbour 14 F7
Ryhall 46 D6
Ryhill 55 G2
Ryhope 69 G1
Rylstone 62 C4
Ryme Intrinseca 9 G3
Ryther 63 J6
Ryton, Glos 30 B4
Ryton, N. Yks 64 E5
Ryton, T. & W. 75 H7
Ryton, Shrops 43 J7
Ryton-on-Dunsmore 34 E3

S

Saasaig 111 G4
Sabden 62 A6
Sacombe 21 H2
Sacriston 68 E2
Sadberge 69 F5
Saddell 76 C4
Saddington 35 H1
Saddle Bow 48 B6
Saddlethorpe 58 C4
Sadgill 67 G6
Saffron Walden 37 J7
Saham Toney 48 E7
Saighton 52 F7
St Abbs 81 H3
St Agnes 2 F4
St Albans 21 F3
St Allen 3 G4
St Andrews 87 J3
St Andrews Major 25 K7
St Annes 61 G7
St Ann's 73 F4
St Ann's Chapel, Corn 4 E4
St Ann's Chapel, Devon 5 G7
St Anthony-in-Meneage 3 F7
St Arvans 29 K8
St Asaph or Llanelwy 51 K3
St Athan 25 J8
St Austell 3 J4
St Bees 66 B5
St Blazey 3 J4
St Boswells 80 D6
St Breock 3 H2
St Breward 3 J2
St Briavels 29 K7
St Brides 26 B6
St Bride's Major 25 G7
St Bride's-super-Ely 25 J7
St Brides Wentlooge 16 D2
St Budeaux 4 E6
Saintbury 34 C7
St Buryan 2 C7
St Catherine 17 J3
St Catherines 84 D4
St Clears 27 G6
St Cleer 3 L3
St Clement 3 G5
St Clether 4 C3
St Colmac 76 F1
St Columb Major 3 H3
St Columb Minor 3 G3
St Columb Road 3 H4
St Combs 99 K2
St Cross South Elmham 39 H2
St Cyrus 93 J5
St David's, Dyfed 26 B5
St David's, Tays 86 C2
St Davids, Fife 86 E6
St Day 2 F5
St Dennis 3 H4
St Dogmaels 26 F3
St Dominick 4 E5
St Donats 25 H8
St Endellion 3 H2
St Enoder 3 G4
St Erme 3 G5
St Erth 2 D6
St Erth Praze 2 D6
St Ervan 3 G2
St Eval 3 G3
St Ewe 3 H5
St Fagans 25 K7
St Fergus 99 K3
St Fillans 85 J2
St Florence 26 E7
St Gennys 4 B2
St George 51 J3
St Georges, Avon 16 E4
St George's, S. Glam 25 K7
St Germans 4 D6
St Giles in the Wood 7 F4
St Giles on the Heath 4 D2
St Harmon 41 J6
St Helen Auckland 68 D4
St Helens, Mers 53 G4
St Helens, I. of W 11 H6
St Hilary, Corn 2 D6
St Hilary, S. Glam 25 J7
Saint Hill 13 G3
St Illtyd 25 L4
St Ishmael's 26 C7
St Issey 3 H2
St Ive 4 D5
St Ives, Dorset 10 C4
St Ives, Corn 2 D5
St Ives, Cambs 37 G3
St James South Elmham 39 J2
St John 4 E6
St Johns, H. & W 33 H5
St John's, I. of M 60 P3
St John's Chapel 68 A3
St John's Fen End 47 J6
St John's Highway 47 J6
St John's Town of Dalry 72 B5
St Judes 60 Q2
St Just 2 B6
St Just in Roseland 3 G6
St Katherines 99 G5
St Keverne 3 F7
St Kew 3 J2
St Kew Highway 3 J2
St Keyne 3 L3
St Lawrence, Essex 23 F3
St Lawrence, I. of W 11 G7
St Lawrence, Corn 3 J3
St Leonards, Dorset 10 C4
St Leonards, Bucks 20 D3
St Leonards, E. Susx 14 D8
St Leven 2 B7
St Ippollitts 36 E8
St Lythans 25 K7
St Mabyn 3 J2
St Margarets 29 H4
St Margaret's at Cliffe 15 K4
St Margaret's Hope 112 C5
St Margaret South Elmham 39 J2
St Mark's 60 P4
St Martin, Corn 4 C6
St Martin, Corn 2 F7
St Martins, Tays 86 E1
St Martin's, Shrops 42 E4
St Mary Bourne 18 F5
St Mary Church 25 J7
St Mary Cray 21 J7
St Mary Hill 25 H7
St Mary Hoo 14 E1
St Mary in the Marsh 15 G6
St Mary's 112 C4
St Mary's Bay 15 G6
St Mawes 3 G6
St Mawgan 3 G3
St Mellion 4 D5
St Mellons 16 D2
St Merryn 3 G2
St Mewan 3 H4
St Michael Caerhays 3 H5
St Michael Penkevil 3 G5
St Michaels, H. & W 32 E4
St Michaels, Kent 14 E5
St Michael's on Wyre 61 H5
St Michael South Elmham 39 J2
St Minver 3 H2
St Monance 87 J4
St Neot 3 K3
St Neots 36 E4
St Newlyn East 3 G4
St Nicholas, Dyfed 26 D4
St Nicholas, S. Glam 25 J7
St Nicholas at Wade 15 J2
St Ninians 85 K5
St Osyth 23 H2
St Owen's Cross 29 K5
St Paul's Cray 21 J7
St Paul's Walden 21 F1
St Peter's 15 K2
St Petrox 26 D8
St Pinnock 3 L3
St Quivox 77 J5
St Stephen 3 H4
St Stephens, Corn 4 D3
St Stephens, Corn 4 E6
St Teath 3 J1
St Tudy 3 J2
St Twynnells 26 D8
St Vigeans 93 H7
St Wenn 3 H3
St Weonards 29 J5
St Winnow 3 K4
Salcombe 5 H8
Salcombe Regis 8 C6
Salcott 22 F2
Sale 53 J4
Saleby 57 J6
Sale Green 33 J5
Salehurst 13 L4
Salem, Dyfed 41 F5
Salem, Dyfed 27 L5
Salen, Strath 89 G4
Salen, Highld 89 H2
Salesbury 61 K6
Salford, Beds 36 C7
Salford, Oxon 31 G5
Salford, G. Man 53 K4
Salford Priors 34 B5
Salfords 13 F2
Salhouse 49 H6
Saline 86 D5
Salisbury 10 C2
Sallachy 94 E2
Salle 49 G5
Salmonby 57 H6
Salperton 30 E5
Salph End 36 D5
Salsburgh 79 F3
Salt 44 A5
Saltash 4 E6
Saltburn 97 F2
Saltburn-by-the-Sea 69 J4
Saltby 46 B5
Saltcoats 77 H3
Saltdean 13 G6
Salter 61 K3
Salterforth 62 B5
Salterswall 53 H7
Saltfleet 57 J4
Saltfleetby All Saints 57 J4
Saltfleetby St Clement 57 J4
Saltfleetby St Peter 57 J5
Saltford 17 H4
Salthaugh Grange 59 G4
Salthouse 48 F3
Saltmarshe 58 B4
Salton 64 E4
Saltwood 15 H5
Salum 88 B4
Salwarpe 33 H4
Salwayash 9 F5
Samala 110 B1
Sambourne 34 B4
Sambrook 43 J5
Samlesbury 61 J6
Samlesbury Bottoms 61 K7
Sampford Arundel 8 C3
Sampford Brett 16 B6
Sampford Courtenay 7 G5
Sampford Peverell 8 B3
Sampford Spiney 5 F4
Samuelston 87 H7
Sanachan 94 D1
Sancreed 2 C7
Sand, Shetld 113 C4
Sand, Highld 100 E2
Sandaig 94 C5
Sandbach 53 J7
Sandbank 84 D6
Sandbanks 10 B6
Sandend 98 E2
Sanderstead 21 H7
Sandford, Dorset 10 A6
Sandford, Strath 78 E5
Sandford, Avon 17 F5
Sandford, Devon 7 J5
Sandford, Cumbr 67 K5
Sandfordhill 99 L4
Sandford-on-Thames 31 K7
Sandford Orcas 9 H2
Sandford St Martin 31 J5
Sandgate 15 J5
Sandgreen 71 G6
Sandhaven 99 J2
Sandhead 70 B6
Sandhoe 75 F7
Sandholme, Humbs 58 C3
Sandholme, Lincs 47 G4
Sandhurst, Glos 30 C5
Sandhurst, Berks 20 C7
Sandhurst, Kent 14 D6
Sand Hutton, N. Yks 64 D7
Sandhutton, N. Yks 63 G1
Sandiacre 45 F4
Sandilands 57 K5
Sandiway 53 H6
Sandleheath 10 C3
Sandleigh 31 J7
Sandling 14 D3
Sandness 113 B4
Sandon, Essex 22 D3
Sandon, Herts 37 G7
Sandon, Staffs 43 L5
Sandown 11 G6
Sandplace 3 L4
Sandridge 21 F2
Sandringham 48 B5
Sandsend 69 L5
Sandsound 113 C4
Sandtoft 58 B6
Sandwich 15 K3
Sandwick, W. Isles 110 C2
Sandwick, Orkney 112 C5
Sandwick, W. Isles 108 D3
Sandwick, Shetld 113 D5
Sandwick, Cumbr 67 G5
Sandwith 66 B5
Sandy 36 E6
Sandycroft 52 E7
Sandygate 60 Q2
Sandyhills 66 A1
Sandy Lane 18 A4
Sangobeg 105 F2
Sannox 76 F3
Sanquhar 72 C3
Santon 58 D3
Santon Bridge 66 D6
Santon Downham 38 D2
Sapcote 45 F8
Sapey Common 33 G4
Sapiston 38 E3
Sapperton, Lincs 46 D4
Sapperton, Glos 30 D7
Saracen's Head 47 G5
Sarclet 107 J5
Sarisbury 11 G4
Sarn, Powys 32 B1
Sarn, M. Glam 25 H6
Sarnau, Powys 42 D6
Sarnau, Dyfed 27 H2
Sarnau, Dyfed 27 H6
Sarnau, Gwyn 51 J7
Sarn Bach 40 C1
Sarnesfield 29 H2
Sarn Meyllteyrn 50 B7
Saron, Dyfed 27 H4
Saron, Dyfed 27 L6
Sarratt 20 E4
Sarre 15 J2
Sarsden 31 G5
Satley 68 D2
Satterleigh 7 G3
Satterthwaite 66 F7
Sauchen 99 G7
Saucher 87 E1
Sauchieburn 93 H5
Sauchrie 77 J6
Saughall 52 E6
Saughtree 74 B4
Saul 30 B7
Saundby 56 B5
Saundersfoot 26 F7
Saunderton 20 B3
Saunton 6 E2
Sausthorpe 57 H7
Saval 102 D3
Savon Street 38 B5
Sawbridgeworth 21 J2
Sawdon 65 G4
Sawley, Lancs 62 A5
Sawley, N. Yks 63 F3
Sawley, Derby 45 F4
Sawrey 67 F7
Sawston 37 H6
Sawtry 36 E2
Saxby, Leic 46 B5
Saxby, Lincs 56 E5
Saxby All Saints 58 D5
Saxelbye 45 J5
Saxilby 56 C6
Saxlingham 48 F4
Saxlingham Nethergate 49 H8
Saxmundham 39 J4
Saxondale 45 H4
Saxtead 39 H4
Saxtead Green 39 H4
Saxthorpe 49 G4
Saxton 63 H6
Sayers Common 12 F5
Scackleton 63 K2
Scadabay 108 B5
Scaftworth 55 K4
Scagglethorpe 64 F5
Scalasaig 82 C2
Scalby 65 H3
Scaldwell 35 J3
Scaleby 73 K7
Scaleby Hill 73 K7
Scale Houses 67 H2
Scales, Cumbr 60 F2
Scales, Cumbr 66 F4
Scalford 45 J5
Scaling 69 K5
Scalloway 113 D5
Scamblesby 57 G6
Scampston 64 F5
Scampton 56 D6
Scarastavore 108 B5
Scarborough 65 H4
Scarcliffe 55 H7
Scarcroft 63 G5
Scardroy 96 A3
Scarfskerry 107 H2
Scargill 68 C5
Scarinish 88 B4
Scarisbrick 52 E2
Scarning 48 E6
Scarrington 45 J3
Scarth Hill 52 F3
Scartho 59 G6
Scaur or Kippford 71 K6
Scawby 58 D6
Scawton 63 J1
Scayne's Hill 13 G4
Scethrog 25 K2
Scholar Green 53 K8
Scholes, W. Yks 54 E3
Scholes, W. Yks 63 G6
Scleddau 26 D4
Scole 39 G3
Scolpaig 110 B1
Scolton 26 D5
Sconser 111 F3
Scoor 88 F7
Scopwick 56 E8
Scoraig 100 F2
Scorborough 58 E2
Scorrier 2 F5
Scorton, N. Yks 68 E6
Scorton, Lancs 61 J5
Sco Ruston 49 H5
Scotasay 108 B5
Scotby 67 G1
Scotch Corner 68 E6
Scotforth 61 H4
Scothern 56 E6
Scotland Gate 75 J5
Scotlandwell 86 E4
Scotsburn 103 F6
Scots' Gap 75 G5
Scots Hole 57 F7
Scotstown 89 K2
Scottas 94 C5
Scotter 58 C6
Scotterthorpe 58 C6
Scotton, Lincs 58 C7
Scotton, N. Yks 68 D7
Scotton, N. Yks 63 G4
Scottow 49 H5
Scoughall 87 K6
Scoulton 48 E7
Scourie 104 C4
Scousburgh 113 C6
Scrabster 107 F2
Scrainwood 75 F3
Scrane End 47 G3
Scraptoft 45 H7
Scratby 49 L6
Scrayingham 64 E6
Screapadal 111 F2
Scredington 46 D3
Scremby 57 J7
Scremerston 81 J5
Scriven 63 G4
Scrooby 55 K4
Scropton 44 C4
Scrub Hill 47 F2
Scruton 68 F7
Sculthorpe 48 D4
Scunthorpe 58 C5
Scurlage 24 C6
Seaborough 9 F4
Seacombe 52 E4
Seacroft 57 K7
Seaford 13 H7
Seaforth 52 E4
Seaforth Head 108 C4
Seagrave 45 H6
Seaham 69 G2
Seahouses 81 L6
Seal 13 J1
Sealand 52 E7
Seamer, N. Yks 69 G5
Seamer, N. Yks 65 H4
Seamill 77 H3
Sea Palling 49 K5
Searby 58 E6
Seasalter 15 G2
Seascale 66 C6
Seathorne 57 K7
Seathwaite, Cumbr 66 E5
Seathwaite, Cumbr 66 E7
Seaton, Cumbr 66 C3
Seaton, Leic 46 C8
Seaton, Devon 8 D5
Seaton, Corn 4 D6
Seaton, Humbs 59 F2
Seaton, Durham 69 F2
Seaton, Northum 75 K6
Seaton Burn 75 J6
Seaton Carew 69 H4
Seaton Delaval 75 K6
Seaton Ross 58 B2
Seaton Sluice 75 K6
Seatown 9 F5
Seave Green 69 H6
Seaview 11 H5
Seavington St Mary 8 E3
Seavington St Michael 8 F3
Sebastapol 29 G8
Sebergham 66 F2
Seckington 44 D7
Sedbergh 67 J7
Sedbury 29 K8
Sedbusk 68 A7
Sedgeberrow 34 B7
Sedgebrook 46 B4
Sedgefield 69 F4
Sedgeford 48 C4
Sedgehill 9 K2
Sedgley 43 L8
Sedgwick 61 J1
Sedlescombe 14 D7
Seend 18 A4
Seend Cleeve 18 A4
Seer Green 20 D4
Seething 49 J8
Sefton 52 E3
Seghill 75 J6
Seifton 32 D2
Seighford 43 K5
Seilebost 108 B5
Seisdon 43 K8
Seisiadar 108 D3
Selattyn 42 D4
Selborne 11 J1
Selby 63 K6
Selham 12 C4
Selkirk 80 C7
Sellack 29 K5
Sellafirth 113 D2
Sellindge 15 H5
Selling 15 G3
Sells Green 18 A4
Selly Oak 34 B2
Selmeston 13 J6
Selsdon 21 H7
Selsey 11 K5
Selsfield Common 13 G3
Selside 62 A2
Selston 45 F2
Selworthy 7 K1
Semer 38 F6
Semington 17 K4
Semley 9 K2
Send 12 D1
Senghenydd 25 K5
Sennen 2 B7
Sennen Cove 2 B7
Sennybridge 25 H2
Sereveton 45 J3
Sessay 63 H2
Setchey 48 B6
Setley 10 E4
Setter 113 D2
Settiscarth 112 B4
Settle 62 B3
Settrington 64 F5
Sevenhampton, Wilts 18 D1
Sevenhampton, Glos 30 E5
Seven Kings 21 J5
Sevenoaks 13 J1
Sevenoaks Weald 13 J1
Seven Sisters 25 G4
Severn Beach 17 G2
Severn Stoke 33 H6
Sevington 15 G4
Sewards End 37 J7
Sewerby 65 J6
Seworgan 2 F6
Sewstern 46 B5
Sezincote 31 F4
Shabbington 20 A3
Shackerstone 44 E7
Shackleford 12 C2
Shader 108 C2
Shadforth 69 F2
Shadingfield 39 K2
Shadoxhurst 15 F5
Shaftesbury 9 K2
Shafton 55 G2
Shalbourne 18 E4
Shalcombe 10 E6
Shalden 19 H6
Shaldon 5 K4
Shalfleet 10 F6
Shalford, Essex 38 C8
Shalford, Surrey 12 D2
Shalford Green 38 C8
Shalmsford Street 15 H3
Shalstone 35 H7
Shamley Green 12 D2
Shandon 84 E6
Shandwick 103 G6
Shangton 45 J8
Shanklin 11 G6
Shanquhar 98 E5
Shap 67 H5
Shapwick, Dorset 10 A4
Shapwick, Somer 16 F7
Shardlow 44 F4
Shareshill 43 L7
Sharlston 55 G2
Sharnbrook 36 C5
Sharnford 35 F1
Sharoe Green 61 J6
Sharow 63 G2
Sharpenhoe 36 D7
Sharperton 74 F3
Sharpness 30 A7
Sharpthorne 13 G3
Sharrington 48 F4
Shatterford 33 G2
Shaugh Prior 5 F5
Shavington 43 H2
Shaw, Berks 19 F4
Shaw, Wilts 17 K4
Shaw, G. Man 53 L3
Shawbost 108 C2
Shawbury 43 G5
Shawell 35 G2
Shawford 11 F2
Shawforth 53 K1
Shawhead 72 D6
Shaw Mills 63 F3
Shearsby 35 H1
Shebbear 6 E5
Shebdon 43 J5
Shebster 107 F3
Shedfield 11 G3
Sheen 54 E7
Sheepscombe 30 C6
Sheepstor 5 F5
Sheepwash 6 E5
Sheepy Magna 44 E7
Sheepy Parva 44 E7
Sheering 21 K2
Sheerness 14 F1
Sheet 11 J2
Sheffield 55 G5
Sheffield Bottom 20 A7
Shefford 36 E7
Sheinton 43 H7
Shelderton 32 D3
Sheldon, W. Mids 34 C2
Sheldon, Devon 8 C4
Sheldon, Derby 54 E7
Sheldwich 15 G3
Shelf 62 E7
Shelfanger 39 G2
Shelfield 44 B7
Shelford 45 H3
Shelley 54 F2
Shellingford 31 H8
Shellow Bowells 22 C3
Shelsley Beauchamp 33 G4
Shelsley Walsh 33 G4
Shelton, Northnts 36 D4
Shelton, Norf 39 H1
Shelton, Notts 45 J3
Shelton Green 39 H1
Shelve 42 E8
Shelwick 29 K3
Shenfield 22 C4
Shenington 34 E6
Shenley 21 F3
Shenley Brook End 36 B7
Shenleybury 21 F3
Shenley Church End 36 B7
Shenmore 29 H4
Shenstone, Staffs 44 C7
Shenstone, H. & W. 33 H3
Shenton 44 E7
Shenval, Grampn 98 B6
Shenval, Highld 96 C6
Shepherd's Green 20 B5

Toab, Orkney 112 C4
Toab, Shetld 113 C6
Tobermory 89 G3
Toberonochy 89 J8
Tobson 108 B3
Tocher 99 F5
Tockenham 18 B3
Tockenham Wick 18 B2
Tockholes 61 K7
Tockington 17 H2
Tockwith 63 H4
Todber 9 J2
Toddington, Beds 36 D8
Toddington, Glos 30 E4
Todenham 34 D7
Todhills 73 J7
Todmorden 62 C7
Todwick 55 H5
Toft, Lincs 46 D6
Toft, Cambs 37 G5
Toft Hill 68 D4
Toft Monks 49 K8
Toft next Newton 56 E5
Toftrees 48 D5
Toftwood 48 E6
Togston 75 J3
Tokavaig 111 G4
Tokers Green 20 A6
Tolland 8 C1
Tollard Royal 10 A3
Toll Bar 55 J3
Toller Fratrum 9 G5
Toller Porcorum 9 G5
Tollerton, Notts 45 H4
Tollerton, N. Yks 63 J3
Tollesbury 22 F2
Tolleshunt D'Arcy 22 F2
Tolleshunt Major 22 F2
Tolpuddle 9 J5
Tolsta Chaolais 108 B3
Tolworth 21 F7
Tomatin 97 G6
Tombreck 96 E5
Tomchrasky 96 A7
Tomdoun 90 C1
Tomich, Highld 96 B6
Tomich, Highld 103 F6
Tomintoul, Grampn 98 A7
Tomintoul, Grampn 92 C2
Tomnavoulin 98 B6
Tonbridge 13 J2
Tondu 25 G6
Tong 43 J7
Tonge 44 F5
Tongham 19 K6
Tongland 71 H6
Tongue 105 G3
Tongwynlais 25 K6
Tonmawr 25 G5
Tonna 25 F5
Ton-teg 25 J6
Tonwell 21 H2
Tonypandy 25 H5
Tonyrefail 25 J6
Toot Baldon 31 K7
Toothill, Hants 10 E3
Toot Hill, Essex 21 K3
Topcroft 39 H1
Topcroft Street 39 H1
Toppesfield 38 C7
Toppings 53 J2
Topsham 5 K3
Torbay 5 J5
Torbeg 76 E5
Torbryan 5 J5
Torcross 5 J7
Tore 96 E3
Torksey 56 C6
Torlum 110 B2
Torlundy 90 C4
Tormarton 17 J3
Tormitchell 70 D2
Tormore 76 D4
Tornagrain 97 F4
Tornahaish 98 B8
Tornaveen 98 F8
Torness 96 D6
Torpenhow 66 E3
Torphichen 86 C7
Torphins 93 H1
Torpoint 4 E6
Torquay 5 K5
Torquhan 80 C5
Torran 111 F2
Torrance 85 J7
Torridon 100 D6
Torrin 111 F3
Torrisdale 105 H2
Torrisholme 61 H3
Torroble 102 D3
Torry, Grampn 98 D4
Torry, Grampn 99 J8
Torrylinn 76 E5
Torterston 99 K4
Torthorwald 73 F6
Tortington 12 D6
Tortworth 30 B8
Torvaig 111 F2
Torver 66 E7
Torwood 85 L6
Torworth 55 K5
Toscaig 94 C2
Toseland 37 F4
Tosside 62 A4
Tostock 38 E4
Totaig 110 E2
Totarol 108 B3
Tote, Highld 111 F2
Tote, Highld 111 F2
Totegan 106 D3
Totland 10 E6
Totley 55 G6
Totnes 5 J5
Totscore 110 E1
Tottenham 21 H4
Tottenhill 48 B6
Totteridge, Bucks 20 C4
Totteridge, G. Lon 21 G4
Totternhoe 20 D1
Tottington 53 J2
Totton 10 E3
Tournaig 100 D3
Toux, Grampn 98 E3
Toux, Grampn 99 J3
Tovil 13 L1
Toward 72 B2
Towcester 35 H6
Towednack 2 C6
Towersey 20 B3
Towie 98 D7
Towiemore 98 C4
Tow Law 68 D3
Town End 47 H8
Townhead, D. & G. 71 H7
Townhead, Cumbr 67 J3
Townhead of Greenlaw 72 C7
Townhill 86 E6
Townshend 2 D6
Town Yetholm 81 G7
Towthorpe 63 K4
Towton 63 H6
Towyn 51 J3
Toynton All Saints 57 H7
Toynton Fen Side 57 H7
Toynton St Peter 57 J7
Toy's Hill 13 H1
Trabboch 77 K5
Trabbochburn 77 K5
Traboe 2 F7
Tradespark, Orkney 112 C4
Tradespark, Highld 97 G3
Trafford Park 53 J4
Trallong 25 H2
Trallwng or Welshpool 42 D7
Tranent 87 H7
Trantlemore 106 D4
Tranwell 75 H5
Trapp 27 L6
Traprain 87 J7
Traquair 80 B6
Trawden 62 C6
Trawsfynydd 51 G7
Trealaw 25 J5
Treales 61 H6
Trearddur 50 B3
Treaslane 111 E2
Trebartha 4 C4
Trebetherick 3 H2
Treborough 8 B1
Trebudannon 3 G3
Treburley 4 D4
Trecastle 25 G2
Trecwn 26 D4
Trecynon 25 H4
Tredavoe 2 C7
Tredegar 25 K4
Tredington 34 D6
Tredinnick 3 H2
Tredomen 25 K1
Tredrizzick 3 H2
Tredunnock 29 H8
Treen 2 B7
Treeton 55 H5
Trefdraeth 50 D3
Trefecca 25 K1
Trefeglwys 41 J4
Trefenter 40 F7
Treffgarne 26 D5
Treffynnon 26 C5
Trefil 25 K3
Trefilan 27 K2
Trefnanney 42 D6
Trefnant 51 K3
Trefonen 42 D5
Trefor, Gwyn 50 C2
Trefor, Gwyn 50 C6
Treforest 25 J6
Trefriw 51 G4
Tregadillett 4 C3
Tregaian 50 D3
Tregare 29 J6
Tregaron 28 A1
Tregarth 50 F4
Tregeare 4 C3
Tregeiriog 42 C4
Tregele 50 C1
Tregidden 2 F7
Treglemais 26 C5
Tregole 4 B2
Tregonetha 3 H3
Tregony 3 H5
Tregoyd 29 F4
Tre-groes 27 J3
Tregurrian 3 G3
Tregynon 41 K4
Trehafod 25 J5
Treharris 25 J5
Treherbert 25 H5
Treknow 4 A3
Trelan 2 F8
Trelawnyd 51 K3
Trelech 27 G4
Treleddyd-fawr 26 B5
Trelewis 25 K5
Treligga 4 A3
Trelights 3 H2
Trelill 3 J2
Trelissik 3 G6
Trelleck 29 K7
Trelleck Grange 29 J7
Trelogan 52 C5
Tremadog 50 E6
Tremail 4 B3
Tremain 27 G3
Tremaine 4 C3
Tremar 3 L3
Trematon 4 D6
Tremeirchion 51 K3
Trenance 3 G3
Trenarren 3 J5
Trench 43 H6
Treneglos 4 C3
Trenewan 3 K4
Trent 9 G3
Trentham 43 K3
Trentishoe 7 G1
Treoes 25 H7
Treorchy 25 H5
Tre'r-ddol 41 F4
Tresaith 27 G2
Trescott 43 K8
Trescowe 2 D6
Tresham 17 J1
Tresillian 3 G5
Tresinwen 26 D3
Tresmeer 4 C3
Tressait 91 K5
Tresta 113 C4
Treswell 56 B6
Tre Taliesin 41 F4
Trethewey 2 B7
Trethomas 25 K6
Trethurgy 3 J4
Tretio 26 B5
Tretire 29 K5
Tretower 25 K2
Treuddyn 52 D8
Trevalga 4 A2
Trevanson 3 H2
Trevarren 3 H3
Trevarrick 3 H5
Trevellas 2 F4
Treverva 2 F6
Trevethin 29 G7
Trevigro 4 D5
Trevine 26 C4
Treviscoe 3 H4
Trevone 3 G2
Trewarmett 4 A3
Trewarthenick 3 H5
Trewassa 4 B3
Trewellard 2 B6
Trewen 4 C3
Trewidland 3 L4
Trewint 4 B2
Trewithian 3 G6
Trewoon 3 H4
Treyarnon 3 G2
Treyford 11 K3
Trickett's Cross 10 B4
Trimdon 69 F3
Trimdon Grange 69 F3
Trimingham 49 H4
Trimley St Martin 39 H7
Trimley St Mary 39 H7
Trimpley 33 G3
Trimsaran 27 J7
Trimstone 6 F1
Trinant 25 L4
Tring 20 D2
Trinity 93 H5
Trislaig 90 B4
Trispen 3 G4
Tritlington 75 J4
Trochry 92 A7
Troedyraur 27 H3
Troedyrhiw 25 J4
Trofarth 51 H3
Trondavoe 113 C3
Troon, Corn 2 E6
Troon, Strath 77 J4
Troston 38 D3
Trottiscliffe 14 C2
Trotton 11 K2
Troutbeck 67 G6
Troutbeck Bridge 67 G6
Trowbridge 17 K5
Trow Green 29 K7
Trowle Common 17 K5
Trows 80 E6
Trowse Newton 49 H7
Trudoxhill 17 J6
Trull 8 D2
Trumisgarry 108 A6
Trumpan 110 E1
Trumpet 33 F7
Trumpington 37 H5
Trunch 49 H4
Truro 3 G5
Trusham 5 J3
Trusley 44 D4
Trusthorpe 57 K5
Trysull 43 K8
Tubney 31 J8
Tuckenhay 5 J6
Tuckhill 33 G2
Tuddenham, Suff 38 C3
Tuddenham, Suff 39 G6
Tudeley 13 K2
Tudhoe 68 E3
Tudweiliog 50 B7
Tuffley 30 C6
Tufton 26 E5
Tugby 45 J7
Tugford 32 E2
Tullibody 86 B5
Tullich 84 C3
Tullich Muir 103 F6
Tulloch 93 J4
Tullynessle 98 E7
Tumble 27 K6
Tumby Woodside 57 G8
Tummel Bridge 91 J6
Tunbridge Wells, Royal 13 J3
Tunga 108 D3
Tunstall, Kent 14 E2
Tunstall, N. Yks 68 E7
Tunstall, Humbs 59 H3
Tunstall, Suff 39 J5
Tunstall, Lancs 61 K2
Tunstall, Staffs 43 K2
Tunstall, Norf 49 K7
Tunstead 49 H5
Tunworth 19 H6
Tupsley 29 K3
Tupton 55 G7
Turgis Green 20 A8
Turkdean 30 F6
Tur Langton 45 J8
Turnastone 29 H4
Turnberry 77 H7
Turnditch 44 D3
Turners Hill 13 G3
Turners Puddle 9 K5
Turnworth 9 K4
Turriff 99 G3
Turton Bottoms 53 J2
Turves 47 G8
Turvey 36 C5
Turville 20 B4
Turville Heath 20 B4
Turweston 35 H7
Tutbury 44 D5
Tutnall 34 A3
Tutshill 29 K8
Tuttington 49 H5
Tuxford 56 B6
Twatt, Orkney 112 B3
Twatt, Shetld 113 C4
Twechar 85 K7
Tweedmouth 81 H4
Tweedsmuir 79 J7
Twelveheads 2 F5
Twemlow Green 53 J7
Twenty 46 E5
Twerton 17 J4
Twickenham 21 F6
Twigworth 30 C5
Twineham 12 F5
Twinhoe 17 J5
Twinstead 38 D7
Twitchen, Shrops 32 C3
Twitchen, Devon 7 H2
Two Bridges 5 G4
Two Dales 55 F7
Two Gates 44 D7
Twycross 44 E7
Twyford, Berks 20 B6
Twyford, Hants 11 F2
Twyford, Norf 48 F5
Twyford, Bucks 35 H8
Twyford, Leic 45 J6
Twyford Common 29 K4
Twynholm 71 H6
Twyning 33 J7
Twyning Green 33 J7
Twynllanan 25 F2
Twyn-y-Sheriff 29 J7
Twywell 36 C3
Tyberton 29 H4
Tyburn 34 C1
Tycroes 27 L6
Tycrwyn 41 L2
Tydd Gote 47 H6
Tydd St Giles 47 H6
Tydd St Mary 47 H6
Ty-hen 50 A7
Tyldesley 53 J3
Tyler Hill 15 H2
Tylers Green 20 D4
Tylorstown 25 J5
Tylwch 41 J5
Ty-nant, Gwyn 41 J1
Ty-nant, Clwyd 51 J6
Tyndrum 84 F1
Tyneham 9 K6
Tynehead 80 B4
Tynemouth 75 K7
Tynewydd 25 H5
Tyninghame 87 K7
Tynron 72 D4
Tyn-y-ffridd 42 C4
Tyn-y-graig, Powys 28 E2
Tynygraig, Dyfed 41 F7
Ty'n-y-groes 51 G3
Ty Rhiw 25 K6
Tyringham 36 B6
Tythby 45 H4
Tythegston 25 G7
Tytherington, Avon 17 H2
Tytherington, Somer 17 J6
Tytherington, Ches 53 L6
Tytherington, Wilts 17 L6
Tytherleigh 8 E4
Tywardreath 3 J4
Tywyn, Gwyn 40 E3
Tywyn, Gwyn 51 G3

U

Uachdar 110 C2
Uags 94 C2
Ubbeston Green 39 J3
Ubley 17 G5
Uckerby 68 E6
Uckfield 13 H4
Uckington 30 D5
Uddingston 78 D3
Uddington 79 F6
Udimore 14 E7
Udny Green 99 H6
Udny Station 99 J6
Uffcott 18 C3
Uffculme 8 B3
Uffington, Lincs 46 D7
Uffington, Oxon 18 E2
Uffington, Shrops 43 G6
Ufford, Cambs 46 D7
Ufford, Suff 39 H5
Ufton 34 E4
Ufton Nervet 20 A7
Ugadale 76 C5
Ugborough 5 G6
Uggeshall 39 K2
Ugglebarnby 65 F2
Ugley 37 J8
Ugley Green 37 J8
Ugthorpe 69 K5
Uig, Highld 110 D2
Uig, Highld 111 E1
Uigshader 111 F2
Uisken 88 E7
Ulbster 107 J5
Ulceby, Humbs 59 F5
Ulceby, Lincs 57 J6
Ulcombe 14 E4
Uldale 66 E3
Uley 30 B8
Ulgham 75 J4
Ullapool 101 G2
Ullenhall 34 C4
Ullenwood 30 D6
Ulleskelf 63 J5
Ullesthorpe 35 G2
Ulley 55 H5
Ullingswick 29 K2
Ullinish 110 E3
Ullock 66 C4
Ulpha 66 D7
Ulsta 113 D2
Ultrome 65 J7
Ulverston 61 F2
Ulzieside 72 C3
Umberleigh 7 G3
Unapool 104 D5
Underbarrow 67 G7
Underriver 13 J1
Underwood, Gwent 16 E2
Underwood, Notts 45 F2
Undy 17 F2
Unifirth 113 C4
Union Mills 60 D4
Unstone 55 G6
Unthank 67 G3
Upavon 18 C5
Up Cerne 9 H4
Upchurch 14 E2
Upcott 29 H2
Upend 38 C5
Up Exe 7 K5
Uphall 86 D7
Uphall Station 86 D7
Upham, Hants 11 G2
Upham, Devon 7 J5
Up Hatherley 30 D5
Uphill 16 E5
Up Holland 53 G3
Uplawmoor 77 K2
Upleadon 30 B5
Upleatham 69 J5
Uploders 9 G5
Uplowman 8 B3
Uplyme 8 E5
Upminster 21 K5
Up Nately 19 H5
Upnor 14 D1
Upottery 8 D4
Upper Affcot 32 D2
Upper Ardchronie 102 E5
Upper Arley 33 G2
Upper Astrop 35 G7
Upper Basildon 19 G3
Upper Beeding 12 F5
Upper Benefield 36 C2
Upper Boddington 35 F5
Upper Borth 40 F5
Upper Breinton 29 J3
Upper Broughton 45 H5
Upper Bucklebury 19 G4
Upper Caldecote 36 E6
Upper Chapel 28 E3
Upper Chute 18 D5
Upper Clatford 18 E6
Upper Dallachy 98 C2
Upper Dean 36 D4
Upper Denby 55 F3
Upper Derraid 97 J5
Upper Dicker 13 J5
Upper Dunsforth 63 H3
Upper Elkstone 54 D8
Upper End 54 D6
Upper Ethie 97 F2
Upper Farringdon 11 J1
Upper Framilode 30 B6
Upper Froyle 19 J6
Upper Gravenhurst 36 E7
Upper Green 18 E4
Upper Hackney 55 F7
Upper Hale 19 K6
Upper Hambleton 46 C7
Upper Hardres Court 15 H3
Upper Hartfield 13 H3
Upper Heath 32 E2
Upper Helmsley 64 D7
Upper Heyford, Northnts 35 H5
Upper Heyford, Oxon 31 J5
Upper Hill 29 J2
Upper Hopton 54 E2
Upper Hulme 54 D7
Upper Inglesham 31 G8
Upper Killay 24 D5
Upper Knockando 98 A4
Upper Lambourne 18 E2
Upper Lochton 93 H2
Upper Longdon 44 B6
Upper Longwood 43 H7
Upper Lydbrook 29 L6
Upper Maes-coed 29 H4
Uppermill 54 C3
Upper Minety 18 B1
Upper North Dean 20 C4
Upper Poppleton 63 J4
Upper Quinton 34 C6
Upper Sanday 112 C4
Upper Sapey 33 F4
Upper Seagry 18 A2
Upper Shelton 36 C6
Upper Sheringham 49 G3
Upper Skelmorlie 77 G1
Upper Slaughter 31 F5
Upper Soudley 30 A6
Upper Stondon 36 E7
Upper Stowe 35 H5
Upper Street, Hants 10 C3
Upper Street, Norf 49 J6
Upper Sundon 36 D8
Upper Swainswick 17 J4
Upper Swell 31 F5
Upper Tean 44 B4
Upperthong 54 E3
Upper Tillyrie 86 E4
Upperton 12 C4
Upper Tooting 21 G6
Uppertown 107 J2
Upper Tysoe 34 E6
Upper Upham 18 D3
Upper Wardington 35 F6
Upper Weald 35 K7
Upper Weedon 35 H5
Upper Wield 19 H7
Upper Winchendon 20 B2
Upper Woodford 10 C1
Uppingham 46 B8
Uppington 43 G7
Upsall 63 H1
Upshire 21 J3
Up Somborne 10 E1
Upstreet 15 J2
Up Sydling 9 H4
Upton, Dorset 10 A5
Upton, Bucks 20 B2
Upton, Notts 56 B6
Upton, Lincs 56 C5
Upton, Mers 52 D5
Upton, Berks 20 D6
Upton, Cambs 36 E3
Upton, Hants 10 E3
Upton, Hants 18 E5
Upton, Cambs 46 E7
Upton, Leic 44 E8
Upton, Ches 52 F7
Upton, Oxon 19 G2
Upton, W. Yks 55 H2
Upton, Notts 45 J2
Upton, Northnts 35 J4
Upton, Norf 49 J6
Upton, Somer 7 K3
Upton Bishop 30 A5
Upton Cheyney 17 H4
Upton Cressett 33 F1
Upton Cross 4 C4
Upton Grey 19 H6
Upton Hellions 7 J5
Upton Lovell 18 A6
Upton Magna 43 G6
Upton Noble 17 J7
Upton Pyne 5 K2
Upton Scudamore 17 K6
Upton Snodsbury 33 J5
Upton St Leonards 30 C6
Upton upon Severn 33 H6
Upton Warren 33 J4
Upwaltham 12 C5
Upware 37 J3
Upwell, Cambs 47 J7
Upwell, Norf 47 J7
Upwey 9 H6
Upwood 37 F2
Urafirth 113 C3
Urchfont 18 B5
Urdimarsh 29 K3
Urgha 108 B5
Urishay Common 29 H4
Urlay Nook 69 G5
Urmston 53 J4
Urquhart 98 B2
Urra 69 H6
Urray 96 D3
Urswick 60 F2
Ushaw Moor 68 E2
Usk 29 H7
Usselby 57 E4
Utkinton 53 G7
Utley 62 D5
Uton 5 J2
Utterby 57 H4
Uttoxeter 44 B4
Uwchmynydd 40 A1
Uxbridge 20 E5
Uyeasound 113 D1
Uzmaston 26 D6

V

Valleyfield 86 D6
Valsgarth 113 E1
Valtos, W. Isles 108 B3
Valtos, Highld 111 F1
Vange 22 D5
Varteg 29 G7
Vatersay 110 B5
Vatten 110 E2
Vauld, The 29 K3
Vaynor 25 J3
Vaynor Park 42 C7
Veensgarth 113 D4
Velindre 29 F4
Veness 112 C3
Vennington 42 E7
Venn Ottery 8 B5
Ventnor 11 G7
Vernham Dean 18 E5
Vernham Street 18 E5
Vernolds Common 32 D2
Verran Island 110 B3
Verwig 27 F3
Verwood 10 B4
Veryan 3 H6
Vicarage 8 D6
Vickerstown 60 E3
Victoria 3 H3
Vidlin 113 D3
Viewpark 78 E3
Village Abberley, The 33 G4
Village, The 33 H2
Vinehall Street 13 L4
Vines Cross 13 J5
Virginia Water 20 D7
Virginstow 4 D2
Vobster 17 J6
Voe 113 D3
Vowchurch 29 H4

W

Waberthwaite 66 D7
Wackerfield 68 D4
Wacton 39 G1
Wadborough 33 J6
Waddesdon 20 B2
Waddingham 56 D4
Waddington, Lincs 56 D7
Waddington, Lancs 61 L5
Wadebridge 3 H2
Wadeford 8 E3
Wadenhoe 36 D2
Wadesmill 21 H2
Wadhurst 13 K3
Wadshelf 55 G6
Wadworth 55 J4
Waen-fach 42 D6
Wainfleet All Saints 57 J8
Wainfleet Bank 57 J8
Wainhouse Corner 4 B2
Wainscott 14 D1
Wainstalls 62 D7
Waitby 67 K6
Wakefield 55 G1
Wakerley 46 C8
Wakes Colne 38 D8
Walberswick 39 K3
Walberton 12 C6
Walcot, Shrops 32 C2
Walcot, Warw 34 C5
Walcot, Lincs 46 D4
Walcot, Shrops 43 G6
Walcote 35 G2
Walcott, Lincs 46 E2
Walcott, Norf 49 J4
Walden 62 D1
Walden Head 62 C1
Walden Stubbs 55 J2
Walderslade 14 D2
Walderton 11 J3
Walditch 9 F5
Waldridge 68 E1
Waldringfield 39 H6
Waldron 13 J5
Wales 55 H5
Walesby, Notts 56 A6
Walesby, Lincs 57 F4
Walford, H. & W 32 C3
Walford, Shrops 42 F5
Walford, H. & W 29 K5
Walgherton 43 H3
Walgrave 35 K3
Walkden 53 J3
Walker 75 J7
Walkerburn 80 B6
Walker Fold 61 K5
Walkeringham 56 B4
Walkerith 56 B4
Walkern 37 F8
Walker's Green 29 K3
Walkhampton 5 F5
Walkington 58 D3
Walk Mill 62 B6
Wall, Staffs 44 C7
Wall, Northum 74 F7
Wallacetown 77 J5
Wallasey 52 D4
Wallingford 19 H2
Wallington, Herts 37 F7
Wallington, Hants 11 G4
Wallington, G. Lon 21 G7
Wallis 26 E5
Walliswood 12 E3
Walls 113 C4
Wallsend 75 J7
Wall under Heywood 32 E1
Wallyford 87 G7
Walmer 15 K3
Walmer Bridge 61 H7
Walmersley 53 K2
Walmley 44 C8
Walpole 39 J3
Walpole Highway 47 J6
Walpole St Andrew 47 J6
Walpole St Peter 47 J6
Walsall 44 B8
Walsall Wood 44 B7
Walsden 62 C7
Walsgrave on Sowe 34 E2
Walsham le Willows 38 F3
Walsoken 47 H6
Walston 79 H5
Walter's Ash 20 C4
Walterstone 29 H5
Waltham, Humbs 59 G6
Waltham, Kent 15 H4
Waltham Abbey 21 H3
Waltham Chase 11 G3
Waltham Cross 21 H3
Waltham on the Wolds 45 K5
Waltham St Lawrence 20 C6
Walthamstow 21 H5
Walton, Bucks 36 B7
Walton, Cumbr 74 B7
Walton, Warw 34 D5
Walton, Somer 9 F1
Walton, Leic 35 G2
Walton, W. Yks 55 G2
Walton, Powys 29 G2
Walton, Shrops 43 G6
Walton, Derby 55 G7
Walton, W. Yks 63 H5
Walton, Suff 39 H7
Walton Cardiff 30 D4
Walton East 26 E5
Walton Highway 47 H6
Walton-in-Gordano 17 F3
Walton-le-Dale 61 J7
Walton-on-Thames 21 F7
Walton-on-the-Hill, Staffs 44 A5
Walton on the Hill, Surrey 12 F1
Walton-on-the-Naze 23 J1
Walton on the Wolds 45 G6
Walton-on-Trent 44 D6
Walton West 26 C6
Walworth 68 E5
Walwyn's Castle 26 C6
Wambrook 8 D4
Wanborough 18 D2
Wandsworth 21 G6
Wangford 39 K3
Wanlip 45 G6
Wanlockhead 72 D2
Wansford, Cambs 46 D8
Wansford, Humbs 65 H7
Wanstead 21 J5
Wanstrow 17 J6
Wanswell 30 A7
Wantage 18 F2
Wapley 17 J3
Wappenbury 34 E4
Wappenham 35 H6
Warbleton 13 K5
Warborough 31 K8
Warboys 37 G2
Warbstow 4 C2
Warburton 53 J5
Warcop 67 K5
Warden 15 G1
Ward Green 38 F4
Wardington 35 F6
Wardle, Ches 53 H8
Wardle, G. Man 53 L2
Wardley 46 B7
Wardlow 54 E6
Wardy Hill 37 H2
Ware 21 H2
Wareham 9 L6
Warehorne 15 F5
Warenford 81 K7
Waren Mill 81 K6
Warenton 81 K6
Wareside 21 H2
Waresley 37 F5
Warfield 20 C6
Wargrave 20 B6
Warham 48 E3
Wark, Northum 74 E6
Wark, Northum 81 G6
Warkleigh 7 G3
Warkton 36 B3
Warkworth 75 J3
Warlaby 69 F7
Warland 53 L1
Warleggan 3 K3
Warlingham 21 H8
Warmfield 55 G1
Warmingham 53 J7
Warmington, Northnts 36 D1
Warmington, Warw 34 F6
Warminster 17 K6
Warmsworth 55 J3
Warmwell 9 J6
Warndon 33 H5
Warnford 11 H2
Warnham 12 E3
Warninglid 12 F4
Warren, Dyfed 26 D8
Warren, Ches 53 K6
Warren Row 20 C5
Warren Street 14 F3
Warrington, Bucks 36 B5
Warrington, Ches 53 H5
Warsash 11 F4
Warslow 54 D8
Warsop 55 J7
Warter 65 F7
Warthill 63 K4
Wartling 13 K6
Wartnaby 45 J5
Warton, Warw 44 D7
Warton, Northum 75 G3
Warton, Lancs 61 H7
Warton, Lancs 61 J2
Warwick, Warw 34 D4
Warwick, Cumbr 67 G1
Warwick Bridge 67 G1
Wasbister 112 B3
Wasdale Head 66 D6
Washaway 3 J3
Washbourne 5 H6
Washfield 7 K4
Washford 16 B6
Washford Pyne 7 J4
Washingborough 56 E6
Washington, W. Susx 12 E5
Washington, T. & W 68 F1
Wasing 19 G4
Waskerley 68 C2
Wasperton 34 D5
Wass 63 J2
Watchet 16 B6
Watchfield, Oxon 18 D1
Watchfield, Somer 16 E6
Watchgate 67 H7
Watendlath 66 E5
Water 62 B7
Waterbeach 37 H4
Waterbeck 73 H6
Waterden 48 D4
Water End, Herts 20 E2
Water End, Herts 21 G3
Waterfall 44 B2
Waterfoot, Lancs 62 B7
Waterfoot, Strath 78 C4
Waterford 21 H2
Waterhead, Cumbr 67 F6
Waterhead, Strath 77 L6
Waterhouses, Staffs 44 B2
Waterhouses, Durham 68 D2
Wateringbury 13 K1
Waterlip 17 H6
Waterloo, Dorset 10 B5
Waterloo, Tays 92 B8
Waterloo, Mers 52 E4
Waterloo, Strath 79 F4
Waterloo, Norf 49 H6
Waterlooville 11 H4
Watermeetings 72 E2
Watermillock 67 G4
Water Newton 46 E8
Water Orton 34 C1
Waterperry 20 A3
Waterrow 8 B2
Watersfield 12 D5
Waterside, Strath 85 J7
Waterside, Strath 77 K3
Waterside, Strath 77 K7
Waterside, Strath 77 L1
Waterstock 20 A3
Waterston 26 D7
Water Stratford 35 H7
Waters Upton 43 H6
Waterthorpe 55 H5
Water Yeat 66 E8
Watford, Herts 21 F4
Watford, Northnts 35 H4
Wath, N. Yks 62 E3
Wath, N. Yks 63 G2
Wath upon Dearne 55 H3
Watlington, Oxon 20 A4
Watlington, Norf 48 B6
Watnall 45 G3
Watten 107 H4
Watterow 8 B2
Wattisfield 38 F3
Wattisham 38 F5
Watton, Norf 48 E7
Watton, Humbs 65 H7
Watton at Stone 21 H2
Wattston 85 K8
Wattstown 25 J5
Wattsville 25 L5
Waunarlwydd 24 E5
Waunfawr 50 E5
Wavendon 36 C7
Waverton, Cumbr 66 E2
Waverton, Ches 52 F7
Wawne 59 E3
Waxham 49 K5
Waxholme 59 H4
Wayford 8 F4
Way Village 7 J4
Weachyburn 98 F2
Wealdstone 21 F5
Weare 16 F5
Weare Giffard 6 E3
Wearhead 67 L3
Weasdale 67 J6
Weasenham All Saints 48 D5
Weasenham St Peter 48 D5
Weaverham 53 H6
Weaverthorpe 65 G5
Webheath 34 B4
Wedderlairs 99 H5
Weddington 44 E8
Wedhampton 18 B5
Wedmore 17 F6
Wednesbury 44 A8
Wednesfield 44 A7
Weedon 20 C2
Weedon Bec 35 H5
Weedon Lois 35 H6
Weeford 44 C7
Week 7 H4
Weekley 36 B2
Week St Mary 4 C2
Weeley 23 H1
Weeley Heath 23 H1
Weem 91 K7
Weeping Cross 43 L5
Weeting 38 C2
Weeton, N. Yks 63 F5
Weeton, Lancs 61 G6
Weir 62 B7
Welbeck Abbey 55 J6
Welborne 48 F7
Welbourn 46 C2
Welburn 64 E6
Welbury 69 F6
Welby 46 C4
Welches Dam 37 H2
Welcombe 6 C4
Weldon 36 C2
Weldrake 58 A2
Welford, Berks 18 F3
Welford, Northnts 35 H2
Welford-on-Avon 34 C5
Welham 35 J1
Welham Green 21 G3
Well, N. Yks 63 F1
Well, Lincs 57 J6
Well, Hants 19 J6
Welland 33 G6
Wellbank 93 F8
Wellesbourne 34 D5
Well Hill 21 J7
Welling 21 J6
Wellingborough 36 B4
Wellingham 48 D5
Wellingore 46 C2
Wellington, Somer 8 C2
Wellington, Shrops 43 H6
Wellington, H. & W 29 J3
Wellington Heath 33 G6
Wellow, I. of W 10 E6
Wellow, Avon 17 J5
Wellow, Notts 55 K7
Wells 17 G6
Wellsborough 44 E7
Wells-next-the-Sea 48 E3
Wellwood 86 D6
Welney 47 J8
Welshampton 42 F4
Welsh Bicknor 29 K6
Welsh End 43 G4
Welsh Frankton 42 E4
Welsh Hook 26 D5
Welsh Newton 29 K6
Welshpool or Trallwng 42 D7
Welsh St Donats 25 J7
Welton, Humbs 58 D4
Welton, Lincs 56 E6
Welton, Cumbr 66 F2
Welton, Northnts 35 G4
Welton le Marsh 57 J7
Welton le Wold 57 G5
Welwick 59 H4
Welwyn 21 G2
Welwyn Garden City 21 G2
Wem 43 G5
Wembdon 8 D1
Wembley 21 F5
Wembury 4 F7
Wembworthy 7 G5
Wemyss Bay 84 D8
Wendens Ambo 37 J7
Wendlebury 31 K6
Wendling 48 E6
Wendover 20 C3
Wendron 2 E6
Wendy 37 G6
Wenhaston 39 K3
Wennington, Cambs 37 F3
Wennington, Lancs 61 K2
Wennington, G. Lon 21 K5
Wensley, N. Yks 68 C8
Wensley, Derby 55 F7
Wentbridge 55 H2
Wentnor 32 C1
Wentworth, S. Yks 55 G4
Wentworth, Cambs 37 H3
Wenvoe 25 K7
Weobley 29 J2
Weobley Marsh 29 J2
Wereham 48 B7
Wergs 43 K7
Wernrheolydd 29 H6
Werrington, Corn 4 D3
Werrington, Cambs 46 E7
Werrington, Staffs 43 L3
Wervin 52 F6
Wesham 61 H6
Wessington 55 G8
West Acre 48 C6
West Allerdean 81 H5
West Alvington 5 H7
West Anstey 7 J3
West Ashby 57 G6
West Ashling 11 K4
West Ashton 17 K5
West Aukland 68 D4
West Bagborough 8 C1
West Barns 87 K7
West Barsham 48 E4
West Bay 9 F5
West Beckham 49 G4
Westbere 15 H2
West Bergholt 38 E8
West Bexington 9 G6
West Bilney 48 C6
West Blatchington 13 F6
Westborough 46 B3
Westbourne, Dorset 10 B5
Westbourne, W. Susx 11 J4
West Bradford 61 L5
West Bradley 9 G1
West Bretton 55 F2
West Bridgford 45 G4
West Bromwich 34 B1
West Buckland, Somer 8 C2
West Buckland, Devon 7 G2
West Burrafirth 113 C4
West Burton, Lancs 62 C3
West Burton, W. Susx 12 C5
West Burton, N. Yks 62 D1
Westbury, Shrops 42 E7
Westbury, Bucks 35 H7
Westbury, Wilts 17 K5
Westbury Leigh 17 K5
Westbury-on-Severn 30 B6
Westbury-sub-Mendip 17 G6
Westby 61 G6
West Caister 49 L6
West Calder 79 H3
West Camel 9 G2
West Challow 18 E2
West Charleton 5 H7
West Chelborough 9 G4
West Chevington 75 J4
West Chiltington 12 D5
West Chinnock 9 F3
West Clandon 12 D1
West Cliffe 15 K4
Westcliff-on-Sea 22 E5
West Clyne 103 G3
West Coker 9 G3
Westcombe 17 H7
West Compton, Dorset 9 G5
West Compton, Somer 17 G6
Westcote 31 G5
Westcott, Bucks 20 B2
Westcott, Devon 8 B4
Westcott, Surrey 12 E2
Westcott Barton 31 J5
West Cross 24 E6
West Cullerley 99 G8
West Curthwaite 66 F2
West Dean, Hants 10 D2
Westdean, E. Susx 13 J7
West Dean, W. Susx 11 K3
West Deeping 46 E7
West Derby 52 E4
West Dereham 48 B7
West Ditchburn 75 H1
West Down 6 F1
West Drayton, Notts 56 B6
West Drayton, G. Lon 20 E6
West End, Beds 36 C5
West End, Humbs 58 D3
West End, Surrey 20 D7
West End, N. Yks 62 E4
West End, Hants 11 F3
West End, Avon 17 F4
West End, Oxon 31 J7
West End, Norf 49 K6
West End Green 20 A7
Westerdale, Highld 107 G4
Westerdale, N. Yks 69 J6
Westerfield, Shetld 113 C4
Westerfield, Suff 39 G6
Wester Fintray 99 H7
Westergate 12 C6
Wester Gruinards 102 D4
Westerham 13 H1
Westerleigh 17 J3
Wester Quarff 113 D5
Wester Skeld 113 C4
Westerton 93 H6
Westerwick 113 C4
West Farleigh 13 L1
West Felton 42 E5
Westfield, Lothn 86 C7
Westfield, Norf 48 E7
Westfield, E. Susx 14 E7
Westfield, Highld 107 F3
West Firle 13 H6
West Fleetham 81 K7
Westgate, Durham 68 B3
Westgate, Humbs 58 B6
Westgate, Norf 48 E3
Westgates on Sea 15 K1
West Gerinish 110 B2
West Ginge 19 F2
West Grafton 18 D4
West Green 19 J5
West Grimstead 10 D2
West Grinstead 12 E4
West Haddlesey 63 J7
West Haddon 35 H3
West Hagbourne 19 G2
West Hall, Cumbr 74 B7
Westhall, Suff 39 K2
West Hallam 44 F3
West Halton 58 D4
Westham, Somer 16 F6
West Ham, G. Lon 21 J5
Westham, E. Susx 13 K6
Westhampnett 11 K4
West Handley 55 G6
West Hanney 18 F1
West Hanningfield 22 D4
West Hardwick 55 H2
West Harptree 17 G5
West Hatch 8 D2
Westhay 17 F6
Westhead 52 F3
West Heath 19 K5
West Helmsdale 103 J2
West Hendred 19 F2
West Heslerton 65 G5
Westhide 29 K3
West Hill, Devon 8 B5
Westhill, Grampn 99 H8
West Hoathly 13 G3
West Holme 9 K6
Westhope, Shrops 32 D2
Westhope, H. & W 29 J2
West Horndon 22 C5
Westhorpe, Suff 38 F4
Westhorpe, Lincs 46 F4
West Horrington 17 G6
West Horsley 12 D1
West Hougham 15 J4
Westhoughton 53 H3
Westhouse 61 K2
Westhouses 55 H8
Westhumble 12 E1
West Hyde 20 E4
West Ilsley 19 F2
Westing 113 D1
West Itchenor 11 J4
West Kennett 18 C4
West Kilbride 77 H3
West Kingsdown 21 K7
West Kington 17 K3
West Kirby 52 D5
West Knighton 9 J6
West Knoyle 9 K1
Westlake 5 G6
West Langdon 15 K4
West Langwell 105 H8
West Lavington, Wilts 18 B5
West Lavington, W. Susx 11 K2
West Layton 68 D6
West Leake 45 G5
Westleigh, Devon 8 B3
Westleigh, Devon 6 E3
Westleton 39 K4
West Lexham 48 D6
Westley, Suff 38 D4
Westley, Shrops 42 E7
Westley Waterless 38 B5
West Lilling 63 K3
Westlington 20 B2
West Linton, Border 79 J4
Westlinton, Cumbr 73 J7
West Littleton 17 J3
West Lulworth 9 K6
West Lutton 65 G6
West Malling 13 K1
West Malvern 33 G6
West Marden 11 J3
West Markham 56 B6
Westmarsh 15 J2
West Marton 62 B4
West Meon 11 H2
West Mersea 23 G2
Westmeston 13 G5
Westmill 37 G8
West Milton 9 G5
Westminster 21 G6
West Monkton 8 D2
West Moors 10 B4
Westmuir, Tays 92 E6
West Muir, Tays 93 G5
West Newton, Norf 48 B5
Westnewton, Cumbr 66 D2
West Newton, Humbs 59 F3
West Norwood 21 H6
West Ogwell 5 J4
Weston, Staffs 44 A5
Weston, Notts 56 B7
Weston, N. Yks 62 E5
Weston, N. Yks 64 E6
Weston, Berks 18 F3
Weston, Lincs 47 F5
Weston, Herts 37 F7
Weston, Ches 53 G5
Weston, Shrops 43 G5
Weston, Northnts 35 G6
Weston, Shrops 43 G8
Weston, Dorset 9 H7
Weston, Ches 43 J2
Weston, Hants 11 J2
Weston, Avon 17 J4
Weston Bampfylde 9 H2
Weston Beggard 29 K3
Westonbirt 17 K2
Weston by Welland 35 J1
Weston Colville 38 B5
Weston Green 38 B5
Weston Heath 43 J6
Weston Hills 47 F5
Westoning 36 D7
Weston-in-Gordano 17 F3
Weston Jones 43 J5
Weston Longville 49 G6
Weston Lullingfields 42 F5
Weston-on-the-Green 31 K6
Weston-on-Trent 44 F5
Weston Patrick 19 H6
Weston Rhyn 42 D4
Weston-sub-Edge 34 C6
Weston-super-Mare 16 E4
Weston Turville 20 C2
Weston-under-Lizard 43 K6
Weston under Penyard 29 L5
Weston under Wetherley 34 E4
Weston Underwood, Bucks 36 B5
Weston Underwood, Derby 44 D3
Westonzoyland 8 E1
West Overton 18 C4
West Parley 10 B5
West Peckham 13 K1
West Pennard 17 G7
West Pentire 2 F3
West Perry 36 E4
Westport 8 E2
West Putford 6 D4
West Quantoxhead 16 C6
West Rainton 68 F2
West Rasen 56 E5
West Raynham 48 D5
Westrigg 79 G3
West Row 38 B3
West Rudham 48 D5
West Runton 49 G3
Westruther 80 E4
Westry 47 G8
West Saltoun 80 C3
West Sandwick 113 D2
West Scrafton 62 D1
West Stafford 9 J6
West Stockwith 56 B4
West Stoke 11 K4
West Stonesdale 68 A6
West Stoughton 16 F6
West Stour 9 J2
West Stourmouth 15 J2
West Stow 38 D3
West Stowell 18 C4
West Street 14 F3
West Tanfield 63 F2
West Tarbert 76 D1
West Thorney 11 J4
West Thurrock 21 K6
West Tilbury 14 C1
West Tisted 11 H2
West Tofts 86 E1
West Torrington 57 F5
West Town 17 F4
West Tytherley 10 D2
West Tytherton 18 A3
West Walton 47 H6
Westward 66 E2
Westward Ho! 6 E3
Westwell, Kent 15 F4
Westwell, Oxon 31 G6
Westwell Leacon 14 F4
West Wellow 10 D3
West Wemyss 87 G5
West Wick, Avon 16 E4
Westwick, Cambs 37 H4
Westwick, Norf 49 H5
West Wickham, Cambs 38 B6
West Wickham, G. Lon 21 H7
West Winch 48 B6
West Winterslow 10 D1
West Wittering 11 J5
West Witton 62 D1
Westwood, Devon 8 B5
Westwood, Wilts 17 K5
West Woodburn 74 E5
West Woodhay 18 E4
West Woodlands 17 J6
Westwoodside 58 B7
West Worldham 11 J1
West Wratting 38 B5
West Wycombe 20 C4
West Yell 113 D2
Wetheral 67 G1
Wetherby 63 H5
Wetherden 38 F4
Wetheringsett 39 G4
Wethersfield 38 C7
Wethersta 113 C3
Wetherup Street 39 G4
Wetley Rocks 44 A3
Wettenhall 53 H7
Wetton 44 C2
Wetwang 65 G7
Wetwood 43 J4
Wexcombe 18 D5
Weybourne 49 G3
Weybread 39 H2
Weybridge 20 E7
Weydale 107 G3
Weyhill 18 E6
Weymouth 9 H7
Whaddon, Wilts 10 C2
Whaddon, Glos 30 C6
Whaddon, Cambs 37 G6
Whaddon, Bucks 35 K7
Whale 67 H4
Whaley 55 J6
Whaley Bridge 54 D5
Whaley Thorns 55 J6
Whalley 61 L6
Whalton 75 H5
Wham 62 A3
Whaness 112 B4
Whaplode 47 G5
Whaplode Drove 47 G6
Wharfe 62 A3
Wharles 61 H6
Wharncliffe Side 55 F4
Wharram le Street 64 F6
Wharton, Ches 53 H7
Wharton, H. & W 29 K2
Whashton 68 D6
Whatcombe 9 K4
Whatcote 34 E6
Whatfield 38 F6
Whatley 17 J6
Whatlington 13 L5
Whatstandwell 44 E2
Whatton 45 J4
Whauphill 70 F7
Wheatacre 49 K8
Wheathampstead 21 F2
Wheathill 33 F2
Wheatley, Hants 19 J6
Wheatley, Oxon 31 K7
Wheatley Hill 69 F3
Wheatley Lane 62 B6
Wheaton Aston 43 K6
Wheddon Cross 7 K2
Wheedlemont 98 D6
Wheelerstreet 12 C2
Wheelock 53 J8
Wheelton 61 K7
Whelford 31 F8
Whelpley Hill 20 D3
Whenby 63 K3
Whepstead 38 D5
Wherstead 39 G6
Wherwell 18 E6
Wheston 54 E6
Whetsted 13 K2
Whetstone 45 G8
Whicham 60 E1
Whichford 31 H4
Whickham 75 J7
Whiddon Down 5 G2
Whigstreet 93 F7
Whilton 35 H4
Whimple 8 B5
Whimpwell Green 49 J5
Whinburgh 48 F7
Whinnyfold 99 K5
Whippingham 11 G5
Whipsnade 20 E2
Whipton 5 K2
Whissendine 46 B6
Whissonsett 48 E5
Whistley Green 20 B6
Whiston, Staffs 44 B3
Whiston, Northnts 36 B4
Whiston, Mers 53 F4
Whiston, S. Yks 55 H4
Whiston, Staffs 43 K6
Whitbeck 60 E1
Whitbourne 33 G5
Whitburn, Lothn 79 G3
Whitburn, T. & W 75 L7
Whitby, Ches 52 E6
Whitby, N. Yks 65 F1
Whitchurch, Oxon 20 A6
Whitchurch, Bucks 20 C1
Whitchurch, Dyfed 26 C5
Whitchurch, Devon 4 E4
Whitchurch, Hants 19 F6
Whitchurch, Shrops 43 G3
Whitchurch, Avon 17 H4
Whitchurch, S. Glam 25 K6
Whitchurch, H. & W 29 K6
Whitchurch Canonicorum 8 E5
Whitchurch Hill 20 A6
Whitcott Keysett 32 B2
Whitebridge 96 C7
Whitebrook 29 K7
Whitecairns 99 J7
Whitechapel 61 J5
White Coppice 53 H2
Whitecraig 87 G7
Whitecroft 29 L7
Whiteface 103 F5
Whitefield 53 K3
Whiteford 99 G6
Whitegate 53 H7
Whitehall 112 D3
Whitehaven 66 B5
Whitehill 11 J1
Whitehills 98 F2
Whitehouse, Strath 76 D1
Whitehouse, Grampn 98 F7
Whitekirk 87 J6
White Kirkley 68 C3
White Lackington 9 J5
White Ladies Aston 33 J5
Whiteley Village 20 E7
Whitemans Green 13 G4
Whitmire 97 H3
Whitemoor 3 H4
White Notley 22 D2
Whiteparish 10 D2
Whiterashes 99 H6
White Rocks 29 J5
White Roding or
White Roothing 21 K2
White Roothing or
White Roding 21 K2
Whiterow 107 J5
Whiteshill 30 C7
Whiteside 79 G3
Whitesmith 13 J5
Whitestaunton 8 D3
Whitestone 5 J2
White Waltham 20 C6
Whiteway 30 D6
Whitewell 61 K5
Whitewreath 98 B3
Whitfield, Northum 74 D8
Whitfield, Avon 17 H1
Whitfield, Bucks 35 H7
Whitfield, Kent 15 K4
Whitford, Clwyd 52 C6
Whitford, Devon 8 D5
Whitgift 58 C4
Whitgreave 43 K5
Whithorn 71 F7
Whiting Bay 76 F5
Whitland 27 G6
Whitletts 77 J5
Whitley, Berks 20 B6
Whitley, Ches 53 H6
Whitley, N. Yks 63 J7
Whitley, Wilts 17 K4
Whitley Bay 75 K6
Whitley Chapel 68 B1
Whitley Row 13 H1
Whitlock's End 34 C3
Whitminster 30 B7
Whitmore 43 K3
Whitnage 8 B3
Whitnash 34 E4
Whitney-on-Wye 29 G3
Whitrigg 66 E3
Whitsbury 10 C3
Whitsome 81 G4
Whitson 16 E2
Whitstable 15 H2
Whitstone 4 C2
Whittingham 75 G2
Whittingslow 32 D2
Whittington, Staffs 44 C7
Whittington, Norf 48 C8
Whittington, Shrops 42 E4
Whittington, Glos 30 E5
Whittington, Derby 55 G6
Whittington, Staffs 33 H2
Whittington, H. & W 33 H5
Whittington, Lancs 61 K2
Whittlebury 35 H6
Whittle-le-Woods 61 J7
Whittlesey 47 F8
Whittlesford 37 H6
Whitton, Powys 32 B4
Whitton, Humbs 58 D4
Whitton, Shrops 32 E3
Whitton, Cleve 69 F4
Whitton, Border 80 F7
Whitton, Northum 75 G3
Whitton, Suff 39 G6
Whittonditch 18 D3
Whittonstall 68 C1
Whitwell, Leic 46 C7
Whitwell, N. Yks 68 E7
Whitwell, Herts 21 F1
Whitwell, I. of W 11 G7
Whitwell, Derby 55 J6
Whitwell-on-the-Hill 64 E6
Whitwick 44 F6
Whitwood 63 H7
Whitworth 53 K2
Whixall 43 G4
Whixley 63 H4
Whorlton, Durham 68 D5
Whorlton, N. Yks 69 G6
Whyle 32 E4
Whyteleafe 21 H8
Wibdon 29 K8
Wibtoft 35 F2
Wichenford 33 G4
Wichling 14 F3
Wick, H. & W 34 A6
Wick, Wilts 10 C2
Wick, Dorset 10 C5
Wick, W. Susx 12 D6
Wick, M. Glam 25 H7
Wick, Avon 17 J3
Wick, Highld 107 J4
Wickam Street 38 C5
Wicken, Cambs 37 J3
Wicken, Northnts 35 J7
Wicken Bonhunt 37 J7
Wickenby 56 E5
Wickersley 55 H4
Wickford 22 D4
Wickham, Berks 18 E3
Wickham, Hants 11 G3
Wickham Bishops 22 E2
Wickhambreaux 15 J3
Wickhambrook 38 C5
Wickhamford 34 B6
Wickham Market 39 J5
Wickhampton 49 K7
Wickham Skeith 39 F4
Wickham St Paul 38 D7
Wickham Street 39 F4
Wicklewood 48 F7
Wickmere 49 G4
Wick St Lawrence 16 E4
Wickwar 17 J2
Widdington 37 J7
Widdrington 75 J4
Widecombe in the
Moor 5 H4
Widegates 4 C6
Widemouth Bay 6 C5
Wide Open 75 J6
Widewall 112 C5
Widford, Essex 22 C3
Widford, Herts 21 J2
Widmerpool 45 H5
Widnes 53 G5
Wigan 53 G3
Wiggaton 8 C5
Wiggenhall St Germans 48 A6
Wiggenhall
St Mary Magdalen 48 A6
Wiggenhall
St Mary the Virgin 47 J6
Wiggington 63 J4
Wigginton, Herts 20 D2
Wigginton, Staffs 44 D7
Wigginton, Oxon 31 H4
Wigglesworth 62 B4
Wiggonby 66 F1
Wighill 63 H5
Wighton 48 E4
Wigmore, H. & W 32 D4
Wigmore, Kent 14 E2
Wigsley 56 C6
Wisthorpe 36 D2
Wigston 45 H8
Wigtoft 47 F4
Wigton 66 E2
Wigtown 71 F6
Wilbarston 36 B2
Wilberfoss 64 E7
Wilburton 37 H3
Wilby, Northnts 36 B4
Wilby, Norf 38 F2
Wilby, Suff 39 H3
Wilcot 18 C4
Wildboarclough 54 C7
Wilden, Beds 36 D5
Wilden, H. & W 33 H3
Wildhern 18 E5
Wildsworth 56 C4
Wilford 45 G4
Wilkesley 43 H3
Wilkhaven 103 H5
Wilkieston 79 J3
Willand 8 B3
Willaston, Ches 52 E6
Willaston, Ches 43 H2
Willen 36 B6
Willenhall, W. Mids 44 A8
Willenhall, W. Mids 34 E3
Willerby, Humbs 58 E3
Willerby, N. Yks 65 H5
Willersey 34 C7
Willersley 29 H3
Willesborough Lees 15 G4
Willesden 21 G5
Willett 8 C1
Willey, Warw 35 F2
Willey, Shrops 43 H8
Williamscot 35 F6
Willian 37 F7
Willingale 22 B3
Willingdon 13 J6
Willingham 37 H3
Willingham by Stow 56 C5
Willington, Durham 68 D3
Willington, Derby 44 D5
Willington, Warw 34 D7
Willington, Beds 36 E6
Willington, T. & W 75 K7
Willington Corner 53 G7
Willitoft 58 B3
Williton 16 B6
Willoughby, Warw 35 G4
Willoughby, Lincs 57 J6
Willoughby-on-the-Wolds 45 H5
Willoughby Waterleys 35 G1
Willoughton 56 D4
Wilmcote 34 C5
Wilmington, Devon 8 D5
Wilmington, E. Susx 13 J6
Wilmington, Kent 21 K6
Wilmslow 53 K5
Wilnecote 44 D7
Wilpshire 61 K6
Wilsden 62 D6
Wilsford, Wilts 18 C5
Wilsford, Wilts 18 C7
Wilsford, Lincs 46 D3
Wilsill 62 E3
Wilson 44 F5
Wilstead 36 D6
Wilsthorpe 46 D6
Wilstone 20 D2
Wilton, Wilts 10 B1
Wilton, Wilts 18 D4
Wilton, N. Yks 64 F4
Wilton, Cleve 69 H5
Wilton Dean 73 K2
Wimbish 38 A7
Wimbish Green 38 B7
Wimbledon 21 G6
Wimblington 37 H1
Wimborne Minster 10 B4
Wimbotsham 48 B7
Wimpstone 34 D6
Wincanton 9 J2
Wincham 53 H6
Winchburgh 86 D7
Winchcombe 30 E5
Winchelsea 14 F7
Winchelsea Beach 14 F7
Winchester 11 F2
Winchfield 19 J5
Winchmore Hill, Bucks 20 D4
Winchmore Hill, G. Lon 21 H4
Wincle 54 C7
Windermere 67 G7
Winderton 34 E6
Windlesham 20 D7
Windley 44 E3
Windmill Hill, Somer 8 E3
Windmill Hill, E. Susx 13 K5
Windrush 31 F6
Windsor 20 D6
Windygates 87 G4
Wineham 12 F4
Winestead 59 H4
Winfarthing 39 G2
Winford 17 G4
Winforton 29 G3
Winfrith Newburgh 9 K6
Wing, Leic 46 B7
Wing, Bucks 20 C1
Wingate 69 G3
Wingates, Northum 75 G4
Wingates, G. Man 53 H3
Wingerworth 55 G7
Wingfield, Beds 36 D8
Wingfield, Suff 39 H3
Wingfield, Wilts 17 K5
Wingham 15 J3
Wingrave 20 C2
Winkburn 56 B8
Winkfield 20 D6
Winkfield Row 20 C6
Winkhill 44 B2
Winkleigh 7 G5
Winksley 63 F2
Winless 107 J4
Winmarleigh 61 H5
Winnersh 20 B6
Winscales 66 C4
Winscombe 17 F5
Winsford, Ches 53 H7
Winsford, Somer 7 K2
Winsham 8 E4
Winshill 44 D5
Winskill 67 H3
Winslade 19 H6
Winsley 17 K4
Winslow 35 J8
Winson 30 E7
Winster, Derby 55 F7
Winster, Cumbr 67 G7
Winston, Durham 68 D5
Winston, Suff 39 G4
Winstone 30 D7
Winswell 6 E4
Winterborne Clenston 9 K4
Winterborne Herringston 9 H6
Winterborne Houghton 9 K4
Winterborne Kingston 9 K5
Winterborne Monkton 9 H6
Winterborne Stickland 9 K4
Winterborne Whitechurch 9 K4
Winterborne Zelston 9 K5
Winterbourne, Berks 19 F3
Winterbourne, Avon 17 H2
Winterbourne Abbas 9 H5
Winterbourne Bassett 18 C3
Winterbourne Dauntsey 10 C1
Winterbourne Earls 10 C1
Winterbourne Gunner 10 C1
Winterbourne Monkton 18 C3
Winterbourne Steepleton 9 H6
Winterbourne Stoke 18 B6
Winterburn 62 C4
Winteringham 58 D4
Winterley 53 J8
Wintersett 55 G2
Wintershill 11 G3
Winterton 58 D5
Winterton-on-Sea 49 K6
Winthorpe, Notts 46 B2
Winthorpe, Lincs 57 K7
Winton, Dorset 10 B5
Winton, Cumbr 67 K5
Wintringham 65 F5
Winwick, Cambs 36 E2
Winwick, Northnts 35 H3
Winyates 34 B4
Wirksworth 44 D2
Wirswall 43 G3
Wisbech 47 H7
Wisbech St Mary 47 H7
Wisborough Green 12 D4
Wiseton 56 B5
Wishaw, Warw 44 C8
Wishaw, Strath 78 E4
Wispington 57 G6
Wissett 39 J3
Wistanstow 32 D2
Wistanswick 43 H5
Wistaston 43 H2
Wiston, W. Susx 12 E5
Wiston, Dyfed 26 E6
Wiston, Strath 79 G6
Wistow, Cambs 37 F2
Wistow, N. Yks 63 J6
Wiswell 61 L6
Witcham 37 H3
Witchampton 10 A4
Witchford 37 J3
Witham 22 E2
Witham Friary 17 J6
Witham on the Hill 46 D6
Witherenden Hill 13 K4
Witheridge 7 J4
Witherley 44 E8
Withern 57 J5
Withernsea 59 H4
Withernwick 59 F2
Withersdale Street 39 H2
Withersfield 38 B6
Witherslack 61 H1
Withiel 3 H3
Withiel Florey 7 K2
Withington, Glos 30 E6
Withington, Shrops 43 G6
Withington, H. & W 29 K3
Withington, G. Man 53 K4
Withington, Ches 53 K7
Withington Green 53 K6
Withleigh 7 K4
Withnell 61 K7
Withybrook 34 F2
Withycombe 16 B6

Withyham 13 H3
Withypool 7 J2
Witley 12 C3
Witnesham 39 G5
Witney 31 H7
Wittering 46 D7
Wittersham 14 E6
Witton Bridge 49 J4
Witton Gilbert 68 E2
Witton-le-Wear 68 D3
Witton Park 68 D3
Wiveliscombe 8 B2
Wivelsfield 13 G4
Wivelsfield Green 13 G5
Wivenhoe 23 G1
Wiveton 48 F3
Wix 39 G8
Wixford 34 B5
Wixoe 38 C6
Woburn 36 C7
Woburn Sands 36 C7
Wokefield Park 20 A7
Woking 20 E8
Wokingham 20 C7
Woldingham 13 G1
Wold Newton, Humbs 59 G7
Wold Newton, Humbs 65 H5
Wolferlow 33 F4
Wolferton 48 B5
Wolfhill 86 E1
Wolf's Castle 26 D5
Wolfsdale 26 D5
Woll 80 C7
Wollaston, Northnts 36 C4
Wollaston, Shrops 42 E6
Wollerton 43 H4
Wolsingham 68 C3
Wolston 34 F3
Wolvercote 31 J7
Wolverhampton 43 L8
Wolverley, Shrops 43 F4
Wolverley, H. & W 33 H3
Wolverton, Bucks 36 B6
Wolverton, Warw 34 D4
Wolveshewton 29 J8
Wolvetton 19 G5
Wolvey 34 F2
Wolviston 69 G4
Wombleton 63 K1
Wombourne 43 K8
Wombwell 55 G3
Womenswold 15 J3
Wonastow 29 J6
Wonersh 12 D2
Wonson 5 G3
Wonston, Hants 19 F7
Wonston, Dorset 9 J4
Wooburn 20 D5
Wooburn Green 20 D5
Woodale 62 D2
Woodbastwick 49 J6
Woodbeck 56 B6
Woodborough, Wilts 18 C5
Woodborough, Notts 45 H3
Woodbridge 39 H6
Woodbury 8 B6
Woodbury Salterton 8 B6
Woodchester 30 C7
Woodchurch 14 F5
Woodcote 20 A5
Woodcroft 29 K8
Wood Dalling 49 F5
Woodeaton 31 K6
Wood End, Warw 34 C3
Woodend, Cumbr 66 D7
Wood End, Warw 44 D8
Wood End, Herts 37 G8
Woodend, Northnts 35 H6
Wood Enderby 57 G7
Woodfalls 10 C2
Woodford, Northnts 36 C3
Woodford, Corn 6 C4
Woodford, Devon 5 H6
Woodford, G. Lon 21 J4
Woodford, G. Man 53 K5
Woodford Bridge 21 J4
Woodford Green 21 J4
Woodford Halse 35 G5
Woodgate, W. Mids 34 A2
Woodgate, H. & W 34 A4
Woodgate, W. Susx 12 C6
Woodgate, Norf 48 F6
Woodgreen, Hants 10 C3
Wood Green, G. Lon 21 H4
Woodhall 68 B7
Woodhall Spa 57 F7
Woodham 20 E7
Woodham Ferrers 22 D4
Woodham Mortimer 22 E3
Woodham Walter 2 E3
Woodhaven 87 H2
Woodhead 99 G5
Woodhill 33 G2
Woodhorn 75 J5
Woodhouse, Leic 45 G6
Woodhouse, S. Yks 55 H5
Woodhouse Eaves 45 G6
Woodhouselee 79 K3
Woodhurst 37 G3
Woodingdean 13 G6
Wooditton 38 B5
Woodland, Durham 68 C4
Woodland, Devon 5 H5
Woodlands, Dorset 10 B4
Woodlands, Hants 10 E3
Woodlands, Grampn 93 J2
Woodlands Park 20 C6
Woodlands St Mary 18 E3
Wood Lanes 53 L5
Woodleigh 5 H7
Woodlesford 63 G7
Woodley 20 B6
Woodmancote, Glos 30 D5
Woodmancote, Glos 30 E7
Woodmancote, W. Susx 12 F5
Woodmancote, W. Susx 11 J4
Woodmancott 19 G6
Woodmansey 58 E3
Woodmansterne 21 G7
Woodminton 10 B2
Woodnesborough 15 K3
Woodnewton 46 D8
Wood Norton 48 F5
Woodplumpton 61 J6
Woodrising 48 E7
Woodseaves, Shrops 43 H4
Woodseaves, Staffs 43 K5
Woodsend 18 D3
Woodsetts 55 J5
Woodsford 9 J5
Woodside, Berks 20 D6
Woodside, Tays 92 D8
Woodside, Herts 21 G3
Woodstock 31 J6
Wood Street 12 C1
Woodthorpe, Leic 45 G6
Woodthorpe, Derby 55 H6
Woodton 49 H8
Woodtown 6 E3
Woodville 44 E6
Woodwalton 36 F2
Woodyates 10 B3
Woofferton 32 E4
Wookey 17 G6
Wookey Hole 17 G6
Wool 9 K6
Woolacombe 6 E1
Woolage Green 15 J4
Woolaston 29 K8
Woolavington 16 E6
Woolbeding 11 K2
Wooler 81 H7
Woolfardisworthy, Devon 6 D3
Woolfardisworthy, Devon 7 J5
Woolfords Cottages 79 H4
Woolhampton 19 G4
Woolhope 29 L4
Woolland 9 J4
Woolley, Cambs 36 E3
Woolley, W. Yks 55 G2
Woolmer Green 21 G2
Woolpit 38 E4
Woolscott 35 F4
Woolsington 75 H6
Woolstaston 42 F8
Woolsthorpe 46 B4
Woolston, Shrops 32 D2
Woolston, Shrops 42 E5
Woolston, Hants 10 F3
Woolston, Ches 53 H5
Woolston, Devon 5 H7
Woolstone, Bucks 36 B7
Woolstone, Oxon 18 D2
Woolston Green 5 H5
Woolton 52 F5
Woolton Hill 18 F4
Woolverstone 39 G7
Woolverton 17 J5
Woolwich 21 J6
Woore 43 J3
Woorton 31 J6
Wooton, Staffs 44 C3
Wooton, Oxon 31 J7
Wootton, Hants 10 D5
Wootton, Beds 36 D6
Wootton, Humbs 59 E5
Wootton, Kent 15 J4
Wootton, Northnts 35 J5
Wootton, Oxon 31 J7
Wootton, Staffs 43 K5
Wootton Bassett 18 B2
Wootton Bridge 11 G5
Wootton Common 11 G5
Wootton Courtenay 7 K1
Wootton Fitzpaine 8 E5
Wootton Rivers 18 C4
Wootton St Lawrence 19 G5
Wootton Wawen 34 C4
Worcester 33 H5
Worcester Park 21 G7
Wordsley 33 H2
Worfield 43 J8
Workington 66 B4
Worksop 55 J6
Worlaby 58 E5
World's End 19 F3
Worle 16 E4
Worleston 43 H2
Worlingham 39 K2
Worlington, Cambs 38 B3
Worlington, Devon 7 H4
Worlingworth 39 H4
Wormbridge 29 J4
Wormegay 48 B6
Wormelow Tump 29 J4
Wormersley 55 J2
Wormhill 54 E6
Wormiehills 93 H8
Wormingford 38 E7
Worminghall 20 A3
Wormington 34 B7
Worminster 17 G6
Wormit 87 G2
Wormleighton 35 F5
Wormley 21 H3
Wormley West End 21 H3
Wormshill 14 E3
Wormsley 29 J3
Worplesdon 12 C1
Worrall 55 G4
Worsbrough 55 G3
Worsley 53 J3
Worstead 49 J5
Worsthorne 62 B6
Worston 62 A5
Worth, W. Susx 13 F3
Worth, Kent 15 K3
Wortham 39 F3
Worthen 42 E7
Worthenbury 42 F3
Worthing, Norf 48 E6
Worthing, W. Susx 12 E6
Worthington 44 F5
Worth Matravers 10 A7
Wortley 55 G4
Worton 18 A5
Wortwell 39 H2
Wotherton 42 D7
Wotter 5 F5
Wotton 12 E2
Wotton-under-Edge 30 B8
Wotton Underwood 20 A2
Woughton on the Green 36 B7
Wouldham 14 D2
Wrabness 39 G7
Wrafton 6 E2
Wragby 57 F6
Wramplingham 49 G7
Wrangham 98 F5
Wrangle 47 H2
Wrangway 8 C3
Wrantage 8 E2
Wrawby 58 E6
Wraxall, Avon 17 F3
Wraxall, Somer 9 H1
Wraxall, Wilts 17 K4
Wray 61 K3
Wraysbury 20 E6
Wrea Green 61 G6
Wreay, Cumbr 67 G2
Wreay, Cumbr 67 G4
Wrekenton 75 J8
Wrelton 64 E4
Wrenbury 43 G3
Wreningham 49 G8
Wrentham 39 K2
Wressle 58 B3
Wrestlingworth 37 F6
Wretham 38 E1
Wretton 48 B8
Wrexham 42 E2
Wribbenhall 33 G3
Wrightington Bar 53 G2
Wrinehill 43 J3
Wrington 17 F4
Writtle 22 C3
Wrockwardine 43 H6
Wroot 58 B6
Wrotham 14 C3
Wrotham Heath 14 C3
Wroughton 18 C2
Wroxall, Warw 34 D3
Wroxall, I. of W 11 G7
Wroxeter 43 G7
Wroxham 49 J6
Wroxton 34 F6
Wyaston 44 C3
Wyberton 47 G3
Wyboston 36 E5
Wybunbury 43 H3
Wychbold 33 J4
Wych Cross 13 H3
Wyche 33 G6
Wyck 19 J7
Wyck Rissington 31 F5
Wycombe Marsh 20 C4
Wyddial 37 G7
Wye 15 G4
Wyke, W. Yks 62 E7
Wyke, Shrops 43 H7
Wyke, Dorset 9 J2
Wykeham, N. Yks 64 F5
Wykeham, N. Yks 65 G4
Wyke Regis 9 H7
Wyke, The 43 J7
Wykey 42 E5
Wylam 75 H7
Wylde Green 44 C8
Wylye 10 B1
Wymering 11 H4
Wymeswold 45 H5
Wymington 36 C4
Wymondham, Leic 46 B6
Wymondham, Norf 49 G7
Wyndham 25 H5
Wynford Eagle 9 G5
Wyre Piddle 34 A6
Wysall 45 H5
Wythall 34 B3
Wytham 31 J7
Wythenshawe 53 K5
Wyverstone 38 F4
Wyverstone Street 38 F4

Y

Yaddlethorpe 58 C6
Yafford 11 F6
Yafforth 69 F7
Yalding 13 L1
Yanworth 30 E6
Yapham 64 E7
Yapton 12 C6
Yarburgh 57 H4
Yarcombe 8 D4
Yardley 34 C2
Yardley Gobion 35 J6
Yardley Hastings 36 B5
Yardro 29 G2
Yarkhill 29 L3
Yarlet 43 L5
Yarlington 9 H2
Yarm 69 G5
Yarmouth 10 E6
Yarnfield 43 K4
Yarnscombe 7 F3
Yarnton 31 J6
Yarpole 32 D4
Yarrow 80 B7
Yarrow Feus 80 B7
Yarsop 29 J3
Yarwell 46 D8
Yate 17 J2
Yateley 20 C7
Yatesbury 18 B3
Yattendon 19 G3
Yatton, H. & W 32 D4
Yatton, Avon 17 F4
Yatton Keynell 17 K3
Yaverland 11 H6
Yaxham 48 F6
Yaxley, Cambs 36 E1
Yaxley, Suff 39 G3
Yazor 29 J3
Yeading 21 F5
Yeadon 62 F5
Yealand Conyers 61 J2
Yealand Redmayne 61 J2
Yealmpton 5 F6
Yearsley 63 J2
Yeaton 42 F6
Yeaveley 44 C3
Yedingham 65 F5
Yelden 36 D4
Yelford 31 H7
Yelland 6 E2
Yelling 37 F4
Yelvertoft 35 G3
Yelverton, Devon 5 F5
Yelverton, Norf 49 H7
Yenston 9 J2
Yeoford 5 H2
Yeolmbridge 4 D3
Yeo Mill 7 J3
Yeovil 9 G3
Yeovil Marsh 9 G3
Yeovilton 9 G2
Yerbeston 26 E7
Yesnaby 112 B4
Yetlington 75 G3
Yetminster 9 G3
Yettington 8 B6
Yetts o' Muckhart 86 D4
Y Fan 41 J5
Y Felinheli 50 E4
Y Ffor 50 C7
Yieldshields 79 F4
Yiewsley 20 E5
Ynysboeth 25 J5
Ynysddu 25 K5
Ynyshir 25 J5
Ynyslas 40 F4
Ynysybwl 25 J5
Yockenthwaite 62 C2
Yockleton 42 F6
Yokefleet 58 C4
Yoker 85 J8
Yonder Bognie 98 F4
York 63 K4
Yorkletts 15 G2
Yorkley 29 L7
Yorton 43 G5
Youlgreave 54 F7
Youlstone 6 C4
Youlthorpe 64 E7
Youlton 63 H3
Young's End 22 D2
Yoxall 44 C6
Yoxford 39 J4
Ysbyty Ifan 51 H6
Ysbyty Ystwyth 41 G6
Ysceifiog 52 C6
Ysgubor-y-coed 41 F4
Ystalyfera 25 F4
Ystrad 25 H5
Ystrad Aeron 27 K2
Ystradfellte 25 H3
Ystradffin 28 B3
Ystradgynlais 25 F4
Ystradmeurig 41 G7
Ystrad Mynach 25 K5
Ystradowen, Dyfed 25 F3
Ystradowen, S. Glam 25 J7
Ythanbank 99 J5
Ythanwells 98 F5
Ythsie 99 H5

Z

Zeal Monachorum 7 H5
Zeals 9 J1
Zelah 3 G4
Zennor, Corn 2 C6

DISTANCES BETWEEN PRINCIPAL TOWNS

ROAD DISTANCES IN KILOMETRES

	Aberdeen	Aberystwyth	Ayr	Berwick-upon-Tweed	Birmingham	Blackpool	Bournemouth	Braemar	Brighton	Bristol	Cambridge	Cardiff	Carlisle	Doncaster	Dover	Dundee	Edinburgh	Exeter	Fishguard	Fort William	Glasgow	Gloucester	Great Yarmouth	Harwich	Holyhead	Inverness	John o' Groats	Kingston upon Hull	Kyle of Lochalsh	Land's End	Leeds	Leicester	Lincoln	Liverpool	Manchester	Newcastle upon Tyne	Norwich	Nottingham	Oban	Oxford	Plymouth	Portsmouth	Sheffield	Shrewsbury	Southampton	Stranraer	Swansea	York
London	810	340	634	544	169	364	161	776	84	196	87	253	484	256	114	698	608	277	418	800	639	175	206	122	407	863	1067	332	927	478	304	156	211	325	298	441	183	196	787	92	351	113	256	241	124	647	312	311
Aberdeen		716	285	293	676	496	908	95	895	793	737	789	356	554	927	108	201	916	790	266	233	753	797	813	707	169	373	586	304	1114	526	666	616	549	547	378	764	610	286	777	990	901	579	642	880	367	795	513
Aberystwyth			510	501	183	246	333	652	378	201	344	169	360	283	454	605	515	323	90	692	515	164	473	452	179	782	990	359	803	523	291	246	320	167	227	439	444	264	663	248	398	357	256	124	323	523	117	330
Ayr				216	465	290	702	230	718	595	575	615	150	378	748	188	117	734	600	214	53	531	647	707	491	320	528	404	341	917	341	481	441	343	341	240	615	441	201	568	792	692	394	433	671	82	589	344
Berwick-upon-Tweed					425	311	663	238	628	566	473	592	140	319	658	182	92	689	632	306	163	512	555	599	501	346	550	298	391	888	251	406	360	352	315	103	528	356	290	521	763	645	306	426	624	254	616	238
Birmingham						198	237	607	262	130	161	166	315	151	283	562	470	253	274	631	470	90	290	269	238	737	945	245	758	452	182	63	145	150	129	333	267	80	618	103	327	227	122	72	206	478	192	209
Blackpool							435	431	460	328	335	336	140	151	478	385	295	454	336	455	295	280	406	443	227	560	769	204	583	652	116	225	206	79	77	208	373	179	443	301	528	425	138	158	404	303	348	154
Bournemouth								843	148	132	248	188	552	378	280	797	707	132	377	867	707	159	367	283	463	961	1165	454	995	330	410	254	336	377	365	558	344	295	853	145	206	84	348	298	50	715	269	433
Braemar									859	737	686	756	291	499	890	84	146	859	742	201	177	674	768	811	632	121	325	494	256	1070	472	608	552	494	492	323	735	568	227	721	945	847	515	575	824	312	777	459
Brighton										220	171	293	568	340	132	782	692	267	468	884	723	214	290	206	501	946	1151	415	1011	465	388	240	295	412	381	525	262	280	871	159	341	77	340	335	98	731	357	394
Bristol											232	72	446	282	299	692	600	122	248	761	600	56	388	307	332	867	1075	375	888	322	312	193	275	259	259	463	356	233	748	119	196	156	259	166	122	608	137	340
Cambridge												288	425	187	201	629	539	354	435	740	58	198	132	108	399	793	998	262	867	538	233	109	137	270	243	370	100	134	727	134	423	200	193	233	211	587	349	241
Cardiff													465	317	383	710	620	195	180	781	620	90	420	396	348	884	1094	393	908	394	354	229	311	272	277	489	388	246	768	174	269	229	288	179	195	628	66	393
Carlisle														229	599	245	154	568	451	315	154	381	497	541	341	422	629	254	443	768	192	332	291	193	192	92	465	291	303	418	642	542	245	283	521	163	439	195
Doncaster															372	443	352	404	375	544	383	241	269	312	269	565	769	76	671	602	47	119	63	138	82	183	237	69	531	233	478	357	29	159	336	391	349	55
Dover																813	723	399	533	914	753	290	298	201	546	977	1181	446	1041	589	418	270	325	439	412	555	280	311	901	206	465	209	370	356	230	761	441	425
Dundee																	90	814	695	204	134	628	711	755	586	212	417	438	299	1014	415	562	505	438	436	267	679	512	188	665	888	790	468	531	768	269	721	402
Edinburgh																		724	605	232	71	538	621	665	496	254	459	348	348	924	325	455	415	348	346	177	589	422	198	575	798	700	378	441	678	200	631	312
Exeter																			370	884	723	179	478	399	454	995	1197	497	1011	198	435	315	398	381	380	586	454	356	858	229	74	190	381	288	169	731	259	462
Fishguard																				782	605	246	563	542	269	872	1080	451	893	568	381	336	410	257	317	529	531	354	753	330	444	404	346	214	346	613	108	420
Fort William																					163	697	813	856	657	106	314	570	127	1081	507	647	607	509	507	407	781	607	79	734	958	858	560	599	837	296	755	510
Glasgow																						536	652	695	496	267	475	409	288	922	346	486	446	348	346	238	620	446	148	573	797	697	399	438	676	135	594	349
Gloucester																							330	286	290	811	1011	272	824	378	256	137	219	203	203	407	299	177	682	84	253	174	203	124	150	552	[illegible]	283
Great Yarmouth																								132	504	834	1038	333	940	676	315	225	206	386	330	452	32	229	800	251	552	319	267	362	330	660	[illegible]	323
Harwich																									507	877	1081	373	983	597	359	217	249	378	385	496	117	241	843	203	473	235	301	354	246	660	[illegible]	463
Holyhead																										763	970	344	784	652	264	262	322	148	200	398	472	275	644	336	528	463	240	166	439	504	[illegible]	303
Inverness																											208	602	135	1193	579	742	687	615	600	431	801	692	185	829	1069	953	632	705	932	402	[illegible]	566
John o' Groats																												806	304	1397	784	947	892	822	805	636	1006	896	393	1033	1271	1157	837	912	1136	610	[illegible]	771
Kingston upon Hull																													697	678	89	195	71	209	153	195	240	145	557	309	571	433	105	235	412	417	[illegible]	60
Kyle of Lochalsh																														1210	634	774	734	636	634	512	908	734	206	861	1085	985	687	726	964	423	[illegible]	637
Land's End																															634	515	597	581	581	785	678	555	1070	441	143	388	581	488	367	930	[illegible]	661
Leeds																																153	109	121	64	148	283	113	494	270	509	393	53	175	373	354	[illegible]	39
Leicester																																	82	161	148	301	192	40	632	117	389	241	100	135	220	494	[illegible]	174
Lincoln																																		190	135	256	169	56	592	200	472	323	74	188	303	454	[illegible]	121
Liverpool																																			56	249	354	158	496	253	455	377	116	93	356	356	[illegible]	159
Manchester																																				212	298	101	494	232	455	356	61	111	335	354	[illegible]	103
Newcastle upon Tyne																																					425	253	375	418	660	542	201	323	521	254	[illegible]	135
Norwich																																						196	768	233	552	299	235	330	311	628	[illegible]	291
Nottingham																																							594	158	430	282	60	132	261	467	[illegible]	124
Oban																																								721	945	845	546	586	824	283	[illegible]	497
Oxford																																									298	124	217	171	103	583	[illegible]	291
Plymouth																																										264	455	362	243	805	[illegible]	536
Portsmouth																																											341	298	34	705	[illegible]	415
Sheffield																																												132	320	423	[illegible]	84
Shrewsbury																																													274	446	[illegible]	214
Southampton																																														684	[illegible]	394
Stranraer																																															[illegible]	357
Swansea																																																404

ROAD DISTANCES IN MILES

	London	Aberdeen	Aberystwyth	Ayr	Berwick-upon-Tweed	Birmingham	Blackpool	Bournemouth	Braemar	Brighton	Bristol	Cambridge	Cardiff	Carlisle	Doncaster	Dover	Dundee	Edinburgh	Exeter	Fishguard	Fort William	Glasgow	Gloucester	Great Yarmouth	Harwich	Holyhead	Inverness	John o' Groats	Kingston upon Hull	Kyle of Lochalsh	Land's End	Leeds	Leicester	Lincoln	Liverpool	Manchester	Newcastle upon Tyne	Norwich	Nottingham	Oban	Oxford	Plymouth	Portsmouth	Sheffield	Shrewsbury	Southampton	Stranraer	Swansea
Aberdeen	503
Aberystwyth	211	445
Ayr	394	177	317
Berwick-upon-Tweed	338	182	311	134
Birmingham	105	420	114	289	264
Blackpool	226	308	153	180	193	123
Bournemouth	100	564	207	436	412	147	270
Braemar	482	59	405	143	148	377	268	524
Brighton	52	556	235	446	390	163	286	92	534
Bristol	122	493	125	370	352	81	204	82	458	137
Cambridge	54	458	214	357	294	100	208	154	426	106	144
Cardiff	157	490	105	382	368	103	209	117	470	182	45	179
Carlisle	301	221	224	93	87	196	87	343	181	353	277	264	289
Doncaster	159	344	176	235	198	94	94	235	310	211	175	116	197	142
Dover	71	576	282	465	409	176	297	174	553	82	186	125	238	372	231
Dundee	434	67	376	117	113	349	239	495	52	486	430	391	441	152	275	505
Edinburgh	378	125	320	73	57	292	183	439	91	430	373	335	385	96	219	449	56
Exeter	172	569	201	456	428	157	282	82	534	166	76	220	121	353	251	248	506	450
Fishguard	260	491	56	373	393	170	209	234	461	291	154	270	112	280	233	331	432	376	230
Fort William	497	165	430	133	190	392	283	539	125	549	473	460	485	196	338	568	127	144	549	486
Glasgow	397	145	320	33	101	292	183	439	110	449	373	360	385	96	238	468	83	44	449	376	101
Gloucester	109	468	102	330	318	56	174	99	419	133	35	123	56	237	150	180	390	334	111	153	433	333
Great Yarmouth	128	495	294	402	345	180	252	228	477	180	241	82	261	309	167	185	442	386	297	350	505	405	205
Harwich	76	505	281	439	372	167	275	176	504	128	191	67	246	336	194	125	469	413	248	337	532	432	178	82
Holyhead	253	439	111	305	311	148	141	288	393	311	206	248	216	212	167	339	364	308	282	167	408	308	180	313	315
Inverness	536	105	486	199	215	348	318	597	75	588	539	493	549	262	351	607	132	158	618	542	66	166	504	518	545	474
John o' Groats	663	232	615	328	342	587	478	724	202	715	668	620	680	391	478	734	259	285	744	671	195	295	628	645	672	603	129
Kingston upon Hull	206	364	223	251	185	152	127	282	307	258	233	163	244	158	47	277	272	216	309	280	354	254	198	169	232	214	374	501
Kyle of Lochalsh	576	189	499	212	243	471	362	618	159	628	552	539	564	275	417	647	186	216	628	555	79	179	512	584	611	487	84	189	433
Land's End	297	692	325	570	552	281	405	205	665	289	200	334	245	477	374	366	630	574	123	353	672	573	235	420	371	405	741	868	421	752
Leeds	189	327	181	212	156	113	72	255	293	241	194	145	220	119	29	260	258	202	270	237	315	215	159	196	223	164	360	487	55	394	394
Leicester	97	414	153	299	252	39	140	158	378	149	120	68	142	206	74	168	349	283	196	209	402	302	85	140	135	163	461	588	121	481	320	95
Lincoln	131	383	199	274	224	90	128	209	343	183	171	85	193	181	39	202	314	258	247	255	377	277	136	128	155	200	427	554	44	456	371	68	51
Liverpool	202	341	104	213	219	93	49	234	307	256	161	168	169	120	86	273	272	216	237	160	316	216	126	240	235	92	382	511	130	395	361	75	100	118
Manchester	185	340	141	212	196	80	48	227	306	237	161	151	172	119	51	256	271	215	236	197	315	215	126	205	239	124	373	500	95	394	361	40	92	84	35
Newcastle upon Tyne	274	235	273	149	64	207	129	347	201	326	288	230	304	57	114	345	166	110	364	329	253	148	253	281	308	247	268	395	121	318	488	92	187	159	155	132
Norwich	114	475	276	382	328	166	232	214	457	163	221	62	241	289	147	174	422	366	282	330	485	385	186	20	73	293	498	625	149	564	421	176	119	105	220	185	264
Nottingham	122	379	164	274	221	50	111	183	353	174	145	83	153	181	43	193	318	262	221	220	377	277	110	142	150	171	430	557	90	456	345	70	25	35	98	63	157	122
Oban	489	178	412	125	180	384	275	530	141	541	465	452	477	188	330	560	117	123	533	468	49	92	424	497	524	400	134	244	346	128	665	307	393	368	308	307	233	477	369
Oxford	57	482	154	353	324	64	187	90	448	99	74	83	108	260	145	128	413	357	142	205	456	356	52	156	126	209	515	642	192	535	274	168	73	124	157	144	260	145	98	448
Plymouth	218	615	247	492	474	203	328	128	587	212	122	263	167	399	297	289	552	496	46	276	595	495	157	343	294	328	664	790	355	674	89	316	242	293	283	283	410	343	267	587	185
Portsmouth	70	560	222	430	401	141	264	52	526	48	97	124	142	337	222	130	491	435	118	251	533	433	108	198	146	288	592	719	269	612	241	244	150	201	234	221	337	186	175	525	77	164
Sheffield	159	360	159	245	190	76	86	216	320	211	161	120	179	152	18	230	291	235	237	215	348	248	126	166	187	149	393	520	65	427	361	33	62	46	72	38	125	146	37	339	135	283	212
Shrewsbury	150	399	77	269	265	45	98	185	357	208	103	145	111	176	99	221	330	274	179	133	372	272	77	225	220	103	438	567	146	451	303	109	84	117	58	69	201	205	82	364	106	225	185	82
Southampton	77	547	201	417	388	128	251	31	512	61	76	131	121	324	209	143	477	421	105	215	520	420	93	205	153	273	579	706	256	599	228	232	137	188	221	208	324	193	162	512	64	151	21	199	170
Stranraer	402	228	325	51	158	297	188	444	194	454	378	365	390	101	243	473	167	124	454	381	184	84	343	410	410	313	250	379	259	263	578	220	307	282	221	220	158	390	290	176	362	500	438	263	277	425
Swansea	194	494	73	366	383	119	216	167	483	222	85	217	41	273	217	274	448	392	161	67	469	369	89	294	267	184	535	664	264	548	285	227	158	209	153	187	319	274	169	461	141	206	182	200	118	161	374
York	193	319	205	214	148	130	96	269	285	245	211	150	244	121	34	264	250	194	287	261	317	217	176	201	288	188	352	479	37	396	411	24	108	75	99	64	84	181	77	309	181	333	258	52	133	245	222	251